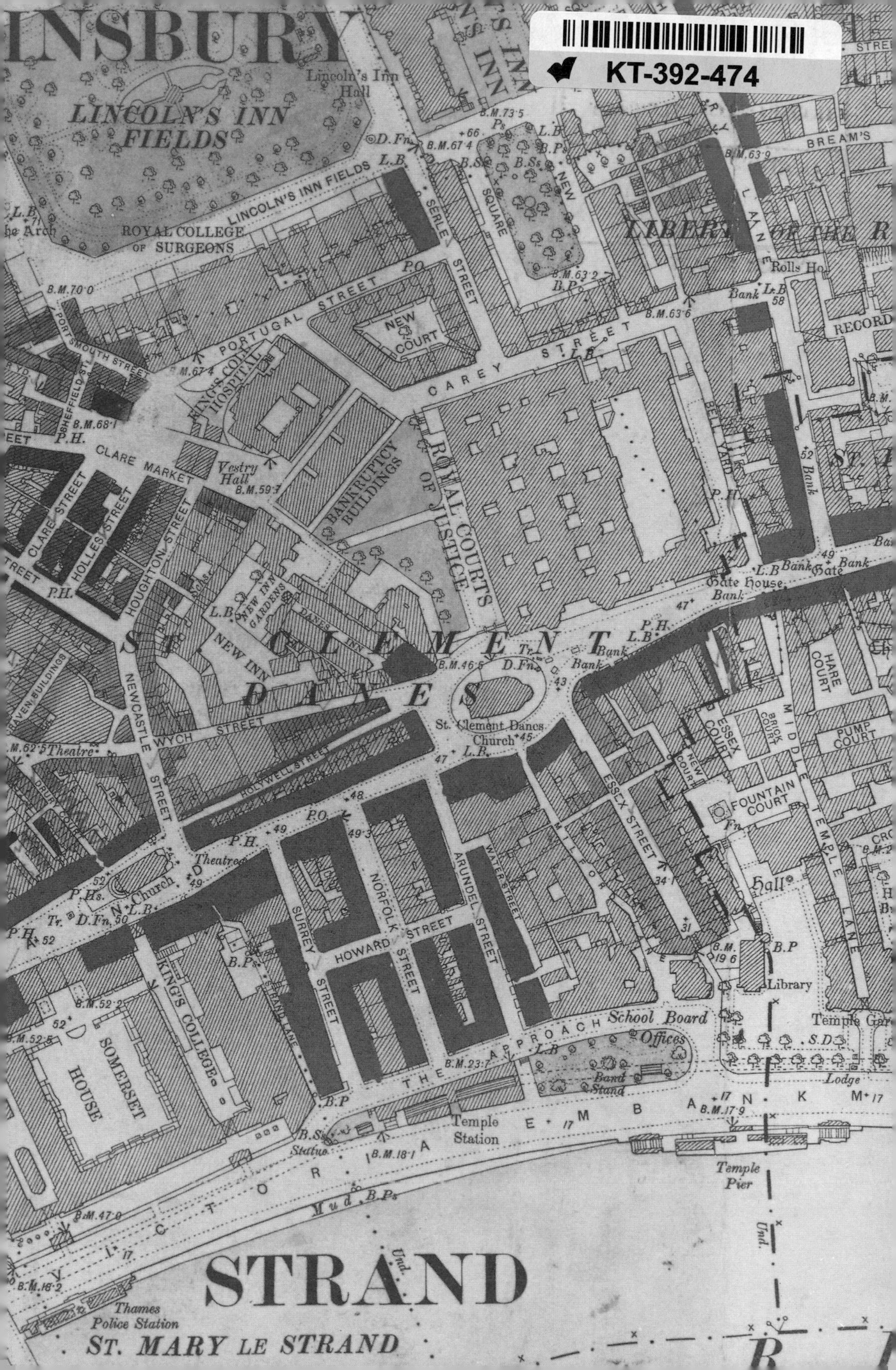

INSBURY
LINCOLN'S INN FIELDS
Lincoln's Inn Hall
LINCOLN'S INN FIELDS
ROYAL COLLEGE OF SURGEONS
SERLE STREET
NEW SQUARE
BREAM'S
PORTUGAL STREET
NEW COURT
CAREY STREET
Rolls Ho.
RECORD
PORTSMOUTH STREET
SHEFFIELD ST.
KING'S COLL. HOSPITAL
CLARE MARKET
Vestry Hall
BANKRUPTCY BUILDINGS
ROYAL COURTS OF JUSTICE
CLARE STREET
HOLLES STREET
HOUGHTON STREET
NEW INN GARDENS
NEW INN
DANES INN
ST. CLEMENT DANES
St. Clement Danes Church
Gate House
HARE COURT
ESSEX COURT
BRICK COURT
PUMP COURT
MIDDLE TEMPLE LANE
NEW COURT
FOUNTAIN COURT
CRAVEN BUILDINGS
NEWCASTLE STREET
WYCH STREET
Theatre
HOLYWELL STREET
ESSEX STREET
MILFORD LANE
WATER STREET
ARUNDEL STREET
NORFOLK STREET
SURREY STREET
HOWARD STREET
Theatre
Church
Hall
Library
KING'S COLLEGE
STRAND LANE
SOMERSET HOUSE
THE APPROACH
School Board Offices
Band Stand
Temple Gar
Lodge
Temple Station
Statue
VICTORIA EMBANKMENT
Temple Pier
Mud
STRAND
Thames Police Station
ST. MARY LE STRAND

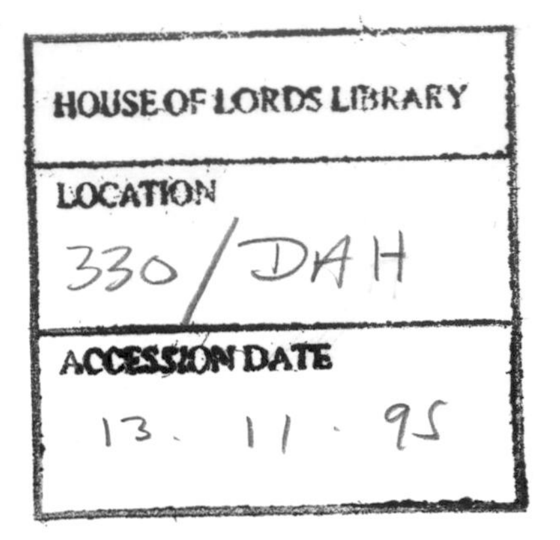

LSE

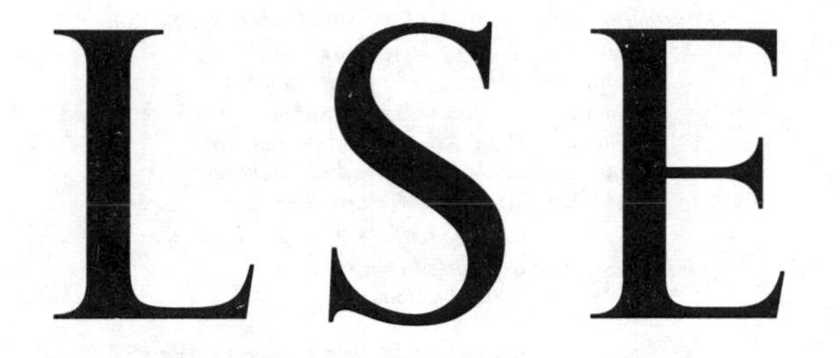

A HISTORY OF THE LONDON SCHOOL OF ECONOMICS AND POLITICAL SCIENCE

1895–1995

Ralf Dahrendorf

OXFORD UNIVERSITY PRESS

1995

Oxford University Press, Walton Street, Oxford OX2 6DP
Oxford New York
Athens Auckland Bangkok Bombay
Calcutta Cape Town Dar es Salaam Delhi
Florence Hong Kong Istanbul Karachi
Kuala Lumpur Madras Madrid Melbourne
Mexico City Nairobi Paris Singapore
Taipei Tokyo Toronto
and associated companies in
Berlin Ibadan

Oxford is a trade mark of Oxford University Press

Published in the United States
by Oxford University Press Inc., New York

British Library Cataloguing in Publication Data
Data available

Library of Congress Cataloging in Publication Data
Dahrendorf, Ralf.
LSE: a history of the London School of Economics and Political Science, 1895–1995 / Ralf Dahrendorf.
p. cm.
Includes bibliographical references and index.
1. London School of Economics and Political Science—History.
I. Title. II. Title: History of the London School of Economics and Political Science, 1895–1995.
H67.L9D33 1995 330'.071'14212—dc20 94–42604
ISBN 0–19–820240–7

1 3 5 7 9 10 8 6 4 2

Set by Hope Services (Abingdon) Ltd.
Printed in Great Britain
on acid-free paper by
The Bath Press, Avon

PREFACE

SOON after I had started work on this history of the London School of Economics and Political Science, one of the more flamboyant alumni of the School invited me to lunch and told me that but for the demands of his job he too would have liked to write this book. He had always wanted to find out what was special about the School. That there was something special he had no doubt. After all, LSE had captured such diverse temperaments as Mick Jagger, Maurice Saatchi, Bernard Levin, and those who appeared with him and Ron Moody on the stage of the Old Theatre in the 1950s for memorable cabarets, as well as the many who had chosen other stages, five Nobel laureates in economics, numerous central bankers and government ministers in the four corners of the world. Surely (this was the suggestion) there must be a way to express in a word the charm of the place which so many call lovingly 'the School' as if there was no other school in the world.

One word? That may be asking too much, and a word would be too little for an answer. One theme perhaps. There is the forever explosive relationship between social science and public policy. Californians worry about the San Andreas Fault and what its violent eruption might do to the peace of their homes; LSE disturbs the peace of mind of those who are directly, or more often indirectly, affected by its doings through another fault line, the Sidney Webb Fault perhaps (no saint he!), though many of quite different persuasions have made their own contribution—the fault line between wanting to know the causes of things and wanting to change things, dispassionate study and committed action, ascetic aspirations and worldly temptations. The very location of the School defines the fault line, at the heart of the polygon which includes the Law Courts and the City, Bloomsbury and Theatreland, Whitehall and Westminster. All critical events in the history of LSE, from George Bernard Shaw's early doubts over the all too academic use of the Hutchinson bequest to Sidney Webb's 'railway speech' of 1911, on to William Beveridge's disputes with Harold Laski and with student politicians, to say nothing of 1968 and all that—all these events tell the same story of two incompatible yet inextricably conjoined human passions at work under one roof.

This is where the architectural as well as the geological metaphor ends, for it was never just the common roof which united the School. LSE may be permanently threatened by quakes of one kind or another, but it is also a place which engenders a special kind of loyalty among its members. LSE matters to those who have come to it. It is not just a few lines in their curricula vitae, an educational experience, or even a first-rate university, but an institution which has laid claim to a part of the hearts and souls of many. This, like much else, has become more difficult in the unfriendly public climate of the 1980s and 1990s, and also in the deteriorating environment of the great capital city, London.

For the major part of its first century, however, the School was a place to work and to play, to spend long days in earnest seminar and corridor talk as well as on frivolous pursuits like lunch-hour dances, to make friends, and for not a few to find their partners for life. LSE creates a common sense of belonging for people who recognize each other wherever they meet, and even if they had been at the School at very different times; what matters is not the 'class of '54' but the School itself. If it was not so public, it might be a sect; if it was not so varied, it might be a party; if it was not so big, it might be a family; as it is, it is simply LSE, the School.

The affectionate use of the adjective-less singular, the School, is almost as old as the institution itself. The abbreviation, LSE, came a little later. It is of course a foreshortened version of the London School of Economics and Political Science. Only the Library is called by its full name, BLPES, and that because it began as the British Library of Political Science before Economics and thus the E were added in 1925. Those who say 'LSE' do not mean to ignore the other subjects taught at the School; the letters have simply become convenient shorthand. The Webbs and others sometimes wrote 'L S of E' in the early years, but a decade after the foundation, L.S.E. had become used within the School, and another ten years later it was understood outside as well, so that soon the dots between L, S, and E could be dropped and LSE became at least as familiar as, say, the LCC, the London County Council, to which the School owed early support, or its later source of funding, the UGC, or University Grants Committee.

This book tells the history of LSE in an utterly unreconstructed manner. It follows for the most part the chronology of events, and it makes much of significant events as well as important and influential personalities. I must ask the indulgence of those who look in vain for their own names in this story, or even for the names of their favourite teachers and friends. The book is short, at least by the standards of a time in which words are computer-processed rather than hand-crafted. The book is also idiosyncratic, if that it not too mild a description. I had fun writing it but there may well be those who think that I should not have undertaken the task at all. I was not only busy with one or two other things while writing it but, more importantly, I am not a historian, and as Director of LSE for one decade of its first century, the years from 1974 to 1984, I was an actor in the story which is told. As a result, this is not the definitive history of the School, nor is it an impartial and objective account. It is in fact itself a testimony to the great tension between knowledge and commitment.

By an enormous effort of self-negation (too enormous for this author) it might have been possible to write a purely factual account like Sydney Caine's splendid *History of the Foundation*. By just following natural inclinations (too natural for this author), a work of self-justification might have emerged, like parts of William Beveridge's intriguing *London School of Economics and its Problems*. Both authors were, of course, Directors of the School for ten and eighteen years respectively, and neither was a professional historian. In trying to steer a course between two such distinguished predecessors, while not making any false claims to a professional expertise, I found it helpful to read Robert Blake's paper on 'Winston Churchill as Historian'. Blake discusses the difficulties of the actor-author and quotes Lewis Namier's advice to Churchill:

> Please do not try to write history as other historians do, but do it your own way. Tell us more how various transactions strike you, and what associations they evoke in your mind.

Churchill took the advice to heart: 'I certainly propose to apply my experience in military and political affairs to the episodes.'

LSE was often split and sometimes torn apart by strong emotions, from the Boer War to 1968 and beyond, but fortunately such divisions did not quite involve its Directors in military affairs, and even the politics of LSE remained for the most part containable. Blake's advice has to be taken therefore *mutatis mutandis*. (Also, this book cannot possibly compare with Churchill's masterly use of the English language, which is not even my mother tongue.) Readers will find, however, that Directors occupy a special place in the book. More is said about them than in earlier accounts of various periods of the School's history, and more understanding is shown for their dilemmas, and for the apparently inevitable cadences in their popularity within a highly-strung institution. By the same token, less is said about other perspectives on LSE, such as those of students, than one would expect from a definitive history.

This, then, is a biography of an institution written by one who was closely involved and remains deeply attached. It places the School in the context of the drama of the twentieth century, but it does so through a mirror, the mirror of the social sciences. These were not exactly invented at LSE but the School brought them together like no other university in Europe, led them to full bloom in all their variety, and then reflected the aches and pains of their maturity, in which professionalization often went hand in hand with an uncertain sense of direction. The need to build a community of the most diverse people in the middle of the distractions of London adds poignancy to the story.

One of the distinguishing marks of LSE was throughout that it never remained silent. There exists an entirely unmanageable quantity of material about the first hundred years of LSE. Lord Beveridge, with his penchant for numbers, calculated that the founders, Sidney and Beatrice Webb, alone had published some five million words or 10,000 pages in the form of books. Much of this was, of course, not concerned with LSE; but once one has added the uncounted reports and drafts and papers, to say nothing of letters and diaries, it would not be surprising to find at least one million Webb-words about LSE. The Beveridges, William Beveridge and Jessy Mair, were not averse to putting large numbers of words to paper themselves, including two books and two dozen or so reports, chapters in books, and articles about LSE. These are but two—two couples rather—of a much larger set of relevant authors. It would be unkind to speak of the logorrhea of those attached to the School (especially for one who is afflicted by the disease himself); but certainly, words, printed and unprinted, were from the beginning the great weapons which LSE people used, weapons of attack and also weapons of self-defence, not least from the often unbearable tension of values and social science.

One result of the abundance of material was that a considerable research effort was required even for this relatively brief history of the School. The ground had been prepared for such an effort, notably by the Archive of the BLPES under the skilful direction of Dr Angela Raspin. Also, a number of authors helped by publishing their own biographies or accounts of aspects of the LSE history in time for me to draw on their works. The Bibliography includes their names and publications. For the rest, I have had the good fortune to be able to rely on an unusual and deeply committed team of research assistants, led with unfailing good humour and a remarkable sense of both relevance and exactness by the former Information Officer of the School, Shirley Chapman. She was joined almost throughout by Lois Reynolds, whose papers uncovered numerous threads otherwise lost in remote textures, and by Fiona Plowman, whose responsiveness and searching intelligence made her invaluable. Without 'the Team' this book would not have been written.

The team not only helped this author but also produced a collection of material on the history of the School which many future authors will find useful. In this way, the generosity of LSE may even be repaid to some extent. The authorities of the School were supportive throughout, beginning with the Directors, Dr I. G. Patel, who initiated the project and encouraged me to write this book, and Dr John Ashworth, and including the Pro-Directors, Professor Robert Pinker, Professor David Bartholomew, and Professor Michael Leifer, the Secretary, Dr Christine Challis, and the Librarians, Mr Chris Hunt and Mrs Lynne Brindley, who provided a home for the History of the School Project.

This is not a definitive, and by the same token it is also not an official history of LSE. None of the officers of the School mentioned so far have read the manuscript before publication. But others, members of the Steering Committee set up by the School, to whom I am much indebted, and personal friends, have read parts or the whole of the text. Their suggestions were invariably helpful even if I did not respond to all of them, so that the text remains my responsibility. My special gratitude for their comments on sections of the book goes to Professor Martin Bulmer, Sir John Burgh, Professor Alan Day, Professor Sir Raymond Firth, Professor John Griffith, Dr Jose Harris, Professor Norman MacKenzie, Dr Peter Mathias, Dr Angela Raspin, and Professor Michael Wise. Professor Harold Edey not only read parts of the manuscript but also provided invaluable help in improving my, and in some cases even the School's, reporting of LSE accounts.

Editors and designers of Oxford University Press have contributed significantly, with the supportiveness and attention to detail for which they are justly renowned, to improve my text and produce a book which gives pleasure as well as food for thought.

Two friends who are not LSE alumni but professional historians, Mr Timothy Garton Ash and Professor Fritz Stern, have been generous in encouraging me while steering me gently away from pitfalls. My wife, as Ellen de Kadt a graduate student and later a teacher in the Government Department, has read and pondered every word of this book about a School which is as close to her heart as it is to mine. We share the hope that our LSE will enrich the lives of people from many walks of life and parts of the world in its second century, as it did in the first.

R.D.

Oxford
September 1994

ACKNOWLEDGEMENTS

As author of this book, I have had the benefit of assistance from many members of the School, their relatives, and friends. I am deeply grateful to all of them. Perhaps the most significant material has come from the reminiscences of LSE people, some written and some through the collection of recorded interviews. Mr Anthony Seldon gave early advice on oral history methodology. Professor Theo Barker has been helpful not only in conducting interviews but in pointing research in the right direction. Professor Eileen Barker and Professor Ben Roberts have also given generously of their time. A number of teachers at the School have interviewed their senior colleagues and thus helped produce what must be a gold-mine of material for future research. To interviewers and interviewed, I am indebted. I have greatly benefited also from the interviews with important economists by Mr Nadim Shehadi, to which he gave me access despite his own publication plans.

Libraries and Archives have provided access and assistance. The following deserve special acknowledgement: the Rockefeller Archive Center, New York (and special help from Erwin Levold); the University of Sheffield Library (and special help from Peter W. Carnell) (Papers of W. A. S. Hewins); the Scottish Record Office, Edinburgh (Papers of Sir Arthur Steel-Maitland); the University of Oxford, School of Geography (Papers of Sir Halford Mackinder); the Alexander Turnbull Library/National Library of New Zealand, Wellington (Papers of W. Pember Reeves); the University of Birmingham Library (Papers of Lancelot Hogben); the University of Kent at Canterbury Library (for supplying copies of oral history tapes). In the case of the first four I can testify personally to the helpfulness of librarians and archivists. Research in New Zealand was done by Professor Michael Hill, now of the Victoria University in Wellington. Professor Alan Musgrave, now of the University of Otago, and Dr Ralph Hayburn made available a recorded conversation on the 1960s 'troubles'. Mrs Wendy Weinberg, an alumnus and Governor of the School, consulted newspaper libraries in Washington, DC.

Important insights were gained from information supplied by members of LSE families. I want to mention with appreciation Dr Edmund Carr-Saunders, Dr Adrian Hogben, and Mr and Mrs Thomas Blanco-White.

Many others have given special assistance to the research team; among them particularly Dr Anne Bohm, Lady Cynthia Postan, Mr John

Raybould, Mrs Kit Russell, and Dr Esther Simpson. The project has benefited also from co-operation with biographers and other scholars working in related areas including especially Mlle Ariane Azema, Dr Maxine Berg, and Professor Susan Howson. Ms Molly Mortimer has been a regular and valuable helper. Ms Laurie Johnston became virtually a member of the team. While pursuing her Ph.D. work she gave much valuable assistance in building up the project's biographical collections, and collating material about 1968 and other subjects.

Staff of the Manuscripts and Special Collections division, indeed all members of the Library (BLPES), have been consistently helpful, as have those in charge of School records. Members of particular departments of LSE have given advice and much time and expertise in a range of fields such as computing, photography, model-building, and technical drawing.

Finally, I want to thank all holders of copyright for the permission to quote published or unpublished material. Formal acknowledgements are due to Macmillan Ltd. for permission to quote extracts from *Autobiography of an Economist*, © Lord Robbins, 1971; to Routledge (Allen & Unwin) for extracts from Lord Beveridge's *The LSE and its Problems 1919–37*; to Robson Books for extracts from Joan Abse's *My LSE*; to the Master and Fellows of Peterhouse, Cambridge for an extract from the Governing Body Minutes and to the Society of Authors on behalf of the Bernard Shaw Estate.

R.D.

CONTENTS

PART III: STRENGTH IN ADVERSITY

The War and Reconstruction, Expansion and Troubled Times, 1937–1995

LIST OF ILLUSTRATIONS

PLATES

(*between* pages 108 *and* 109, 268 *and* 269, and 428 *and* 429)

ENDPAPERS

FIGURE

PART I

AN EPIC OF CLARE MARKET

Foundation and Early Development

1895–1919

1

THE FOUNDING YEARS

A PLAN IS HATCHED

Borough Farm

When Sidney Webb came down for breakfast at Borough Farm near Godalming in Surrey on 4 August 1894, he had surprising news for the small group assembled to share a frugal late morning repast. His wife Beatrice was there, of course, as well as Graham Wallas, house guest of the Webbs at their summer retreat, and George Bernard Shaw, who had cycled over from a friend's home in nearby Tilford. The three men had known each other for more than ten years as fellow radicals and intellectual protagonists of reform, not least as members of the 'junta' of the political club-cum-'think-tank' founded in 1883 under the name Fabian Society. On that August morning, Sidney Webb told his friends of a letter which he had received on the previous day from a Derby solicitor informing him that the gruff though loyal Fabian supporter Henry Hunt Hutchinson had taken his own life. In his will he had appointed Webb one of his executors and stipulated that the major part of what turned out to be £20,000 which he had left should be applied 'at once, gradually and at all events within ten years to the propaganda and other purposes of the said [Fabian] Society and its Socialism, and to advancing its objects in any way [the executors] deem advisable'.[1]

Money has a strange effect on men of modest means and extravagant ideas, and £20,000 was a tidy sum in 1895. It is the equivalent of nearly one million pounds a hundred years later. However, before the breakfast party allowed their imagination to run away with the pot of gold, their thoughts, not unnaturally, turned to its originator, the 'foolish old fellow', as his beneficiaries sometimes unkindly described 'Old Hutch'.[2] Born in 1822, Henry Hunt Hutchinson had spent much of his life in a job which his father had held before him, as clerk to the magistrates in Derby, his home town. Apparently, the uninspiring position made him increasingly restless. At one point, in 1878, he tried to escape it and went to California, where two of his sons had prepared the ground by the purchase of a ranch for the men to which an embroidery shop run by the women was added later. But his grandson Bertram (who was to become an LSE student, and a pacifist, during the Second World War) remembers stories that things in California 'did not work out very well'. While three of the five children stayed, Hutchinson and his wife returned to England, a loveless couple it seems. 'After his death, . . . [his widow] would sit in front of his portrait . . . addressing his memory in uncomplimentary terms.'[3] Hutchinson grew old, found that he had cancer, and became more and more weary of life. He had joined the Fabian Society in 1890, not least in order to find friends to discuss his manuscripts on Herbert Spencer and other subjects of political theory. He made generous contributions but he also had frequent rows, notably with George Bernard Shaw, whom he found rude and hard to take. Eventually, on 26 July 1894, he had shot himself to put an end to what had become an unbearable life.

It did not take the breakfast party at Borough Farm long to agree that Hutchinson had been unduly parsimonious to his family and especially to his wife. His will might not stand up to legal scrutiny if challenged. Moreover, Hutchinson's daughter Constance had been named an executor along with Sidney Webb; both were also Trustees of the fund to be set up with three other Fabians—Edward Pease, William de Mattos, and William Clarke—to join them. Thus it was decided to set aside a portion of the money for the Hutchinson family. In the event, Constance Hutchinson died a year later and left the residue of her estate to the main fund. Her father had expressed the wish to see his voluminous manuscripts on political philosophy published. Edward Pease, co-founder and indispensable secretary of the Fabian Society as well as Hutchinson Trustee, went through them and concluded that 'there was nothing fit for publication'.[4] A part of the money was later used to establish the Hutchinson Silver Medal awarded annually to a student of the London School of Economics 'for excellence of work in research'.

With the Hutchinson problems out of the way, the conversation at Borough Farm became lively. After all, there still remained at least £10,000, so that there was much to talk about. What exactly were the most urgent purposes of the Fabian Society and what should be done to 'advancing its objects in any way [the Trustees] deem advisable'?[5] Given the temperaments of those assembled, it seems a little unlikely that Sidney Webb simply 'told us' that some of the money would be used 'to found a School in London along the lines of the École Libre des Sciences Politiques in Paris'.[6] In any case, a 'long discussion' ensued. Someone (though it is hard to think who—Beatrice, perhaps?) appears to have suggested rather disingenuously that the money be spent on ordinary Fabian Society activities but this was soon discarded. A more likely proposal (by Shaw?) was that 'a big political splash might be made with [the bequest]—all the Fabian executive might stand for Parliament! and I.L.P. [Independent Labour Party] candidates might be subsidized in their constituencies.'[7] 'Old Hutch' would probably have approved of the proposal; but Sidney Webb thought that 'collectivism' and 'radicalism' were already such strong and widespread sentiments among thinking and working people alike that no further 'splash' was needed to promote them. With the quiet determination which was so characteristic of him he steered his companions to another idea. Rather than fritter away the funds for ephemeral political purposes, a permanent institution should be created. It would bolster the cause by knowledge. Research was wanted and economic study; in short, a London School of Economics and Political Science.

Webb probably did not invent the idea on the spur of the moment he received the letter from Derby. Janet Beveridge tells us (though without giving chapter and verse) that 'the project to create a London School of Economics was in [Sidney's] mind even before he met Beatrice in 1890'.[8] Certainly, Webb had been impressed by research-driven academic institutions like the Massachusetts Institute of Technology (MIT) which he had visited in 1888 ('the *Economics* course is the best I ever heard of'[9]), and the German Technische Hochschulen which became one of the inspirations of his Technical Education Board (TEB), as well as Imperial College in London. Also, he was familiar with the École Libre des Sciences Politiques, the Paris school of political science which was the object of much interest in London in the early 1890s, and to whose name only 'London' and 'Economics' had to be added. But the plan for what came to be known as LSE, if not yet its emergence to reality, was hatched on that August morning in Surrey. A few weeks later, on 21 September 1894, Beatrice Webb noted in her diary about her husband's intentions: 'His vision is to found, slowly and quietly, a "London School of Economics and Political Science"—a centre not only of lectures on special subjects, but an association of students who would be directed and supported in doing original work.'[10] It is not clear that Beatrice was entirely happy with this use of the Hutchinson bequest, and certain that George Bernard Shaw was not; but Graham Wallas supported his friend, who in any case was not easily deflected.

This is, at least, what may have happened on 4 August 1894. It is probably not far from the truth though the 'future historian' invoked by two of the participants of the Borough Farm breakfast finds himself in a slight quandary. Or is it appropriate that the momentous event of the invention of LSE should be shrouded in a little mystery? Graham Wallas even wanted the latter-day researcher to establish its precise date when he reminisced on the occasion about which he had otherwise little doubt. The letter from Derby is likely to have reached Borough Farm on 3 August. Webb immediately sent a note to his trusted confidant Edward Pease: 'Don't tell anyone all this—we ought to keep it as quiet as we can, lest there be huge claims from everyone.' Was he himself the one to tell 'all this' to the ever-talkative G.B.S? Certainly Wallas's story, told in 1922 and again, with slight variations, in 1925, did not impress Sidney Webb. In his own 'Reminiscences' written thirty-five years after the event, in 1929, Webb compares the Borough Farm breakfast to the founding myths prevalent in India, and takes issue with Graham Wallas's account of the 'invention' of the School at 'Borough Farm, a couple of miles south-west of Godalming, early in the morning of a certain day in August, 1894'. 'Any future historian', says Sidney Webb, 'might quickly detect here the presence of myth. The map of Surrey would show that Borough Farm is near Milford, and has no connection with Godalming, which is remote and unfriendly.'[11] But his memory had temporarily deserted the great fact-finder, for in 1894, his own letterhead read 'Borough Farm, Milford, Godalming'. Moreover in 1932, three years after the 1929 'Reminiscences', when faced with William Beveridge for an interview about the origins of the School, Webb was in a less mystifying mood and confirmed Graham Wallas's story. We have independent evidence that the Webbs, Graham Wallas, and George Bernard Shaw were at or in the case of Shaw near Borough Farm in early August 1894. We also know that apart from scribbling away at their respective projects they occasionally met for a 'ravenous plain meal' over which they talked at length.[12] We even have the drawing by the Fabian activist and artist Bertha Newcombe, also a house guest of the Webbs at the time, which shows the protagonists seated at what the present owner of Borough Farm calls 'the sand pit'.[13] With such circumstantial evidence and in the absence of contrary facts there is little reason to doubt Wallas's account of the event.

Bertha Newcombe's drawing, published in 1895 by the *Sketch*, is also a reminder of the youth of the Borough Farm quartet. By the standards of a later age at least, the four children of the later 1850s were young; a century later, not one of the School's seventy-three professors would be younger than the 35-year-old Sidney Webb of 1894 (the youngest, and he an exception, being 37). They were also full of plans and ideas. None of them followed a conventional career. Sidney Webb had actually abandoned two possible careers, one in a City broker's office and one as a civil servant, in order to devote himself to a lifetime of social reform by research and by political activity. Beatrice joined him in the memorable intellectual partnership which was to last for more than fifty years. Graham Wallas was then an easily distracted classics master at a North London school and would soon become the first politics professor at the London School of Economics. Shaw, 'drifting in London's intellectual bohemia',[14] had haltingly embarked on a life of journalism and before long of playwriting as well as public activity in innumerable campaigns and organizations.

Such language must not be misread. The four Fabians were by no means freelancers and drifters without a purpose. They were serious, if anything—with the partial exception of Shaw—a little too serious even when they were young. They were all untempted by most worldly pleasures and had a strong puritanical streak. They were early representatives of a post-Gladstonian England which more often than not turned against a market-oriented, free-trade liberalism. Gladstone's last government had fallen in March 1894; now Lord Rosebery of the 'efficiency' movement was Prime Minister, and Joseph Chamberlain the social imperialist began to dominate public debate as well as political reality. In this climate of change, the Webbs and their friends formed a vision of the future which had more to do with Comte and Saint-Simon, and with the positivism of Herbert Spencer, than with the Gladstonian half of John Stuart Mill. They dreamt of an organized, well-run society rather than the free-for-all of 'Manchester liberalism', of the hegemony of well-trained benevolent experts rather than entrepreneurs. Moreover, they were convinced that the first step of the road to reform was to find out facts; 'facts shall make you free'. That is why they, Sidney Webb at any rate, wanted a London School of Economics and Political Science, and that is what they wanted this School to promote.

Graham Wallas Declines the Directorship

In the months that followed the Borough Farm breakfast, the plan was taken forward 'slowly and quietly', as Beatrice Webb had anticipated. At times it progressed very slowly indeed. There were several reasons for the delay. One was simply that Sidney Webb was extremely busy. On one occasion early in 1895 when pressed about the LSE plan, he replied in exasperation: 'I have only two hemispheres to my brain and ten fingers!'[15] He certainly needed the two hemispheres. Since March 1892 he had been the Progressive Liberal member of the London County Council (LCC) for Deptford, and had taken a major interest in its education committees. In

March 1895, a difficult re-election battle was due. Beyond that the Fabian Society was not the only organization to attend to, and of course there were the ever-present research projects; in 1894, the Webbs' *History of Trade Unionism* had appeared, to be followed by *Industrial Democracy* three years later. Still, LSE was never far from Webb's mind. One 'evening we sat by the fire [Beatrice noted in her diary] and jotted down a list of subjects which want elucidating—issues of facts which need clearing up. Above all, we want the ordinary citizen to feel that reforming society is no light matter, and must be undertaken by experts specially trained for the purpose.'[16]

Webb had been right to suspect that it would take some time to settle the legal and technical questions concerning the Hutchinson estate; the process was not finally concluded until February 1895. In the mean time, the Hutchinson Trustees met on two occasions, in September and in November 1894. They had in fact become Sidney Webb's little planning group for LSE, their impressive-sounding name concealing perhaps just how little it was. Constance Hutchinson attended once but in October 1894 went to the United States, where she died in 1895. William Clarke rarely turned up for meetings. Like William de Mattos, Clarke was an early Fabian, though the 'journalist of depressed disposition' left the Fabian Society and died in 1901. De Mattos attended meetings for a while but in 1895, 'after some scandals about his attitude to sexual reform',[17] emigrated to Canada, where his tracks vanish after he boarded a ship at Vancouver. On the other hand, Sidney Webb made the reliable Edward Pease Secretary of the Hutchinson Trustees and occasionally invited Graham Wallas to their meetings.

While Webb thus had no difficulty persuading the Hutchinson Trustees of his plan, the Fabian Executive was another matter. They had, of course, no formal standing in the question, but Webb wanted them on his side, if only because of the explicit mention of the Fabian Society in Hutchinson's will. Slowly but surely—though never entirely—Sidney Webb wore down the resistance of his fellow Fabians to what had by now become a fixed idea in his mind. On 8 February 1895 the Minute Book of the Hutchinson Trustees records agreement on the decision to spend the major portion of the money on education, and more particularly on the London School of Economics and Political Science. Its structure is outlined in some detail, including subjects of lectures ranging from 'the working of democratic machinery' to 'railway economics' and 'the history of the regulation of wages by law' ('chosen rather', as Webb endearingly reveals, 'because I know of suitable men available'). Five hundred pounds was set aside to start the 'experiment' in October. 'It would make a great public sensation, and would, I am convinced, "catch on". If not, it need not be continued.'[18]

Sydney Caine concluded later: 'We can thus identify 8 February 1895 as perhaps the most important date in the School's history, the date when the decision was taken to create it.'[19]

Webb certainly had the idea to found a School of Economics and Political Science in London, but he could not have done the day-to-day work of setting it up himself. For one thing he was preoccupied with other affairs. The LCC and its Technical Education Board was but one practical step towards his wider concern with social reform. More importantly perhaps, Webb had created a pattern for his life from which he would have found it hard to escape even then; in fact he never broke with it until his death. LSE had a place in his scheme of things but Webb was not to be constrained by becoming its director. Thus a director had to be found if the plan was not to join scores of good intentions on the scrap heap of reforms. Not unnaturally, Webb picked the member of the breakfast party and fellow Fabian, Graham Wallas.

Graham Wallas pondered the question for what to Webb seemed an inordinate amount of time. Six weeks went by from the 8 February green light by the Hutchinson Trustees to Wallas's negative decision on 23 March. The reasons for the decision remain unclear, especially since Wallas was to join the School as a lecturer of politics when it opened, remain in post until his retirement in 1923, and continue to teach and help in other ways to his death in 1932. From all accounts Graham Wallas was an impressive and much loved teacher, not quite a Laski perhaps, of somewhat more sedate and painstaking temperament, but an effective preacher and devoted mentor of his students. He had become a Fabian through his Oxford friend Sydney Olivier, who in turn was Sidney Webb's Civil Service colleague during Webb's stint in the Colonial Office; through Olivier, Webb and Wallas met in 1885. The 'Three Musketeers', as they were dubbed in Fabian circles, contributed major pieces to the *Fabian Essays* of 1889. Wallas's essay on 'Property under Socialism' contrasted the idealism of socialists 'fifty years ago' with the current temptation 'to undervalue the ideal'. This he regretted, for he did not belong to the union of fact-finders assembled at the early School. Yet the Webbs liked him. Beatrice praised his utter unselfconsciousness and lack of vanity and even called him 'a loveable man', which was generous in view of the fact that Wallas left the Fabian Society in a huff in 1904 because he objected to Shaw's conversion to imperialism. In a more backhanded way Beatrice Webb described Wallas's 'enthusiasm, purity of motive, hard if somewhat mechanical work'.[20] His *Great Society* and *Human Nature in Politics* certainly do not read as well as his lectures must have sounded.

In retrospect, one should perhaps be grateful that Wallas declined the offer after long weeks of reflection. Had he accepted, it is quite possible that the LSE would never have come about at all. Wallas's biographer, Terence Qualter, describes his writings in terms which might also be applied to his life's work in general. Initially Wallas was often exciting and dazzling. 'But longer acquaintanceship leaves one curiously flat. There is a sense of waiting for something to happen. But it never does. Wallas was magnificent in suggesting what ought to be done, and was the obvious inspiration of dozens of books which followed him, but his own positive contributions were slight.' This is a harsh judgement by a biographer who—as sometimes happens—seems to have fallen out with his subject three-quarters of the way through telling the story. If one wants to find out about Graham Wallas's political thought, one is therefore well advised also to read Martin J. Wiener's *Between Two Worlds.* Still, there is truth in Qualter's statements, and even more in the conclusion that 'Wallas's own contributions to administrative reform were slight. He was, as always, much more adept at inspiring and preaching than in actually doing things.'[21]

Given the close relationship between the men, it seems almost certain that Graham Wallas had initially indicated to Webb his readiness to accept the directorship. In the weeks after the Trustee decision of 8 February 1895, Webb was too involved in his election campaign to give much thought to what happened. On 2 March he was narrowly re-elected in the Deptford constituency and could think about the LSE plan again, only to find that Wallas declined the offer of the directorship. 'There remains a little mystery whether Wallas changed his mind or whether, as he afterwards implied, he had never been intended to do more than hold the job until someone else was available.'[22] Either way, unless someone else could be found quickly, the plan to found a London School of Economics and Political Science would falter.

TWO YOUNG MEN IN A HURRY

Enter Hewins

On Sunday 24 March 1895, Sidney Webb wrote an urgent letter from his Westminster home at 41 Grosvenor Road to an Oxford don:[23]

Dear Hewins
The project of the 'London School of Economics'—necessarily dormant during [London County Council] elections so far as I was concerned—was, I hoped, progressing. But Graham Wallas now decides after all that he cannot undertake the Directorship!

This is an unexpected blow. I write in haste to ask whether it would be possible for you to come and see me any day during the coming week. We should be very pleased to put you up. It is now a matter of serious import whether the scheme can be carried through. I am still keen on it, and if it should be possible for you to help to a greater extent than we contemplated it might still be done.

Yours very truly

SIDNEY WEBB

The letter came as a surprise, not least to its addressee—though Hewins, as we shall see, thrived on such surprises—and it changed the course of the foundation significantly. Now the School was to find its other founder, the one who was to put ideas into practice and thus create its reality. Hewins could not have been at the Borough Farm breakfast for he was emphatically not 'one of us'. He was not a Fabian, let alone a socialist; in later years he became a Conservative Member of Parliament. In fact, Sidney Webb hardly knew him. He and Beatrice had had a chance encounter with him in the Bodleian Library at Oxford where the young don had saved them from the fangs of a not very user-friendly librarian. They also knew of him as the author of a book on seventeenth-century trade and (unsuccessful) applicant for the Tooke Chair of Economic Science and Statistics at King's College London in 1891. Hewins in turn had heard of the plan to establish a School of Economics and had actually agreed to contribute a course of lectures; in fact, it was his characteristically impatient question which had led Webb to write that he only had 'two hemispheres to my brain and ten fingers'.

If the 35-year-old Sidney Webb was young by the standards of a later age, the recipient of his letter of 23 March was young by any standard. William Albert Samuel Hewins was 29 years old at the time, and a tutor in economic history at Pembroke College, Oxford, as well as an active University Extension lecturer teaching economics to adult audiences all over the country. He had floated with ease through a conventional school and university career which was facilitated by an open scholarship at Pembroke College. But exposure to his father's industrial ventures in the Black Country left a deeper impression on the young man than Oxford conventions. He had briefly intended to seek ordination as a Church of England minister but then decided that economics was perhaps a more effective tool for pursuing social reform in the modern age; an economics at any rate which is aware of its 'practical value'. Although he got his first degree in mathematics, the motive force of Hewins's work was application. It was certainly not ideology.

In 1892, Hewins married Margaret Slater, who also came from a family of industrialists and with whom he had three children. His private life remained throughout just that; he kept his personal feelings to himself though he found it easy to express emotions about great men and important ideas. He was intensely loyal to the great men he encountered. To his equals 'he was an agreeable, sympathetic, instructive and elevating companion, producing the impression alike of a keen and exact student, and of an uncommon personality, strong, purposive and independent'.[24] Yet even between the lines of this appreciation by an obituarist who had known Hewins from his Oxford days, one can sense why it was that early students at LSE described him as 'remote' and even regarded him as 'cold and fish-like'.[25] Apart from his mother, who called him Albert, no one seems ever to have addressed him by one of his first names, William Albert Samuel. In his case this was more than the public-school style of the time. His wife called him 'my dearest', and everyone else 'Hewins' or, like Beatrice Webb, 'W. Hewins'. Miss Mactaggart, the early administrator of the School, remembered many years later that 'he worked far too hard and had too much to do', and also that he was unable to delegate work to others. 'He did not leave anything to anybody; nothing was done by anyone else.'[26]

In those March days of 1895 Hewins was eager enough to be a part of the new venture. He telegraphed on receipt of Webb's letter that he would arrive in Paddington on the following Tuesday afternoon at 4 p.m. Sidney Webb may have considered telegrams a slight indulgence, or he was not in quite as much of a hurry as his correspondent; in any case his letter of reply told Hewins on Monday that 'we shall be very glad to see you as your telegram proposes'. Hewins duly arrived, and had tea with Beatrice Webb while Sidney was attending to LCC business. In the evening, guests came for dinner at Grosvenor Road, so that intimate conversation was difficult. But on Wednesday morning, Sidney Webb and Hewins talked things over, from the Director's salary to the needs of a 'fair trial' for the new School. In the afternoon, Hewins returned to Oxford.

Two days later, on Friday 29 March 1895, Sidney Webb wrote again. He reported that on Thursday the Hutchinson Trustees had met and had solemnly decided 'to go ahead with the proposal of a London School of Economics and Political Science on the lines which you and I worked out together. They attach great importance to securing your cordial assistance and, indeed, unless you can undertake the Directorship and carry out the scheme, it will probably have to be abandoned.'[27] Such drastic action turned out to be unnecessary. Hewins replied on Saturday 30 March telling Webb that 'you will be glad to know that I accept [the Trustees'] invitation to undertake the Directorship of the proposed new School and that I will at once set about the work of organization'.[28] He was as good as his word. His appointment took effect on the following Monday, 1 April 1895, eight days after he had first heard of the opportunity.

Hewins did not slacken his pace. Within six months he found rooms for the School, designed the syllabus of its courses, gathered support for the new venture *urbi et orbi*, and attracted over two hundred students so that the first academic year of the London School of Economics and Political Science could start on 10 October 1895. From the moment of his appointment, in other words, things moved anything but 'slowly and quietly'. In fact they moved with what to us in a more bureaucratic age appears as breathtaking speed. To be sure, the end of the nineteenth century was what Germans called a *Gründerzeit*; a time of founders. But the same can be said for the 1960s, at least in the academic world. In Britain seven new universities were founded within three years. Albert Sloman, the first Vice-Chancellor of the new University of Essex, has described the process in his 1963 Reith Lectures, *A University in the Making*. He traces the progress of the university from initial inquiries through the promotion of the idea, the application to the University Grants Committee, the process of academic planning, the early appointments and the architects' site plan to the admission of the first students and concludes: 'I am inclined to think that a period of less than three years between establishment and taking the first students would be inadvisable, even if it were possible.'[29] We like to speak of the ever-accelerating pace of history in the twentieth century but, so far as deliberate human endeavours are concerned, the opposite seems closer to the truth. For every year of the foundation of the University of Essex, that of LSE required one month.

The reasons are many, but they are epitomized by the fact that in 1895 there were no committees, councils, or boards which had to be consulted. Even the Hutchinson Trustees, when they met on 28 March 1895, consisted of Sidney Webb as chairman, Edward Pease as secretary, one other member, William de Mattos, and, on this occasion, Graham Wallas as a guest. Thus, the foundation of the London School of Economics was the work of two men, Sidney Webb and W. A. S. Hewins—two young men, though even their age was not particularly remarkable in 1895. They had an idea and the means to put it into practice, which is what they did. The prehistory of their venture was over; the 'great romance' of the London School of Economics began.

The Story of the Foundation (1)

The story of the foundation of the LSE has been told many times. Sir Sydney Caine, the sixth Director of the School, published most of the relevant documents in his *History of the Foundation of the London School of Economics and Political Science*, and added a characteristically cool, calm, and collected account of events. Friedrich von Hayek, for twenty years professor at LSE and later Nobel laureate in economics, paid much attention in his memoir on the occasion of the fiftieth anniversary to the academic climate in which the School was established. Janet (Lady) Beveridge, as Jessy Mair, was Secretary of the School for most of the period of the directorship of William Beveridge who became her husband after their departure for Oxford, and completed a more romantic account of the foundation—*An Epic of Clare Market*—just before her death in 1960. This 'labour of love' is devoted to the 'parents' of the School, Sidney and Beatrice Webb, and their intriguing relationship, as well as to some of the 'godparents' among Fabians and others. The founders themselves have told their story, though perhaps not as fully as one would have wished. Sidney Webb's account has to be read through the eyes of his wife in her diaries and autobiography, except for the brief 'Reminiscences' of 1929 and the transcript of the Beveridge interview in 1932. Hewins devoted a few pages of his *Apologia of an Imperialist* to the early days of the LSE, though the main objective of his autobiography is, as the title suggests, the story of tariff reform. Graham Wallas wrote a short piece in 1922 for the *Students' Union Handbook* which provides the first published evidence of the inception of the School at Borough Farm. There are other reminiscences by students and staff of the early days as well as later accounts based partly on hearsay and partly on documents. In more recent years a number of dissertations have been written in several languages about aspects of the early history of the School.

The breakfast apart, there is nothing mysterious about the story of the foundation, although it was later embroidered by authors who wanted to make their own points or were carried away by nostalgic affection. Appropriately, most accounts begin with the money which turned vague ideas into a viable project. However, no one observed that while Hutchinson's money unlocked the founders' imagination only a fraction of it was needed to make the foundation possible. When Hewins later drew up the School's first annual budget on the back of an envelope, it came to just over £714. This sum, to be sure, erred on the low side. It did not even include Hewins's own salary of £300 let alone the foreseeable needs of a growing school with its staff salaries, scholarships, research facilities, and the rest. In actual fact, £2,400 was needed to keep the School going during its first year. But even that was not only much less than the Hutchinson money might have provided, but included such other sources as tuition fees, which rapidly rose from 10 per cent to 25 per cent of the School's income; donations from individuals like Bertrand Russell and soon Shaw's 'lady' Charlotte Payne-Townshend; specific contributions, for example for railway studies; and public funds. Tapping these was the work of Sidney Webb, who at least in financial terms may be said to have founded the School twice over.

When Webb was first elected to the London County Council in 1892, he found there a committee set up under the Technical Instruction Acts of 1889 and 1891. The remit of this committee was anything but clear though it had funds for 'technical education' at its disposal, which stemmed from the customs duty levied on beer and spirits under the 1890 Local Taxation (Customs and Excise) Act. The funds were colloquially referred to as the 'whisky money'. Sidney Webb redefined and reorganized the relevant committee into the Technical Education Board (TEB), got himself elected as chairman, and from 1895 included LSE in the list of eligible recipients of financial support. Clearly this helped though equally clearly it would have been difficult to set up the School entirely on 'whisky money'. Webb wrote years later that 'not more than between three and four thousand pounds' of the Hutchinson money had actually ever been used by LSE,[30] and no doubt he knew what he was saying. But of course the Hutchinson money was free money, entirely at his (and his fellow Trustees') disposal. At the very least, the Hutchinson bequest provided a welcome cushion of security for the new scheme. It also concentrated the minds of those who pursued it.

Meanwhile, W. A. S. Hewins was able to concentrate his mind on other matters. When Hewins began as Director of an as yet non-existent LSE on 1 April 1895 he saw three tasks for himself which were: to harness support for the new idea, to give it academic substance, and to see to the practical requirements of implementation. He tackled the tasks in this order of importance, though not necessarily in this sequence. Almost at once Hewins sent letters to dozens of economists and other social scientists all over Europe informing them of the new venture. Without exception they replied with encouraging comments and offers of help. Even Alfred Marshall, who was later to fall out with Hewins and with LSE, wrote on 7 June 1895: 'I was very pleased to see that so much progress has been made on what seems so excellent a plan.' If there were doubts, they concerned less the project itself than its chance of realization, as in the letter by the Manchester historian Thomas F. Tout on 9 June 1895: 'I am afraid it will be a long while before we are ripe for anything like the École Libre des Sciences Politiques in this half-educated country.'[31] Hewins soon managed to prove such sceptics wrong, if only by persuading scholars like William Acworth, Arthur L. Bowley, Herbert Foxwell, William Cunningham, and Halford Mackinder to come and teach at the new School.

Public support, academic substance, and practical issues merged when Hewins began to look for rooms for the planned lectures and wrote to the Society of Arts and the Chamber of Commerce. Sir Henry Trueman Wood of the Society of Arts raised a question which was soon to become only too familiar: 'Is the new school to be a genuine teaching institution? or is it to be founded and worked in the interests of any particular set of social economists? In the latter case, I doubt if the Society of Arts could wisely meddle with it.'[32] Hewins went to see the Committee and persuaded it of the School's scholarly neutrality. The Chamber of Commerce proved even more reluctant, though in the end its council passed a resolution which stipulated the condition: 'Lectures should deal with principles and with historic facts rather than with the advocacy of the particular view of any one party or thought.'[33]

It was not very difficult for Hewins to embark on this particular journey of persuasion. As Miss Christian Mactaggart observed in her memoir, he 'had accepted the post [of Director] only on the understanding that the lecturers should be appointed only because of capacity to deal with the subject to be lectured on with nothing whatever to do with the colour of their political opinions'. So Hewins went and told the Chamber of Commerce on 27 June 1895 that 'the School would not deal with political matters and nothing of a socialistic tendency would be introduced.'[34] In this he had the full support of Webb, though both were soon to be reminded that they were not alone in the world.

A Significant Distraction

On 1 July 1895, four days after Hewins's Chamber of Commerce address, George Bernard Shaw had a day of unfortunate mishaps. He was caught by rain coming back to London from his country place in Limpsfield, 'so that I reached home so encrusted with mud that I looked like the beginning of a very rough sketch of myself in clay for the statue that will no doubt some day be erected for me'. (In the event, the plan to erect a full-size statue Shaw in Adelphi Gardens fell victim to a 'confused' and 'discordant'[35] Memorial Fund meeting in 1951, so that the Rodin bust is the best we have got.) Once home on that wet July day, Shaw discovered that some urgently expected tickets had been mislaid or had not arrived and lost much time trying to chase them. When he finally set off to join the Webbs for dinner at Grosvenor Road his bicycle pedal came loose on the way, and as he screwed it back on another shower drenched him. Not surprisingly he abandoned the expedition at that point and returned home, where he proceeded, clearly not in the best of moods, to write a letter addressed to Beatrice Webb but intended for Sidney ('Please shew him this letter and allow it to rankle') about what he called, 'the Hutchinson business':

At the last [Fabian Society] executive discussion of it Webb quite terrified me by shewing an appalling want of sense of the situation; and my most violent efforts failed to wake him up: he kept making exactly the speeches he makes to the Chamber of Commerce and the County Council (TEB) about it. When [the two Fabian Executive members] Bond Holding and Martin began to stare with all their eyes, he first attempted to prove that Acworth was the greatest living authority on railways; then that his object was 'abstract research' (meaning a kind of research in which Bosanquet and Miss Dendy [of the Charity Organisation Society] would arrive at exactly the same conclusions and produce the same information as [Graham] Wallas and [the feminist socialist] Miss Stacy); then, feeling rather hurt at the evident scepticism of Martin as to the feasibility of this academic ideal, he hinted that the bequest had been left to him to dispose of as he thought fit, and that the executive had nothing to do with it. The general impression left was that the Hutchinson Trustees are prepared to bribe the Fabian by subsidies for country lectures and the like to allow them to commit an atrocious malversation of the rest of the bequest; and that as the executive is powerless, the best thing to do is to take the bribe, and warn future Hutchinsons to be careful how they leave any money they may have to place at the disposal of Socialism. This won't do. First, Hewins must be told flatly that he must, in talking to the Guild of St Matthew and the other Oxford Socialists, speak as a Collectivist, and make it clear that the School of Economics will have a Collectivist bias. Any pretence about having no bias at all, about 'pure' or 'abstract research', or the like evasions and unrealities must be kept for the enemy. Second, Hewins must be told that he MUST get somebody else than Acworth and Foxwell to put in the bill. It is easy to say where are the men: the answer is that that is precisely what Hewins is being paid out of Hutchinson's money to find out. Third, the Collectivist flag must be waved, and the Marseillaise played if necessary in order to attract fresh bequests. If the enemy complains, it must be told that the School has important endowments the conditions of which are specifically Socialist; and that if the enemy wants specifically individualist lectures, he must endow too. Meanwhile the Fabian must be kept in mind of the fact that Acworth was chosen by the LCC; and Webb positively must not declare that he is an authority on the subject, or indeed that he is anything but hopelessly wrong and invincibly stupid on it, which is the solemn truth.[36]

So there it is, the great bogeyman which was to haunt the School even before its foundation stone was laid! What was it to be, an academic institution committed solely to advancing and imparting unbiased knowledge, or a place with a mission, and one of collectivism if not socialism at that? In one sense the answer was never given, for it was impossible to give. At least some of the founders held strong views, and however solemnly they protested their readiness to abide by the judgement of 'facts', they did not abandon hope that their views would prevail in the end. Moreover, the dilemma of value and fact has never been resolved in the social sciences, not even by the great Max Weber, who at the time of the establishment of LSE began his advocacy of a value-free social science while also preaching German nationalism. In theory, it may be possible to hold politics as a vocation and social science as a vocation apart, but in practice the history of LSE shows that the two will clash whenever they are given half a chance. It is an achievement to live with the dilemma, tolerating occasional lapses to one side or the other but above all turning the tension between scientific inquiry and political commitment to creative advantage.

Hewins's views on these matters were fairly, though by no means entirely, clear. After all, as a young man at Oxford he had started, in 1886, a Social Science Club, the object of which 'was quite definitely to find a way to the solutions of social difficulties by practical investigation'.[37] Sidney Webb might easily have used such language. Webb in fact placed much emphasis on value-free social science, more than any of his associates including his wife, who probably lost at least some of her interest in the new School when she discovered what it was to be, and even more what it was not to be.

However, Webb was not to be deflected from his plan. At an early point, he turned for support to the suave lawyer–politician Richard Burdon Haldane, an unlikely friend perhaps but one who was to help in many of the ventures in which Webb was involved. Indeed, he had already helped bring Sidney and Beatrice together when her ailing father was still an obstacle to their marriage. Haldane, born in 1856, belonged to the Borough Farm generation but lived in a different world. He enjoyed society life, or perhaps simply life. 'As for the Sidney Webbs,' he wrote a little patronizingly in his autobiography, 'whatever be their failings socially, they are splendid workers, and I should be proud to feel that I had given up so much for a cause as they have.'[38] Still, he supported the cause in many ways, not least by his legal opinion on the alleged 'atrocious malversation' of the Hutchinson money. Not so, he said, and argued with considerable ingenuity

> that the expression 'propaganda and other purposes of the said Society and its Socialism' do not confine the trust to the precise basis and rules of the Society as these exist now, but that, so long as the propaganda of the Society are Socialistic,

any objects within its practice, if that is not directly contrary to these rules, will be proper trust purposes.[39]

It is unlikely that G.B.S. was satisfied by such language, but in fact the Fabian Society accepted the 'bribe' and sponsored a series of 'country lectures' from the Hutchinson bequest; and Webb and Hewins continued to attract funds from 'the enemy', in which category Shaw had generously included not only the Chamber of Commerce but also the LCC and its TEB with Webb as chairman, though presumably not Charlotte Payne-Townshend, who was in due course to become his wife. Shaw and Webb remained close, though they certainly stood on different sides when it came to 'the Hutchinson business' and continued to quarrel over it for years. Shaw never took much interest in LSE; instead he devoted an almost inordinate amount of time and energy to 'the Fabian', as the Society was called by its members. While Webb wrote slightly more of the early Fabian Tracts than Shaw, Webb's were more likely to be entitled *Facts and Figures for Londoners* (Tract nos. 8 and 10), *Questions for School Board Candidates* (no. 25), and *Municipal Tramways* (no. 33), whereas Shaw's were *A Manifesto* (no. 2), *What Socialism Is* (no. 13), *Report on Fabian Policy* (no. 70), and of course the infamous (though innocuously titled) *Fabianism and the Fiscal Question* (no. 116), which nearly split the Society in 1904. For over twenty years the Fabian Society could rely on G.B.S. producing the right text at the right time and defending it effectively.

Webb went in a different direction altogether (and his wife was torn between the two). He not only put his heart—or was it, as always, his head?—into the London School of Economics and Political Science but also formed a clear view of his—and its—objectives. We have seen that his memory was not flawless, but there is no reason to doubt the thrust of his recollection (in 1903) of a conversation with the City dignitary and one-time chairman of the Chamber of Commerce, Sir Owen Roberts:

I remember going to Sir Owen Roberts in 1895 to try to interest him in the School, and telling him very candidly the ideas that I had on the subject. I said that, as he knew, I was a person of decided views, Radical and socialist, and that I wanted the policy that I believed in to prevail. But that I was also a profound believer in knowledge and science and truth. I thought that we were suffering much from lack of research in social matters, and that I wanted to promote it. I believed that research and new discoveries would prove some, at any rate, of my views of policy to be right, but that, if they proved the contrary I should count it all the more gain to have prevented error, and should cheerfully abandon my own policy. I think that is a fair attitude.[40]

If this sounds a little holier than thou, it is none the less an attitude which has marked the best of the London School of Economics over the years.

The Story of the Foundation (2)

While such debates were waging, Hewins got to work on the academic programme of the new School. He filled many pages with notes in his neat and slightly clerkish handwriting; he also continued to solicit advice. Whereas the Webbs had, at their fireplace in Surrey, listed subjects for which they knew likely teachers, Hewins went about the project in a more systematic fashion. Six main disciplines were to be taught (he noted on 24 April): economics ('including economic theory and economic history'); statistics; commerce ('commercial history and commercial geography, tariff systems of Europe, means of transport, railways etc.'); banking, currency, and finance; commercial and industrial law; political science, including public administration. This is not light years away from the present core of LSE, and it was a surprisingly modern list at the time.

By July 1895 Hewins's plan and Webb's realism had been merged into a printed prospectus. The eleven pages of it make impressive reading. Hewins was now able to state proudly, yet without exaggeration: 'The London School starts with the cordial co-operation of the leading economists and students of political science in the United Kingdom, and with the support of the Society of Arts and, on its commercial side, of the London Chamber of Commerce.' There was no precedent to go on for what was announced, certainly none in the United Kingdom, and the repeated references to the École Libre des Sciences Politiques in Paris as well as Columbia College in New York and one or two other foreign institutions sound more like invocations than genuine models. The truth is that the man who had founded a Social Science Club in Oxford in the 1880s now turned the club into a rudimentary university of the social sciences and thereby broke new ground.

The prospectus for LSE betrays no particular bias; certainly not a political bias. However, it contains one significant statement by emphasizing that of course 'the study of economic and political theory' will have a prominent place, but 'the special aim of the School will be, from the first, the study and investigation of the concrete facts of industrial life and the actual working of economic and political relations as they exist or have existed, in the United Kingdom and in foreign countries'.[41] When the Cambridge sage Alfred Marshall read this, he must have begun to wonder about the new institution. The edge is clear enough. Theory, it is suggested, has no real bearing on concrete facts, and ultimately concrete facts matter more than theory. From the start there is what the Fabian historian, Norman MacKenzie, likes to describe as the positivist bias in the academic affairs of the School.

The lectures and classes offered for the first year are now spread over nine subjects instead of four. Hewins himself would lecture under the heading of Economics on 'the state in relation to industry and commerce'. Four years before Webb had asked him to set up LSE, Hewins had applied for the Tooke Professorship of Economic Science and Statistics at King's College on the Strand; but the electors had regarded the 26-year-old as a little too inexperienced. Now Hewins enlisted the man appointed at the time, Cunningham, to lecture on 'the economic effect of alien immigrations'. The Revd William Cunningham was a Doctor of Divinity and came to be known as the 'national economist' because he disliked cosmopolitan liberals as much as the Roman Church. When he left King's College in 1897 to become Archdeacon of Ely, the Tooke Chair was offered to Hewins, who gladly accepted. The directorship of LSE was a part-time job and certainly carried no full salary. The lectures by Hewins and Cunningham were to be accompanied by elementary and advanced classes, so the prospectus tells us. Indeed, in Hewins's view, these classes were to be the academic core of the School.

The other eight subjects were all to be taught by unusual and remarkable men. Arthur L. Bowley announced a course in statistics. W. M. Acworth was to lecture under the heading of commerce, but in fact on railway economics, a subject on which despite Shaw's jibe he really was the leading authority. Railway economics was important for the early School for practical as well as academic reasons; its teaching brought in students, money, and wider support. Other lecture courses and classes were announced by H. J. Mackinder on commercial geography, by Hewins on commercial history, by J. E. C. Munro on commercial and industrial law, by H. S. Foxwell and George Peel on banking and currency, by Edwin Cannan on taxation and finance, and by Graham Wallas on political science.

'The lectures and classes will commence on October 10.' The idea was that all lectures and most classes should be given in the evening between 6 p.m. and 9 p.m. Men and women would be equally welcome. Students would not be prepared especially for any degree, but attendance would be useful for Civil Service examinations as well as those of the Institute of Bankers, the London Chamber of Commerce, and others. There would be three terms, and a fee of £3 would be charged for the year. Applicants were invited to write in specifying the lecture courses in which they were interested so that they could be sent tickets. 'Tickets are not transferable. Cheques and Post Office Orders must be made payable to W. A. S. Hewins.' Not all would be able to afford the fee; some scholarships were made available for students of ability. 'By this means the opportunities of scientific training afforded by the School will be brought within the reach of all who are likely to profit by them.'

The straightforward, almost bureaucratic language must not deceive us: these were innovative, indeed revolutionary ideas. So was the intention to combine lectures and classes with preparation for and assistance of research—'original research' as the prospectus said. The publication of a series of books had already been arranged with Messrs Longman, Green & Co., and not surprisingly among the first titles there would be books by W. A. S. Hewins and Sidney Webb. The collection of publications for a library was announced, although by July 1895 it was as yet unclear where its funding could be found. An 'information department' was to be set up 'to assist British students and foreigners visiting England for the purpose of investigation'. Even the earliest School had some of the makings of a graduate school of social research.

When Hewins sent out the prospectus he already had an address for the School, 9 John Street, Adelphi. The building in which two ground-floor rooms had been rented, has been pulled down since along with that whole side of the street to make way for the huge office block called Adelphi, and the street has been renamed John Adam Street. The area south of the Strand named Adelphi after the three brothers Adam (the *adelphoi*) was clearly a good address. It was also convenient, with the Society of Arts across the road, and the City, the Law Courts, Westminster, and Whitehall in walking distance. Judging from the Scandinavian visitor Rasmussen's words of praise, it even had a special meaning. 'Adelphi was not only a dream of antique architecture; it was just as much a finance-fantasia over risk and profit; the financier was an artist and the artist a financier.'[42]

Hermione Hobhouse, a granddaughter of Beatrice's sister Margaret and a member of one of the important LSE families, had the Adam brothers in mind when she published this description of Adelphi; for there was little finance and even less art in the physical conditions of the early LSE. The Hutchinson Trustees had approved the lease of the rooms in May, at a rent of £100 a year. Lectures were to be held there as well as across the road at the Society of Arts and in the Chamber of Commerce in Eastcheap in the City. The Hutchinson Trustees continued to act as a management committee for the incipient LSE, though it is safe to assume that this meant no more than Sidney Webb and Edward Pease with W. A. S. Hewins in attendance as an interested guest.

When Hewins wrote to his mother at the end of May 1895 that he had been busy with his new duties but 'the preliminary arrangements for starting the School in October are nearly all completed', he was clearly a little too sanguine.[43] For one thing, the Hewins family had to move to London, which they did at the end of August when they found a house at 26 Cheyne Row. On the other hand, later that summer Hewins was able to give a string of upbeat newspaper interviews to the *Citizen*, the *Westminster Gazette*, the *Echo*, and others. They make interesting reading, if only because the more familiar names of the Webbs are rarely mentioned in them, and Hewins was clearly popular with the journalists. Here was an Oxford man who went beyond the limitations of the old universities, as Arnold Toynbee had done before him. 'Such men are in advance of the University' (*Echo*, 8 Oct. 1895). They, or rather the LSE, also offered opportunities to gifted students who had hitherto not found a place. 'Now even the Board-school boy . . . may consider himself a Prime Minister *in posse*; and it is necessary that a more democratic (perhaps also a more efficient) machinery should be evolved for producing the perfect public servant and social philosopher' (*Westminster Gazette*, 27 Sept. 1895). The two are probably rarely the same person. A *Speaker* writer (15 June 1895) warned the founder that his creation 'must avoid the extremes of popularity and of pedantry' which as we know were to be ever-present in the new institution. However, the same writer called the as yet non-existent LSE 'one of the great English institutions of the new age' (*Echo*, 8 July 1895). Hewins himself, when he spoke to the *Westminster Gazette* (27 Sept. 1895), took up a theme which seems to have worried him throughout his life: 'I don't believe the twentieth century will look on social questions from the same point of view as the nineteenth. I believe we have got to the end of the period of mere enthusiasm, and are settling down to careful investigation.' Hewins was certainly a careful investigator, but he was also an enthusiast. Indeed, when he wrote his *Apologia* more than thirty years later, after decades in which he almost seemed to have forgotten the London School of Economics, he found it possible to say:

When I think of the first days of the School of Economics at No. 9 John Street, Adelphi, and contemplate the great organisation which has grown from those beginnings, I can only feel that I was privileged along with my colleagues, to take part in a great romance. Difficulties appeared from day to day only to be overcome. Although we represented different schools of thought and were on different sides of politics, I cannot remember any incident which disturbed the harmony of our relations during those early years or which interfered in any way with the rapid progress of our great undertaking.[44]

When the Hutchinson Trustees met on 30 September 1895 to review the arrangements for the first academic year of the London School of Economics and Political Science, Hewins was able to report success. The money was there, and lecture halls had been rented. The lecturers were in place, indeed one or two were added late to the list, including Bertrand Russell, who had just returned from Germany and offered a course on German Social Democracy. 'Are you quite sure of getting your students?' The *London* reporter who asked this question need not have been anxious; Hewins had left little to chance. Scores of people of all ages and classes had made use of the invitation to buy tickets. On 10 October 1895, the beginning of Michaelmas Term, the London School of Economics and Political Science embarked on its adventurous journey.

Hewins in his *Apologia* remembered the first students calling on him at 9 John Street, where they found 'an almost unfurnished room where there was one bureau and two chairs, one for myself and one for my visitor'. This did not deter the curious and the determined. Over 200, or even, according to Hewins's later account, over 300 students attended lectures and classes during the session 1895/6. Moreover, four weeks after the beginning of the first term, on 9 November 1895, Hewins was obviously pleased to note: 'The hope that the School would appeal to various classes of students has been fulfilled. Amongst those who have joined the School are men and women engaged in business, in municipal and public work, graduates of universities (English and foreign), civil servants, teachers (both in secondary and elementary schools), clerks, journalists, and working men.'[45]

It does not detract from the achievement of the founders to stand back and wonder what the London School of Economics and Political Science really amounted to in its first year. It was obviously a flimsy institution; indeed more an arrangement and a promise than a university. To be sure, a handful of research students were recruited during the first year, a number of classes turned into proper seminars, books were bought and manuscripts published, but beyond that the School consisted of little more than a series of evening lectures given in different places to diverse audiences, tenuously held together by a common name. The lecturers were distinguished but busy academics who, with the exception of the Director, held positions elsewhere. The audiences were a motley crowd ranging from 18- to 70-year-olds; some would drift away after a few lectures whereas others took a more serious and scholarly interest. The subjects taught depended largely on the specialist interests of teachers, though they were all if not applied social science then social science with a view to application. The rest was hope. The 'student from No. 9, John Street' who reminisced in the *Clare Market Review* about 'those far-off romantic days when about thirty excited enthusiasts used to crowd joyfully into a space intended to hold fifteen, happily confident that the social movement was come at last, and they were in and of it',[46] saw things through the rose-tinted glasses of 1907 when the School was well and truly established. In October 1895 an important step had been taken, but almost everything was still to do before even the expression 'the School' had any justification. Fortunately Sidney Webb and W. A. S. Hewins provided the mixture of vision and dogged hard work which was needed.

MEN AND IDEAS

Founders and Supporters

Two young men founded the London School of Economics. They were utterly different and yet perfectly matched. W. A. S. Hewins could never have invented LSE; his was not a visionary or even creative mind. But without him the School would not exist, for Sidney Webb could not have done what Hewins did, not just because he was busy with other things, but because he needed others to organize what flowed from his famous forehead. Beatrice organized his life, and Hewins organized LSE with all his remarkable energy and capacity to put things and people together.

But the two founders could not have done it alone, nor were they living in a vacuum. At least a handful of other names have to be added to the founders' roll of honour: the unfortunate Hutchinson as the ghost at the party, the other three members of the Borough Farm quartet, the great facilitator R. B. Haldane, and of course the first teachers at the newly established School, Arthur Bowley and Halford Mackinder in particular, who both were to play a major role for three decades and more, as well as the first administrators, notably the formidable Miss Mactaggart.

More importantly, all these people were highly sensitive to the intellectual and political currents of the time; they would hardly have thought of a school of economics and political science if that was not the case. These currents were variegated and strong, but they all had in common that they were reactions to the age of Gladstone. The 'five *E*s' which made up the field of intellectual forces in which LSE came into being—Education, Economics, Efficiency, Equality, Empire—were associated with the great or at least the fashionable names of the time. Whether they add up to one coherent philosophy may be doubtful. The 1890s were more a time of dissolution than of construction. 'The former clear objectives were gone, and as yet nothing took their place.'[47] Perhaps the London School of Economics provided one focus for defining the new objectives and preparing people to achieve them.

Sidney Webb, the senior founder, was central to this story. Curiously, there is no full-scale biography of him. Whereas the glamorous-seeming Beatrice Potter-Webb continues to attract authors, Sidney left most of the guild of biographers cold and those who tried despaired of an abundance of material without a central theme. 'Your satellite Wallas and your sun Webb, I being a mere comet,'[48] George Bernard Shaw wrote to Beatrice once, implying perhaps that she is the planet around whom everything revolves. In fact of course the planet revolves around the sun. Beatrice Webb's latest biographer, Carole Seymour-Jones, describes her at the time of the reluctant marriage as 'a woman of ability but lacking in confidence, who was cast adrift from her natural social milieu and was in need of both an ideology and a husband'.[49] Sidney Webb provided both, and more. Born in 1859 to an Evangelical mother and a political father, he grew up in uninspiring if financially secure circumstances in London. The parents of Sidney and his two siblings, Charles and Ada, 'plainly belonged to the anonymous lower middle class, living in genteel but cramped and ugly style',[50] though learning good manners from the dominant mother. Webb never spoke much of his family; indeed he always remained a very private man. From the age of 15, he spent two years in the house of a Lutheran pastor in northern Germany. On returning to London, the 'shy and gauche' cockney boy got a job as a colonial broker and embarked on an extraordinary career of evening courses and qualifying examinations which showed his single-mindedness and capacity for sheer hard work as well as his desire to learn and to find out about things. He eventually got into the Civil Service, more precisely the Colonial Office, of which he was much later to be the responsible cabinet minister. The young clerk decided to take yet another degree, this time in law at London University, and soon after set out for a free-lance career in social research and reform. The deliberateness of this progression in the life of a 30-year-old alone is awe-inspiring.

The story of the marriage of Sidney and Beatrice Webb in 1892 leaves a bitter-sweet taste, however effective their lifelong partnership was to become until it even generated a little affection. Socially, Sidney and Beatrice were worlds apart, and it mattered. The upper-class lady encountered a man of whom Bertrand Russell's grandmother declared, having watched him cope uncertainly with his several knives and forks at dinner, that he was 'not quite'. 'Not quite what?' asked Russell (as his biographer Caroline Moorehead reports). '"Not quite a gentleman in mind or manners", Lady Russell replied.'[51] Beatrice and Sidney had both gone through unhappy experiences of the heart when they met, though in Beatrice's case the much-heralded love affair with Joseph Chamberlain was probably no affair at all but a crush by a young woman on a famously desirable man who played with his admirer's affection without ever responding in kind. Sidney, not unnaturally, wanted to love his wife but Beatrice reminded him firmly of their pact, which does not become more endearing by its frequent quotation: 'Let me have your head only—it is the *head* only that I am marrying.'[52]

It was, to be sure, a remarkable head on whom no one who met him failed to comment. 'He had a fine forehead, a long head, eyes that were built on top of two highly developed organs of speech (according to phrenologists) and remarkably thick, strong, dark hair,' wrote George Bernard Shaw about the first encounter with his then 21-year-old friend, and added an observation on what was going on inside the head: 'He knew all about the subject of debate; knew more than the lecturer; knew more than anybody present; had read everything that had ever been written; and remembered all the facts that bore on the subject.'[53]

Norman and Jeanne MacKenzie have done more than anyone else to give Sidney Webb his proper place in the history of the LSE and beyond. They had harsh things to say about the partnership of two people 'who were over thirty when they met and perennially middle-aged thereafter'. They were 'more often respected than genuinely liked', which is perhaps not surprising if it is true that 'the springs of their energy . . . may well be found in a need to escape from a temperamental mood of despair—with oneself, and with the state of society—a need to force the world into a more regulated and thus more acceptable frame'. It goes with this approach that 'the Webbs had never been unduly sensitive on the question of liberty, for they had always believed that the individual should be subordinate to society—provided that the society was morally just'.

But the MacKenzies also praise the Webbs as 'starters of things', as latter-day Saint-Simonians, and then single out Sidney's 'bold-mindedness'.[54] This was coupled moreover with untiring attention to detail. Webb not only designed new schemes but he saw them through, or at any rate he saw to it that people were appointed to see them through. In the foundation of LSE, Beatrice's role is peripheral. To be sure, she was present not just at the creation but at every stage. She hosted breakfasts and dinners of consequence; she took part in discussions about the scheme; she listened, and of course she recorded events. If the early history of LSE—as that of the Fabians—is as much recorded history as it is history in the making, this is owed to Beatrice Webb's gifts as an observer and a diarist. And what a talent it is to write history as one makes it, and at times make it by writing about it! But somehow one feels that Beatrice's heart was not in the new foundation. LSE was truly Sidney's creation, or rather, that of the unusual team of Webb and Hewins.

Hewins brought into the early School those lecturers who were 'appointed only because of capacity to deal with the subject to be lectured on', and they were a highly respectable group of academics. Webb added somewhat less professional, though equally important people. There were the two other participants of the Borough Farm breakfast of course, Shaw and Graham Wallas. Shaw was hardly a 'mere comet' even in relation to LSE; in the early days he was always around in one capacity or another; he may have irritated people but he also gave them pleasure and food for thought. Yet as we have seen he can hardly be described as a founder of LSE. As his biographer Michael Holroyd puts it, 'Shaw's maverick figure does not find a part in the undramatic plans Webb had made for revolution through research'.[55] Shaw was not the only maverick around; H. G. Wells had his own ambivalence towards Sidney Webb and what in his *roman-à-clef The New Machiavelli* he ironically called Oscar Bailey's 'Westminster School of Politics and Sociology'.[56]

Edward Pease played a more continuous part, not only because he was the eternal Secretary of all relevant trusts and clubs and societies, but also because he lived to the age of 97. Pease was born in 1857, the son of a fundamentalist Quaker. He came to London and found work as a clerk in a trading firm but soon got more and more involved in organizations of social reform, including the Fellowship of the New Life which preceded the Fabian Society. Pease was devoted to Webb and in his notes about the early days of the School describes him as 'the sole force' in the place while characteristically adding: 'I of course knew all that went on, but I do not claim any initiative.'[57]

This is a short list of people. R. B. Haldane must be added to it, another contemporary of the generation of the later 1850s (he was, like Shaw, born in 1856), Scotsman, barrister, and rapidly rising in the Liberal Party, in which he was to have a distinguished political career. Haldane's reliable and friendly presence helped the School through many a crisis and turned out to be crucial for the admission of LSE to the University of London six years later. The philosopher–mathematician as well as lifelong political activist Bertrand Russell not only agreed to lecture at the fledgling institution but also donated his Cambridge fellowship stipend from Trinity College to LSE for research scholarships. The wealthy but emotionally starved Irish lady Charlotte Payne-Townshend allowed herself to be persuaded by the Webbs to join the Fabian Society, by Hewins to help pay for the building at 10 Adelphi Terrace for the School's library, and by G.B.S. to become Mrs Charlotte Shaw. If one adds up the names, little more than a dozen people were involved in the actual foundation of LSE. Why did they do it? Individual temperaments and idiosyncrasies apart, what was it about the spirit of the time that motivated such a bold venture?

The Five Es *(1): Education*

The answer is not straightforward or simple. Given the numerous currents and cross-currents of the time it is in fact all the more remarkable that the founders of LSE should have discovered and pursued with such single-minded determination their own route to the destination they wanted. Even the attempt to summarize the dominant currents as five big *E*s is a gross simplification of the mood of the 1890s. The desire to turn advanced *Education* to practical purposes was widespread and focused particularly on London. The development of *Economics* had accompanied but not really influenced the rise of industry. Now that a combination of entrepreneurial fatigue and Continental competition cast doubt on the inexorable quality of this rise, economists were called to the front or even volunteered. The quest for national *Efficiency* became the 'political catchcry' of the post-Gladstonian period. Its other theme came to be social reform in the name of *Equality* or, as it was mostly called at the time, collectivism, to replace the individualism of Victorian heydays. All these strands were sometimes tied together and sometimes torn asunder by the overarching theme of the *Empire*, which from the Irish Question to the Boer War and on to tariff reform dominated public and political debate. If we want to understand what the intellectual air was like which the founders of LSE were breathing and thus what ideas informed the new School itself, we must come to terms with the five *E*s and those who lived by, for, and with them. As we do so, we shall discover that all of them represented a reaction against Gladstone, against Cobden and Bright, against individualism and *laissez-faire*, cosmopolitanism and free trade and against the rule of intelligent amateurs who had emerged from real and would-be upper-class families, from the public schools, from Oxford and Cambridge.

Every one of the five *E*s has in recent years found its historian, though perhaps the full range of the late nineteenth-century craving for education still needs to be written up. Its source was in part the desire by public-spirited university teachers to reach out and lecture the masses, in part the need perceived by the business community and by some politicians to have a new and more relevant kind of university built around the applied sciences of the natural as well as the social world. Both groups looked abroad for examples and, as so often happens in Britain, they cited overseas developments with limited knowledge but to great effect. This may be a little unfair, considering the German experiences of Sidney Webb, Bertrand

Russell, and R. B. Haldane, the information about the École Libre des Sciences Politiques gathered by several committees at the time, and the close and frequent contacts of many with American universities. Still, such examples were taken out of context; they were used not to promote vocational or even technical training for everybody, but to apply the most advanced opportunities of research-led education to new groups of people and new areas of study. Even the reformers remained true to British type and opened doors for the few while leaving the many ill-housed and locked in.

What came to be called 'university extension' work, that is lecturing by university teachers to adult audiences all over the country, was begun by Cambridge dons before it caught on in Oxford. But in Oxford it soon became a movement rather than a practice. As the *Oxford Magazine* put it in 1883: 'A new faith, with Professor Green for its founder, Arnold Toynbee for its martyr, and various societies for its propaganda, is alive amongst us.'[58] T. H. Green, the self-appointed Hegelian, agreed with his master that the individual must be subordinated to the higher purposes of the state, but Green was in effect a moralist rather than a believer in the inexorable march of history towards a Prussian paradise. Arnold Toynbee died young, but left a legacy of commitment to social reform. On it, Arthur Acland built the growing group of young lecturers who were prepared to go out and teach in Cornwall and Nottinghamshire, Wales and Lancashire under often difficult conditions 'not merely . . . [for] diffusing a taste for intellectual culture but also . . . [for] relieving the monotony and elevating the tone of material pursuits, by nurturing an enthusiasm for the moral ideal'. These are the words of L. L. Price, one of those to whom the Committee Minute of 1889 applies: 'The Life of an earnest University Extension Lecturer must be one of self denial and devotion.'[59]

The organization of extension lectures and classes proved difficult at first. Arthur Acland's attempt to link the group of willing lecturers with the Co-operative Societies ran aground on the primarily practical interests of co-operators. It was only when Michael Sadler, a Student of Christ Church, took over as Secretary of the University Extension Lectures Subcommittee in 1885 that a viable organization emerged. Local committees were set up and a menu of short and extended courses, travelling libraries, syllabuses, classes, and certificates was offered. Between 1885 and 1891 the number of students attending extension lectures rose from 3,000 to 20,000; in 1890/1 192 courses were delivered at 146 lecture centres. If there was a disappointment for the idealists of Oxford, many of whom were members of the Oxford Economic Society, it was in the small number of working men who attended the courses. Attendance resembled more the later Open University mixture (which also disappointed its founders Jennie Lee and Harold Wilson) of people already on the way to semi-professional jobs, white-collar people, an aspiring middle class by education rather than money.

This was not true for a venture in which the Oxford Committee joined the Revd Samuel A. Barnett in 1884, when a settlement in the East End of London was set up. It was named after Toynbee, the young martyr for the cause, who had collapsed and died after an extension lecture. Other settlements preceded or followed Toynbee Hall, but this particular combination of education and social work based on a residential centre remained the most successful of its kind. Moreover, it became almost an extension of LSE—or was it, for a while, even the other way round? Certainly all those involved in the foundation of the School, without exception, had played their part in the good works of Toynbee Hall. A few years later, Beveridge, Carr-Saunders, and Tawney became Residents and helped develop what Barnett hoped would become a working-class college. It did not, but generations of lecturers, many from LSE, were involved in community work and exposed to the poverty and later the racial discrimination of East End life through Toynbee Hall.

Toynbee Hall is still active and very much alive, but the University Extension Movement did not last. When Hewins joined it in 1888 as a very young man, indeed (as Alon Kadish reminds us) 'half a year after obtaining his degree in mathematics and without having been required to give a trial course', its work had gone on for little more than five years. In 1889 Hewins was asked to organize economic studies at Toynbee Hall but refused. Instead he took charge of the successful summer schools of the Oxford Extension Committee. By 1891 when the Tooke Chair came up, much of the enthusiasm for extension teaching was gone. When Hewins finally left for LSE, 'it was because of the promise of the new School as an institute of higher education that would use Extension methods and ideas to succeed where the Extension failed'.

One can speculate why it failed and whether the inevitable conflict between idealized expectations and real needs proved fatal. Clearly, 'Oxford's faith in its mission was not entirely rational',[60] and perhaps the more practical approach by Cambridge could be sustained longer. Above all, however, extension teaching was an informal harbinger of more formal developments which began to take shape at the same time. Not just LSE emerged, but also transformed university colleges in cities like Birmingham, Sheffield, and Bristol, later Reading, in which Oxford (and Cambridge) dons came to be unwelcome intruders from yesterday's world. However much fun Oxbridge snobs tried to make of these 'Lilliputian universities', they were there to stay and had powerful champions. R. B. Haldane was proud when he became the first Chancellor of the University of Bristol and gave an address on 'The Civic University'. That was in 1912, but in the 1890s he had already decided: 'My great question was how to extend University Organisation in England.'[61]

The key to many of these developments lay in the University of London, which had been designed when it gained corporate status in 1836 as an examination board rather than a teaching institution. In 1894 the Gresham University Commission reported to Parliament and proposed the creation of a Gresham University in London. Whether the name of the distinguished City man Sir Thomas Gresham, who had died more than 300 years earlier, helped or hindered the subsequent debate is not clear. The Report had good things to say about university extension courses but only as an instrument of 'bringing under the direct influence of University study many students who would otherwise have remained outside that influence'. The core proposal was to create a full-blown university which 'so far as its teaching functions are concerned, should be limited to London and its neighbourhood'.[62] It was this latter point which triggered the seemingly unending parliamentary and public battles of the next four years. Not even Haldane's legendary ability to bridge apparently unbridgeable gaps sufficed to get the London University Bill through Parliament. Too many opposed what they saw as 'an eclipse of the University as an imperial examinations board with worldwide influence and its replacement by a predominantly local institution'. But Haldane, helped in this case by Sidney Webb, did not give up. A second attempt in 1898 led to a dramatic confrontation in Parliament, a dubious bargain over the Irish vote, and in the end an act to set up a University of London Commission. Haldane's biographers Eric Ashby and Mary Anderson summarize the scene:

> Four years of negotiation over the Gresham Report, drafting, seeking to compromise between conflicting views, and occasional log-rolling, had deepened Haldane's understanding of academic affairs; it had also whetted his appetite for more. He had demonstrated to himself and to his colleagues his flair for the strategy and tactics of policy-making, his rare ability to turn those earnest discussions in Beatrice Webb's drawing room and the talk among carefully chosen companions round his own dinner table into practical political decisions.[63]

It is not just a footnote to add that when the reconstituted University of London was finally created in 1900/1 it provided an essential anchorage for the London School of Economics. It is on the other hand a footnote to mention that there is also a Gresham College in London, sustained by City

Corporations, though this became little more—and probably less—than the lecturing arrangement of LSE in its first year.

The Five Es *(2): Economics*

Evidently the first two big *E*s, Education and Economics, were closely linked from the beginning. Economists were not the only extension lecturers, but they provided the core of the Oxford initiatives. In 1894 the British Association for the Advancement of Science held its annual meeting in Oxford and devoted much time to education. The Report of the Committee chaired by the Tooke Professor W. Cunningham on 'Methods of Economic Training in this and other Countries' pulls no punches: 'While fully recognizing the great energy with which individual teachers in this country have sought to develop the study of this subject, your Committee cannot but regard the condition of economic studies at the universities and colleges as unsatisfactory.'[64] But what exactly does 'economic studies' mean? For Alon Kadish the detailed description of the Oxford Extension Movement is but the prologue to the main theme of his book, *The Oxford Economists in the Late Nineteenth Century*, and these, while almost inevitably split between theorists and historians as well as Manchester liberals and 'new liberals', were largely to be found on the social and historical side of the great divide. Even the early Hewins, who was still advocating free trade and free competition, 'saw the future in terms of an increase in working-class self-help through corporate action, namely trade unions, co-operatives, and friendly societies'.[65]

Economics, like other social sciences, was far from a fully developed, clearly delineated, and professionally organized discipline when LSE was founded in 1895. It was still possible to mix theological exegesis, moral prescription, historical description, and political preference, and call the concoction economics. To quote A. W. Coats, 'many, possibly most, college and university teachers of the subject were amateurs or part-timers'. 'There was no sharp dividing line between the experts and the laymen.'[66] But perhaps this condition went better with the built-in dilettantism of the Oxford tutorial system than with the intentions of the London School of Economics. Here, a more serious issue rose to the surface of scholarly and public debate. Gerard M. Koot, the historian of this issue and its impact on LSE, has rightly observed in one of his articles on the subject that the LSE of the Webbs and W. A. S. Hewins 'sought to mould economic history and applied economics into an alternative economics to Marshall's more theoretical vision of the subject' which dominated Cambridge at the time.[67] The debate, in Kadish's terms, waged between 'Marshallians and

Dissenters', and its subject has remained one of lively divisions to the present day.

Is economics a theoretical science in the strict sense of the term, or is it a discipline of social and historical study which seeks to understand rather than predict? In purely professional terms, the question has been answered: no one would be a serious candidate for the Nobel Prize in Economics who does not accept the theoretical claims of the discipline. Every professor, or at least every student, of economics has to be able to read technical journals which would have made little sense to the Revd William Cunningham (though Hewins with his degree in Mathematics could probably have deciphered them). Theory means that universal statements about economic processes can be made which abstract from specific historical experiences, though predictions follow from them which can be tested by observable events. History, or even sociology by contrast, means that economic phenomena are studied like political movements or social changes without generalizing intent, and often in the light of the value preferences of the scholar. The German epistemologist Wilhelm Windelband introduced, in the 1880s, the distinction between 'idiographic' and 'nomothetic' statements and scholarly disciplines. 'Idiographic' study emphasizes the specific and the unique, whereas 'nomothetic' propositions try to establish laws with general and in principle universal validity. Nomothetic Marshallians and idiographic dissenters dominated the economic debate of the 1890s.

Emotions ran high during this debate. Professor C. F. Bastable emphasized in his address to the Oxford meeting of the British Association in 1894 that 'as soon as we get thoroughly accustomed to contemplating economic conditions in their actual forms as the special products of social life, it is but a matter of course to notice the remarkable differences and equally remarkable resemblances that different instances of the same economic institution or function will present'.[68] This is not the language of contemporary economics, though Professor Cunningham would have agreed, for he told an audience at the time that 'Professor Marshall, instead of accepting the description of medieval or Indian economic forms as they actually occur, sets himself to show that the accounts of them can be so arranged and stated as to afford illustrations of Ricardo's law of rent'.[69] Marshall's defenders, like John Maynard Keynes, had a kinder view of the great man: 'He was conspicuously historian and mathematician, a dealer in the particular and the general, the temporal and the eternal, at the same time.'[70] But, as tends to happen, the temporal fell away and the eternal remained.

Robert Heilbroner, in contrasting the 'Victorian world' with the 'underworld of economics', later took up the theme. 'All the while that Marshall and his colleagues were refining their delicate mechanism of equilibrium, a few unorthodox dissenters were insisting that it was not equilibrium but change—violent change—that characterized the real world and properly formed the subject for economic inquiry.'[71] This is a theoretical as much as a methodological point, and it is related to another one made by A. W. Coats: 'The trouble was that, as economics became recognized as a distinct intellectual expertise, its practitioners stressed the scientific and technical aspects of the subject; but its growing abstractness and complexity made it seem both less relevant to current affairs and less intelligible to the layman.'[72] What people—members of chambers of commerce, politicians, administrators, intelligent laymen—wanted at the time of the foundation of the London School of Economics was economic analysis which would help them understand what was going on around them and identify what could be done about it. What they got was a discipline of economics which was deeply divided between the historical and the theoretical approach. 'In addition to methodology, [the dispute] revolved around nearly opposite views of the economist as a detached expert or a practical adviser, different estimates of the role of history in social science, diverse judgements on the value of pure theory, and serious disputes on matters of policy.'[73]

Who won the great battle of minds? There cannot have been many who shared W. J. Ashley's sense of triumph when he declared in 1907 that the theoretical project of 'English economics as a system of thought' had now 'become a closed chapter in intellectual history'. In retrospect, the opposite was the case, and not just because 'the dissenters sorely lacked a Marshall to help to establish their academic independence'.[74] Theoretical economics was simply the stronger project; it also lent itself to the formation of what came to be called a scientific community. When Alfred Marshall celebrated his eightieth birthday in 1922 he received an address more in the nature of a panegyric: 'Through you, British economists may boast among their foreign colleagues that they have a leader in the great tradition of Adam Smith and Ricardo and Mill, and of like stature.'[75] Some of the 'foreign colleagues', like Schumpeter and Taussig, signed along with thirty-two other leading economists, who included the LSE luminaries Arthur Bowley and Edwin Cannan as well as William Beveridge, by then Director of the School.

However, this is only one end of the story. After all, it could be argued that the whole dispute was not, and certainly not only, about two approaches to economics—one theoretical and one historical—but about two equally legitimate disciplines which can happily exist side by side—economics and economic history. The study of economic history was developed to great and in some respects singular distinction in Britain, not least at LSE. Then there is that other famous victory with which Alon Kadish concludes his study of nineteenth-century economists: 'The foundation of the London School of Economics to a large extent saved the careers of some of the dissenters and allowed them to develop their own independent lines of enquiry.'[76]

But is this true? It is certainly not the whole truth. Yes, Hewins and Cunningham gave the first lectures in the subject; but there was also another professor, Edwin Cannan, who was soon to take over most of the teaching of economics, and who according to his colleague and successor Friedrich von Hayek 'created the tradition which, more than anything else, determined the intellectual climate in the central department of the School'.[77] When Cannan died in 1935 Lionel Robbins remembered how in the early 1920s 'he dominated our horizon'. 'Besides being a great teacher of Economics in the narrow sense, [Cannan] was also a social philosopher of the lineage of David Hume, Adam Smith, and Bentham.'[78] Cannan was unorthodox in his ways yet largely orthodox in his economics. He taught at LSE for thirty-one years but never moved from his beloved Oxford, where he had a singular personal library and also served as City Councillor, so that his lectures had to finish at 6.57 p.m. sharp to enable him to catch the 7.30 train from Paddington. Cannan had his own mind; he never simply followed fashion, and he did believe in the first tenet of LSE—application. Truth is not a value in itself, he said; there must always be a practical aim. Or, as he put it in his book on 'local rates', the first to appear in Hewins's LSE series in 1896: 'We do not study such subjects from a love of truth in the abstract or to while away a wet Sunday afternoon, but because there are practical controversies about them and we hope that we may learn something which may be of assistance in these controversies.'[79]

The Five Es *(3): Efficiency*

Application could mean many things, but in the late 1890s and early 1900s one subject dominated most others, the third of our big *E*s, Efficiency. When in 1894 Lord Rosebery succeeded Gladstone, this marked a watershed in British political life even if the break was not immediately apparent. Rosebery was to become the main advocate of 'national efficiency', and national efficiency in turn epitomized the end of Gladstone's world. Rosebery's Government did not last long, though Lord Salisbury, who succeeded him a year later, while coming from the other side of the House of Commons, shared his views up to a point (an important point which had to do with democracy vs. expertocracy, as we shall see). In any case,

Rosebery rapidly dissociated himself from the Liberal Party. Efficiency, like Empire, cut across the political divide, but whereas Empire created new divisions, efficiency for a while seemed to unite the nation. This was actually the whole point; the very idea of efficiency was alien to democratic politics.

Efficiency provided first of all, as G. R. Searle put it in his book *The Quest for National Efficiency*, a 'political catchcry', that is 'some phrase or slogan which sums up the hopes and fears of the hour, though in a maddeningly imprecise way'.[80] The fears were that Britain was falling behind. Neither its industrial organization nor its military strength kept pace with competitors in Europe and America. Lack of systematic development and application of modern science, and an educational tradition aimed at gentleman amateurs, were two of the reasons for this decline. Also, national resources were squandered by poverty and destitution. The hopes were therefore that Britain would follow another model. To quote G. R. Searle again:

> If one were to sum up its meaning in a single sentence, one might describe the 'National Efficiency' ideology as an attempt to discredit the habits, beliefs and institutions that put the British at a handicap in their competition with foreigners and to commend instead a social organization that more closely followed the German model. It was contended by one journalist that the key to the internal policy of the German Empire was 'this central idea of national efficiency', just as the key to British national life was to be found in 'the idea of personal liberty'.

And if this were not shocking enough, there were those who wanted to go one step further and adopt Japanese ways. 'Far from rejecting the values of an autocratic, military state, the Fabian leaders, Rosebery and Haldane were thus actually attracted by the streak of silent, calculating ruthlessness they detected in the Japanese national character.'[81]

Rosebery was of course hardly a Fabian leader, nor was Haldane more than a follower, but the Fabians certainly joined 'new liberals' and Chamberlain Conservatives in their advocacy of efficiency. Germany, if not the German model, has come up on several occasions already in this story. Webb lived in Germany as a youth and spoke the language well. Bertrand Russell thought the model of German Social Democracy worth studying. R. B. Haldane brought back from one of his many trips the idea of the Technische Hochschule in Berlin-Charlottenburg and proceeded to set up a 'London Charlottenburg' in the form of the Imperial College of Science and Technology. But the efficiency dream went further. It meant Bismarck's combination of social policy and military strength. Others, starting with Thorstein Veblen's *Imperial Germany and the Industrial*

Revolution (1915), have since described the same combination as one of political oppression and the continued domination of a feudal military-industrial class. The realities which the efficiency party praised did not please those who preferred the idea of personal liberty.

But around the turn of the century this was apparently a minority view. There were those who opposed Lord Rosebery's coalition; G. K. Chesterton for example, or L. T. Hobhouse, who after a stint with the *Manchester Guardian* was to become a Professor of Sociology at LSE. True Conservatives like Lord Salisbury or Lord Derby opposed the 'efficient' notion that a 'Council of Science' should provide Parliament with the necessary expertise and pointed out that the House of Commons would not wear such expert attacks on its sovereign powers. However, the majority of the political class applauded Lord Rosebery's speeches in Chesterfield in 1901, or in Glasgow in 1902. There he described efficiency as 'a condition of national fitness equal to the demands of our Empire—administrative, parliamentary, commercial, educational, physical, moral, naval, and military fitness—so that we should make the best of our admirable raw material'.[82] Human raw material, that is, the stuff from which an Imperial Race can be bred by efficient administration and, if need be, even eugenics.

'We believe', noted Beatrice Webb at about this time in her diary, 'in a school of administrative, political and economic science as a way of increasing national efficiency, but we have kept the London School honestly non-partisan in its theories.'[83] Such siren sounds reveal slightly sinister undertones in the light of Lord Rosebery's campaign. It is obvious that Rosebery—and the Webbs—were not liberals in the sense of being prepared to live with what H. G. Wells called the 'muddle' of the real world and to hope for progress to emerge from diversity and conflict rather than organization. By the same token they did not much like the rough-and-tumble of democratic politics. 'Honestly non-partisan' could and to some extent did mean anti-party. Political parties were at best a convenience. The Webbs were often accused of using them all—literally, all: Progressive, Liberal, Conservative, Independent Labour, Imperialist, even Irish—if it suited their purpose. In fact this was not opportunism, but the all-too honestly anti-partisan approach of an ideology of efficiency.

For a while it looked to some as if Lord Rosebery might carry the day, especially since many observed how close his views were to those of Joseph 'Joe' Chamberlain. However, the two did not get on. Many did not get on with 'the mysterious Rosebery', as Beatrice Webb described him. 'He is first rate at "appearances"', but he is no leader of a group. 'All he has yet done is to *strike attitudes* that have brought down the House at the time, and left a feeling of blankness a few days later.'[84] The 'rhyming' Liberal MP Wilfrid Lawson put it even more unkindly:

> Lord Rosebery was a most wonderful man,
> He had every species of scheme on his shelf.
> But 'efficiency' still formed the gist of his plan,
> And 'efficiency' meant nothing else than himself.[85]

Hewins found Rosebery rather distasteful, but Webb insisted that the Director keep up good relations with a man who indeed turned out to be useful to the School. In 1899 and 1900 he and Haldane organized fund-raising occasions for LSE in the City, and when the Bishop of London, Dr Mandell Creighton, died in 1901, Rosebery succeeded him as largely though not entirely honorific President of LSE. When the new School building was officially opened in 1902, Rosebery performed the ceremony as Chancellor of the reconstructed University of London.

The Five Es *(4): Equality*

Efficiency, it will be noted, became a national movement a few years after the foundation of LSE, although it was close to its intentions. Equality, on the other hand, was very much a founding idea of the School. Or should it be called Collectivism? Or indeed Fabianism? Beatrice Webb happily merged these divergent strands into one—her, or rather their—tissue. After the first encounter of the Webbs with Hewins, and in order to emphasize the difference between him and them, she described her own views and Sidney's: 'We were democratic collectivists, believing in the eventual triumph, in so far as social environment is concerned, of the principle of equality between man and man; if only by the roundabout way of the "inevitability of gradualness".'[86] So there it is, the Fabian triptych of method, values, and objective: the allusion to Quintus Fabius Cunctator, who would defeat the enemy in the end by not rushing things at first; the opposition to a century of individualism and its denial of both solidarity and the benevolent role of the state; the search for ways to bring about a social condition of equality. That she naturally speaks of 'man and man' shows that Beatrice lived before a later age of 'political correctness'; in any case, she was not politically correct in that she had little time for the women's movement and initially even opposed women's suffrage.

The Fabians are arguably the most overwritten episode in the political-intellectual history of Britain. Are they also overrated? They were, as G. M. Trevelyan put it in his *British History in the Nineteenth Century*, 'intelligence officers without an army'. 'This little band of prophets', Anne

Fremantle called them. 'But they influenced the strategy and even the direction of the great hosts moving under other banners'.[87] Still, overwritten they were, for several reasons. One is that the Fabians themselves were mainly preoccupied with writing in virtually every genre available to this noble human activity—tracts, speeches, treatises, novels, plays, diaries, memoirs, ditties. The latter were mostly about the Webbs and rarely complimentary even if composed by fellow Fabians like G. D. H. and Margaret Cole:

> O that Beatrice and Sidney
> would get in their kidney
> a loathsome disease
> Also Pease.[88]

They did not, but instead all lived to the age of 90 and more, which did not exactly diminish their influence. Moreover, the Fabians were lucky in their historians, from the founders themselves and notably Edward Pease, through Anne Fremantle, who got quite carried away by the 'Fabian' election victory of 1945 ('Among the dense crowd of Socialists were the Fabian Prime Minister, Clement (now Earl) Attlee, and 229 other Fabians'),[89] to the thorough, shrewd, and amusing MacKenzies and beyond.

But perhaps the most complicated reason for the Fabian hype has to do with the fact that many of the things which they said actually happened. In some cases they happened because they made them happen; Sidney Webb (though not '*the* Fabians') actually wanted LSE and made it possible. In other cases, including the huge project of the welfare state, the connection is more tenuous. Keith Middlemas's *Politics in Industrial Society* gives a sobering account of the disjunctions between radical projects and practical actions; the two wars stimulated more social reforms in this century than all Labour governments put together, though Labour and even Fabians prepared the ground. But we do not like the complexity of real history. The authors of ideas prefer to think that they are directly responsible for realities which correspond to their speeches or writings, and the rest love simple causal explanations, not to say conspiracy theories. The fallacy which logicians call *post hoc ergo propter hoc* is almost irresistible. The Webbs recommended universal insurance in their Minority Report for the Poor Law Commission. Its implementation many years later must then be their achievement. This is how myths are born which, if repeated often enough, sound almost like history: 'The tools forged by the Fabians to do the job they had set themselves—that of socializing Britain—did just that job. These tools were three in number: the Labour Party, the London School of Economics, and the weekly magazine, the *New Statesman*.'[90]

The Fabian Society made its first real splash when it published, in December 1889, a series of essays edited by George Bernard Shaw and written by Shaw, Webb, William Clarke, Sydney Olivier, Annie Besant, Graham Wallas, and Hubert Bland. The first edition, which was sold out after a month, sported a dramatic olive-green cover with a reproduction of a lithograph probably by Walter Crane. Under the banner heading 'Socialism' a substantial man in top hat and City clothes, held firmly around his expanding waist by a belt inscribed 'Privilege', steps down a ladder from a luscious apple tree. The ladder is called 'Capital' and at the bottom of it, two rather threatening-looking young men named 'Labour' await him, one with an axe, the other with a shovel. One wonders whether the apple tree will survive the treatment. The *Essays* themselves, which were to sell 30,000 copies at the considerable price of 6*s*. (or at least £12 a century later) in the first four years, were altogether more serious. Apart from numerous specific analyses and proposals, they came back time and again to the point that individualism had failed. The assumption (wrote Webb) 'that absolute freedom in the sense of individual and "manly" independence, plus a criminal code' would bring about the good society had clearly been refuted.[91] Something else was now needed, the opposite of individualism; that is collectivism, socialism, democracy. One of the *Essays* authors, the Hutchinson Trustee William Clarke, put it in terms of practical politics a few years later:

> It may be expected that the closing years of the century will witness an attempt at the completion of the evolution of modern Liberalism by shedding altogether the 'moderate' or Individualist wing, which is visibly declining and must start to disappear. The new party of progress in England can scarcely avoid being largely Collectivist, since it will probably be the outcome of a union of radicalism with the labour movement, nearly all the rich and socially influential classes gravitating steadily to the Conservative party.[92]

Concepts have their own history. Who would, a century later, gladly call themselves 'collectivists', or even 'socialists'? In the 1890s almost the opposite was true. People ran away from Spencer and Darwin, Bentham and Mill as they did from Cobden and Bright and from Gladstone. T. H. Green was 'in', as were the historical economists and other collectivists, Oxford or otherwise. A number of authors have later traced this sea change, some with an eye on LSE like Peter Clarke in his *Liberals and Social Democrats* or Stefan Collini in *Liberalism and Sociology*. The much-vaunted 'new liberalism' should in fact more appropriately be described as a kind of socialism—and who minded? 'I called myself a Socialist', G. K. Chesterton wrote in his autobiography, 'because the only alternative to being a Socialist was not being a Socialist. And not being a Socialist was a perfectly ghastly thing. It meant being a small-headed and sneering snob, who grumbled at the rates and the working-classes, or some hoary horrible old Darwinian who said the weakest must go to the wall.'[93] Political correctness after all, but one also wonders about cycles of the public mood. Is there not, a century later, and after a revival of libertarian individualism, a hankering after a new collectivism, only this time called communitarianism?

Like most of the Fabians, the Webbs were no serious philosophers. Their deeper beliefs, if deep they were, displayed an almost indiscriminate eclecticism. A bit of T. H. Green and Hegel, a little Christianity, a touch of romanticism, vegetarianism as an ideology, and various other snippets produced a viable philosophy of life. This meant of course that they were not Marxists either. Every now and again Sidney Webb is liable to sound historicist: 'The main stream which has borne European society towards Socialism during the past one hundred years is the irresistible progress of Democracy.'[94] Webb means democracy in Tocqueville's sense, of course, that is, equality; but he emphatically does not mean that nothing needs to be done to channel and gradually direct the stream of history. Norman and Jeanne MacKenzie summarized his and other Fabians' dream in one sentence: 'It was, indeed, something very like a Positivist state inspired by a Religion of Humanity and governed by a disinterested élite.'[95]

In two respects the Fabians were special. One had to do with their unbending belief in the 'gradualness' of change (albeit 'inevitable') and the programme of 'permeation'. For them this was not like the 'long march through the institutions' advocated by some of the Maoist student revolutionaries of 1968 who wanted to destroy institutions in the process, but it was an attempt to win by sheer competence, by superior expertise. This is why research was needed; the hard slog of collecting facts and writing them up in reports which then had to be published and disseminated. This is also why a caste of experts had to be trained; the vanguard, not to say the *nomenklatura* of the Positivist State. In the case of Sidney Webb the process of permeation and applied expertise had to begin at home, as it were, in the municipality in which people live. In fact, this is where it had already begun. 'It is the municipalities which have done most to "socialize" our industrial life,' he wrote in the *Fabian Essays*[96] and it is entirely consistent that he should have stood for election to the London County Council where he could promote his 'gas-and-water socialism' long before he considered standing for Parliament. It is consistent also that he wanted the *London* School of Economics within a new and 'predominantly local' University of *London*.

The other special feature of the Fabians was the Society itself. *Educate, Agitate, Organize!* commanded one of their slogans, and organize they certainly did. From its foundation out of a split in the Fellowship of the New Life in 1884 to the creation of the Fabian Research Department in 1913 and the ascendancy of that other couple, G. D. H. Cole and Margaret Postgate-Cole, the Society was dominated by the 'junta' of the Webbs and Pease, with Olivier and Wallas often added and G.B.S. much in evidence. Not many observers or participants noticed what Anne Fremantle describes as 'the most outstanding characteristic of the Fabians', 'their humour—that is to say, their capacity to take themselves lightly, to laugh at themselves', nor would her statement go unchallenged that the Fabian Society 'owes much of its perennial vitality to the warm bonds between its members'.[97] Humour and light-heartedness were the preserve of sceptical and even dissident Fabians who, from H. G. Wells to Anthony Crosland, regarded the orthodoxy as all too drab and dreary as well as self-important. Debate at the meetings of the Society was often heated and sometimes vicious. Indeed, this was the attraction. For many years the Fabian Society was the place 'where it was at'. Important subjects were discussed by informed people whose views provided a benchmark for others though not always a prescription. Permeation also worked, not just in terms of the number of lectures given and tracts published, or even of members of 'the Fabian' in important positions. The Fabian Society was in tune with the mood of the 1890s and continued to play a major part in the 1900s.

The Five Es *(5): Empire*

Many of the early students of LSE were undoubtedly motivated by Fabian hopes. After all, Sidney Webb was teaching at the School, as was Graham Wallas (though Fabians were a distinct minority among the early lecturers). Moreover, the desire to train an administrative élite of experts in economic and social matters linked Fabians with the Chamber of Commerce, to say nothing of Lord Rosebery and the entire 'national efficiency' movement. Yet if one were to single out one intellectual current which provided much of the tension, creative or otherwise, of LSE in its first fifteen years, it would not be Equality, or even Efficiency by itself, but Empire. There were, yet again, surprising combinations. Arnold White's populist book *Efficiency and Empire* (1901), added that other *E*, Eugenics, to the list, and Rosebery liked it. We shall presently find that Empire and Equality, or at least social reform, also were far from incompatible. One could detect the shadow of Bismarck, or rather of Bismarck as seen by those who did not have to live under him, loom from over the North Sea: industrial efficiency, social reforms, domination by an administrative élite, military strength, imperial ambitions, the whole gamut of the 'German model'. However, its British exponent was not a Bismarck in either style or temperament though he may have had similar ambitions, and was described at times as 'a secondhand Bismarck'.[98] This was Joseph Chamberlain, whose importance for Britain at the time of the foundation and early history of LSE, and for the School itself, it is hard to exaggerate.

To describe the British political scene from the 1890s on as being in a state of flux is a considerable understatement. In fact it was in a state of turmoil and underwent a profound transformation. In the fifty years between the fall of Lord Rosebery's Government in 1895 and the election of Clement Attlee in 1945 there were only two periods of 'normal' majority government by a single party—the Liberals after 1906 and the Conservatives after 1924—which add up to ten years at the most. Even in these periods the parliamentary majorities were challenged from within. All other governments were either formed by minorities in Parliament or by coalitions including the 'national governments' during the two wars and after the Great Depression. In the second half of this period the underlying transformation had much to do with the uncertain progress of Labour and the certain decline of the Liberal Party, but in the first half, and particularly in the years before 1914, turmoil had more to do with imperialism and its precarious relationship with social reform.

The old certainties had become shaky a decade before 1895 when Gladstone embraced Home Rule for Ireland and thereby split the old Liberal Party several ways. His most significant loss to what were to be the Liberal Unionists was his former President of the Board of Trade, Joseph Chamberlain. No other public figure of the time attracted the unorthodox as strongly, which meant of course that he was a thorn in the flesh of the Establishment. The youthful Beatrix Potter fell hopelessly in love with him without sharing any of his views. When the young officer Winston Churchill 'looked out of my regimental cradle' he found that 'Mr. Chamberlain was incomparably the most live, sparkling, insurgent, compulsive figure in British affairs'. He was all these things, but in reviewing J. L. Garvin's *Life of Joseph Chamberlain*, Churchill also noted a little more sceptically that 'all our British affairs to-day are tangled, biased, or inspired by his actions'.[99]

These began with a successful industrial career as a screw manufacturer in Birmingham which gave Chamberlain material independence. There followed the 'gas-and-water radicalism' (for a socialist he was not) of his years as Mayor of Birmingham from 1873 (when he was nearly 40). Once in national politics, he gave this radicalism widely audible expression in his 'unauthorized programme' of 1885, which did not endear him to either Whigs or Tories. Soon afterwards the Liberal Unionist phase began, mostly in coalitions with Tories, and notably as Colonial Secretary during the Boer War. Finally, the great conversion happened not just from 'Fiery Red' to 'True Blue' or even from 'a ruthless Radical and, if you challenged him, a Republican' to 'a Jingo Tory and Empire Builder',[100] but above all from free trader to tariff reformer and protectionist.

In fact the contrasts miss the point. Chamberlain's fatal attraction lay in the combination of imperialism and social reform. G. M. Trevelyan is not the only one who doubts that this misalliance actually worked. 'Chamberlain would indeed have liked to run Imperialism and Social Reform together. But he had been forced to choose between the two.'[101] Perhaps wanting to run them together and indeed having plausibly stood for both was enough even if it did not make the slogan 'Tariff Reform Means Work for All' wholly plausible. 'The democracy' (as Joseph's son Austen described the people) 'want two things; imperialism and social reform.'[102] The mixture turned out to be the most explosive political cocktail of the age. Bismarck, and in his own very different way Chamberlain, represented its more harmless version. But Chamberlain's son Neville was to encounter (and fail to cope with) the variety which almost blew European civilization to pieces; it was called National Socialism. In his thorough and evocative study, *Imperialism and Social Reform*, Bernard Semmel returns time and again to the dream of an 'Imperial race' which will make the nation able to survive the world. H. G. Wells, one of the believers, went so far along this road in his *Anticipations* that his most recent biographer, Michael Coren, sees him helping 'create an intellectual climate in the 1920s and 1930s' which may not have led to Hitler and Stalin but 'certainly gave credibility to the atrocities of the dictators'.[103]

However, in the early years of the century British imperialism was above all about economics. Indeed one can hear echoes of the European debate of the 1950s, 1970s, and again 1990s. Semmel plausibly distinguishes between two kinds of imperialism. One was industrial and unrestrainedly protectionist; it proposed to draw a ring fence around the Empire with preferential tariffs within and significant barriers without. Chamberlain, the former industrialist, after his 'conversion' subscribed to this view, as did Hewins, whose father and father-in-law had both been in the declining iron industries of the Midlands. Whereas this faction generally ended up with the Conservatives and was prepared to use economic means for political purposes without even excluding war, another group remained closer to the

Liberals and formed the Liberal Imperialists. These were the people involved in commerce and finance who saw politics as an instrument for advancing economic interests especially of the City, and they were more likely to advocate the use of taxes than tariffs for financing social reform. Halford Mackinder, the second Director of LSE, came to be a major proponent of this school, having undergone his own conversion from liberalism and free trade. His imperialist argument was, as we shall see, geopolitical—what is needed is a 'controlling centre' of world-wide economic activity.

When imperial preferences were finally introduced in 1932 the author was again a Chamberlain but the reasons were entirely defensive and had little to do with the pre-war movement. Joseph Chamberlain's son Austen continued the struggle after his father had suffered a debilitating stroke, but in 1913 he wrote a despondent letter to his stepmother: 'I have done my best, but the game is up. We are beaten and the cause for which Father sacrificed more than life itself is abandoned!'[104] He was right, but what changes had been wrought in the process! Imperialism had alternately united and divided all political groups. Unlikely alliances were forged not just between Empire and Efficiency but also between Empire and Equality, and of course between Empire and Economics, to say nothing of Empire and Education at least at LSE. Then the alliances fell apart; Chamberlain and Rosebery did not get on; the Webbs were always suspicious of Chamberlain's enthusiasms, and of course Shaw's unlikely conversion to the cause in 1904 led some of the founders to leave the Fabian Society. When Joseph Chamberlain died in 1914 the badly shaken political parties and their leaders must have breathed a sigh of relief, though by that time the agenda of politics had changed. Germany, from being a model of efficiency had become a great threat for the selfsame reason.

Liberal or Not?

No other subject dominated informal conversation at and around the early LSE more than the conjunctions and disjunctions of imperialism and social reform. Even 'Sidneywebbicalist' efficiency had to take second place;[105] it was in any case more suited to instruction than to conversation. All the time people wondered where they belonged in the scheme of public affairs. Whig and Tory had ceased to make much sense. There were many Radicals and Progressives, to be sure, the latter above all in the London County Council. But then came the Limps (the Liberal Imperialists) and the Tory Imperialists too. The Labour Party was as yet more an idea than a reality, which it became only after 1906; and then it was divided between Labour

Representationists and Lib-Labs. Thus the School was right at the centre of the new currents of the time. It was an exciting place. The very things which concerned an intelligent public most were happening at LSE, and if they were not literally happening, they were thoroughly studied and consciously taught and also debated with passion and zeal.

Yet for a liberal, the scene surveyed in this chapter is not a source of great pleasure. I do not mean the 'new liberal', of course, but rather the Gladstonian, unreconstructed or not. Almost every current of innovative intellectual and political debate around the turn of the century runs against his preferences. Education was certainly in need of expansion and reform, but a purely functional training of administrative and managerial élites can destroy the best possibilities of creative thinking and civilized learning, quite apart from the fact that it patronizes the many who leave school without any qualification. So far as Economics is concerned, it would be good to see a hundred flowers bloom, or at any rate three, of which one combines theoretical rigour with systematic data management ('econometrics'), one keeps us informed of the varieties of the history of economic institutions and processes, and one seeks to apply our obviously incomplete knowledge to contemporary economic problems. And Efficiency? We have learnt in the twentieth century that there are limits to amateurism let alone dilettantism, but if philosopher-kings are a bad idea because they ascribe a monopoly of truth to fallible mortals, then expertocracy is an abomination which adds greyness and control by computer files to the fallacy of certainty. Equality is fine as long as it enables citizens to go their own ways on the basis of common rights upheld by the rule of law, but collectivism disenfranchises the citizen as soon as he or she is born. Finally, Empire is not just a gigantic distraction but also the enemy of the ultimate liberal dream, which must be one of openness and multilateral institutions if not a world civil society.

Thus LSE was not exactly born in a liberal age. This is not to say that it was merely the creature of a passing *Zeitgeist*. For one thing one must resist the temptation of identifying an institution, a university at that, with particular ideas or intellectual currents. Good universities are never of one piece, and those which have tried to be that did not survive or became third-rate training schools for some *nomenklatura* or other. LSE certainly never was of one piece. Students and staff alike were searching for answers and not simply bringing them along as they came. Every creed was held and taught by someone. Hewins is not Webb, and Cannan is not either; to say nothing of later configurations like Laski and Robbins and Hayek and Beveridge. The mood of the foundation was one thing, but the currents which flowed through John Street and Adelphi Terrace and later Clare Market and Houghton Street were varied and many.

Then there is the context of the founding years to consider. True liberals are no relativists. They are what Ernest Gellner has called Enlightenment Puritans, who may not have incontestable grounds for claiming the universal validity of their beliefs but who hold and apply them wherever, whenever. Of course there are some times in which this is easier than in others. In the post-Gladstonian world it was difficult to be just a liberal, and for good reasons. My father grew up in post-Bismarckian pre-First World War Germany. He came from an unskilled labourer's home and did not turn against his origins when he had worked his way up through journalism and politics. Like his father before him, he became a social democrat and fought all his life both for democracy and for social improvement. He would have appreciated many of the ideas which went into the foundation of LSE, and it is not difficult for his son to appreciate them through his eyes even where another generation has different preferences. Much that is thoroughly bad flowed from the prevailing ideas of the turn of the century, but also much that is good and necessary. This includes the advancement of the social sciences as well as the creation of the welfare state. It entails above all the fact that the School was the place where practice met theory in the whirlpool of the currents of the time.

TO CLARE MARKET

Administering by Trial and Error

W. A. S. Hewins was at least as good at what later came to be called public relations as he was at day-to-day management. Two terms into the first academic year he published a *Brief Account of the Work of the School.* The number of students had risen to more than 300. Seventy of them had decided to take 'the whole or part of the three years' course of study' in economics which was here advertised for the first time in detail. Success was also reported for the 'research department' and the collection of the Library, though 'it is impossible to proceed further with this branch of the work until the School removes to more commodious buildings'.[106] A summer school was announced, along with the continuation of university extension courses.

But Hewins was by now exhausted. Already on 22 March 1896 Sidney Webb had sent him a note expressing, on behalf of the Hutchinson Trustees, appreciation of the 'tact and energy' with which he had tackled 'the specially arduous work of the first year' and offering him an additional

£100, but also urging him that 'you must take a good holiday whenever you can get it'.[107] Hewins did not, and just before the beginning of the second year Beatrice Webb notes on 5 October 1896: 'Found Hewins in a state of nervous collapse threatening severe illness, sent him away with his wife and child and took over the work of preparing for the coming term.'

This she did not enjoy, nor did Sidney. Beatrice had herself had a somewhat miserable summer, and now 'poor Sidney trudges over there directly after breakfast and spends the morning with painters, plumbers, and locksmiths, would-be students intervening to whom he gives fatherly advice', which was clearly not his favourite activity.[108] However, by November Hewins was back and began to tackle the three problems which were to occupy him throughout the next years: establishing the Library, finding 'more commodious buildings' for the School, and getting it recognized by the emerging new University of London. For all this he needed help, though as we have heard he was not much good at using such help. The first 'Secretary' of the School was Sidney Webb's invention, the Fabian propagandist Harry Snell, later Lord Snell. But Snell and Hewins did not get on at all. 'It is quite possible that Snell had views as to how the School should be run that differed from those of Mr. Hewins who was not a socialist.'[109]

Snell himself put the problem in somewhat kinder terms when he thought back to his early days at the School many years later. The School, he recorded, owed a great deal to Hewins. 'He was, I think, a Conservative in politics, and there was, so far as I remember, little that was forward-looking in his teaching. But he was approachable and industrious, and his students liked him. I certainly had a great regard for him.' However, the dynamic personality was Sidney Webb, who 'believed in an eight-hour day for everybody except himself'. The greatest concern of the early days was to lay the 'scientific foundation' of the School in such a way as to allay all suspicions that it was a propagandist organization.[110] This succeeded thanks to the quality of the staff and the enthusiasm of the students.

So why did Sidney Webb call Snell one day 'to see me and at the end of less than ten minutes conversation I had lost my job as secretary to the Director of the School'? Miss C. S. Mactaggart's suspicion of a clash in the politics of the Director and his Secretary was probably only one reason. She herself was not altogether displeased for she was soon to succeed Lord Snell, though not before she had removed one other rival, and for some years without the title of Secretary. In the first instance, someone had to be found to put order into Hewins's rather disorganized affairs. Hewins, so an early student reports, had a habit of putting appointments into a book which he never opened to check but only to put in new dates, so that students arrived in vain. The same student, and no doubt others, therefore welcomed the appointment of 'J. McK.', the one-armed Scottish-Australian John McKillop, who quickly coped with the prevailing 'tendency to chaos'. 'He took charge of the appointments and reduced the daily round of work to order, routine details were organized, and the once murmuring student faces were wreathed in smiles.'[111]

One person who did not smile was Miss Mactaggart who, like J.McK., had been an occasional student during the first year of the School, and was asked to stand in for a sick colleague as Lady Superintendent, a kind of domestic bursar. McKillop had recommended her, apparently even known her in Australia (there were few women around whom McKillop did not know) but she soon spoke in scathing terms about him. He was, she remembered, 'a rather brilliant man with an extraordinary capacity for being wrong'. The formidable lady did not despair, however. 'As a matter of honest truthful fact I'—and at this point of her confession she adds the rather baffling parenthesis '(not willingly)'—'gradually ousted poor old McKillop from his work as secretary but he remained secretary—and librarian—while I was secretary and registrar.' 'Poor old McKillop'—when Miss Mactaggart wrote this in 1934, she knew, of course, that his end had not been happy. 'As a matter of fact, he hated me like poison.'[112] This was hardly surprising. He also spent many a lunch with Edward Pease 'at an A.B.C. when he usually discussed his perennial feud with Miss Mactaggart' which in the end, in 1909, made him resign. 'Very foolish of him,' thought Pease. 'A few years later he went to the bad, exactly how or why I don't know, and died as a tramp in a workhouse.'[113]

The British Library of Political Science

All this began in 1896/7 and did not bode well for the Director's managerial talents. In fact the administration of the School was only sorted out properly when Halford Mackinder assumed the directorship. But the little domestic difficulties did not prevent progress on all three fronts on which the School tried to move forward. The first of these was the Library. It was essential both as a research tool and as a visible indication that the School was there to stay. In February 1896, a 'big house' had become available around the corner from John Street, at 10 Adelphi Terrace. The Hutchinson Trustees decided to rent it although it was both too large and too expensive. Early students comment not just on its size but on its extensive 'cellars' or 'basement', which suggests that some of the action took place underground. Apart from the ground floor there were three other floors, and the first was well suited for a library. More importantly, Charlotte Payne-Townshend had agreed to rent the upper floors for herself, which meant that she not only relieved LSE of a part of the cost but was also at hand for further attempts at persuasion by George Bernard Shaw to marry him and by Sidney Webb to support the new Library.

In the early spring of 1896 a prospectus about the 'Proposed Establishment of a Library of Political Science' was issued. It sets out in considerable detail the many studies which cannot be conducted in England in the absence of a proper library. 'It is, for instance, impossible for an English student to find, in any library in this country, the material for a precise knowledge of the Referendum, or the Second Ballot, or of existing arrangements for the control of the Liquor Traffic.' It was to be, one notes, a library of *political science* (though economics came to be added later). The Library was to be established 'in connection with the London School of Economics and Political Science', but as a separate trust. 'It would at once become an inalienable part of London's educational endowments.'[114] But public money was not available for the purpose. Ten thousand pounds was needed to put up a new building, and at least £6,000 to refurbish an old one. Nothing specific was said in the appeal text about the cost of purchasing books and documents.

It is not clear that the appeal was ever formally launched or even widely publicized. Hewins certainly was reluctant to go public before the University of London question had been sorted out. However, letters were written to potential supporters. Not all of them seemed persuaded by the project. The ageing Herbert Spencer replied rather belatedly (on 24 March 1897) and in ill temper. He did not like the idea of free libraries any more than that of free bakeries. It was not so much that he thought there is no such thing as a free lunch, but there should be no such thing; 'food for the mind should no more be given gratis than food for the body'. All this was 'socialistic, and I am profoundly averse to socialism in every form'. In any case, who wants 'the idea that political science is based upon an exhaustive accumulation of details of all orders, derived from all sources'? 'I hold contrariwise that political science is smothered in such a mass of details: the data for true conclusions being relatively broad and accessible.'[115]

The clash of minds—Herbert Spencer's and Sidney Webb's—could not have been greater; but Webb remained unimpressed. The more difficult feature of the proposed appeal was that it looked a little confused. A new building would surely serve the School as much as the Library, and even the refurbishment of an old one was no longer the issue once 10 Adelphi Terrace had been rented. Moreover, initial funding was available, notably through the gift of £1,000 by Charlotte Payne-Townshend. Her relation to the foundation of the Library was not unlike that of the Hutchinson bequest to the School itself; both provided a helpful stimulus as well as a useful cushion for realizing plans which might, however, well have come to fruition without them. Actually, the Hutchinson Trustees voted £1,500 for the Library, though only a part of the money was spent. Other donations came from known benefactors of the School and even from the Webbs themselves. Thus Sidney's temporary concern with 'painters, plumbers, and locksmiths' to fit up Adelphi Terrace was really all that was needed. The new Library—by then it was called *British* Library of Political Science—was duly opened on 9 November 1896, with a governing body of five Trustees, the three Fabian 'junta' members Webb, William Clarke, and, of course, Edward Pease, the inevitable R. B. Haldane, and Charlotte Payne-Townshend. Hewins became its first Director though McKillop acted as Librarian until B. M. Headicar succeeded him in 1910.

The Webbs had first met Charlotte Payne-Townshend in the autumn of 1895, at about the time of the opening of LSE. 'We, knowing she was wealthy, and hearing she was socialistic, interested her in the London School of Economics.' To such cynicism Beatrice soon added a bit of scheming. She suggested that they take a house in the country together in order 'to bring her more directly into our little set of comrades'. 'To me she seemed at that time, a pleasant, well-dressed well-intentioned woman; I thought she would do very well for Graham Wallas!' It soon emerged that she was not just pleasant and well-dressed, but had her own intentions which betrayed 'certain volcanic tendencies'.[116] Graham Wallas bored her whereas Shaw was fun. Michael Holroyd has told the story of their turbulent courtship, with sometimes one and sometimes the other taking the initiative, and neither, though G.B.S. even less than Charlotte, being able to bear the absence of the other for very long. On 1 June 1898 they finally got married, with Graham Wallas as a witness, and the *Star* reported:

> Miss Payne-Townshend is an Irish lady, with an income many times the volume of that which 'Corno di Bassetto' used to earn, but to that happy man [G.B.S.], being a vegetarian, the circumstance is of no moment. The lady is deeply interested in the London School of Economics, and that is the common ground on which the brilliant couple met. Years of married bliss to them.[117]

The years turned out to be decades, until Charlotte's death parted them in 1943. Meanwhile the Library was thriving. It remained separate from the School until the last of its Trustees, Edward Pease, died in 1955; but the distinction was more formal than real. Sydney Caine, in his *History of the Foundation*, raises doubts about Beatrice Webb's explanation of the

Library's legal independence; she invoked Sidney's discovery that under the Literary and Scientific Institutions Act of 1843 such institutions were exempt from local rates. While this sounds much like Webb, and he did in fact obtain exemption from rates, Caine believes that the point was 'more to give a basis for an appeal to a different, and wider, body of supporters than those who could have been counted on to contribute to the acquisition of a collection of books by the London School of Economics alone'.[118] In any case, the Library went from strength to strength, was renamed (still with a difference!) British Library of Political and Economic Science (BLPES) in 1925, and became in due course the leading social science book and document repository in the world.

Passmore Edwards Hall

In the mean time, Hewins and Webb turned their minds to the next stage, a building of its own for the School. Once again it was Sidney Webb's manifold connections which did the trick. As a member of the LCC he knew, of course, of the plan to redevelop the area between Holborn and Lincoln's Inn in the north and east, Covent Garden in the west, and the Strand in the south. This was but five minutes' walk away from the elegant Adelphi-Adam quarter yet one of the festering slums which Charles Booth's *Map Descriptive of London Poverty* had shown up, around 1890, in dark colours of destitution. Following plans first made in 1892 all this was now to be changed; old buildings were being demolished even as the School began its work at the Adelphi; by the beginning of the new century the crescent of the Aldwych had emerged, soon to be followed by the grand new thoroughfare which when completed after Queen Victoria's death was named Kingsway.

While all were impressed, some did not much care about such modernization, as a little exchange between the chairman of a session of the Sociological Society, Patrick Geddes, and the forever heretic Fabian speaker H. G. Wells showed in 1906. Wells had, under the heading 'The So-Called Science of Sociology', spoken of Utopia, and Geddes wanted him to be more specific: 'I would only ask him whether these Utopias might not be brought down definitely to geographical position and civic use; whether we are not all discussing a Utopia of some kind by making the Kingsway, for instance.' Wells was not one to agree such concreteness. 'In reply to Professor Geddes I would say that the plan of making the Kingsway was Utopian until it was made, then it became a melancholy fact.'[119] Melancholy? In any case, the meeting took place in the new building of LSE on Clare Market, which would not have existed without the Kingsway redevelopment. Thus the School had benefited from Utopia coming down to earth and being put to academic if not civic use.

Getting the site required all Webb's skills, especially since he was in a sense contracting with himself, the Chairman of LSE with the Chairman of the Technical Education Board of the LCC. In the end he managed to wrest 4,300 square feet in Clare Market from the Council's Improvement Scheme property 'on permanent loan' for a peppercorn rent. The next question was the funding of the building itself. At the time there was a philanthropist about town who was known to be prepared to give grants for bricks and mortar, especially libraries and hospitals (although *Who Was Who* records that he also 'erected eleven drinking fountains, and placed thirty-two marble busts of eminent men by eminent artists in public buildings'); his name was John Passmore Edwards. He was a West Country man, born in 1823, self-educated, and therefore, contrary to Herbert Spencer, much interested in free libraries. After varied business ventures he had made a fortune out of technical publications like *Building News* and *Mechanics' Magazine* before he bought the halfpenny daily, *Echo*. In the end he sold most of his share to Andrew Carnegie and devoted his time to numerous causes of reform, for a while, from 1880 to 1885, as a Liberal MP, but then as a discriminating and deliberate benefactor. There were over seventy Passmore Edwards buildings when he died in 1911, some of which, like the one in the Epilepsy Colony of Chalfont St Peter, Buckinghamshire, still display the characteristic style of the man and his time. When Sidney Webb first approached John Passmore Edwards early in 1899, things did not go well. Sidney was a little too pedantic; he came along with a detailed Trust Deed for an LSE Building Fund, whereupon the benefactor not unreasonably 'took fright' and 'said it was nonsense and verbiage and unnecessary'.[120]

But Webb did not give up, which was as well because in the end he got what he wanted, though the correspondence between him and Passmore Edwards makes painful reading. In some ways it is quite typical of the relationship between a benefactor and an academic beggar for institutional money; as time went on, Webb needed a little more for slightly different purposes under changed legal conditions, all of which the philanthropist did not like though he had fortunately heard it before. In other ways, Passmore Edwards was special. On 16 August 1899 he suddenly decided that he did not like the name, Passmore Edwards *Hall*. 'The word Hall means nothing in particular and is best known in its connection with music halls . . . There are more dancing halls in the country than music and education halls.' So why not call the building 'institute'? Sidney Webb was horrified. The building is not, he replied the next day, 'an "institute" in itself';

in any case, universities do not have institutes but 'Oxford, Cambridge and most other Universities have Halls within them'.[121]

Passmore Edwards Hall it remained then, and the donor paid up all of £11,500 plus a history scholarship and various bits and pieces. However, he remained involved in detail though fortunately his preferred architect, Mr Maurice B. Adams, was also the one whom Sidney Webb had arrived at 'by the statistical method'. Edward Pease remembered almost correctly, judging from an article in the *Building News* (Mr Edwards's paper, of course) on 29 June 1900 that the 'plan was selected from two [it was actually three] designs submitted because it gave 80% useful wall space out of the total and the other [two] gave only 50%.'[122] When the foundation-stone was laid on 2 July 1900, building work had already begun. Still, a formal ceremony was organized, with five speeches and the printed announcement that 'the proceedings will take not more than fifty minutes'. 'I remember the laying of the foundation stone,' Professor Arthur Bowley tells us in his splendid memoir written in 1945.

I drove in a hansom cab from Paddington through the slums round Drury Lane, Great Wild Street and the rest—and across the scene of devastation that was afterwards Kingsway. My cabman said, 'this is a rum place to come to'. It was raining heavily and Bishop Creighton [Bishop of London and President of LSE] solemnly laid the foundation stone in the name of the Trinity, there to rest forever. The next day the masons moved it, to put it in its right place. When in the next building phase, Passmore Edwards Hall was demolished it was moved again, but it can be found by the curious in the wall outside the Library window that looks into the court behind Houghton Street.[123]

The curious can still find it there, though the Library has moved and the stone is now behind the ground floor. When the foundation-stone was laid, the child was christened Passmore Edwards Hall. Miss Mactaggart felt unhappy about the name and suspected the hand of McKillop. But then she felt unhappy anyway because, while the rest were enjoying themselves in the rain, she had to make tea for the reception that was to follow at 10 Adelphi Terrace. While Sidney Webb showed much interest in the drains and other practicalities of the new building, many thought back with nostalgia to the 'beautiful fireplace and handsome ceiling' at Adelphi Terrace, which was now and until its demolition in 1936 to be occupied entirely by the Shaws. Passmore Edwards Hall? 'We hated it,' Miss Mactaggart stated with customary firmness, and Edward Pease called it 'ugly'.[124] When Sydney Caine wrote his *History* over sixty years later he commented on the outside appearance, which must 'seem to us absurdly fussy and very remote in feeling'.[125] How the times, and the tastes, change! Passmore Edwards

Hall would, thirty years after Caine's condemnation, be a listed building and much admired for its *art nouveau* design.

There were the usual hitches about completing the building. Money ran out and an additional £8,000 had to be found. This is where the funds raised by the otherwise unsuccessful Mansion House appeal by Rosebery in 1901 and notably Lord Rothschild's donation of £5,000 came in handy. LSE thanked Rothschild later by inviting him to become President of the School in succession to Lord Rosebery; he was, helpfully, Rosebery's brother-in-law. Webb was eager to see the work completed before the beginning of the session 1901/2 and was disappointed when the formal opening by Lord Rosebery, who was by then Chancellor of London University, had to be deferred until May 1902. But at last the School had its own home, which Passmore Edwards Hall was to be for the next twenty years. This mattered in more ways than one. More space was badly needed. Concentration on one site saved time and added to collegiality. A proper building opened the opportunity for daytime use, at first for research and later for teaching as well. Moreover the simple fact of having a visible, tangible domicile called the London School of Economics and Political Science gave the School a promise of permanence.

Incorporation in London University

This promise was enhanced by less tangible though equally significant legal and organizational changes. We have left the story of the University of London at the time of the 1898 'Haldane' Act and the appointment of a Commission. The Commissioners duly drafted a constitution for the hybrid but workable structure of the University of London as it existed for nearly a century; that is until its breakup in the 1990s. There were to be internal and external students, thus satisfying local needs as well as the lingering national and imperial ambitions of some. Academic staff were to be university appointees who were organized in faculties but had their actual place of work in one of the schools or colleges of the University. The University would (increasingly) channel public funds to its constituent parts. Its structure would recognize the dual components, faculties and schools, as well as the two functions, academic and financial. A part-time Vice-Chancellor and, more importantly at least for three-quarters of the century, a full-time Principal would preside over the whole and be accountable to a Senate for all academic and a Court for all financial matters.

For LSE, two issues were of critical importance. One was that a Faculty of Economics be created to give its teachers academic recognition, and the other that the School itself be accepted as a college of the new University. On 16 March 1900, Webb reported to the Hutchinson Trustees that the School's application to be a school of the University had been successful. Soon after, the Faculty of Economics and Political Science (including Commerce and Industry) was created and ten LSE lecturers became 'recognized teachers'. A year later, the B.Sc.(Econ.) and the D.Sc.(Econ.) were established—'the first university degrees in the country devoted mainly to the social sciences'.[126] All this did not come about without a struggle. Hewins had several anxious conversations with the Secretary of the University Commission, Bailey Saunders. Webb had to persuade others in the University and beyond that LSE really was an academic and not a propagandist institution.

Hewins and Webb also had to fend off attacks on the financial basis of LSE, the soundness of which was one condition of incorporation. Some of these originated from 'friends' within the LCC and especially from Ramsay MacDonald, 'always strangely hostile to Sidney', and Will Crooks; 'together they made a dead set against the School,' wrote Janet Beveridge in her *Epic of Clare Market*.[127] The bitterness of Ramsay MacDonald had its origin (Norman MacKenzie suspects) in his 'disappointment about the Hutchinson money'. Like Shaw he would have preferred 'something like a working-men's university'[128] in which he, the Hutchinson lecturer on propaganda tours, might have found a special place. In 1898 the issue had flared up once more, with threats by MacDonald to take Webb to the courts for malversation of funds.

Time and again, Haldane, friend of the Webbs, supporter of the School, and of course inventor of the reconstituted, enlarged University of London, had to come to the rescue. But by the time the Passmore Edwards Hall rose towards completion, all was settled. No one would quarrel with Arthur Bowley's comment forty-five years later: 'Only those who were familiar with the small and tentative beginnings of the School—small in numbers and income, but great in the far-seeing vision of its promoters—can appreciate its enormous success in obtaining equal recognition with the old Colleges of the University, and in the establishment of a Faculty of Economics.'[129]

In the mean time, one little matter of a more domestic nature had been got out of the way as well. What exactly was LSE in legal terms? Was it a part of the Hutchinson Trust? Surely not. Should a Royal Charter be sought? There is no evidence that this was considered. Sydney Caine has described what took place so clearly and succinctly that his account can speak for itself:

In fact recourse was had to the simpler process of incorporation under the Companies Acts as a company limited by guarantee. The necessary application to the Board of Trade was signed by seven members of the School's Advisory Committee (Webb, Alfred Comyns Lyall, William Garnett, Charlotte F. Shaw, Jervaise Athelstane Baines, W. Pember Reeves and Edward W. Whittuck) and on 13 June 1901 the 'Incorporated London School of Economics' was duly registered. The Memorandum and Articles empower the School to continue its established activities and broadly to do everything appropriate to the development of study and research in the social sciences. It gave the School the usual powers to hold property and by placing all powers of administration and control in the hands of the members of the corporation as Board of Governors, with power to appoint such officers and committees as they thought fit, gave it in fact a highly flexible operating constitution which has proved very convenient in later years of growth and adaptation. [Bishop] Dr Mandell Creighton was elected first President of the Board and Webb was elected Chairman.[130]

Some Governors never cease to be surprised when once a year a meeting of the Board turns into an Annual General Meeting of a company and they briefly become directors, though the shock is alleviated by the fact that on 14 June 1901 the Board of Trade directed that the Incorporated LSE was registered 'without the addition of the word "Limited" to its name', and on 2 August 1957 the School was even allowed by the Registrar of Companies to delete the 'Incorporated' before LSE. For later Directors, the quality of the Articles of Association was more important and in particular Article 3 (A) iii, which defines one of the 'objects of the Corporation', 'to provide *for all classes and denominations without any distinction whatsoever*, opportunities and encouragement for pursuing a regular and liberal course of education of the highest grade and quality in the various branches of knowledge dealt with by the institution'.

Life at the Early School

While all this was going on in what might be called the *bel étage* of LSE history, life on the ground floor and even in the basement grew and flourished. In 1899 Hewins published a *Brief Report on the Work of the School.* The thirty-two pages of this Report recount a remarkable success story. In its first three years, LSE had amply achieved what it set out to do. More than 1,000 students had been registered already to whom over 400 were added in 1898/9. Many of them were graduates of other universities, a significant proportion women; students had come from sixteen countries. Since degrees were not given, 'there was no reason why students should join the School except genuine interest in the subjects taught and a desire to learn'. New teachers had been added to those of the first year while most of the latter carried on. Lawyers like C. A. Montague Barlow and E. A. Whittuck are listed; G. Lowes Dickinson in Political Theory, A. J. Sargent in Economic History, and above all a galaxy of Occasional Lecturers including A. V. Dicey, F. Y. Edgeworth, and George Peel as well as such future permanent teachers as L. T. Hobhouse, and Miss Lilian Tomn, later Mrs Knowles. The rapid expansion of the Library is given due attention.

All this had its cost, between April 1895 and March 1899, a mere £9,000. (Thus Hewins's figures: expenditure for the full first four years actually added up to £10,500.) Nearly 50 per cent of the income was public money from the TEB, more than 10 per cent fees, a significant portion private donations, including Mrs Shaw's share of the rent; the Hutchinson fund had ceased to matter except on special occasions. On the expenditure side, almost all lecturers, as well as the Director, were part-timers, and many gave their time on an unpaid, voluntary basis.

One striking new element in the 1899 Report is the emphasis on 'commercial education'. Sidney Webb had lectured on the subject at the International Congress on Technical Education in 1897, where he described LSE as 'the beginning of a "High School of Commerce"',[131] and Hewins had often insisted on a close fit between the School's teaching and the needs of those responsible for British business. The 'High School of Commerce' is nothing other than a Business School in later terminology. A separate section of the Calendar emphasizes the thrust. It announces courses on 'chartered companies', 'the policy of different states in relation to means of transport', 'markets and dealing', 'modern company law and its connexion with the development of English commerce', 'the principles of local taxation', and the like. In this context the otherwise surprising comment has its place that students are attracted by 'personal negotiation' rather than advertisement. This applies to the 'railway students' sent by four of the main railway companies and later, when Haldane was Minister of Defence, to the 'Army Class'.

An undertone of the 1899 Report suggested that the unqualified internationalism of the early days was now somewhat mitigated. The Director was still proud of the list of countries represented among his students and staff, but a more consciously British approach had crept into the statement of aims. After all there was much 'uncertainty as to how far arrangements which had proved successful in other countries would be suitable to England'. This was Hewins's language, which led two years later, in the *Brief Description of the Objects and Work of the School* of March 1901, to a kind of 'mission statement':

The conception of Higher Commercial Education adopted by the School was that of a system of higher education especially adapted to the needs of 'the captains of industry and commerce', a system, that is, which provides a scientific training in the structure and organization of modern industry and commerce, and the general causes and criteria of prosperity, as they are illustrated or explained in the policy and the experience of the British Empire and foreign countries.[132]

One wonders what Bernard Shaw thought when he read this, but then perhaps he had other things to do. Sidney Webb for his part probably did not mind. When the subject of commercial education came back to LSE twenty years later, it was Webb's children, the teachers of the School, who objected; but that is a matter for another chapter of the LSE story.

Was the School then, in its early days, a University as we understand it today, or was it more a professional school along the lines of the later business schools? Alon Kadish (in his unpublished study of 'The City, the Fabians, and the Foundation of the London School of Economics') tells us that it was both, and in that he is surely right. But he also argues that Sidney Webb's early approaches to the London Chamber of Commerce and Hewins's interests led to the same conclusion. Here one must be allowed to wonder. It is certainly true that the LSE 'embodied from its very inception a different approach to the study of economics, and to its relation to current problems and needs', but the result of this turned out to be curiously paradoxical. It was the Fabians and notably Sidney Webb who insisted on the need for unbiased research, who sought to hire research students and promoted publications which would 'permeate' general consciousness, whereas Hewins did not hesitate to advocate the training of an Imperial commercial élite. As has happened not infrequently since, a strenuously though not always successfully value-free left found itself confronted with an unashamedly opinionated right, though at this time as on some later occasions the two had one belief in common which dominated for many years, indeed for decades. The Rt. Hon. James Bryce (so Kadish reminds us) put it into words when in an address to the School's students in 1900 he stated that possibly 'Plato had a school like the present in his mind when he said all kings should be philosophers, and philosophers kings'.[133] Popper came too late to prevent some of the consequences of this fallacy, though fortunately he not only came, but came to LSE.

Happily we do not have to rely entirely on official reports to get an idea of what it was like at Adelphi Terrace in the early years of the School. The accounts by Professor Bowley, Professor Wallas, Miss Mactaggart, and a number of the first students give some of the flavour. Lectures and classes were one thing, and obviously not all lecturers were equally popular with everybody. Even in the John Street days one student who made no secret of his preferences observed 'Hewins in the front parlour lecturing to a dozen or so mostly *men* on economics; Wallas in the back parlour lecturing to twenty or thirty mostly *women* on Poor Law'.[134] 'In those great days,' another student remembered, 'Dr Cannan was apt to be overshadowed by one or two of his more brilliant colleagues.' These remain unnamed but the same student refers to 'the apostolic fervour of A' and 'the golden-throated eloquence of Z' and thus reminds all students to the present day of one or the other of their teachers.[135]

But formal lecturing was only a part, and perhaps a small part of the life of the School. The Webbs began to give their much appreciated 'At Homes' at Grosvenor Road where members of the School, the Fabian, and London society met. More relevant for most was the afternoon tea hour arranged by Miss Mactaggart at which staff and students—day as well as evening students—met to gossip and sometimes to hold serious discussions. The latter were formalized when in 1897 the Students' Union was founded, to begin with as the Economic Students' Union. The first chairman, C. M. Knowles, was to become Lilian Tomn's husband. The Union met fortnightly for debates on topical subjects and soon became a focus for social activities, for 'from the beginning, students showed themselves gregarious and argumentative, eager to take advantage of every chance of meeting each other'.[136] Dining and dancing soon became a vital part of student life. LSE was a sociable place which helped form a strong *esprit de corps* as well as many lasting alliances.

One man of special importance joined LSE in 1896, the Head Porter Edward Dodson, who was to stay in the job for twenty-seven years. He soon became indispensable, especially for the students. 'The presiding genius', wrote one, 'was the Head Porter, Dodson, an old seaman, who knew everybody and was reputed to know everything.'[137] Arthur Bowley adds his experience to the list of virtues.

> Dodson brought from the Royal Navy an adaptability for all tasks, a stentorian voice and a strong sense of discipline. He saw that the students kept such rules as there were, knew which were industrious and which idle, and, I daresay forecast their examination results. He was the mentor of the students and the adviser of the staff.[138]

Dodson also began a tradition, for the porters have, more than any other group, kept the real School (as distinct from its more nebulous intellectual being) on the straight and narrow. They have throughout shown the self-confidence not to join any particular group or take sides in disputes, not even the side of the Director. As a result they commanded universal trust.

Members of the School throughout the decades will understand why Governor Clinton on his inauguration as US President in 1993 invited only one member of his former Oxford College, the head porter, to the grand occasion.

Dodson contributed much to the atmosphere of 'domesticity' which one of the students of the early years of the century remembered. The Director and the Secretary were (according to this student) 'remote' and 'aloof' figures. 'Neither of these is really essential to the picture, but the Porter's Lodge at the entrance, with Dodson within it, the first sight on entering and the last on leaving, is essential.'[139] This was after the move to Clare Market, and Arthur Bowley similarly recalls that 'in those days there was a quasi-domestic atmosphere', though he adds that 'as time went on [it was] gradually dissipated'.[140] What remained was the other charm of the School, 'the feeling that one was embarked on an adventure, that one had within one's grasp the key to esoteric wisdom that was to solve the riddle of social justice'. The student who wrote this for the fiftieth anniversary collection of reminiscences in 1945 tended to be rather critical of his teachers ('from Cannan and Foxwell I learnt nothing') but the sense of adventure pervaded all.[141] From its earliest days, LSE was not just another academic institution, but had a special cachet which had something to do with the intensity with which the events of the times penetrated its walls and the deliberateness with which those within tried to influence events.

This meant, of course, that the School was more immediately affected by incidents and trends of twentieth-century history than others, and that it was severely tested by them. The first such test came even before the century began, by the South African War—the Boer War—from October 1899 to May 1902. The list of early speakers at the Economic Students' Union includes the name of Alfred (later Viscount) Milner, who apart from the Colonial Secretary, Joseph Chamberlain, was the central figure in the events which unfolded to the dismay of some and the delight of others. Milner as High Commissioner tried to enforce Chamberlain's plan to incorporate the gold-rich provinces of Transvaal and the Orange Free State into the British Empire. The Boers resisted and in the early actions revealed what came to be seen as the 'inefficiency' of British forces as well as leaders; but then the massive deployment of half a million British troops at a cost of over £200 million began to wear down the Afrikaners with their force of 60,000 men. They sued for peace, became a part of the Empire, and Milner could proceed, with the help of his young Oxford advisers, to 'reconstruct' the Boer states primarily by creating an 'efficient' mining industry. 'Joe's War' was won, at a cost of more than 70,000 lives, of whom

26,000 died in special camps in which the Boers were rounded up: 'concentration camps', as they were called.

Not surprisingly the Boer War led to bitter divisions at home. It was one of those incidents by which 'old friendships were broken and families driven to savage quarrels'.[142] After Queen Victoria's death in 1901 Britain was not a happy country. More surprisingly, perhaps, the divisions affected the political left as well as the rest. Anne Fremantle offers a rather neat description of the Fabian stances:

> The Boer War actually split the Fabians four ways. There were the out-and-out pacifists, generally dubbed pro-Boer; there were those who opposed this particular war as imperialist; there were the bitter-enders, who had been against getting involved, but once in, felt England must go on; and there were the blood-letters, who, with a moral fervour, approved of the war as salutary, and of these, G.B.S. was passionately one.[143]

The Fabians dealt in their own messy way. An early resolution by J. A. Hobson to dissociate the Society from the war was carried, but the Executive promptly arranged a postal ballot which came down narrowly in favour of not making any official pronouncement. It may well be that any other decision 'would have destroyed the Society',[144] but when Shaw added his tract on *Fabianism and the Empire* to the official cop-out, many significant names left the Society.

LSE, of course, did not have to take a stance. Indeed it was clearly not supposed to do so. However, this did not spare it many of the same divisions, with notable non-Fabians like Edwin Cannan and Bertrand Russell strongly advocating the pacifist or even pro-Boer position. Then there were the students. Perhaps for the first time, a pattern emerged which was to become familiar in later crises. 'At the School the staff was very largely imperialistic, and though probably the students, if polled, would have shown a handsome majority on the other side, among actual frequenters of the class-room and Union, "pro-Boers" were apt to be much out of fashion.'[145] Certainly, the Director and the Chairman were among the supporters of the war, though the enthusiasm of Hewins and the indifference of Webb made their alliance another illustration of the tenuous combination of imperialism and social reform.

WEBB AND HEWINS

Synergy without Intimacy

Occasionally the founders still worried about the survival of their creation even when it was four years old and, in Janet Beveridge's words, a 'lusty child'. 'I got nervous last night', Sidney Webb wrote to 'his Bee' on 28 April 1899, 'about the London School and the University, and all the complications, thinking that it would all collapse like a pack of cards.' Webb was actually a bit of a worrier and needed to be reassured, which Hewins promptly tried on this occasion, though Webb still confessed to his wife: 'And if it does collapse, well then it must. We shall have done our best.'[146] Janet Beveridge makes much of the 'two serious attacks'[147] which the School suffered in its third and its seventh year. The first was the attempt by Ramsay MacDonald and others to withdraw the TEB grant, and the second the doubts expressed by members of the new University Senate about the impartiality and academic credentials of the School. However, between them Webb and Hewins mastered both, the latter by Webb's carefully drafted if oddly apologetic letter to the Vice-Chancellor Dr Archibald Robertson on 3 January 1903. The passage about the 1895 conversation with Sir Owen Roberts has been quoted earlier. Webb went on to lean over backwards in denying a Fabian bias, almost to the point of admitting the opposite inclination. 'By far the largest gifts have, as a matter of fact, come from persons who are Unionist in Politics, and Individualistic, Free Trading monometallists in Economics.' And of course 'Hewins was then as now a strong Imperialist and Churchman, and at any rate, "anti-Liberal"'. Fortunately, the first Director also got some less backhanded praise from his chairman: 'Here I must name Hewins, whom I chose as first Director of the School, and who did, at the start, all the organizing and nearly all the teaching, throwing up all his Oxford and other connections, and putting inexhaustible energy and ability into the work of building up a new institution, for what was at first merely a nominal salary.'[148]

As we know from his *Apologia*, Hewins responded in kind:

> We were making a genuine attempt to build up an institution which would deal with economics and political science in a manner worthy of these subjects . . . Outside the School I was often warned by uneasy economists of the danger of my association with Sidney and Beatrice Webb. Nothing could have been more absurd. I have worked with many different colleagues, but never with any two people so free from prejudice or so absolutely loyal and devoted to the main objects of what we were doing and so indifferent to the ordinary prejudices to be found in academic circles as Mr. and Mrs. Webb.[149]

There are several reasons why the venture of building up a new institution (as they both put it) succeeded. It benefited, in the apt phrase used in later Calendars (and probably coined by Sydney Caine) from 'the conjunction of a need with an opportunity'. Circumstances were favourable; a significant demand for social science education could be tapped; thanks to Henry

Hunt Hutchinson and Charlotte Payne-Townshend and others the wherewithal was found; influential persons like R. B. Haldane were prepared to lend their support. But when all is said and done, no deconstruction of the history of LSE can detract from the fact that its foundation was the work of two unusual men, Sidney Webb and W. A. S. Hewins.

The two were very different, yet in this particular venture totally complementary. Both had their paradoxical side. Webb was the pedant as visionary, Hewins the enthusiast as administrator. Webb's qualities were much underestimated until Norman MacKenzie brought them to light. Hewins is even less recognized, though also harder to understand. Beatrice, who was so good at brief characterizations, ran into difficulties with Hewins and in the end called him 'one big paradox'. 'He is disinterested with regard to money, he is ambitious of power.' 'But the most characteristic paradox of his nature is the union of the fanatic and the manipulator.'[150] The 'fanatic' is Bernard Shaw's term; the manipulator hers. What she meant was that he had ways of making others do what he wanted them to do. Nearly two years after that diary entry of January 1901 Beatrice Webb added two other characteristic observations. Hewins 'inspires confidence in men of affairs and has, in fact, more the business than the academic mind'. He is also not as precise, let alone pedantic as the Webbs (and Miss Mactaggart) would have liked, indeed Beatrice describes him as 'slovenly' and 'dilatory' at times: 'But there is usually method in his carelessness and things left undone or mistaken are usually matters about which his judgement has been overruled or to which his aims are slightly different from those of the Governors.'

'With such a character it is difficult to be intimate, however much it may excite one's admiration, liking and interest.'[151] Beatrice may have craved intimacy, but for Sidney—and for Hewins—its absence was the very basis of their synergy. Sydney Caine, and the dissertation author whom he quotes, E. J. T. Brennan, are led by Beatrice's account in *Our Partnership* to regard the relationship as one based on a division of labour, with Hewins looking after the 'internal organization of the School' and Sidney after 'the whole financial side' (Beatrice meanwhile 'roping in influential supporters').[152] Closer study and a critical reading of Beatrice Webb's numerous texts reveals this as only a part of the truth. If Hewins knew one thing, it was how to mobilize support for the project (and he did it again later for others), and Sidney never ceased to take an interest in matters of internal organization. Certainly much of the initial funding had come in through Sidney's connections, but especially in the early years financial viability and academic achievement were inextricably linked. In any case Hewins was always brought in to persuade those who had money as well as political doubts. In so far as there was a division of labour between the two, it was one of temperament rather than activity. Moreover one must not forget that for Sidney Webb the School was one of many concerns which actually shrank in relative importance as the years went by, precisely because Hewins was there and put all his considerable energy into the School at least for the first seven years.

The symbiosis of the two was remarkable. They met all the time. Even after six years they had lunch at Grosvenor Road every Tuesday, dinner with some group or other at least once a week, and that in addition to meetings of the Trustees or the Governors or some other committee. And when there was an emergency involving LSE, they would turn naturally towards each other, though the reservation 'involving LSE' is important; there is no evidence that either turned to the other with worries or even delights caused by different subjects. In that sense there was no intimacy between them. In 1898 the Webbs went on a tour of the world which took them away from London for nine months. A stream of letters kept them in touch though it is well to remember that there was no air mail at the time, so that letters to and from the West Coast of America, let alone New Zealand and Australia, took a little while to reach their destination. For these nine months in its third and fourth years Hewins ran LSE by himself.

Still, the correspondence gives us an idea of what the two men were talking about when they met and how they saw each other. Fairly soon after the Webbs' departure, Sidney wrote to Hewins from Boston (on 14 May 1898). He had evidently been talking to many people already about LSE and found that 'the School lacks a little in status on this side'. This was hardly surprising after a mere two and a half years but it led Sidney to a series of practical suggestions. The prospectus should be sent to '*all* the American economic professors, economists and historians' (the emphasis on *all* is Sidney's), and 'an elaborate "Calendar"' would be even better. Provision must also be made for students to get some kind of 'certificate of study and research'. Oh, and great care has to be taken not to give the wrong kind of political impression; 'we ought not to let it be imagined that the School is especially for study of *Labour* questions'. (One cannot help feeling that Webb was beginning to go further than necessary on this issue, which is as much a comment on his tactical sensitivities as on the substantive concerns of his environment.) The Library, Webb added, should acquire American journals and monographs. Perhaps universities in the United States can be asked for free copies; otherwise second-hand sets might be available. 'All kindly greetings to yourself and Mrs Hewins. This is a business letter!'[153] There were no others.

Hewins's tone was a little more emotional, as was his wont, though the substance remained strictly business-related. His letters to Webb were evidently written as much to clear his own mind because Webb could do little from a distance. They were also written because the Director needs someone to talk to, and this cannot easily be a colleague or subordinate. 'We have had rather anxious times at the School,' he wrote after he had received Sidney's Boston suggestions on 30 May, only to add quickly, 'or perhaps I should say, I have, because most people think the School moves with a sort of triumphant momentum, independent of the will of man, and it is not wise to shake confidence. But I have been very anxious.' There follows a blow-by-blow account of the discussions with the TEB about continuation of their grant as well as a more light-hearted description of 'Miss P.-T.'s' preparations for marrying Shaw, though this is also seen as a possible threat if she should decide to give up Adelphi Terrace. 'The TEB episode and now Miss P.-T.'s marriage illustrate the insecure basis on which at present the School rests.'

Hewins then takes up Webb's remark about the School lacking status and responds to it with a veritable outburst, not against Sidney Webb or anyone else in particular but about his deeper beliefs in education, the role of the School, and public life in Britain. 'The only way I know of creating status is to grow a number of able men, to identify them and their work with the School, and so accumulate great traditions.' The issue is not to train students of economics, at least not now. The first task is 'to inspire public policy, not by drafting programmes, but by presenting ideas clothed in the living garments of reality,' as the Webbs had done by their *Industrial Democracy* and the campaign of speeches and meetings that followed it. One day there may be an elaborate system of degrees in economics and political science which will then attract good students without any special effort, but for the moment the greatest need is to help those who seek a public career, of whom there are more in England than anywhere else. They require not just academic training; there is rather 'the necessity of a closer union between the School and the public life and aims of the times, than you get in the analogous institutions of foreign countries'.

What a reversal of roles! Here is the 'socialist' Sidney Webb recommending caution in showing too great an interest in labour questions, and there the 'conservative' Hewins arguing against too academic a definition of the School. This does not sound like a division of labour which confines the Director to matters of internal organization. They were both strong-willed and influential men—modest as Hewins may sound when he writes a little later that 'small men as we may be, we can do much to concentrate the general interest in the School'. The outburst is not yet finished. Hewins praises Oxford for having captured the minds of so many men of state and of business. He then adds the ominous remark:

> If we are to attach any importance to the signs of the times, it is not inconceivable that in the near future the work done in building up the present British Empire will be put to the test. Whenever the struggle comes, it will either become impossible to carry on the School, or the School will have to become a greater force then we have yet dreamt of.'

The Boer War? And what kind of image of LSE does this betray? As if to underline the point, Hewins gives a beautifully ironic account of the lying in state and funeral of Gladstone, who had died on 19 May 1898. Lord Samuel had asked Hewins on behalf of the Liberal Party whether the School would accept a Gladstone memorial endowment, whereupon Hewins suddenly became almost Webb-like and insisted that this was possible only 'if it was of a genuinely national character', for 'the School, like the rain, must fall equally on the just and the unjust, know no distinction of parties but equally inspire all in the faith that the strongest arguments will win the day'. The endowment went to Oxford.

Hewins had still not done, though his emotion was spent. He went on to describe the work of the School as well as his own ('I rarely get home before 10 p.m.') and put Webb's mind at rest over details concerning the Calendar and the Library. 'I am afraid I have bored you with a very long letter. With kind regards to Mrs Webb, believe me, yours sincerely, W. A. S. Hewins.' Later letters to Webb the world-traveller are more businesslike throughout, though they do contain the occasional 'I miss you horribly' and some characteristic touches of self-deprecation: 'I feel fairly fit, but if I don't survive until your return you must raise money on my death. I don't suppose I should be worth new buildings, but a small lectureship or studentship, or at worst some new shelving would not be out of the way.'[154]

The secret of the relationship between Webb and Hewins lay in the way in which the common project of LSE reined in the enormous differences between them and made them productive for the cause. The pedant and the enthusiast, the upstart and the Oxonian, the municipal socialist and the conservative Imperialist, the undramatic planner of 'revolution through research' and the sometimes melodramatic organizer of new 'movements' had surprisingly little in common and yet they were bound together by the School. The deeper bond may have been of a different order and has perhaps emerged in this story as it unfolded. The two men needed each other. Webb needed Hewins because despite his obsession with detail he was not a doer, and Hewins needed Webb because despite his enthusiasms he was not an inventor. Whether this meeting of needs could have sustained a life-long relationship in the absence of a real meeting of minds is a question which in the event was not put to the test.

A Rapid Departure

Hewins left LSE in exactly the way in which he had arrived on the scene—quickly, with almost clinical neatness, and in order to help, nay enable a great man to set in train a project which was also a movement. Early in November 1903 Hewins wrote to Joseph Chamberlain to say that he had considered his request to become 'secretary of the proposed Tariff Commission'. The responsibility was heavy, but Chamberlain's confidence encouraged him that 'I can be of real service to the Imperial cause and hasten the adoption of your policy': 'I have therefore decided to accept your invitation, and I propose to begin work at once. I am today resigning the Directorship of the School of Economics and my other academic posts so that I can henceforth devote all my powers to the work of the Commission.'[155]

Chamberlain was obviously pleased but also rather taken aback by his tempestuous friend, as his immediate reply shows:

> But I confess I am a little concerned to find that you think it would be necessary to break your connection with the School which is really your creation and which has done so much good work. I do not like the idea of destroying one good thing in attempting to build up another, and I had hoped that it might have been possible to combine the two.[156]

But that was not Hewins's way, nor was he concerned about giving up the Tooke Professorship for a temporary position. There is some doubt about the exact dates, though Hewins quotes from his own diary when he says that Chamberlain formally

> invited me to become Secretary in a letter I received on Tuesday, Nov. 17th, 1903. I had informed the Webbs that my resignation of the School was probable, on Monday, Nov. 9th, and on Nov. 16th that it was certain. I sent in my resignation on Nov. 18th and it was accepted by the Governors on the 20th to take effect on Dec. 25th.[157]

Hewins had crossed the Rubicon and never looked back.

Again, the decision had not come entirely out of the blue. Hewins's Imperial predilections had long been known. On 15 May 1903 Joe Chamberlain had made his famous Birmingham speech in which he announced to the world his 'conversion' from free trade to protection and Imperial preferences. Hewins was one of many who were thrilled. He tried to persuade the Webbs but found them rather less impressed. 'I venture to suggest', Sidney wrote on 30 May, 'that it will be important to keep the School out of the stormy controversy that is going to arise; and as you and I are already both "suspect" on the subject of Imperial protection, the less we say the better.'[158] But Hewins was not to be restrained. In June he agreed to write regular articles to promote Chamberlain's cause. His closest correspondent and former research student Firth warned him on 15 June that his stance 'might frighten away some money from the School'.[159] The problem was not just money. Alfred Marshall used the occasion to cut the link finally when he wrote to Hewins on 14 July: 'And so I trust you will forgive me if I state bluntly and categorically that I dissent from your economic arguments, and even from your statements as to fact, to an extent which I had not anticipated in the very least.'[160] Sidney Webb's concern grew during the autumn of 1903 and by the end of October Hewins's position at the School had become all but untenable.

The successes of Hewins the tariff reformer did not quite match those of Hewins the founder of LSE. Beatrice Webb may not be the most reliable witness; she had, perhaps understandably, been ungracious about Hewins leaving LSE in order to form an alliance with the man who had jilted her. Hewins 'is an ideal henchman for Joe—fanatical, well-informed, unveracious and devoted to his cause',[161] she wrote in her diary. Still, her impressions when she met Hewins and his wife for the first time after a long interval of silence in January 1906 are not improbable. Hewins was then obviously disappointed by the Liberal election victory. He seemed also 'irritated against Joe'. 'Poor Hewins, with his grand castles in the air that he has been, for the last three years, inhabiting—now lying in ruins about him!'[162]

The Imperialist had his moments of triumph. But there was much disappointment too. Hewins tried three times unsuccessfully to enter Parliament (Shipley, 1910; Middleton, 1910 and 1911) before he was elected for Hereford City in 1912. His six years as an MP were unremarkable, although he briefly served as a Parliamentary Under-Secretary to the Colonial Secretary Walter Long before losing his seat after boundary changes in 1918. Three further attempts to win Swansea West for the Conservatives failed. Meanwhile he rarely went to LSE though on a few occasions he accepted invitations to lecture and in 1910 he came to the rescue of an embattled Sidney Webb with a gentlemanly letter to *The Times.*

In 1914 Joseph Chamberlain died, having been incapacitated by a stroke for some time; 1914 was also the year in which Hewins was received into the Roman Catholic Church. Was he yet again in need of a firm lead and framework in which to develop his talents? During the last ten years before his sudden death at the age of 66 in 1931 he was not seen around much, certainly not by his old LSE friends. Polite and, as always, respectful letters were exchanged between Webb and Hewins from time to time, the last one when Webb thanked him for his *Apologia of an Imperialist* in 1929, but to all intents and purposes the School faded away for Hewins and Hewins lost touch with its story.

2

GROWTH AND GROWING PAINS

A MEETING OF MINDS

Mackinder and Webb

W. A. S. Hewins's sudden departure soon after the beginning of the academic year 1903/4 left an obvious gap. Unless Sidney Webb was going to take charge himself (as he had done briefly at the beginning of the second year of the School and was to do again after the end of the First World War), a new Director had to be found almost instantly. Fortunately one was at hand, having been a lecturer since the first week of the first term in 1895—the geographer Halford Mackinder. Sidney Webb picked him and persuaded him, but the appointment was no longer in his gift nor even in that of the Trustees or other School committees. Now that LSE had become a part of the University of London the University Senate had to approve and formalize the appointment, which was as well since on this occasion the School wanted Mackinder confirmed as University Lecturer in Physical Geography to supplement his meagre Director's salary of £400. The Senate took both decisions with commendable expeditiousness on 16 December 1903 so that Mackinder could assume the directorship immediately when Hewins's contractual notice ran out a week later.

To say that Mackinder had been at the School since its first term is in a noteworthy sense both right and wrong. It appears that he was never just in one place. As one reads different accounts of this period in his life, one wonders whether it is possible that a man is simultaneously a Reader in the University of Oxford, the Principal of the University College of Reading, and Director of LSE. In fact he was all these, and the geographical link invoked by his biographer E. W. Gilbert, who said that 'for a time Mackinder held office in all three Thames valley towns simultaneously', does not really explain how it could be done. (It is almost more relevant 'that he then owned three dress suits, one kept at Oxford, one at Reading and another in London'.)[1] Nor were the three jobs all that Mackinder did at the time. In 1899 he went on an extended, difficult, and successful expedition to Mount Kenya. In 1900 he stood unsuccessfully for Parliament as a Liberal at Warwick and Leamington. One begins to appreciate why his student and admirer Martha Wilson Morse would write to Jean Ritchie after his death: 'No work on Sir Halford will ever be finished. Like Beethoven, Sir Halford is too mighty, too universal.'[2] The likeness may seem a little remote. Mackinder himself preferred to say that he was 'a pluralist'.

Two other characteristics of Mackinder are more relevant. One is his predilection for new initiatives. 'Mackinder told me that he liked beginnings,'[3] Gilbert recalled, and Mackinder's most comprehensive biographer Brian W. Blouet gives this preference a psychological twist: 'Driven by a necessity to keep proving himself to himself,' Mackinder 'always felt the need to accept the next challenge'.[4] The other significant fact is that geography was his life's passion, so much so in fact that his very first letter after the election to the directorship of LSE on 16 December 1903 was written to tell the President of the Royal Geographical Society that the post 'will give me great power on behalf of Geography'.[5]

Mackinder, and the Royal Geographical Society, had looked for an opportunity to establish geography firmly at least since March 1895. But where should it be? In Oxford? In London? On 11 December 1895, a few weeks after the opening of LSE, John Scott Keltie of the Royal Geographical Society had written to Hewins enquiring whether LSE would be willing to co-operate in developing an institute or school of geography in London. Hewins replied sympathetically. A few weeks later, in February, Mackinder produced a prospectus for a new London School of Geography. The proposal was to create a place for lectures and evening classes as well as for research by a number of geographers. The historian of LSE cannot suppress a sense of *déjà vu* in reading the document, which in part looks like a carbon copy of the LSE prospectus produced by Hewins nine months earlier, all the way to the French reference which, in the case of geography, suggests that the new school should be formed 'somewhat on the lines of the Paris *École des Chartes*' (to which Mackinder added, a little shamefacedly, 'and the lately established London School of Economics'). 'Mr Mackinder's Estimate', submitted a little later, reads like a description of 9 John Street with its two rooms and a clerk.

More than sympathetic interest was needed to make the plan real. The LSE story had shown the way. But Mackinder did not have a Hutchinson bequest; more importantly, Webb's Technical Education Board of the LCC, when approached, did not find it possible to support the venture; and the Royal Geographical Society, while sympathetic, decided in the end that it would be cheaper to set up the School of Geography in Oxford. This duly happened in 1899, and the Oxford University Reader of Geography became its first director. He was of course none other than Halford Mackinder who eventually resigned from Oxford in 1905.

Why then should he be Director of LSE? Mackinder, born in 1861, was four years older than Hewins. His father had been neither a vicar nor an industrialist but a medical officer in Lincolnshire. Halford Mackinder had read Physical Science at Oxford, though his pluralism showed early, for he also read Modern History, held a scholarship in geology, became President of the Union, and was called to the Bar (Lincoln's Inn) before he was 27. More relevant was perhaps that he became a friend of Michael Sadler and fell for the challenge of University Extension teaching. He is said to have given 600 extension lectures within three years and travelled 30,000 miles in the process, but then he was an 'outstanding' lecturer who throughout his life held his audiences spellbound. In reading reports of Mackinder's appearances, one can see why his kind of synthetic geography with its mixture of visual components and imaginative argument would go down well in Barnsley and Skipton as well as Salisbury and Taunton. (Somebody has appropriately produced a map of England marking the 'Towns in which H. J. Mackinder gave courses of Oxford University Extension Lectures'.[6]) Obviously, the paths of Mackinder and Hewins crossed frequently during these Oxford days.

When Mackinder met the Webbs they got on well, though Beatrice's first diary entry about him shows her in one of her less charitable moods. 'He is a coarse-grained individual (Bertrand [Russell] says brutal) but with a certain capacity for oratory, and strong picturesque statement. If he got his foot on the ladder he might go far towards the top, especially as there is an absence of able young men.' To which she adds mysteriously (because she had heard of the Mount Kenya expedition?): 'Signs in him of negroid blood?'[7] Whatever was meant by such signs, Mackinder and the Webbs saw quite a lot of each other during that summer of 1902 in which the School had taken possession of the new building. They spent time in the Webbs' new summer retreat in Gloucestershire, talking about the School, the recently introduced B.Sc.(Econ.), and then increasingly about a new plan (in Sidney's language); another beginning (in Halford Mackinder's).

Sidney was still pondering ways of following up his much-discussed article of the previous autumn about 'Lord Rosebery's Escape from Houndsditch', that is about the way in which a new political philosophy of efficiency and expertise could help the country get away from the ('Houndsditch') tailors' patchwork of 'Gladstonian rags and remnants'. His suggestion to Rosebery had been to bring together 'a group of men of diverse temperaments and varied talents, imbued with a common faith and a common purpose, and eager to work out, and severally to expound, how each department of national life can be raised to its highest possible efficiency'.[8] Nothing to that effect had happened. Then Beatrice conceived the idea that Sidney might do worse than start with his own acquaintances and bring them together in a dining club. Sidney liked the prospect, and so did Mackinder, with whom he and Beatrice discussed it. The plan was that Beatrice would preside over the initial dinner, but would then withdraw and leave the men to their own resources. One day in the summer of 1902 when they went out for a ride (a bicycle ride to be sure), Beatrice Webb explained all this to Mackinder, who recalls it in one of his autobiographical fragments. 'Her husband, who was riding just ahead of us, threw back over his shoulder—"I will give your club its name—the Co-efficients".'[9]

The Pentagram Circle

And so one of the more intriguing clubs of that age of apparently endlessly clubbable men (not women) was born. In later years the Coefficients Club came to be much written about, for it lends itself to a multitude of interpretations. What exactly was it? A brains trust? A conspiracy? The nucleus of a new party or a future government? The mystery started early. 'The "dining club" takes shape; and I am asked by the half a dozen who have nominated themselves the first members to invite you pressingly to join.'[10] Did Sidney know what he was doing when he wrote this note to H. G. Wells on 12 September 1902? In any case, Wells joined. Like other early members he later wrote about the Coefficients in his autobiography, though many years earlier, in 1911, he had given a more disrespectful account in what he called his 'queer confused novel', *The New Machiavelli.* 'In those days there existed a dining club called—there was some lost allusion to the exorcism of party feeling in its title—the Pentagram circle.' The Druids were at work in Edwardian London. Still, Wells's description probably gives as good a sense of the club as any:

> We were men of all parties and very various experiences, and our object was to discuss the welfare of the Empire in a disinterested spirit. We dined monthly at the Mermaid in Westminster, and for a couple of years we kept up an average attendance of ten out of fourteen. The dinner-time was given up to desultory conversation, and it is odd how warm and good the social atmosphere of that little gathering became as time went on; then over the dessert, so soon as the waiters had swept away the crumbs and ceased to fret us, one of us would open with perhaps fifteen or twenty minutes' exposition of some specially prepared question, and after him we would deliver ourselves in turn, each for three or four minutes. When every one present had spoken once talk became general again, and it was rare we emerged

> upon Hendon Street before midnight. Sometimes, as my house was conveniently near, a knot of men would come home with me and go on talking and smoking in my dining-room until two or three. We had Fred Neal, that wild Irish journalist, among us towards the end, and his stupendous flow of words materially prolonged our closing discussions and made our continuance impossible.[11]

There was little love lost between H. G. Wells and not only 'Fred Neal' (probably J. L. Garvin) but also that other Irishman George Bernard Shaw, and other Coefficients as well who are recognizable in Wells's story. The restaurant where the club met was actually the Ship Tavern rather than the 'Mermaid', and it was not on 'Hendon Street' but on Whitehall, though later the club moved to St Ermin's Hotel.

Just in case anyone should wonder what this story has to do with the history of the London School of Economics, the answer is simple. The twelve founding members of the Coefficients included all three Directors of the School from 1895 to 1919, W. A. S. Hewins, Halford Mackinder, and William Pember Reeves. In addition of course Sidney Webb was there and also—how not?—R. B. Haldane. Bertrand Russell and H. G. Wells had at least some association with the School, so that the 'strangers' were at first a minority. They included Leopold S. Amery, recent *Times* correspondent to South Africa and Conservative Imperialist, who together with Mackinder served as secretary of the club and produced brief but informative reports. The others, like everyone else, were there as 'experts' of one kind or another, Lieutenant Carlyon Bellairs, RN, as a naval expert, Sir Clinton Dawkins, KCB, as a financial expert, The Rt. Hon. Sir Edward Grey, Bt., MP, as an expert on foreign policy, and L. J. Maxse, Esq., as an expert on journalism. Had there been an LSE board of directors, these might well have been its members.

The group was constituted as planned, at a dinner arranged by Beatrice Webb on 6 November 1902, and held its first formal meeting at R. B. Haldane's apartment on 8 December. (We know about these meetings because their printed minutes, or reports, have been found in the papers of several participants.[12]) 'How far and on what lines are closer political relations within the Empire possible?' was the question, and W. P. Reeves opened the discussion. He suggested that 'a closer union' would require 'two bodies, a permanent advisory committee and a periodical conference between the Imperial ministers and the ministers of the self-governing Colonies'. The smaller permanent body should represent expertise whereas the larger body might be the nucleus of an 'Imperial Cabinet'. The question of 'the power of the purse' was raised as well as that of 'an adequate Imperial court of appeal', and some discussion was held of policy matters like 'Inter-Imperial Commerce and Imperial Defence'. The similarity to the European debate half a century later is almost uncanny.

The first meeting showed little disagreement whereas the second (which took place at the Ship on 19 January 1903) 'gave evidence of a considerable divergence of opinion among members'. It was introduced by Professor Hewins and dealt with the question: 'How far, and upon what conditions, is preferential trade within the Empire attainable or desirable?' Hewins presented a double-barrelled argument, with the barrels not necessarily pointing in the same direction. On the one hand, shrinking industries and narrowing markets required protection if Britain was 'to support the large armaments required to maintain our international position'. On the other hand, the German example of the *Zollverein* had shown that 'political unification could never be really complete without economic unification'. Other Coefficients were worried about trade diversion and artificial commercial units though all agreed, as tends to happen when the substance of an issue is controversial, that there was inadequate information and that therefore 'some closer investigation of the whole character of our present trade with the Colonies' was needed. The LSE members must have been pleased.

The Coefficients seemed also of one mind on a rather more significant issue:

> There was a general agreement among members that in practice the Empire was for them an object in itself, an ideal that had gradually grown up in their minds, which it did not occur to them to refer to any standard, but which was to them in itself the principal standard by which they judged political issues. To use the words of one member, he 'no more asked himself why he became an Imperialist than he asked himself why he fell in love'.

Thus begins the report of the sixth meeting held on 15 June 1903 and introduced by Mr Amery on the question: 'For what ends is a British Empire desirable'. As usual, the report is signed by L.S.A. (Amery) and H.J.M. (Mackinder), but it is in more ways than one economical with the truth. (It shows incidentally how misleading apparently straightforward documentary evidence can be.) For one thing, this was the meeting at which Bertrand Russell resigned from the club, or rather was 'flung out' of it in H. G. Wells's words. 'Hewins, Amery and Mackinder declared themselves fanatical devotees of the Empire. "My Empire, right or wrong", they said. Russell held against that there were a multitude of things he valued before the Empire. He would rather wreck the Empire than sacrifice freedom.'[13] For another thing, this was in fact not the sixth, but the seventh meeting because there is a report of a rather whimsical evening on 28 May, attended by Grey, Maxse, Russell, Bellairs, and Reeves at which Amery 'entered the ante-room with an opera hat and the cares of Empire on his brow', only to tell the others that he had to go and dine with Lord Roberts, the promotor of the National Service League. The others stayed and listened to Mr Maxse denouncing 'the subversive and heretical pilot [scheme?] just wantonly sent up by Mr Chamberlain'.

This was a reference to Joseph Chamberlain's Birmingham address of 15 May which (as Mackinder put it in an after-dinner speech in 1931) 'dropped a bomb among us', the Coefficients.[14] 'I remember,' writes L. S. Amery,

a little dinner at Edward Grey's house three days after that pronouncement, when I made an impassioned appeal to him not to let a small theoretical issue prevent his joining in a great Imperial movement which might revivify the whole of our political and social life. But Grey, though I think moved, would not be persuaded. Fear of a new idea, added to the pull of old party ties, was too strong. On the other hand I did succeed, after several talks, in persuading Mackinder. Not to his personal advantage, I fear, for he would almost certainly, if he had stayed with his party, have risen to high office in the 1906 Parliament.[15]

Mackinder became a Liberal Imperialist and later a Unionist, allied with the Conservatives. The Coefficients continued their evening discussions and in due course tackled every one of the five *E*s. Empire was what they were all about. Economics, at least Hewins-type economics, accompanied them throughout. Efficiency hardly needed mentioning given their name. Equality was introduced, unsurprisingly, by Sidney Webb at the seventh meeting on 9 November 1903, which was devoted to the question: 'How far is it possible by legislative regulation to maintain a minimum standard of National well-being?' The usual rapporteurs, L.S.A. and H.J.M., tell us that 'very few objections from the individualist standpoint . . . were suggested', but they also report an interesting twist to the discussion:

It was urged that just as it was necessary to have legislation in order to enable the individual to maintain a higher standard, so a State which endeavoured to set up a higher standard required to be protected against the competition of others, at any rate for a time till the system of regulation should have had its fullest effect.

The Empire was never far from their minds. Mackinder's own first contribution, however, was devoted to the fifth *E*, to Education: 'What should be the limits of national free education?' The report of the tenth meeting on 21 March 1904 on this subject is rather disappointing. 'There was general agreement among members as to the necessity for a considerable extension of higher Education,' but no useful ideas were advanced as to how to go about it. Moaning and groaning about a decline in the quality of professionals and the ignorance of parents who were demanding free education irrespective of the talents of their children dominated the evening.

The Coefficients went on for several years. They grew in size as they declined in importance. Viscount Milner joined them in 1905, and surprised the older hands by his openness and intelligence. Later reports betray a change of mood, above all of the urgency of taking a view. 'It is scarcely necessary to add that no conclusions were reached at the end of a very interesting evening,' writes H.G.W. [Wells] on 17 April 1905, thereby indicating that the Coefficients were about to become just another club. Mackinder is even more laconic when he confines his report of the twenty-first meeting to a single sentence: 'The discussion which ensued [on 'A Possible First Step towards Revolution'!] dealt with the general tendencies of the time as disclosed by the discussions at the twenty previous meetings.' The club was not actually discontinued until at least 1907, but it 'petered out', as its protagonists tell us, and as is also evident from Sidney Webb's letter to his 'dearest one' on 18 June 1907:

> Last night I went to the Co-Efficients—9 present, including Lord Milner. I propounded the Poor Law Reform Scheme, not saying that was your scheme, though implying it, and not much alluding to the [Poor Law] Commission. They were all favourably impressed and made little criticism, except deprecating the expense (Milner), and fearing medical opposition (Mackinder and Newbolt) fear of stereotyping medical practice (Bellairs). No useful suggestions were made.[16]

By then, the Liberal election victory of 1906 had taken place and had put most of those whom we have encountered so far in this story—the Coefficients, the Fabians, and even LSE—in a political no man's land for a while, though some familiar men, notably Haldane, helped them regain their bearings.

'I find my thoughts lingering about the Pentagram Circle.'[17] H. G. Wells is not the only one. When Sydney Caine discovered the minutes of the Coefficients in 1973, he read into them Webb's dream. 'The strong links of the Coefficients with the LSE have been noted; and the atmosphere of objective discussion and of the bringing together of diverse opinions is wholly familiar to those who know the real character, rather than the popular reputation, of the LSE.'[18] Alas! the Coefficients were more an example of sometimes vicious clashes of strongly held partisan views. 'The shadow of Joseph Chamberlain lay dark across our dinner-table . . . More and more did his shadow divide us into two parties.'[19] Wells saw himself aligned with Reeves and Russell against Maxse, Bellairs, Hewins, Amery, and Mackinder. Mackinder himself noted that 'on the issue of Protection and Free Trade we divided hopelessly. We tried for a time to hold together, but we were carried apart as a majority and a minority into new intimacies of comradeship.'[20] This also answers Bernard Semmel's question: 'Why did not the Coefficients succeed in becoming the brains trust of a new social-imperial political party as the Webbs had hoped it would?'[21] Because they were two parties at least and probably three, and in any case the story of real politics was not written over a glass of port in the Ship Tavern. The Coefficients were neither an academy, an Imperial College of Economics and Political Science, nor a future political movement, but simply a club, a Pentagram Circle which on its better days, or rather nights, gave its members pleasure 'not so much by starting new trains of thought as by confirming the practicability of things [they] had already hesitatingly'—and in some cases not so hesitatingly—'entertained'.[22]

The Circle had one other benefit; its members got to know each other rather well, and some, notably Wells and Amery, have imparted their knowledge to posterity. The liberal internationalist and the conservative imperialist did not have a great deal in common, but they were equally outspoken about their fellow Coefficients. Neither of them liked the 'sham expert officialdom of [the 'Bailey'-Webbs] to plan, regulate, and direct the affairs of humanity'.[23] Both wondered whether Edward Grey was really up to his job, indeed 'doubted whether he fully understood what was happening in international affairs'. The two autobiographers did not have much time for Haldane. 'Haldane was steeped in university and metaphysical German,' thought Amery. 'Whether Haldane's metaphysics really contributed much to that "clear thinking" which he incessantly preached, I always doubted.'[24] So did Wells, who found him 'intellectually unsympathetic'. 'The "Souls", the Balfour set'—another club in which Haldane may well have felt more at home—'in a moment of vulgarity had nicknamed him "Tubby". He was a copious worker in a lawyer-like way and an abundant—and to my mind entirely empty—philosopher after the German pattern.'[25]

Geographer of Central Europe

Both Wells and Amery respected Mackinder. Amery in particular found him 'a more forceful personality and a more powerful brain than either Grey or Haldane'.[26] The word 'imagination' is often used when people refer to Mackinder, the 'intelligent and imaginative man' (in the words of his biographer Blouet).[27] Mackinder's Reading colleague W. M. Child linked this imagination with other virtues. 'He had a way of blending dreams and hard sense, subtlety and simplicity, and he never seemed to know when he passed from one to the other.'[28] His public manner was 'commanding' but in private he seemed rather shy and at times insecure. Bertrand Russell describes (with perhaps a tinge of snobbishness?) a dinner given by the Webbs for Arthur Balfour in 1905. 'Poor Mackinder made a bee-line for Balfour, but got landed with me, much to my amusement. It was a sore trial to his politeness, from which he extricated himself indifferently.'[29] When he married Emilie Catherine ('Bonnie') Ginsburg at Virginia Water in Surrey in 1889, the society occasion was widely reported and he looked a rather dashing young man. But the marriage did not bring happiness. Their only child died as soon as he was born on New Year's Day 1891. Up to the turn of the century his wife helped in some of his ventures, but then she succumbed to symptoms of stress and the couple separated without ever making the full break. Indeed, many years later, in 1937, they re-established some kind of relationship, and Halford Mackinder even left Bonnie a small bequest in his will. Blouet thinks that 'the failure, like any failure, haunted Mackinder'. Others have ascribed his self-imposed isolation—'few people could write of the inner man from firsthand knowledge'—to his lack of personal fulfilment. However, while Mackinder may have had 'few close friendships and few disciples',[30] he was much admired by those who heard and watched him and he almost always commanded respect.

He was also a commanding figure, especially in his proper field, geography. 'Mackinder is a figure who cannot be ignored in the history of modern geography.'[31] In Amery's words, 'his real strength lay in his insight into the bearing of geography on history'.[32] Mackinder saw his subject not just as a synthesis of physical and human or social factors but also of current and historical phenomena, and indeed of the study of visible facts and intuition about their meaning. For Mackinder, the science of geography was an art form as well, and certainly a part of the humanities.

> Geography is the imaginative understanding of the great regions of the earth's surface, the power of visualizing wide areas which may be the field of long campaigns; the power of extending what you see before you or beyond the horizon, and far beyond it, the power of looking upon the map of a large area, and carrying away in the mind, not simply a picture of the map, but a picture of the country; and, again, not merely that, but the power of seeing into it with the mind's eye, and of perceiving the inter-related facts which go to make up its geography—the relief of the land; the play of the winds upon that relief; the resulting rainfall as distributed in quantity and through the seasons; from the rainfall and relief, the system of drainage of the country; from the relief and the rainfall, the distribution of soils as the product of substance of the rocks; and from all these factors, the economic value of the country in its different parts; in other words its productivity; in other words its value for supply. You take to pieces the country and you reconstruct it in your imagination. You put, as it were, X-rays through it, rather than dissect it, so that you see the interplay of the parts.

Inspired words indeed (as E. W. Gilbert comments), which demonstrate the qualities of the extempore lecturer as well as the 'view of geography as a synthesis'.[33] It is thus, in the words of a historian of the Royal Geographical Society, a 'subject of vast scope',[34] so vast that one wonders whether it can be taught, let alone learnt. Yet Mackinder devoted much of his life to precisely this task. The Oxford School, and the LSE department (set up as a joint department with King's College in 1921/2) are institutional success stories, and his students in the early LSE years like Hilda Rodwell Jones (after her marriage in 1920 Hilda Ormsby) or Alice Bottomley or Horace Wilson were full of praise for the 'inspiring lecturer' and the 'sweep and scope of what he said'.[35]

Mackinder's contributions to defining and establishing the subject of geography are worth mentioning in their own right, but it is above all important to recognize them because he is so largely remembered for his more problematic contribution to what came to be called 'geopolitics'. In April 1904, the seminal paper which Mackinder had read only weeks after his appointment as Director of LSE, on 25 January 1904, to the Royal Geographical Society, on 'The Geographical Pivot of History', was published. It is an amazing *tour de force*, exhibiting, as the author says, 'human history as part of the life of the world organism'; no less. The 'Columbian age' of exploration and sea power is over; 'it ended soon after the year 1900'. Now that the world is known, and everywhere ruled, 'we shall again have to deal with a closed political system' and turn our attention 'from territorial expansion to the struggle for relative efficiency'. This gives the 'heartland' of 'Euro-Asia', that is Russia and Poland, a special, indeed pivotal importance. The new struggle between 'Columbians' and 'post-Columbians' has deep historical roots. 'It is the Romano-Teuton who in later times embarked upon the ocean; it was the Graeco-Slav who rode over the steppes, conquering the Turanian.' L. S. Amery, who was present on the occasion, disagreed with this Romano-Greek contrast and while he accepted the notion of 'the pivot state, Russia',[36] he drew quite different conclusions from those of Mackinder. Roger Louis has traced Amery's advocacy of 'air power combined with sea power' to concern about 'redressing' the power of the European 'heartland' *à la* Mackinder.[37]

This is important because taken by itself Mackinder's paper was oddly inconsequential. Was it prediction or exhortation, and if the latter, to what end? Even before the 'Pivot' lecture, a German geographer, Joseph Partsch, had been asked by Mackinder to write a book on Central Europe; when it appeared, it argued 'that this centre of Europe is great enough, and favoured enough by position, climate, nature, and confirmation, to hold its independent place for ever among the great powers of the world'.[38] Mackinder's later writings did not make any clearer what he had in mind, though he renamed the 'pivot state' and called it 'heartland', and he also moved it further and further to the west until by the 1940s it included Berlin though no longer the vast spaces beyond the Urals. Mackinder left no doubt about the need to contain the 'landsmen'. Even so, an ambiguity remains in his famous advice to the Versailles peacemakers in the 1919 book, *Democratic Ideals and Reality*:

When our Statesmen are in conversation with the defeated enemy, some airy cherub should whisper to them from time to time this saying:
Who rules East Europe commands the Heartland:
Who rules the Heartland commands the World-Island:
Who rules the World-Island commands the World.[39]

Halford Mackinder lived to a ripe old age; thus he came to see the excesses committed in the name of 'geopolitics' by Germany. He had warned of such possibilities, and of his own belief in democratic ideals there had never been any doubt. Yet when the octogenarian received the Charles P. Daly Medal of the American Geographical Society in 1944, he found it necessary to defend himself against the accusation—levelled against him in *Reader's Digest*, no less—that German expansionism owed much to 'three links in a chain', Albrecht Haushofer, Rudolf Hess, Adolf Hitler, and that he, Mackinder, had inspired Haushofer. The defence was somewhat disingenuous. 'Of the second and third [in the chain] I know nothing,' he said, and whatever Haushofer 'adapted from me' was taken 'from an address I gave before the Royal Geographical Society [i.e., the 'Pivot' address] just forty years ago, long before there was any question of a Nazi party.'[40] He then refers to a book by a German exile, Hans W. Weigert, on *Generals and Geographers*, for his defence.

Mackinder's biographer Blouet gives him the benefit of the doubt. But he does not refer to the book by James Trapier Lowe which appeared eight years before Blouet's, on 'Mackinder's philosophy of power' (*Geopolitics and War*). Lowe tries to establish three points. The first is that Haushofer made the study of Mackinder's works one of his main concerns and praised his 'teacher' many times. (Mrs Haushofer actually translated the popularized theory of Mackinder's geopolitics by his former student James Fairgrieve.) The second point is that Hitler's ADC Rudolf Hess was a student of Haushofer's and introduced the professor to Hitler at Landsberg officers' prison where he was detained (in comfort) after the Munich *putsch* of 1923. And thirdly, Hitler after Landsberg had a plan, a project, which

Hitler before Landsberg had not. 'There is little or no doubt that [Mackinder's] philosophy and provocative theory of world power greatly influenced Haushofer when he was the theoretician of the Third Reich, and that through him Hitler was given a positive program of territorial expansion, which Hitler proclaimed to the world in the pages of *Mein Kampf*.' The book was written of course in Landsberg prison.

Lowe is not accusing Mackinder, whom on the contrary he describes as similar to 'the Cato of old', the great exhorter, though he also says: 'Mackinder provided the warning, and Hitler the sword.'[41] Others disagree. 'The imputation that Mackinder's imperialism was the inspiration for and theoretical basis of German aggression is quite false.'[42] Thus the Liverpool geography lecturer Gerry Kearns. One of Mackinder's successors at LSE, Professor Michael Wise, adduces evidence to show that Hitler's mind, if it was ever influenced by ideas of others, was made up before Landsberg, and that Haushofer had absorbed many strands of thought other than Mackinder's. Mackinder had sought peace and freedom in the face of threats which needed to be exposed.

Unintended consequences are not just the subject but also the bane of social science; many a theory has been used for purposes which its author abhorred. Perhaps that is what happened to Mackinder's 'vision'. Blouet concludes that 'Mackinder tended to be isolated by his vision.'[43] Can anyone be surprised? The vision is really a threat. It does not tell Britain and its friends what to do, but what to fear. It is a positive vision only for an expanding Germany, or else for a Russia which 'commands the world'. This leaves Britain with two options, containment or benign neglect. Either way, Imperialism becomes a curiously defensive notion, an attempt to gather the threatened periphery of the 'Columbian age' and help it survive against the mainstream of history. This weakness was particularly evident in Mackinder's initial and purely political imperialism of 1899. After his 'conversion' to tariff reform in 1903 he had of course become explicitly protectionist, though the leopard does not really change its spots. In 1906 he published the little pamphlet on 'the underlying principles rather than the statistics of tariff reform', *Money-Power and Man-Power* (written, as it says on the title-page, 'for a quiet hour either before or after the elections'). This is in essence a disquisition on power and the need to back it up by human strength as well as commercial success. 'It appears to me that the Free Importers attach too little significance to our Man-Power, and too much to our Money-Power—too little to our power of doing, and too much to our power of buying.'[44] But despite the final flourish—'There is such a thing as the power to do good'—the defensive undertone of the argument is unmistakable. Sometimes one wonders whether Mackinder regretted not having been born in the 'Heartland'.

However, he was British and clearly felt strongly about the Empire, the role of Britain in it, and the need to educate people to 'think imperially': 'Our national education must have an Imperial aim—conscious, clear, and effective.'[45] Amery records a curious episode from the summer of 1903. Under the influence of Joe Chamberlain's 15 May speech, a number of organizations were set up to promote tariff reform. One was of course the Tariff Reform Commission of which Hewins became the Secretary. Another, of a more campaigning nature, was the Tariff Reform League, which was soon also looking for a 'first rate organizing secretary' with 'imagination and drive' as well as 'a complete grasp of the economic and Imperial case'. 'Precisely the right man was available in the shape of Mackinder whom I persuaded to offer his services.'[46] Fortunately for LSE, one of the sponsors of the League, the proprietor of the *Daily Express*, C. Arthur Pearson, thought that Mackinder was too strong-willed or otherwise objectionable, and by the time the question of inviting him after all was reopened, Mackinder had accepted the directorship of the School.

THE 'HEAD BEAST' TAKES CHARGE

Orderly Growth

Halford Mackinder came to the directorship unexpectedly after Hewins's sudden departure in December 1903. Even after he had given up the principalship of Reading that same autumn and the Oxford readership two years later, he regarded the LSE post as part-time and leaving him free to teach, to write, and to campaign for his political beliefs. He served for little more than four years, a shorter period than any other Director during the first century of LSE. Yet when he took over, he became very much the 'head beast' (according to Bertrand Russell, who had of course also described him as 'brutal'[47]) and took charge. What is more, the School needed someone to take charge because Hewins's style, with his convenient lapses of memory and limited interest in matters of internal organization, was beginning to grate on people and lose its original charm.

When Mackinder looked back on his LSE days, he found only one achievement worth noting. 'During my time the School changed over from a wholly evening establishment to a day College of the University of London and was placed on the Treasury Grant List.'[48] This was clearly important, even crucial, yet the more immediate effect of the new broom was to bring some order into the School's affairs. Mackinder shared one unfortunate habit with Hewins: 'He was not very punctual. He never came when he said he would.' That unfortunate deficiency apart, 'he had an enormous grip on things'. 'He always saw the way things were going—even if they were going the way he did not want them to go—things in the Senate, things in his own household.' 'He was difficult to work for—but really splendid to work with.' This is Miss Mactaggart speaking, who thought later that Mackinder's services to the School were not properly appreciated. 'He had never had justice done to him; he did not leave any kind of popularity behind him.'[49]

The tone set by Janet Beveridge's 'organic' description of Mackinder's directorship has been adopted by historians of LSE ever since. 'His period of office came in the lull which followed the earlier struggle for existence'; 'it was a period on the whole of peaceful progress and consolidation'; and then: 'Its teething troubles over, the Lusty Child was entering on a normal adolescent growth.'[50] Perhaps the metaphor tells more than its author intended. There may be normal as well as abnormal patterns of adolescent growth, but so far as growth goes, adolescence is surely the least normal period of all. This was certainly true for the School. In the session 1902/3 (Hewins's last full session) 'the work of the School was doubled'. Student numbers increased from 542 to 1,002, fee income from £374 to £1,034, and 10,000 volumes were added to the Library in a single year. During the next session (when the directorship changed hands at the end of the first term) student numbers increased by a further 30 per cent to 1,300, and the number of university graduates by 50 per cent to 150.

These and other figures can be found in Mackinder's first Director's Annual Report, which he produced as early as May 1904. It is also the first Report of its kind, and despite Mackinder's reference to 'discussing the position with my predecessor' one can sense his disapproval at the lack of regular reporting, which made it necessary for him to give account of Hewins's last session as well as his own beginnings. Mackinder is also critical of the 'carefully considered, although somewhat expensive, system of making known the advantages offered by the School'; 'such advertisement cannot, in a commercial sense, pay'. It did pay, of course, in every other sense, for without Hewins's talent for public relations the School might never have got off the ground. But Mackinder was not ungenerous. Advertising offered only one explanation of the rapid growth of the School; others were 'the unexhausted impulse derived from the new building and the University' and 'the programme of work arranged by my predecessor'.

Mackinder then proceeded, unusually, to announce his own programme of work under the innocuous heading 'Session 1904/05', but in fact six months before the beginning of that session. The three-year undergraduate degree course will have to be completed. For the business school activities, notably with respect to railway and insurance studies, special advisory committees will be set up in order to provide some guarantee of continuity. A systematic programme of distinguished evening lectures by outsiders is to be put in train. It is to be Mackinder's answer to the old bogeyman:

> In this manner we might further secure the expression of opinions of all schools of thought in Economics and Policy. It would be undesirable in the highest degree to limit in any way the free expression of opinion in such a school as ours, and the best method, therefore, of securing an atmosphere of truly scientific discussion, and of avoiding mere propagandism, would seem to be to provide opportunities for the exposition, in succession, of the results of the experience and the opinions of the leading administrators of all shades of thought.

The apparent urge for self-restraint, even self-denial, on the part of these passionately opinionated men is remarkable. It is almost as if they were happiest when views were expressed, at least at LSE, which were contrary to their own convictions. Perhaps such discipline pays in terms of the quality of teaching and research; it certainly imbues lectures and classes and the whole atmosphere of the place with a live tension absent in less worried institutions, but it also betrays a puritanism of the mind which it cannot be easy to live with even for its protagonists.

Mackinder's programme of May 1904 contains two further projects of which one was to be an unqualified success whereas the other did not come to fruition in his time. He wanted to develop the academic side of the School and he did this by turning the Library into an attractive research centre in such subjects as sociology and history (which he singled out for special mention). Then he complained about lack of space. The School had barely moved into Passmore Edwards Hall, but already 'the space at present available in our building is inadequate'. Mackinder the metropolitan geographer could not fail to notice that the opening of the Kingsway in 1905 and other attendant developments had made LSE more accessible to a growing number of Londoners. Moreover, the changes around Clare Market offered their own opportunities. As usual, Mackinder's imagination came into play: 'I think that our chief entrance ought to be placed upon the new Kingsway.' It was not, as everyone knows. When the massive expansion of the School occurred almost twenty years later, it turned to Houghton Street, which in Mackinder's time was still made up of a row of old houses on one side and St Clement Danes Grammar School on the other (and, if Lady Beveridge is to be believed, an organ-grinder in the middle). When shortly after Mackinder's resignation as Director the School was offered sites between Clare Market and the Aldwych development no action was taken. A somewhat refurbished and slightly extended Passmore Edwards Hall remained the home of the London School of Economics.

Academic Developments

In all other respects the subsequent Reports by Halford Mackinder are success stories. So far as the academic work of the School is concerned, the 1904/5 Report contains most of the relevant elements. The School then had over 1,400 students. Of these, 132 were University graduates, a number that was quickly to grow to 150 and beyond. Nearly one hundred were University of London undergraduates and this number too increased rapidly in subsequent years. For the rest—the bulk of those described as involved in 'work of lower than research standard'—there were (in 1904/5) over 600 students sent by the railway companies, for the most part clerical staff. More than 200 came from insurance offices. Another identifiable category were the fifty-four civil servants. Sixty, later eighty and more, came from abroad, and the Director finds it necessary to make a statement about them which he repeats in his 1904/5 Report:

Criticism has sometimes been directed to this characteristic of the School, but provided that the number of foreigners is not excessive—and at present it does not appear so—there are two great advantages in their presence. They carry the reputation of the School over the world to the advantage of all its students, British as well as foreign, and they enrich its student life, both in the Common Room and in the Union.

In 1904/5 these students were taught by forty-one lecturers, more than half of whom belonged to the permanent staff and had University appointments. They gave nearly 1,900 lectures and classes—nearly fifty per lecturer, and more by the permanent faculty—and now taught both in the afternoon and in the evening, as well as sometimes in the morning and even on Sundays. It is of interest to quote the list of subjects dealt with from Mackinder's 1904/5 Report:

Political Economy, including Economic History and the History of Economic Theory. Statistical Method and the Incidental Mathematics. The British Constitution, including the Colonies and Dependencies. Constitutional Law and History. Public Administration, including Public Finance. The Political and Economic Position of the Great Powers. International Law. Comparative Ethics. The History of Political Ideas. Ethnology and Social Institutions. The Original Sources of History, Mediaeval and Modern, together with Palaeography and Diplomatic? Geography—Economic, Political and Historical. Accountancy and Business Methods. Currency, Banking and the Money Market. Industrial

Organisation. Foreign Trade. Industrial and Commercial Law. Transport—Law and Economics. Insurance—Law and Economics. Library Economy and Bibliography. Educational Organisation.

Few of these subjects had not been available at least *in nuce* in the early days of the School, though the increasing role of law and the addition of 'ethnography and social institutions' deserve mention.

During the next three years, the remaining years of Halford Mackinder's directorship, several changes took place. The number of graduates and undergraduates grew. One railway company, the London and North-Western, pulled out of the scheme. The insurance scheme never really took off. But there was compensation. 'It is in connection with the Army, however, that, during the past Session [1906/7], the School has made the experiment which has attracted most public attention.' Strictly speaking, it was actually not the School that made the experiment but R. B. Haldane, who had become Secretary of State for War in the 1906 Liberal Government, and who released some thirty military officers for 'a course of instruction in preparation for administrative duties'. They were soon to be called 'Haldane's Mackindergarten', with a more apposite and witty use of the nursery metaphor than the earlier notion of Viscount Milner's kindergarten of 'young men who had served with him in South Africa and who went on to serve the nation in prominent places'.[51] The LSE Army Class was interrupted by the First World War but was reinstated in 1924 and it continued until 1932.

Mackinder was very active in the University, so much so that his biographer Blouet concluded: 'Mackinder's most important task was to bring the School fully into the life of the University of London.'[52] The Director himself found it necessary, in his last Report for 1907/8, to explain why he sometimes 'had to leave undone duties at the School itself because of the competition of my duties at South Kensington', the seat of the University at the time. He thought that 'the easy and influential relations which have been established with the other Schools of the University' in this way, 'justify the energy which has been sacrificed', and proceeded to give a poignant example. 'Perhaps the most noteworthy [event of significance during the Session 1907/8] is the conclusion of a concordat with University College for the delimitation of the aims of the two institutions, and for the prevention of harmful and vexatious competition.' New institutions always have territorial worries, and in any case the academic community loves cartels. The Agreement is wonderfully absurd at least in terms of the advancement of knowledge. It stipulates for example 'that the teaching of the Department of Political Economy at University College be specially developed in the directions required by the Faculty of Arts, leaving to the London School of Economics and Political Science the teaching required for the Faculty of Economics and Political Science.' In history, on the other hand, the advantage is on the other side, except that both UC and LSE 'will be free to teach Constitutional History', and indeed 'that as regards 19th Century History, the School of Economics and Political Science will be free to teach any form of History'.

No doubt this was a triumph for Mackinder, though perhaps not the most important one. The School which he left in 1908 had a small but solid postgraduate and research element. Indeed, in comparative terms it was not so small. In 1905/6 there were 181 postgraduates in the whole of England and Wales, twenty-seven at Oxford, thirty-six at Cambridge, forty-nine in other universities—and sixty-nine at the London School of Economics. Clearly Mackinder's heart was in this part of the work of the School. On the other hand the infant LSE would not have been viable as a rudimentary graduate school alone. To be viable, both financially and in terms of its recognition by an outside world which was not exactly waiting for a purely academic institution, it required the more practical element provided by railway and insurance students, the Army Class, and civil servants, as well as the dwindling proportion of those who, while doing 'work of lower than research standard', did so out of curiosity and the desire to learn rather than for any vocational purpose.

Governance and Finance

To whom did the Director report all this? One of Mackinder's first acts as Director was to give the governance of LSE a more enduring structure. In purely administrative terms, as we have seen, Miss Mactaggart had a clearer run than before, though the difficult McKillop was still around not just for Library but also for budget matters. From among the academic staff, Mackinder created a Professorial Committee which included such teachers as Bowley and Cannan, Dicksee and Foxwell, Sargent and Wallas. The Professorial Council was the precursor of the later Academic Board and also the Appointments Committee. But for many years the School's governance focused on the Court of Governors and, as a part of Mackinder's reforms, its management committee (later Council of Management, in line with the language of the Articles of Association), together with the Finance and General Purposes Committee.

Lord Rothschild had succeeded Bishop Creighton and Lord Rosebery as a very non-executive 'President' of the Governors and thus the School; he was the last to hold the office, which was discontinued in 1908. Sidney Webb, on the other hand, was the highly executive Chairman of the Governors. He was assisted by a Vice-Chairman in the person of Sir John Cockburn, an Australian reformist politician who had decided to settle in England and who remained Vice-Chairman for twenty-six years. Of the remaining thirty-two Governors some are familiar names: the Fabians Hubert Bland and Edward Pease; R. B. Haldane and Lord Rosebery; Mrs Bernard Shaw and Mrs Sidney Webb; Bertrand Russell, W. Pember Reeves. Sidney Webb was one of two Governors appointed by the LCC; two represented the Senate of the University of London; in addition there were five chairmen and general managers of railway companies. The Governors of LSE have always remained a large and somewhat motley crowd, though also one which showed remarkable devotion to the cause of the School and to its immediate as well as long-term needs.

Most of the needs, at least in so far as they were of interest to the Governors, were of course financial. The time has come to look a little more closely at the School's finances, especially since this leads to another one of Mackinder's success stories. The brief survey of figures is made easier by the fact that during the period in question, and indeed for many years to come, we can—unbelievably for us ninety years on—largely discount inflation and take nominal figures as real. After the initial eighteen months or so, the School had for five years, from 1897 to 1902, an annual expenditure of about £2,500. This is not including one-off expenditure, of course, notably for building, equipment, and library purposes. After 1902, when the annual grant from the University of London began, expenditure rose first to about £5,500 (1902–4), then to about £7,000 (1904–6), and further to about £9,000 in 1906/7, and around £10,000 in the subsequent three years to 1910. After 1910, a further significant increase in expenditure occurred.

These are rounded figures intended to give an indication of the dimension of expenditure and of its changes. Very roughly, it can be said that the School's expenditure had doubled by comparison with the early years when Mackinder came, and quadrupled when he left. Over this period the structure of expenditure remained more or less the same. About 65 per cent went on academic salaries, and around 13 per cent on administrative salaries. The remainder was distributed between administrative expenses, including some repairs and upkeep (around 10 per cent); the Calendar, which continued to swallow funds despite Mackinder's doubts about advertising (with other printing and stationery about 5 per cent), the British Library of Political Science (less than 3 per cent), scholarships and prizes (around 3 per cent), the refectory (2 per cent), and research (3 per cent). In later years, of which more will have to be said, maintenance expenditure and a much larger share for the Library changed this picture significantly.

Where did the money come from which the School spent in its early years? The School had no endowment other than some small and designated trusts like Mrs Bernard Shaw's trust for a research studentship or the Whittuck Trust for teaching commercial law. Thus it had to rely on three sources, all of which fluctuated in sometimes unpredictable ways: fees, grants from bodies like the Technical Education Board of the LCC, the University of London (after 1901), and donations. Fees depended of course on student numbers, which is why the railway, insurance and army classes were so important. Fee income went up to more than £1,000 in 1902/3, and doubled in 1904/5. Donations were naturally subject to major fluctuations, £764 in 1901/2 (though this excludes the Whittuck 'endowment' of £1,500), £1,461 in 1902/3, £526 in 1903/4, £992 in 1904/5, and so on. It is not irrelevant to notice that donations fell to an all-time low of £300 in Mackinder's last year, 1907/8, which is one of the reasons for the 'financial crisis' of which more will have to be said.

Grants from public bodies, on the other hand, remained fairly stable throughout. The TEB of the LCC gave the School £1,200 (in one year £1,500) 'whisky money' a year, and later raised this subsidy by about 10 per cent. From 1902, the University contributed both by stipends to teachers and by a maintenance grant. At the time, this was clearly the kind of income source which the School would wish to cultivate and extend. It was therefore 'a matter of great importance [which] affects the whole future of the School' that the Director was able to tell the Governors on 6 July 1905 of new developments which had taken place nationally. The Treasury Committee which preceded the later University Grants Committee had decided during that year not to exhaust its resources because 'there are, and will be, institutions coming within the spirit of the Reference to which assistance ought to be given on a similar footing':

> As an example of a new and striking foundation which in their opinion deserves special encouragement, they would mention the London School of Economics. This College, which is a School of the New Teaching University of London, is established in a new building at Clare Market. Its function is to teach Political Economy, not only generally, but in its application to current problems. Its special feature is the very large amount of original research work done by its students under the guidance of the teachers. It was founded in 1895, but its development has been so rapid that although the fees paid by the Students and the income from local sources have not yet reached the minimum amount prescribed by the reference, they will probably shortly do so. The Committee think that work of the kind done by this institu-

tion, although of a somewhat novel and special nature, deserves encouragement by the State.[53]

Those were the days! Not surprisingly the Director took this cue and by June 1906 was able to write to the Rt. Hon. H. H. Asquith, MP, Chancellor of the Exchequer, that the School had now achieved the minimum of fee and grant income prescribed by the Reference for the University Colleges Committee and wished therefore to apply for a share of the state grant. The Director's letter is a low-key but confident and impressive assertion of the School's achievement. As was Mackinder's way, he placed special emphasis on the fact that 'the School undertakes an unusually large amount of the highest kinds of University work—namely, those which are known as Postgraduate and Research Work', and that it had seventy-eight research students. Interestingly, there is no reference to the applied and practical nature of teaching at LSE except for a passing mention of 'the public services' expected of LSE in the peroration. The School, it says, needs money for buildings, for the Library, for research studentships, for additional teaching, 'and we are convinced that, unless more funds are available, it will be impossible for the School to fulfil its present brilliant promise, and to perform adequately the public services now expected of it'.[54]

The effect of the application was immediate and pleasing. In 1906/7 the School received its first grant from the Exchequer: £250. This sum rose quickly to £2,000 by 1910, and £13,500 by 1919. Halford Mackinder makes little in his Reports or even his sketchy Reminiscences of this achievement, yet it must count among his most important.

To be sure, the School was riding on the crest of a wave. When King Edward VII visited the new University of Sheffield in April 1909 he remarked that 'the great development of the University movement is a remarkable feature of the march of education'.[55] Walter Runciman, the President of the Board of Education, stated in his Introductory Report on the Universities and University Colleges in Receipt of Treasury Grant 1908/9: 'In every direction there are signs of active interest in university affairs, and the future is full of hope.' But institutions have been known to miss such tides of support. LSE did not miss it and thereby made sure that future crises would never threaten its survival.

It is only fair to add that the Director's efforts to add new buildings to the Passmore Edwards Hall were not a total failure. The extension of the Hall provided above all much needed space for the social purposes which are an indissoluble part of any university, and more particularly of one in the centre of a metropolis. The Refectory was opened in 1907, and so were the Common Rooms. Sidney Webb seemed to worry that the School would get too sociable. 'It is *quite impossible*,' he wrote to McKillop on 17 December 1906, carefully emphasizing the words himself, 'for the School to have a *licence*—it would lead to the forfeiture of the building to the L.C.C. and loss of grant—there must be no question of it.'[56] But the School could circumvent such rulings by forming a club. 'I have no hesitation', the Director wrote in his 1907/8 Report about the new facilities, 'in expressing my opinion that as a consequence of the new opportunities a very important and fresh sense of corporate life has been established.' Students of the time confirm the opinion. While the officers of the Army Class 'kept themselves to themselves', all others mixed freely. 'Looking back,' Shena Potter (later Lady Simon of Wythenshawe) wrote in 1960, 'I feel that these meal times discussions were some of the most important features of my years there.'[57] In any case, LSE had settled down in its location. A fleeting thought about South Kensington, where the University then was and where Imperial College was emerging, quickly led the Director to the conclusion: 'The situation of the School building, beside the Law Courts and the Record Office, and midway between the City and Westminster, could hardly be bettered.'[58]

THE NEW SCIENCE OF SOCIETY

A Discipline for Wealthy Amateurs

Among Halford Mackinder's lasting achievements there is one which has so far not been mentioned. He used all his multifarious connections with the University to attract to LSE a major benefaction to support sociological studies. The generous donor was Martin White, whose name continues to adorn the Martin White Chair, which was held successively by L. T. Hobhouse, M. Ginsberg, T. H. Marshall, D. V. Glass, and D. G. MacRae. James Martin White was a Scottish landowner who enjoyed life's pleasures but also took a serious interest in public affairs and the study of society. Born in Dundee in 1857, he belonged to Webb's generation. For one year, 1895/6, he served as a Gladstonian Liberal MP for Forfarshire, but the experience of Parliament only strengthened his belief that public servants needed a better education; they needed to know what began to be called sociology.

At the time few people had a clear idea of what exactly this newfangled term was supposed to mean, and it was used to describe almost anything from the study of sex to social work. The latter connotation is of interest in this story because before long it came to be combined, yet not merged, with the prevailing conception of sociology at LSE. Since the 1860s one strand of social reform had developed around an association called the Charity Organisation Society (COS). The COS, associated with the name of C. S. (later Sir Charles) Loch, and those of the Bosanquets, Charles and Bernard in the early days and Helen later, aimed at helping the poor by direct and practical action rather than by political agitation and institutional change. Advocates of the COS shared the desire to find out relevant facts and even a general sense of social needs with the Fabians and their friends, but the two had little in common otherwise and were engaged in often bitter disputes. The difference between the individual and the institutional approach was real—it still is—and much of LSE seemed to favour political action rather than the education of social workers.

Shortly after the foundation of the School in 1897, Bernard Bosanquet tried to do for his concerns what Halford Mackinder had attempted for geography, that is copy the success of LSE and set up another school like it, the London School of Ethics and Social Philosophy. LSESP rather than LSEPS! The venture failed, though from its ruins there emerged in 1903 a School of Sociology and Social Economics led by Edward Johns Urwick. Urwick, another Oxford graduate attracted by the evident need to do something about social problems, had worked at Toynbee Hall, where he preceded Beveridge as sub-warden from 1900 to 1903, before he embarked on an academic career. Jose Harris regards Urwick as 'a neglected figure in Edwardian intellectual history', a saintly man who helped 'build bridges between the speculative studies of Bosanquet and Hobhouse and the more empirical disciples of Booth and Rowntree'. The School of Sociology was one such bridge. 'Very soon after its establishment, therefore, the School of Sociology was an academic success, both in terms of the quality of its lecturers and of the popular demand generated for its product.'[59] This product has been described as 'a proper professional training for social work'.[60]

Martin White sympathized with this approach and was to be instrumental nine years later in arranging the marriage between the then ailing School of Sociology and the thriving LSE, but in the early years of the century sociology for him meant above all the set of ideas represented by his friend and mentor, Patrick Geddes, who along with his student and younger colleague Victor Branford is sometimes described as the founder of British sociology. The description is more organizational than intellectual. It ignores not only the author of the *Principles of Sociology*, Herbert Spencer, but above all the great moral philosophers of the Scottish Enlightenment, Adam Smith, Adam Ferguson, and John Millar, who are now widely acknowledged to have invented the sociological mode of thought. Sir Patrick Geddes (he was knighted in 1932) was by contrast a man of many parts, 'biologist, sociologist, educationist, and town-planner' (as his son Arthur Geddes describes him in the *Dictionary of National Biography*). Born in 1854, he too came from Scotland. In his younger years he tried and abandoned careers in banking and in the arts, went to study zoology in Paris, and discovered what was to become his lifelong interest in town planning. To his friend Martin White's dismay he did not become the first Martin White Professor of Sociology in London, though the benefactor endowed a (part-time) chair in botany at Dundee for him instead. In 1892 he had bought Outlook Tower, the 'World's First Sociological Laboratory'.[61] From all accounts, it must have been a cramped and grubby little library-cum-seminar building. For him, however, this was just one of many interests. 'His civic improvements at Edinburgh, his betterment of Indian cities, his efforts for residential halls for British students in French universities with the view of international understanding, and, perhaps most intimately of all, his gardens, were all the expression of the idealism that was the dominant note in a noble character.'[62]

Victor Branford came to Outlook Tower when he was still a banker and chartered accountant but soon fell for the kind of sociology which he learnt there. When Martin White decided to give some money to advance the study of sociology in 1903, initially about £1,000 went to the University of London for a series of lectures, but a smaller sum was used to set up the Sociological Society, for which Branford and White wrote the prospectus in June 1903. The Society was an immediate success. When its President, the lawyer and political theorist James Bryce, presented his First Annual Report he appended a map of Britain which showed how the 408 members were distributed all over the country (as well as, thirty-two of them, the Colonies and foreign countries). Among the early members we find public and literary figures like H. H. Asquith, Hilaire Belloc, the Bishop of Stepney, foreign scholars like Émile Durkheim, Ferdinand Tönnies, as well as familiar LSE names including H. S. Foxwell, H. J. Mackinder, Bertrand Russell, Mr and Mrs Shaw, Graham Wallas, and Mrs Sidney Webb. Sociologists all?

The preparatory meetings of the Society took place at the Royal Statistical Society, but for its first formal session on 16 May 1904 and for nearly all subsequent meetings the Sociological Society moved to the 'School of Economics (University of London)'. This was clearly appropriate though it did not necessarily help the process of definition. Nor did the telegram address which the founders chose and which tops their early letterheads: *Volkwissen, London*. No doubt postal confusion with others was thus avoided, but 'folk knowledge', German folk knowledge at that, seemed a strange concept to choose. Closer definition of disciplines of scholarship was both an obsession of the time and a hopeless project as long as many if not most social scientists cultivated their synthetic ambitions. Hewins had used the label of economics to deal with the universe of things social from a historical point of view, and Mackinder the geographer had read into maps both past events and current social structures. Now James Bryce saw the Sociological Society as one 'which surveys with the eye of science the whole field of human activity'. Victor Branford, in his inaugural lecture for the Society, offered a no less ambitious definition of its task:

> The first task of Sociology—as pure science—is thus the deliberate, systematic, and ever-continuing attempt to construct a more and more fully-reasoned social theory—a theory of the origin and growth, of the structures and functions, of the ideals and destiny of human society. The second task of Sociology—as applied science—is the construction of principles applicable to the ordering of social life, in so far as concrete problems can be shown to come within the range of verifiable knowledge.[63]

Both Bryce and Branford, and other members of the Society as well, expressed their dismay at the late and limited development of sociology in Britain. In fact it was not only late and limited but also strangely uneasy and unconvincing, as its historians, Philip Abrams and R. J. Halliday, have pointed out. The Sociological Society did not really help. If anything, it strengthened the notion that sociology was a 'movement' rather than a discipline of scholarship. In his inaugural lecture Branford also stated that 'as affecting the genesis of Sociology, the main features of the century were, in the first place, the creation of the Biological Sciences as definite systems of study, and in the second place the growth of the conception of a Science of History'. How does sociology fit in this picture? It does not, at least not clearly. On the one hand, there are hints that it may add a third, a twentieth-century triumph of knowledge to those of the nineteenth century, but on the other hand sociology is seen as some kind of combination of the two, biology and (philosophy of) history. Even the hybrid word, sociology, with its Roman and Greek roots, which has bothered purists since Auguste Comte, is now regarded as 'a convenient memento of the twofold nature of human society'.[64]

It is thus difficult not to agree with Philip Abrams when he speaks (in his *Origins of British Sociology*) of 'a desperate piecing together of intellectual interests whose real tendency was to fly off in a dozen different directions'. Abrams rather enjoys poking fun at those who 'took the decisive part in institutionalizing sociology in the Edwardian period' and who were 'wealthy amateurs with careers elsewhere, academic deviants, or very old men'.[65] Two more professional groups of people—the social workers assembled in the Charity Organisation Society (and later the School of Sociology) and the empirical researchers following in the footsteps of and indeed including Charles Booth and B. Seebohm Rowntree—while members of the Sociological Society, took little part in its work and never determined its culture. Thus the 'alliance between town-planners, eugenists, charity organizers and workers in the various social settlements', to which Halliday refers,[66] remained very 'temporary' indeed. The sociological debate was dominated by biology and history, or to use contemporary terms (and thereby add two further *E*s to our list), Eugenics and Evolution, with a little Ethics thrown in for good measure.

Eugenics or Sociology?

The Sociological Society published the lectures which it sponsored as well as informative minutes of the subsequent discussions in three magisterial yearbooks, entitled *Sociological Papers*. Whether these were widely read is at least uncertain. (My own copy carries the ex-libris of the then Archbishop of York, Cosmo Gordon, but the holy man never even leafed through the volumes, whose pages were left uncut.) Both the 1904 and the 1905 volumes begin with a section on eugenics, introduced by the venerable 'biometrician' Francis Galton, who was then 82 years old. In his statements Galton shows a remarkable mixture of actuarial pedantry and religious fervour. His twin thesis is simple. Eugenics—the term he chose after dismissing his earlier proposal, 'stirpiculture'—'the science which deals with all influences that improve the inborn qualities of a race', needs to be developed as a research project but also applied as a practical proposition in order to 'raise the average quality of our nation to that of its better moiety at the present day'. The purpose of eugenics is first to establish and then to enhance people's 'civic worth'. Research should be statistical and aim, for example, at 'a "golden book" of thriving families'. Application, on the other hand, should not be enforced but attempted by persuasion. To be sure, people should, in due course, get 'Eugenic certificates', later called 'honour-certificates', but above all eugenics 'must be introduced into the national conscience, like a new religion'. In the 1905 *Papers* Galton goes even further in this direction: 'In brief, eugenics is a virile creed, full of hopefulness, and appealing to many of the noblest feelings of our nature.'

It is not easy ninety years later, after Auschwitz, Gulag, and more recent experiences of 'ethnic cleansing', to discuss questions of eugenics dispassionately. Yet this needs to be done because they were so much in the forefront of the thinking of many great minds, and by no means immoral ones, in Edwardian times and beyond. In fact, C. P. Blacker is quite right to point to the 'earnest moral tone' of the early Eugenics debate.[67] Blacker not just became the historian but was also a protagonist of that thread of the history of LSE—and of social science—which leads from the early interest in eugenics through the foundation of the Eugenics Society in 1907 and its later progeny, the Population Investigation Committee, to a modern demography which merges with history as much as biology. This is fortunately a story of increasing sophistication. For the research suggested by Galton was, to put it mildly, simple in nature. He gives us an example of what he had in mind in 1904, 'A Eugenic Investigation' in the form of a list of some 'thriving families' of Fellows of the Royal Society. This demonstrates for example that Darwin had a 'fa fa' (paternal grandfather) who was a 'physician, poet and philosopher', a 'fa bro' who died at the age of 20 but was 'of extraordinary promise', and that Francis Galton was in fact his 'fa ½si son', to say nothing of the 'me fa's' (maternal grandfathers), 'me bro's', and others who were all distinguished Wedgwoods.

Galton uses anthropological evidence to prove that marriage is regulated in many societies, and that such regulation is sanctified by religion. What then is the objection to 'a whole-hearted acceptance of Eugenics as a national religion'? The debates reported in both the 1904 and 1905 volumes of the *Sociological Papers* display an enormous variety of views. The most sceptical commentators are Continental scholars and medical practitioners. The former wonder whether we really want 'to produce a few Grecian Gods and Goddesses in the sacred circle of the privileged few' (Max Nordau) or even whether intelligence is all there is to excellence (Ferdinand Tönnies). The latter, the doctors, suggest that it is easier to influence the environment anyway and that therefore 'hygienics' and better school meals might do most of the things that need doing at the moment.

Some other discussants are not quite so reticent. The Hon. V. Lady Welby gets more insistent by the year on women's need 'to take a truer view of their dominant natural impulse towards service and self-sacrifice'. George Bernard Shaw pleads for polygamy as the answer to all questions of eugenics. 'What we must fight for is freedom to breed the race without being hampered by the mass of irrelevant conditions implied in the institution of marriage', which means 'freedom for people who have never seen each other before and never intend to see one another again to produce children under certain definite public conditions, without loss of honour'. H. G. Wells makes his familiar unsavoury point that the best are hard to find, but in any case, 'the way of Nature has always been to slay the hindmost. . . . It is in the sterilisation of failures, and not in the selection of successes for breeding, that the possibility of an improvement of the human stock lies.' Sir Richard Temple is more optimistic so far as identifying the best is concerned. 'It would appear that a beginning has been made, as regards men, in the Rhodes Scholarships.'

A number of comments come to mind as one tries to re-create the debates of the early years of the twentieth century. The directness of many comments is striking. People said exactly what they thought and as they thought it. It is sometimes argued at the end of the twentieth century that ours is a time in which 'anything goes'. (Paul Feyerabend's use of this phrase in his *Beyond Method* rang in the 'postmodernist' age even for science.) In fact, however, the late twentieth century is remarkably 'politically correct', especially as we approach the boundary between social and genetic factors, sociology and biology. There was also in the beginning of the century an almost unbounded optimism, at least among intellectuals, both about what research could uncover and what could be done by its application. After all, why not 'improve the stock'? But then such optimism presupposes a simplicity which we have lost. When Galton tries to argue that eugenics has nothing to do with moral judgements, on which people differ according to time and place, but only with such undisputed values as 'health, energy, ability, manliness and courteous disposition', the combination of naïve relativism and naïve dogmatism strikes a later generation as truly breathtaking. One wonders whether a less optimistic age is not in the end a kinder age for humans.

Above all, a later reader is struck by the lack of sophisticated social science, indeed of any social science in some of the statements recorded in the *Sociological Papers*. In part this lack of sophistication was a British problem, certainly at the time with which we are here concerned. There was none of the extensive investigation and analysis of 'social facts' which had begun elsewhere. In the United States, cities, immigrants, and other phenomena had become the subject of important surveys, and from Sumner's anthropology to Veblen's economic sociology there were general analyses of note. In Germany, Max Weber had confidently defended both the methodology and the 'basic concepts' of sociology and offered an impressive sociological explanation of early capitalism by relating the propensity to save and to invest to the readiness of Calvinist bourgeois to forgo immediate gratification. There was never a reference to biological, racial superiority, nor did Émile Durkheim in France invoke lunacy or genetic disabilities when he explained suicide in terms of anomie, the social fact of a lack of norms.

A paper by Durkheim was actually discussed by the Sociological Society, in the absence of the author, on 20 June 1904 ('On the Relation of Sociology to the Social Sciences and to Philosophy'). It did not state Durkheim's position as clearly as he had done in his *Rules of Sociological Method*, in which the term 'social fact' first appears. Still, he made the point that sociology is a discipline of specialisms and not therefore a branch of philosophy. The discussants raised characteristic objections: sociology offers little that cannot be reduced to psychology (E. Reich); sociology is essentially an approach and not a science (S. H. Hodgson); sociology uses other sciences for its own purposes and is therefore a derived science (J. M. Robertson). Even Hobhouse's defence of Durkheim seemed weak when he argued that the French sociologist had only tried to describe where we are today and not where we might go in future. For many of the early supporters of sociology there was paradoxically (so it would seem) no such thing as society.

Towards Institutionalization

However, this survey of sociology at the time of its establishment at LSE is still incomplete. Indeed, eugenics—rephrased in the third volume of the *Sociological Papers* as 'Biological Foundations of Sociology'—was by no means the only subject in which the Sociological Society took an interest. Patrick Geddes is represented in each of the three volumes with his own subject 'Civics', which advocates a 'civic worth' of another kind. Civics for him is 'applied sociology', and notably 'the application of Social Survey to Social Service'. This is playing with words, a game which Geddes enjoyed, though his concerns were real enough; they had to do with cities, good cities, *Eu-topia* as he chose to call the best combination of 'folk', 'work', and 'place'. In 1906 he proposed to that end a 'Civic Museum' in which all 'civic facts' are assembled, whether geographical or economic, architectural or political, in order to develop 'our civic consciousness, our civic conscience, our active citizenship'.

Other names appear in the *Sociological Papers* which will reappear in these pages. Edward Westermarck gave one of his anthropological exposés every year. L. T. Hobhouse chaired many of the meetings of the Society and contributed to most of them. Younger interested lay and professional people attended as well. The Preface of Janet Beveridge's *Epic of Clare Market* begins with the sentences:

> I went to the School of Economics for the first time fifty-five years ago in 1904 to attend one of the earliest meetings of the Sociological Society. Development of sociology in the School was a major interest of several fellow Scots from the same part

of Scotland as my own, in particular Victor Branford and Martin White. It was an interest also of David Beveridge Mair, my first husband and father of my children. He was a member of the Sociological Society from its beginning. Through this I was brought into friendly contact with outstanding men in this new field of social science, notably Francis Galton.

In fact David Mair contributed to the discussion on 'Sociology as an Academic Subject' and suggested that 'sociology is a serious study, but the oftener it is lightly undertaken the better'. (Halford Mackinder added his own warning note on the likelihood of the 'higher syntheses' of sociology to this discussion: 'No man was really fit to deal with them until he was forty years of age.') Janet does not mention William Beveridge although he was present too at these early meetings of the Sociological Society. In fact the young man, barely 26 years old at the time, commented rather rudely on an admittedly opaque and very elaborate paper given by J. S. Stuart-Glennie on 6 April 1905 which postulated 'laws' of history, of which Beveridge said that 'it seems quite easy to prove that they are either untrue or unimportant'. A year later, Beveridge himself gave a very different, that is to say strictly empirical paper on 'The Problem of the Unemployed'. For some, like J. A. Hobson, the theorist of imperialism, this paper was if anything too empirical: Beveridge 'has carried his analysis to the beginning of an understanding of the vitals of this issue' but no further. Others, like A. L. Bowley, disagreed. The great statistician called it 'the best paper I have heard of this kind, and I think it goes very nearly as far towards the root of the matter as one can get in a short half hour'. Beveridge and the theorists, Beveridge and the empiricists: an enduring subject, as we shall see.

The Sociological Society lived on, but the *Sociological Papers* gave way, after 1906, to the *Sociological Review*, which has appeared more or less regularly to the present day. Martin White remained close to the Society though his attention had shifted to the University of London. By 1907, he had given £2,250 to endow one temporary lectureship in sociology and one in ethnology. The latter went to A. C. Haddon, who held it for five years from 1904 to 1909. The former lectureship was offered to the Finnish social anthropologist Edward Westermarck, who accepted it gladly and remained connected with the School for twenty-six years. In 1907 his post was endowed for five years in the first instance, an endowment made permanent in 1911 for the part-time professorship which Westermarck combined with his continuing obligations in the University of Åbo (Helsingfors). In 1907, also, Martin White gave the University of London £10,000 to endow a permanent chair in sociology, the Martin White Professorship. Its first incumbent, L. T. Hobhouse, had already lectured at the School since 1904; indeed he had been a temporary lecturer in 1896/7. To these generous benefactions Martin White added funds for bursaries and scholarships in sociology.

Taking it all together, the Sociological Society, the support for Geddes, the London chairs, the lectureships, the scholarships, it is not too much to say that one man, Martin White, established the discipline of Sociology in Britain. One is reminded of the way in which the Ford Foundation sixty years later—and therefore an institution rather than one person—established the study of International Relations. Moreover, sociology came to life at LSE. 'Mackinder was perhaps a critical figure.' Abrams thinks of Mackinder the scholar who could provide 'a badly needed counterweight to Geddes', but he also refers to Mackinder the administrator. 'His active support was critical in establishing the teaching of sociology at the LSE.'[68] Mackinder himself was characteristically modest when in his 1907/8 Report he commended the munificence of Martin White, by then a Governor of LSE, and thanked the University for placing 'its Department of Sociology in our School'. 'The number of students who have attended the lectures of Professors Westermarck and Hobhouse during the past Session makes it evident that the School has gained a material accession of resources—a further evidence, if it were necessary, of the value of intimate relations with the University.'

Hobhouse and Westermarck

The inauguration of the Martin White Professorships of Sociology on 17 December 1907 became a major event. The Principal of the University, Arthur W. Rücker, recounted the story of the benefaction. Then the Vice-Chancellor, Sir William Collins, expressed his pleasure at being able to renew his 'old connection with the School of Economics as a former President of the Students' Union'. The medical man had been an Honorary President for a year in 1902/3 so that his knowledge of the 'monotechnic institute' LSE was rather slight, and his prejudices sounded familiar. 'It is to be sincerely hoped that if your efforts do not succeed in making politics scientific they may not result in making science political.' Fortunately, Sir William found more felicitous words to introduce the two lecturers of the day.[69]

L. T. Hobhouse rose first, and he spoke about 'The Roots of Modern Sociology'. He took a little while to get to his main point, as he was wont to do, but when he got to it, he caught the imagination of his audience. One root of sociology is political philosophy, which for 200 (by now nearly 300) years has analysed 'general conceptions that underlie society, which constitute the generic essence of the social bond'. Another root is 'philosophy of history', which has taught us the variety of social options and the gradualness of progress. Furthermore, 'speculation upon social matters has been influenced by the contemporary state of the sciences of nature', though we must beware of facile conclusions, for 'the last word of biology is the first of sociology', no less, yet no more either. Sociology is about institutions, and these can be studied in their own right. Adam Smith was the first to undertake that task with a view to 'a general science of society', and John Stuart Mill followed in his footsteps. Since then, various specialisms like demography, anthropology, social history have developed. We now 'have a body of social sciences, a number of social specialisms amassing a vast material'. What is needed is to bring them together and produce the synthesis of a General Sociology. This synthesis will be based on one idea which links all the specialisms, Evolution. With regard to evolution, sociology has two tasks, one, to find out what actually happened, and the other, to identify what is good and desirable. 'Thus, for the completion of our task we need both a science and a philosophy, and it is only through the union of the two that we can bring the certainty and precision of systematic thought to bear upon the problems of practical life.'

Leonard Trelawny Hobhouse was so important for the history of the School, and so unusual among its early heroes, that he deserves a little more attention. He was born in 1864, a year before Hewins, into a distinguished family of public figures of which his father, a vicar, was an 'undistinguished member'.[70] He read Greats at Oxford, got involved in Toynbee Hall and extension work, and became a part of Oxford's radical if not socialist groups. However, he did not actually join the Fabian Society nor did he ever show any gut sympathies for the bureaucratic fantasies of the Webbs or even the Coefficients. Above all, he had no time for any brand of Imperialism. He was horrified by the Boer War, unimpressed by Joe Chamberlain, and, at the time at which Hewins left the School to become Secretary of the Tariff Commission, Hobhouse accepted the secretaryship of the Free Trade Union at the other end of the tariff debate spectrum, and joined the School. Despite several attempts to offer him a safe seat in Parliament, Hobhouse was never tempted but remained a journalist and a scholar. The former career, notably with the *Manchester Guardian*, preceded the latter, at LSE where he taught until his death in 1929.

Hobhouse was a liberal, of sorts. The qualification is necessary because his liberalism had acquired a strong (T. H.) Green flavour ever since his Oxford days. Stefan Collini has shown, in his masterful study of *L. T. Hobhouse and Political Argument in England 1880–1914*, that Hobhouse was a man of many ambiguities. It would be difficult, for example, to place his liberalism in Isaiah Berlin's terms of 'positive' and 'negative freedom'. Hobhouse certainly did not confine himself to defining liberty as the absence of constraint. His political coordinates saw his contemporaries arranged on a more elaborate scale: 'Communist—Theoretical Socialist [i.e. Fabian]—ordinary Labour [i.e. Ramsay MacDonald]—Good Liberal [i.e. Hobhouse]—Bad Liberal [i.e. Mackinder?]—ordinary Tory [i.e. Hewins?]—Diehard.' The Good Liberal was in many ways a Liberal Collectivist; in any case one for whom the common good was the lodestar of individual development. Hobhouse not only supported social policies like pensions and a minimum income, but his New Liberalism gave a special role to the community. In his organic scheme of things 'liberty becomes not so much a right of the individual as a necessity of society'.[71] In *Morals in Evolution* Hobhouse takes the hypostasis of society even further. There is not just individual, but also social salvation. 'Society has a soul to be saved, and its salvation is in the justice, humanity and freedom realized in its inward and outward relations.'[72]

Collini not only doubts whether it is illuminating to call Hobhouse a Liberal but also whether he really was a sociologist. In any case, 'Hobhouse had the role of sociologist thrust upon him'. The offer of the Martin White Chair 'came at a critical stage in his life. At the age of 42 he had resigned his short-lived editorship of the *Tribune*, with no obvious career in front of him, a succession of rather unsatisfactory jobs behind him, and a wife and three young children very much with him.' To make matters worse, he was in the middle of 'one of his more frequent depressions' so that he was quite uncertain whether to accept the Chair. 'With this conspicuous lack of enthusiasm Hobhouse finally accepted, and in September 1907 became Britain's first professor of sociology.'[73]

All this is perhaps more a comment on Hobhouse's temperament, the ups and downs which accompanied him throughout his life, than on his sociology. One may doubt the lasting intellectual quality of his work. What Collini describes as Hobhouse's ambiguities are often signs of a lack of rigour and clarity of thought. Even a friendly obituarist commented on his 'impetuosity' and 'often diffuse' style. 'Emotion always coloured the white light of his thought.'[74] One wonders whether he ever really felt at home in the community of scholars, including the School. But none of this can detract from the strong sociological flavour of his philosophy of progress, or from his contributions to LSE. The non-Fabian, non-Coefficient semi-Liberal introduced a welcome fresh element into the academic life of the School. The Webbs, who had liked the early Hobhouse of *The Labour Movement*, remained on good terms with him. Hobhouse not only taught sociology, but helped administer the Ratan Tata Fund and thus build up applied social science at the School. It may be true that his later, 'post-war writing played a series of rather repetitive variations on the themes of his earlier work',[75] but Hobhouse remained a valued citizen of the School. What is more, through his—in many respects his only—disciple Morris Ginsberg, Hobhouse's combination of sociology and social philosophy lived on.

The other inaugural lecturer of that 17 December 1907, Edward Westermarck, was of a different ilk. Westermarck enjoyed exploring and explaining the intriguing and seemingly endless variety of ethnographic facts; he was also a worldly man with a great sense of fun. His lecture on 'Sociology as a University Study' was shorter and clearer than that of Hobhouse, if arguably less profound. Why is sociology a latecomer among the sciences? 'Sociology is a young science, because it is a difficult science.' It raises complex questions of value and fact. It forces the scholar to become free-floating. 'The sociologist must, so far as possible, cut himself off in thought from his relationships of race, country, and citizenship,' says the (Swedish) Finn in Britain with an argument which anticipated Karl Mannheim's later sociology of knowledge. Westermarck does not seem to find such abstinence too difficult so that he is quite unworried about the old LSE bogeyman, whom he puts to rest in his own inimitable way: 'Finally, it has been argued that sociology suggests socialism, and in fact sociology and socialism are even now frequently confounded with one another. Such a confusion is quite human. I am told that the Sultan of Turkey has prohibited the importation of dynamos into his country, because he is afraid of dynamite.'

Westermarck agreed with those who thought that sociology was often 'too vague, too constructive, too full of far-reaching but unproved generalisations, to deserve the name of science'. True, 'sociology is the science of social phenomena in the widest sense of the word', but: 'What we want at present is not text-books on sociology as a whole, but sociological monographs.' These would provide insights of practical as well as theoretical usefulness. 'The Indian mutiny might in all probability have been prevented by a little greater insight into native ideas and beliefs.' Westermarck's own monographs on marriage, ritual, and belief in Morocco and elsewhere are examples. Morris Ginsberg has tried, in a generous assessment of Westermarck's life and work after his death in 1939, to find a theoretical niche for the social anthropologist, but his place in the Hobhouse–Ginsberg tradition is quite tenuous. He provided data, notably about the 'simpler peoples', which the social philosophers used to substantiate their claims not just for moral evolution but for moral progress. Westermarck himself remained an agnostic and moral relativist, so that he had less reason to be shattered by 1914, let alone the dark clouds on the horizon of the inter-war years.

Westermarck's lively and amusing *Memories of My Life* were written in Swedish but underline the statement in his inaugural lecture: 'I have never felt myself as a foreigner in this country'. He liked Britain though he resisted some of its temptations, like going to the Derby with Martin White, which he did not feel 'sufficiently English' to enjoy. 'I knew, of course, that one horse ran better than another, but it was a matter of complete indifference to me as to which of them all was the fastest.'[76] White, on the other hand, was so obsessed with the Derby that he wrote an angry letter to Beveridge in 1926 complaining that Governors' meetings had been scheduled on Derby Day. 'We are doing what we can for sports for the students—should we deprive ourselves of the less active open air sports—we who are older?'[77] Westermarck and White got on well; they drank 'Waterloo brandy' in London clubs, went motoring on the Continent, and celebrated the end of term with champagne dinners. A photograph of the two men on the verandah of Westermarck's house in Tangier taken in 1907 makes one understand how people during their motoring expedition across Spain could mistake 'my friend Martin White for King Edward, his wife for the English Queen, and me—I was wearing a black astrakhan cap—for a Russian Grand-Duke'. One somehow feels that Westermarck need not have worried 'how far I have succeeded in realizing the founder's [i.e., Martin White's] intentions'.[78]

THE REIGN OF THE WEARY REFORMER

Departure and Arrival

Halford Mackinder did not much like Sidney Webb's comment that he had been able to run the School 'with two fingers of one hand',[79] nor was he pleased by the implied criticism of the 'pluralism' of his interests. When Friedrich von Hayek wrote the School memoir for the fiftieth anniversary of LSE in 1945, the then Director, Alexander Carr-Saunders, sent Mackinder an advance copy for comment, and the 84-year-old responded plaintively that the text might be construed to say: 'If only he had not been a "pluralist" . . . he might have run the Sch of Econ at a spectacular rate which would have accomplished in his time many things only achieved later.' On the contrary, Mackinder said, 'through the years 1887–1908 there ran through my life a rope of continuity in diversity'.[80]

Hindsight? There are worse ways of assessing achievement. In any case,

Mackinder's biographer Blouet has probably got it right when he describes the School years as 'a partially settled period in Mackinder's life', which was as much as he could hope for. 'The years as director of LSE (1903–1908) covered Mackinder's most productive time as a writer and editor.'[81] They were also a success story at LSE itself. Mackinder left the School better organized and academically more solid than he had found it. He had also integrated it firmly into the University of London as well as the national system of university finance.

What is more, Mackinder did not leave the School. The directorship may have been an episode in his life, but LSE was not. He resigned from his administrative responsibilities because he wanted to go on to politics and other things, but he remained a reader, and from 1923 a professor of geography. After failing to get himself elected in a by-election at Hawick in 1909, he became Liberal Unionist MP for the Glasgow Division of Camlachie in 1910 and again in 1918; served as High Commissioner to South Russia in 1919/20; received a knighthood in 1920; was appointed chairman of the Imperial Shipping Committee in the same year (until 1939), and a member of the Privy Council in 1929. At a farewell dinner given for him by the Students' Union as he was about to leave the directorship in 1908, he gave a speech which those present found memorable. Mackinder emphasized the scholarly, scientific objectives of LSE. 'It was in a scientific spirit, and with a scientific object, he said, that the School was organised, and he appealed to the students to maintain that attitude, not only in their work but in their social life.' The latter proposition, the scientific social life, sounds a little ominous, but the students liked it anyway. Their magazine, the *Clare Market Review* (then in its fourth year and one of the achievements of the Mackinder years) also enjoyed the occasion for another reason. 'From an editorial point of view, the present transitional condition of the School is eminently satisfactory. There is always something new to comment on, and, generally speaking, something to applaud.'[82]

The Governors, and Sidney Webb as their Chairman, did not find the condition quite so satisfactory. Mackinder told the Webbs of his intentions on 19 May 1908. 'For the last months we have known that his resignation was imminent, he intending to devote himself to the affairs of the Empire, in preparation for Parliament and office in the next Conservative ministry.' Their views may rarely have met, but the Webbs, and especially Sidney, appreciated Mackinder's 'competence', which is another way of saying that Webb felt he could leave the School to its Director and concentrate on other matters. How to find another man of the same ilk was the question. Several names were considered, including H. A. L. Fisher and G. M.

1. The Borough Farm Quartet, August 1894: George Bernard Shaw, Beatrice and Sidney Webb, Graham Wallas.

2. The School's first homes at the Adelphi. A view of John Street and (*inset*) the door to No.10 Adelphi Terrace.

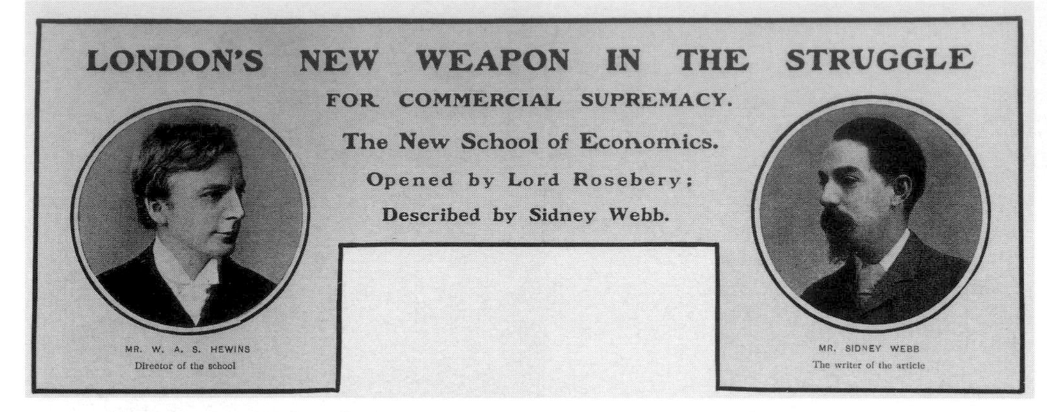

Two young men in a hurry...

3. *Above*: banner heading to an article by Sidney Webb in *The Sphere*, 31 May 1902
Below: as politicians

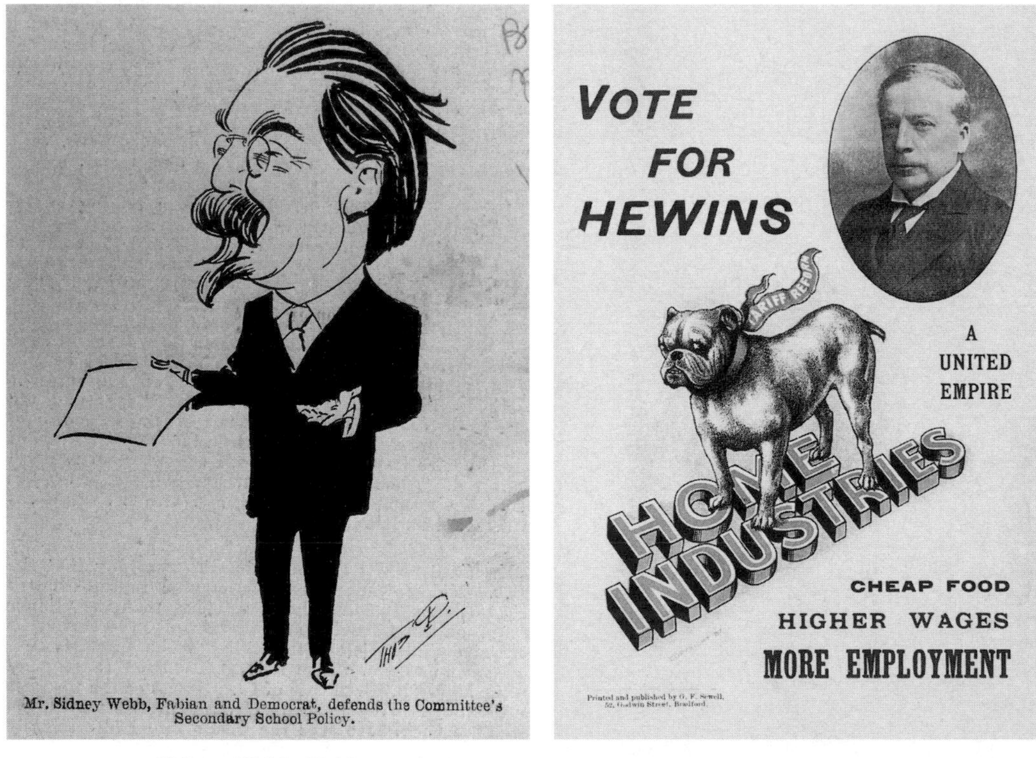

4. Sidney Webb, Fabian and Democrat, 1907

5. W.A.S. Hewins, Conservative and Imperialist. Election poster, 1910

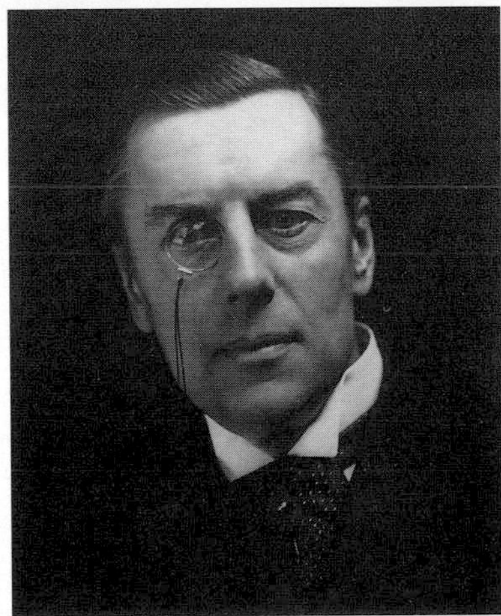

6. **Early supporters and friends . . .**

(*clockwise from top left*) Henry Hunt Hutchinson (1822–1894): Charlotte Payne-Townshend (Shaw) (1857–1943): Richard Burdon (later Lord) Haldane (1856–1928); Lord Rosebery (1847–1929); Joseph Chamberlain (1836–1914)

7. The first School Annual Dinner, Frascati's Restaurant, London W1, February 1898. Those present include W. A. S. Hewins, Beatrice and Sidney Webb, John McKillop, Charlotte Payne-Townshend, Bernard Shaw, A. L. Bowley.

8. The School moves to Clare Market. Benefactor John Passmore Edwards (1823–1911) *top right*.

9. Professor Sir Halford Mackinder (1861–1947). Second Director (1903–8) and long-term teacher of geography.

10. William Pember Reeves (1857–1932). Third Director (1908–19) and family *c.*1908 *l–r* Amber, Beryl, Maud, and Fabian.

11. The first LSE hockey team *c.*1911 with the Director and (*on his r*) Vera Powell, her future husband Percy Anstey (*standing extreme l*), H. B. Lees-Smith (*seated second l*), Professor A. L. Bowley (*standing extreme r*), Professor A. J. Sargent (*seated extreme r*), N. M. Mazumdar (Students' Union President 1912–13) *in front*

12. In the Staff Common Room. *At centre* Professor E. Cannan (Political Economy) with *l–r* Professor L. R. Dicksee (Accountancy), B. M. Headicar (Librarian), Professor A. J. Sargent (Commerce), Professor Lilian Knowles (Economic History)

13. In 1911 Sidney Webb retired as Chairman of the Court of Governors. He had ten successors before 1995.

Lord Robbins (1968–74) and Sir Huw Wheldon (1975–85) are shown in Plate 43.

Rt Hon Russell (Lord) Rea (1911–16)

Sir Arthur Steel-Maitland (1916–35)

Josiah (Lord) Stamp (1935–40)

Sir Otto Niemeyer (1940–57)

Edward (Lord) Bridges (1957–68)

Sir Morris/Mr Justice Finer (1974)

Sir John Burgh (1985–87)

Sir Peter Parker (1988–)

Trevelyan, and according to Janet Beveridge at least Trevelyan was asked but refused. This was awkward because an early appointment seemed desirable. Webb and his Fabian friends then thought of the Birmingham economist W. J. Ashley, but alas! he was a Fair Trader and, as Edward Pease tells us: 'The first two Directors were both Protectionists, & the controversy between Free & Fair Trade was acute at that period. When Mackinder resigned Webb said we really must appoint a freetrader: otherwise the City will think we are Fair traders in disguise.'[83]

All this happened within a matter of days; Webb, as we have seen, was never one to dither in the face of decisions. In the end he went for someone he thought he knew, the old friend, William Pember Reeves, 'New Zealand Fabian', Coefficient, Governor of the School since the early years and as Agent-General of New Zealand 'a good administrator, with experience, with colonial connexions, the right opinions' (thus Beatrice Webb)[84]—though perhaps only a mild, not to say wet free-trader judging from his contributions to the Coefficients and the little book *A Council of the Empire*. Sidney Webb wrote to him—Reeves was basking in the Portuguese sun at Sintra—on 23 May in his characteristic manner. 'Mackinder is retiring *at once* from the Directorship of the London School of Economics. The question is, would *you* be prepared to accept the position if we offered it to you?' The salary is a mere £700 but there are other attractions. Beatrice, Mrs Shaw, and R. B. Haldane support the proposal, and the University Senate would no doubt endorse it. '*The matter is urgent.*' Unless Reeves accepts, the post will have to be advertised, '& the door opened for three month intrigues &c., & the new session opened without a Director'. 'Can you *telegraph* to me . . . after taking a few hours to think it over?'[85] The answer was yes. 'We are quite content and he begins with enthusiasm,' Beatrice Webb noted in her diary.[86]

Born in 1857, Reeves was the last of the Webb generation to take up a leading position at LSE. When he was appointed, he already had more than one life behind him. Indeed Keith Sinclair does not reach LSE until page 311 of his 344-page biography of Reeves, which is not an accident. Willie, later Will Reeves (in his youth and political days very much a man of first names, with the middle name, Pember, his domineering mother's family name, a mere P.) had a chequered educational career. Neither remote sheep stations nor Christ Church at Oxford gave him much to crow about, and the eventual training for the law was essentially a means to an end. The same is true for journalism, though Will Reeves loved writing and his uninspired but sympathetic poetry understandably made him much loved in his country:

Though young they are heirs of the ages,
Though few they are freemen and peers,
Plain workers—yet sure of the wages
Slow Destiny pays with the years.
Though least they and latest their nation,
Yet this they have won without sword,
That Woman with Man shall have station,
And Labour be lord.

The volume, *New Zealand and other Poems*, was published in London in 1898 when Reeves's political career was almost over, though the verses show what he stood for—citizenship, equality, the rights of labour. First as a journalist, then from 1887 as a New Zealand MP, to begin with in opposition and since 1890 in what would now be called Lib-Lab coalitions (though party names meant little at the time and Reeves was technically a Liberal), as Minister notably of Education and of Labour. His Industrial Conciliation and Arbitration Act of 1894 became a model for others and, despite many changes, its traces still live on in New Zealand as does the name of William Pember Reeves, which is even today a starter of good conversation.

Reeves was not a stable man. Throughout his life acts of courage and great physical stamina alternated with phases of depression and frailty. Health became his almost continuous preoccupation though he lived to the age of 75. Reeves's political career came to an unhappy end in 1896. He was, as Keith Sinclair puts it, 'much more concerned with reform than power' so that his enemies and even more so his uncertain allies had for some time looked for ways to get rid of him. Eventually, he received the most golden handshake on offer, the Agent-Generalship in London, where he arrived in March 1896. Contrary to the sneers of his erstwhile electors in New Zealand, Reeves's London life was by no means one of red carpets and candlelight dinners in aristocratic company, but he did find friends in radical circles. His wife, Maud, followed him soon and brought the two daughters, Amber and Beryl, born in 1887 and 1889, as well as the little boy whom they had named Fabian. For a while they were much sought-after guests as well as hosts in London's political-intellectual society, and Amber was to make her name even into literature as H. G. Wells's *Ann Veronica*, though she almost broke her father's heart in the process.

But despite the fact that Reeves was much liked and often rumoured to be destined for greater things, he was not happy. In part this was his temperament, to be sure, but the worries about his health and the moods of despair were not helped by his sense of rootlessness. Sinclair quotes the telling interview which Reeves gave in 1905:

Conversely, I have for the last ten years lived in London with my eyes turned half the time to the Antipodes. Thus all my life I have been, as it were, looking across the sea. Without ceasing to be a New Zealander I have also become an Englishman. Yet in talking over affairs with English friends our point of view seems almost always not quite the same. On the other hand I do not look at things quite as I should if I had never left New Zealand. It is a detached kind of position.[87]

The pathetic feeling left Reeves above all weary. As early as 1897, an uncharitable G.B.S. had noted: 'The Agent-General, snoring on the sofa, is going to wake—yes: there he goes.'[88] Seven years later, Beatrice Webb made the more devastating comment: 'W. P. Reeves and his wife are always at hand—the same in opinion, but he has grown stale in English politics and is settling down to a certain plaintive dullness of spirit and aim.'[89] It must have appeared as a new lease of life to Reeves to be asked to do another job. He resigned as Agent-General and accepted two directorships at once; those of the New Zealand National Bank and of the London School of Economics.

The School which the new Director found was not without problems. Historians of the period have noted that LSE was 'going through a period of acute financial crisis in the late 1900s', and even added that 'the School's third director, William Pember Reeves, had been appointed in 1908 specifically to raise funds and to arouse confidence among City business men'.[90] There is both truth and exaggeration in such statements. Fundamentally the School's finances were safe and sound since the Treasury grant had been secured. At the same time Mackinder had overspent, or rather overborrowed, in order to complete the extension of Passmore Edwards Hall and the erection of 'shedifices' on the vacant site next to it. Consequently, his last paper to the Governors, somewhat misleadingly entitled 'Memorandum from the Director on the Policy of the School in the Immediate Future', is a three-page explanation of the £2,000 debt and further needs amounting to another £1,000. Reeves may not have agreed with Mackinder's assessment that 'so far as ways and means are concerned the position is for the moment not very difficult'. He certainly made a point of saying in his first Annual Report that 'the most difficult questions which I, as a newcomer, have had to cope with have been financial'. But he also reported that already, after less than a year, 'we are now not far off [the] very desirable goal' of 'financial equilibrium'. An appeal for donations and certain economies (dramatically described as 'heavy sacrifices') had been necessary, but by 1910 no more needed to be said about the passing crisis.

Instead, Reeves was able to report, throughout his early years, steady growth. The numbers of both undergraduates and research students increased. The Army Class (though not the Insurance Class nor the course for librarians) flourished. New courses were offered, and many outside lecturers supplemented the home team. Reeves was neither very personal nor very poetic when he reported 'certain impressions' after his first year in 1908/9. His account is none the less of interest:

I fully expected to find [the School's] life busy and its students distinguished by zeal and intelligence. So far from proving a disappointment, what I have seen has surpassed my expectations. The abilities of the students as a body are well above the average, and this is not only true of those engaged in research or studying for degrees. The officers of the Army class are men of undoubted capacity. The railway clerks and officials attending lectures on Transport throw themselves into their work in a manner that is more than creditable when it is remembered that their studies come at the end of the duties of a long day. The number of graduates and undergraduates goes on growing, and this is gratifying; but the general high level of intelligence among other students is to me even more cheering. The energy, capacity, and courtesy of the lecturers are only what might be looked for from gentlemen of their high academic standing. I am none the less grateful for it. Nor can I pass over the loyal services rendered by the staff of the School during a year which has had to be marked by various changes and experiments in organisation. Speaking generally, I have formed a very high estimate of the value to modern education of the work of the London School of Economics. My belief is that expansion in the future can only be checked by limited financial resources. Should funds be forthcoming for further building and equipment I feel certain that our work could speedily be extended and the number of our students rise in proportion.

A Sense of Belonging

In the event this was not to happen during Reeves's decade at the helm of the School, nor could the new Director foresee how non-financial developments were to check the expansion of the School when the World War began. However, to begin with things continued to look up. When John McKillop finally resigned in July 1909, two changes became possible at once. One was that Miss Mactaggart at last got the title of what she had long been in fact—the Secretary and Registrar of the School. The other change was that the road was now open to the appointment of a full-time librarian. On 1 March 1910 B. M. Headicar was appointed to that position, which he held for the next twenty-three years. A new porter (later Clerk of Works), Charles Wilson, who was to be with the School for over fifty years, had been added to Dodson's staff earlier, in 1905; George Panormo, who came in 1912, was to stay for forty years.

On the academic side, a number of students and staff whose names were to become familiar if not famous entered the scene, even apart from those associated with the Sociological Society. Two distinguished international lawyers, Professor Alexander Pearce Higgins and Sir John Macdonell, joined the staff in 1908. Edwin Cannan finally became a professor of economics in 1907, and A. J. Sargent in 1908. Professor Seligman succeeded A. C. Haddon in ethnology in 1910. More significantly, women made their mark in the academic life of the School. The President of the Students' Union in 1907/8 and the Hutchinson Silver Medallists in 1907, 1908, and 1912 were all women. In 1910 the Director reported with pleasure 'the steady relative increase in the number of women students as compared with men'. The proportion—35 per cent in 1909/10, almost 40 per cent in 1910/11—was unusually high for mixed university institutions but distorted by the large number of LCC teachers on a special course; only about 30 per cent of those on the general register were women.

So far as teachers are concerned, there can be little doubt that the general air of discrimination did not stop at the doors of LSE. Of the sixty-one women teachers of the University who presented a petition to the ('Haldane') Royal Commission on University Education in London in 1910 demanding more involvement in academic governance, only a handful came from LSE (though fewer from other non-women's colleges). Still, studies by Margherita Rendel, Gillian Sutherland, and others on the problems faced by women academics do not ring quite true for the School. Women, Gillian Sutherland suggests, were less likely to get jobs in subjects related to power, like geography and administration. At least two of Halford Mackinder's students provided exceptions. Lady Alice Bottomley taught geography from 1908 to 1913. In 1912, Hilda Rodwell Jones-Ormsby was first appointed to an assistantship in geography.

Then there is the extraordinary story of economic history which leads Maxine Berg to conclude:

> The L.S.E. proved to be by far the most fertile environment for the development of economic history. Moreover it was a particularly attractive place for women scholars for two reasons: first, the position of women at the L.S.E. in relation to their position at other colleges; second, the particular role there for economic history in relation to social policy.[91]

The names resound in the history of the School. Lilian Knowles (née Tomn), the formidable Conservative with a capital C, became a lecturer, in 1903. Eileen Power, the most remarkable academic of the group, arrived as a Shaw research student in 1911, and then stayed on the School staff till her death in 1940. Vera Anstey was also a student before the war, and joined the staff later. Others with a more remote association can be cited, including Beatrice Webb, of course. Maxine Berg argues that these women academics might have become economists, had it not been for Alfred

Marshall's infamous attitude to women at Cambridge. Many of them in fact got their first degrees at Girton College.

Talking about women, power, and LSE it is hard to omit Miss Christian Mactaggart, who for many years was the real seat of power at the School. More than that, she was followed by another powerful woman as Secretary; indeed the secretaryship has been occupied by women for nearly two-thirds of the first hundred years of LSE.

The liveliest account of life at the School before the First World War was written by a woman, Mary Stocks. As Mary Brinton, she came to the School in 1910, in the same year as Vera Anstey, who was then still Vera Powell. Like Mrs Anstey, Mrs Stocks was soon to be a lecturer—one in commerce, the other in history—at the School. Sixty years later Mary Stocks remembered that she 'clocked into [LSE] with such easy informality in 1910'. She liked the mixture of age, culture, and background of her fellow students though she notes that the School 'was preponderantly male'. (This was to change—in her perception—in 1912 when the Diploma in Social Studies under Professor Urwick produced 'Urwick's harem'.) Lectures and classes were, as always, a mixed blessing; not everyone could be as 'brilliantly stimulating' as Lilian Knowles, who led Stocks to specialize in economic history for her B.Sc.(Econ.). But in any case, 'the centre of the school's social life was the refectory' on the top floor of the Passmore Edwards building. 'There, staff and students mixed informally. There were only two fixed stars in the whole galaxy. At the head of a centre table sat Mr Pember Reeves, the somewhat withdrawn Director, with Miss Mactaggart the Secretary on his right hand.'[92] Otherwise, students might well find themselves sitting next to the great of the time, Edwin Cannan, Lilian Knowles, Graham Wallas, L. T. Hobhouse. Talking to them, Mrs Stocks wrote many years later, was much more educational than to listen to the great on radio. LSE (as she put it in an earlier 'Reminiscence') 'was as personal as an Oxford College and far more in touch with the world outside'.[93]

'The LSE was in those days very politically-minded,' and so was Mrs Brinton-Stocks, who became secretary of the Union when the first Indian, Nandlal Marreklal Muzumdar, was elected President in 1912. In the Union's Clare Market Parliament she clashed with Hugh Dalton, who was then still at Cambridge though he was to join the School shortly. In a basement in Clement's Inn Passage the Webbs were conducting their anti-Poor Law campaign; another building in the same little street housed the militant suffrage campaign of Christabel Pankhurst. Then Mary Brinton met John Stocks at Oxford, and through him another crowd, Tawney and

Gilbert Murray and William Temple, though until she had completed her studies the couple remained 'engaged'. 'Nowadays I suppose we should have got married immediately and I should have commuted from Oxford to the School of Economics fortified by the "pill".'[94]

Much of this story did not look very different from the other side of what was evidently a rather low fence: the perspective of teachers. Edward Westermarck had the anthropologist's eye for detail, especially in a foreign country. His first lecture in 1904 attracted an audience of only six, 'but its variety made up for its small proportions, as it was representative of the same number of nations and the lecturer himself made a seventh'. Westermarck praises the brightness of students and their propensity for discussion, but he too adds that 'the restaurant'—he clearly means the refectory—'is one of the School's most important institutions'. However, he perceived more stratification, at least at lunch-time. 'After the meal, the teachers withdraw to their common-room, whilst many of the students amuse themselves with a dance.' Things are different in the evenings when groups mix more freely and often repair afterwards to a meeting of one of the numerous societies of the School. Westermarck concludes with a rare compliment from a Continental: 'Englishmen have shown their wisdom in making food and drink important ingredients in their social life, and academic circles form no exception to this rule.'

In one respect Mary Stocks's story is unusual: she found her husband in Oxford. More characteristically, Percy Anstey, Vera Anstey's husband-to-be, had been President of the LSE Students' Union in 1908/9. From its earliest days onwards, the School was not just a matchmaking place but one of extraordinary intimacy, or perhaps inclusiveness. Again, we can trust the anthropologist Edward Westermarck to uncover the fact and some of the reasons: 'Thanks to the domestic arrangements and other circumstances the teachers at the School find it convenient to spend a comparatively large portion of their time within its walls, and consequently come into closer contact with each other than is the case with the professors at many other universities, Helsingfors, for example.'[95] What is true for teachers is no less true for students. It was notably true for those who after their working days chose to come to LSE in order to learn and debate, the evening students who by virtue of that fact alone must have found themselves somewhat set apart from their colleagues at work. It would be wrong to call the School a 'total institution', though malevolent outsiders might want to apply Erving Goffman's description of prisons and lunatic asylums to it; LSE was at all times far too open to the world to deserve that label. But for all its openness it was always more than just another university; for many who came to it it became almost a universe of life, an experience which provided a lasting bond with those who shared it.

There is no shortage of examples to prove the point. It would perhaps be misleading to cite the founders, Mr and Mrs Webb, as evidence, let alone Mr and Mrs George Bernard Shaw. The first three Directors did not involve their wives much in the School, though Mrs Hewins occasionally gave parties which were fondly remembered, and Maud Reeves not only came with Beatrice Webb to lunch at the School and pursue her interest in women's rights and in the lot of those whose family lives revolve *Round about a Pound a Week* (as her best-selling 1913 book was entitled) but was an occasional lecturer from 1912 to 1916. The story of William Beveridge and Mrs Jessy Mair alias Janet Beveridge is more relevant, as is in a different way that of a later Director who found his first wife (Vera Banister, who died in 1981) as well as his second wife (Ellen de Kadt, a teacher in the Government Department) at the School before he began to write the history of this remarkable institution. He was not the first Director to marry within the charmed circle; Sydney Caine married his fellow student Muriel Harris in 1925.

There are in fact whole dynasties of LSE connections. Beatrice (Webb-) Potter's brother-in-law Leonard Courtney became the first Honorary President of the LSE Students' Union and surrounded himself with fellow students in later life. R. H. Tawney married William Beveridge's sister Jeannette. Alexander Carr-Saunders, the fifth Director of the School, was actually a cousin of Maud Reeves. And for the children of LSE names it was almost *de rigueur* to come back to the School. Amber Reeves was merely registered for a short period, but Alice Bottomley's daughter (as well as her grandson's wife) completed her course. Jessy Mair's daughter Lucy became a distinguished anthropologist who taught at the School for over forty years. A. L. Bowley's daughter Marian got her B.Sc.(Econ.) and her Ph.D. at LSE. Graham Wallas's daughter May got a Ph.D. in economics in 1926 and was on the staff from 1928 to 1945. The list is long and gets longer every year.

In his 'Reminiscences', Arthur Bowley emphasized another aspect of the inclusiveness and sense of belonging which the School gave its members. 'It has been my frequent experience to be greeted anywhere in England or abroad by someone who says, I lectured at the School of Economics, or I was a student, or my father was a student, or more rarely I attended your lectures; and such must have been a more frequent experience for those whose subject was more popular. Thus I have found friends in all the countries I have visited.'[96] Bowley is too modest about his subject, statistics, but certainly right in observing that the School has created a fellowship, national and international, which sustains its members. This is not so much a fellowship of 'the class of '29', that is those who were actually at the School together—though this matters too—as one of all those who had any association with the place. Having been at LSE provided instant recognition, and much to talk about even among strangers.

Not all talk is nice, to be sure. It would be quite wrong to describe the intimacy of the School as a source of unbridled pleasure; at all times it was also one of endless and often vicious gossip. William Pember Reeves discovered this the painful way when his daughter Amber, a girl with 'a slender nimble body very much alive, and a quick greedy mind', fell in love and eloped with the man who described her thus—H. G. Wells. Wells's seemingly insatiable sexual appetite was exceeded only by his indiscretions as an author. In *Ann Veronica*, which he based on his relation with Amber Reeves, he combined both and upset the parents more than words can tell. Wells had little time for the pompous Victorian father, though he professed to like the 'subtle and interesting'[97] mother but in any case his feelings made little difference to his behaviour. Amber survived the episode and found a forgiving, gentle husband, George Rivers Blanco-White. The daughter of the affair, Anna Jane got her B.Sc.(Econ.) at LSE where she also met her husband Walter Eric Davis. The father and Director, Will Reeves, found it harder to cope. 'To someone so proud and thin-skinned, the wide notoriety of the affair, which he helped to spread, was a calamity.' The School did not help. 'At the L.S.E. the students were talking of nothing else but the current literary gossip. As he took his seat at lunch in the Refectory, one student thought, he looked like "a condemned man".'[98]

LIFE AFTER WEBB?

The Railway Speech

On 17 September 1910 Sidney Webb was invited to speak to trade unionists on the occasion of the formal opening of the new offices of the Amalgamated Society of Railway Servants in Euston Road. Webb gave many such speeches, often four or five a month (as a list drawn up by Beatrice for December 1910 shows). On this occasion, his wife accompanied him. As usual Webb spoke from notes, though he later regretted having left behind for the journalists present some scraps of paper including figures which were estimates rather than facts. He pointed to the large number of railway workers who had to work for more than twelve hours on their working days, and the more than 100,000 who were earning less than

£1 a week. He also attacked the recent House of Lords ruling in the Osborne case according to which union dues could not be used to support political campaigns. To make his case, Webb employed rather strong language; he called the low earnings of railwaymen 'a scandal' and referred to the 'colossal ignorance' of the Law Lords with respect to trade union history.[99]

In fact, Webb got a little carried away by the audience and the rousing reception for him and his wife. At one point he encouraged the union to take stronger action than hitherto in defence of its members. 'Every penny earned at the expense of working excessive hours, or by paying less than an adequate wage—is robbery. (Hear, hear.)' Webb also attacked railway directors and their honesty in interpreting arbitration awards. Beatrice Webb, in her reply to the vote of thanks, went even further and said unkind things about one particular railway director, her own father, who never 'tried to improve the men', which made her conclude 'that there was something very rotten in the whole capitalistic idea of Labour'.

Understandably, the audience loved it all. They applauded warmly, proceeded to lunch, heard a number of toasts, and after a further vote of thanks (so the *Railway Review* tells us on 23 September 1910) 'the proceedings concluded with the singing of "Auld Lang Syne"'. Obviously the official railway interests were less happy when they read all this. 'Does Mr. Webb demand a legal minimum wage of 24s. per week for lamp lads, lad porters, van boys, messenger lads, and the many other grades in which boys are employed?' However, even the *Railway News* began its attack on Webb's statistics with the sensible statement: 'Such speeches usually call for scant notice and pass quickly to the lumber of forgotten things.'[100]

This is just what appeared to be happening which was as well in view of the importance of the Railway Course for the finances of LSE. Then out of the blue, and more than four weeks later, the national press carried, on 18 October 1910, a letter addressed to 'Dear Mr. Webb' and signed by (Lord) 'Claud J. Hamilton' on behalf of three LSE governors. All three were railway bosses, Lord Claud Hamilton, MP, chairman of the Great Eastern, Mr Oliver R. H. Bury, and Mr James C. Inglis, general managers of the Great Western and the Great Northern Railways respectively. Lord Claud Hamilton explains his late response by 'absence from England' and then proceeds to take 'strong exception' to Webb's statements about the Osborne judgement and the working conditions of railwaymen. His attempt to refute Webb's case begins with the statement, 'I am bearing particularly in mind the fact that you occupy the responsible position of chairman of the Board of Governors of the London School of Economics' where railway economics is taught, though he is cautious enough to mention that the Committee of Governors which oversees this teaching is actually chaired by him, Lord Claud Hamilton. Still, he insists, 'it is in my view impossible to dissociate your attitude in this matter from your position as chairman of the Board of Governors of the London School of Economics. . . . I cannot avoid the conclusion that your public utterances must in some sense be reflected in the teaching in the class-rooms of our school.'[101] Since he will not be involved in misleading the young, he feels compelled to resign as a Governor of the School and sever all connections with it.

Sidney Webb was in Scotland at the time and more than a little upset about the letter. On the day of its publication, Beatrice wrote and advised him 'to let the matter blow over', and 'to sit still and say as little as possible'. However, in London all hell broke loose when *The Times*, on 19 October, published a leading article endorsing the railway directors' views. From its title—'The Degradation of Economics'—to its final flourish—comparing Sidney Webb as chairman of LSE with the absurd idea of the French Socialist leader Jean Jaurès as chairman of the École des Sciences Politiques—the article exemplified that exquisite viciousness of which the old *Times* was capable when its vested interests were threatened. Two insinuations hurt particularly. *The Times* invoked Mrs Webb ('whom it is impossible to avoid mentioning in this connexion') to prove that since the '"Crusade" on behalf of the Minority Report on the Poor Law, the Chairman of the School of Economics has preferred agitation to science'. And the article cited Mr Hewins, who, it suggested, had resigned his directorship on joining Mr Chamberlain's movement because 'he knew that, if he did not do so, the school might suffer'. Others were only too pleased to grasp the opportunity to vent their long-standing resentment against the School and the Webbs. 'The London School of Economics is one of those peculiar educational bodies which covers the political doctrines it propagates among its students with the decent if not always efficacious mantle of apparent impartiality,' wrote the *Globe*, gleeful at the apparent revelation of the true LSE.

But *The Times* had gone too far. Citing Hewins in particular turned out to have been a mistake. On 22 October the paper had to publish a dignified and unambiguous letter by Hewins which left no doubt that he had resigned in 1903 for one reason only: 'It was quite impossible to combine the very arduous work of the director of the School of Economics with the work of the Tariff Commission.' Actually, this was generous by Hewins in view of the discussions which had preceded his resignation, but then he was always a gentleman. Beyond the resignation issue Hewins pointed out that the Governors are in fact not responsible for the academic work of the School. Even so, if anyone expressing a controversial view would have to resign, 'resignations would be exceedingly general'. The School had laid down in its Articles that no one must be under any disability in respect to social, religious, and political opinions. What is more, Sidney Webb had at all times been a guardian of the best academic values. 'I never had a suggestion from him which was incompatible with the principles which I have just laid down.'

Two days later, Halford Mackinder added his voice to that of Hewins in the letter columns of *The Times*. 'I am in full agreement with Mr. Hewins.' Like Hewins, Mackinder differed deeply from Sidney Webb in his political views, but 'I bear testimony to the fact that on no occasion did he seek to bias the teaching given under my charge.' 'Impartial and virile work' is accomplished not by muzzling teachers but by allowing all shades of opinion expression; in this respect, LSE is a model for others. The third director, Reeves, expressed his views in a letter to the *Globe* on the same day. Webb had spoken, he said, as a private person, not to students, and not at the School. 'You will allow me to add that your statement that the School of Economics exists for the purpose of impressing on "callow youth" any peculiar political propaganda is comically at variance with facts.'

Meanwhile, Sidney Webb was not quite so sanguine. He drafted a long letter to another governor, Sir Alfred Lyall, who had asked him for an explanation. In it, he called his railway speech 'somewhat indiscreet', but also blamed misreporting while bending over backwards in expressing his respect for the Law Lords, and mildly correcting some of his own figures ('I wish now that I had said 50,000, instead of 100,000').[102] Indeed, Webb expressed his appreciation of the railway barons and stressed that he had always been an opponent of strikes. In his published response to Lord Claud Hamilton on 24 October, Webb did not go quite so far; yet while his friends described it as a 'dignified reply', others found it a 'singularly lame defence'.[103] (Beatrice was in a slightly defensive mood too; when she saw Reeves during the critical October days, she not only told him that it was 'better to avoid *all* controversy' but also advised him 'to give up his speechmaking'.)[104] Webb's reply emphasized first of all that his speech had no connection with LSE, and that the reports gave a false idea of it. He then pointed out to Lord Claud Hamilton that despite their fundamental differences of opinion they had worked well together. 'But your letter raises a larger issue—nothing less, in short, than that of freedom of thought and freedom of speech in university administration.'[105] In this respect the School has an unblemished record. Throughout its history, every shade of opinion has been represented, which is the right way to produce impartiality. Mr Hewins's letter was the best proof of this record.

The Founder-Chairman Resigns

From the beginning the Webbs suspected that the controversy had what is called nowadays a subtext. Nor was the top layer of this subtext very submerged. In her first letter on the matter to Sidney on 17 October, Beatrice Webb had remarked, 'I think it probable that Lord George has something to do with it and that it is a determined effort to oust you from the London School of Economics.' Lord George Hamilton, Lord Claud's brother, had been chairman of the Poor Law Commission and was somewhat bruised by his encounters with Beatrice Webb as a member, and with Sidney, who had drafted many a paper with her. Two days later Beatrice sent her husband the *Times* leader. 'The *Times* article which I send you (please keep it as I want to put it into my diary) lets the cat out of the bag. It is the Minority Report agitation which has led to this attack. It is another indirect reply from Lord G. Hamilton—an attempt to injure us as he cannot fight us.'

The Royal Commission on the Poor Law and the Relief of Distress was arguably the most important event in Beatrice Webb's professional life. No less than one-third of her long book, *Our Partnership*, is devoted to it. The relevant chapter begins by noting two outstanding events which happened at about the same time: 'the advent of the Liberal Government in December 1905' and 'my appointment, in November of the same year, to serve on the Royal Commission'. She was the only woman on a Commission of twenty, mostly experts and a few politicians, charged to enquire into the working of the Poor Law (which was then 60 years old), as well as into other means of combating distress. Initially, she actually liked the chairman, Lord George Hamilton. 'This experienced politician and attractive *grand seigneur* combined exceptional personal charm and social tact with an open mind and a willingness to give free play to the activities of his fellow-commissioners.'

With Beatrice Webb the chairman needed all his social graces. Even after the first Commission meeting she complained about the procedures which he had suggested. As time went on—and the Poor Law Commission sat for over three years, reporting in February 1909—differences became more substantial. Indeed, from 1907 onwards Beatrice Webb pursued her own political agenda and line of argument to such an extent that she all but ceased to be a useful member of the Commission. The Commission agreed to demand a radical overhaul of the old Poor Law. Its main proposal was a scheme of national assistance administered by municipal authorities which was regarded as courageously reformist when it was first published (and which was not implemented until some twenty years later). Meanwhile Beatrice Webb, with the help of Sidney, produced a Minority Report which epitomized the social and political thinking of the Webbs.

The Minority Report scheme was based on the seemingly simple assumption that poverty has many causes and each of these needs its own remedy. The simplicity evaporates, of course, once one begins to think of implementation. The establishment of a 'national minimum' to define poverty was still a viable proposition. Beyond that, however, what followed from the Minority Report was a welfare state with specialized public agencies for every need: health and housing, education and unemployment, disability and old age. There were many reasons to be sceptical. Some undoubtedly preferred the status quo to the 'socialism' of a welfare state of this kind; others abhorred the bureaucratic nightmare and the rule of unaccountable 'experts'. But the Webbs were unmoved by such opposition. As soon as they had recovered from the exhaustion which followed the preparation of the Minority Report, they decided on what was for them a new course of action: propaganda and campaigning. In May 1909 the National Committee for the Promotion of the Break-Up of the Poor Law was set up. The negative connotation of the 'break-up' soon turned out to be an encumbrance, and it was renamed National Committee for the Prevention of Destitution. Thousands of members were recruited, hundreds of speeches made, and for a while the euphoria of success prevailed. Beatrice Webb, forever conspiratorial, noted in her diary on 27 September 1909:

> We are, in fact, creeping into the public mind much as the School of Economics crept into the University, into the City, into the railway world [*sic*! but of course a year before the great controversy—R.D.], into the civil service, into the War Office—before anyone was aware of it. Our little office, wedged between the Fabian Office close on its right, and the London School of Economics a few yards to its left, is a sort of middle-term between avowed Socialism and non-partisan research and administrative technique.

Where there are subtexts, it is seldom a question of one alone. The Webbs were beginning to work themselves into a corner from which it was hard to see an escape. They deserted, for a while, their favourite obsession with facts and figures, but they were not yet prepared to throw in their lot with a political party. Perhaps they were looking for a way out even then, for by the end of 1909 they had begun to plan a long trip 'to the East' which was in fact to take them around the world and away from Britain for a whole year. Planning the venture took a while but by the time the 'bomb [was] thrown into the School by the railway magnates led by Lord Claud

Hamilton'[106] the decision to embark on the journey around the world in June 1911 had been taken.

The Minority Report campaign had its year but not much more. Even without the trip, this was hardly surprising. Since the most progressive government conceivable at the time was already in power, it was hard to see who would do what the campaigners wanted. Without hope of success campaigns lose their lifeblood. At the same time their campaigning mode had lost the Webbs many society friends in London: Lord George Hamilton scored social points where it hurt Beatrice most, with Balfour and even with Haldane. During this time the London School of Economics also moved in some ways beyond the Webbs. Hewins was undoubtedly right when he concluded his *Times* letter with the statement that no one should 'underrate Mr. Sidney Webb's services to the School of Economics. There are few men who have contributed more to the success of that institution or to that of the University of London.' A later Director, William Beveridge, also had a point when he ended his comment on the railway affair with the observation that 'it remained true of the Governors, as of many other gatherings, that where ever Sidney [Webb] happened to sit, there was the Chair—the source of ideas and practical expedients and reconciling compromises'.[107] Yet by 1911 Webb had grown a little tired of the School, and the School had begun to raise him to a pedestal of respected irrelevance.

It was Beatrice who noted in the middle of the railway row how much Sidney valued his '*authoritative* connection with the School', meaning one of position rather than just association, but like Sidney she also knew that 'we shall have to drop into the background of the School's life'. Ostensibly the storm about the railway speech blew over. The 'four errant knights of individualism'[108] (Douglas Owen had joined the other three though he continued to teach Marine Transport to the Army Class) resigned as Governors on 16 December 1910, but the Railway Course continued with great success, and on 13 June 1911 two new 'railway' Governors were elected, Mr F. H. Dent of the South Eastern and Mr W. Guy Granet of the Midland Railway (the latter to stay only for a few months). Yet, even in his public reply to Lord Claud Hamilton, Sidney Webb had mentioned that he expected 'to be away from England on a prolonged tour of the East, which would, in any case, have necessitated some arrangement of chairmanships among the governors'.

Initially he had thought that the trusted and loyal vice-chairman, Sir John Cockburn, could act as chairman in his absence. But as the world tour drew nearer, a more drastic rearrangement seemed inevitable; Sidney Webb offered his resignation from the chairmanship of the Court of Governors and the Council of Management. On 25 May 1911 the Council resolved 'that owing to his approaching prolonged absence from this country the resignation of Mr. Webb be accepted with deep regret and in the hope that on his return he will resume his active work in connection with the School'. The Rt. Hon. Russell Rea MP (later Lord Rea) was elected chairman. Sidney Webb did resume some of his activities for the School after his return and even played a critical role at the end of the First World War. Yet his resignation marked the end of the era in which no major decision could be taken without him. There was, for everyone to see, life after Webb.

The Ratan Tata Department

The Webbs left London in June 1911 and returned in May 1912. Their tour was exhausting, eventful, and yet possibly pointless except as a way out of an impasse. Norman MacKenzie put it more kindly: 'In retrospect the journey seems like a long ante-room to the next phase of the partnership. When the Webbs arrived back in London they would have to make another fresh start.'[109] Yet there were moments of significance during the tour, even for LSE. Most important among them was the encounter in Bombay, during the last lap of their trip in the spring of 1912, with Ratan Tata, the second son of the founder of the industrial and commercial Tata empire. Like his brother Sir Jamsetji, Sir Ratan spent much time on philanthropy. He had been interested for some time in setting up a School of Social Studies, and had tried to persuade the Institute of Science in Bangalore to accept it. But Bangalore declined, perhaps out of fear of the subversive effects of social science, and Tata was persuaded to give the money to the University of London instead. In April 1912 Sidney Webb 'clinched the deal' not just for London but for the London School of Economics.

The 'deal' had changed significantly from the original School of Social Studies. Through the good offices of Sir Ratan's secretary, B. J. Padshah, contact had been established between the benefactor (who was then living in Twickenham) and the new Principal of the University of London, Sir Henry Miers. The Principal in turn had got in touch with Professors Hobhouse and Urwick, who had produced a scheme to endow 'research into the principles and methods of preventing and relieving destitution of poverty'. This sounded like the language of the Poor Law Commission though the scheme left the controversial issue open by insisting that work 'should not be confined to questions of private philanthropy but should extend to legislative and administrative measures dealing with poverty, pauperism, and the causes and effects thereof'. The scheme was essentially a research proposal though its authors insisted on its recognition by the

University of London and connection with its sociological work and saw it as 'a branch of what may become a distinct Sociological Department of the University'.[110] Webb may not have liked all this; in any case he 'wrote a long memo';[111] but Tata accepted the proposal and made available funds for the Ratan Tata Foundation at LSE for a period of five years.

In terms of the LSE end of the deal, it was no accident that the Ratan Tata Foundation became a driving force in the *ménage à trois* which historians of the School's Department of Social Science and Administration have described. At the end of 1912, Martin White and Sidney Webb jointly proposed to their fellow Governors 'to establish a new department to continue "the work admirably carried on since 1903 under Mr. C. S. Loch of the Charity Organisation Society"'. Thus the merger of the School of Sociology and LSE took place, with financial assistance from Tata research funds. Like all mergers between weaker and stronger partners, it had its traumas; but Jose Harris plausibly argues that it would be wrong to exaggerate the intellectual and political differences between them. On the need for professionalization in social services as well as the desirability of 'positive' research the two sides agreed. The statement of intent in the LSE Calendar for 1912/13 combines the two strands, practical training and class teaching, throughout. 'It is regarded as essential that the academic teaching should be intimately connected with the lessons drawn from the students' experience in practical work.' It helped in all this that the two and soon three academic heads of the new department—Hobhouse, Urwick, and Tawney—came from different traditions and held different views but managed to run the new department without unnecessary friction.

In so far as there were problems in the *ménage à trois*, they arose from the third partner, the Ratan Tata Foundation, and even then they may have been unintended. In fact, the family metaphor is not quite the right description of the relationship in the early years; for while the School of Sociology and the LSE Department of Social Science and Administration were merged, the Ratan Tata Foundation merely supported a major research project in the department. It was only when the first five-year period of such support was drawing to a close that thought was given to drawing the Foundation closer to the department. In the end this took place in a dramatic manner. In his Report on the Session 1916/17 the Director admits that 'it is with some regret that I see a department which we have, without question, made very successful, transferred to the University'. What had happened was that the Ratan Tata Foundation had renewed, indeed increased, its support but insisted on the School's department becoming the Ratan Tata Department under the direct auspices of the University.

In practice, this change—occurring as it did, during the darkest days of the war—did not mean as much as would appear at first sight. The department did not actually leave the School. Contact between LSE and the Tata family remained close. Sir Ratan Tata died in 1918, but even his successors might well have given major support for an enlarged Department of Social Science and Administration to the School, had it not been for the collapse of the Tata business empire in the early 1920s. The department survived and became a critical part of LSE. Jose Harris lays much stress on the 'broad, speculative, intuitive and humane' philosophy which informed the department from the days of Urwick and Tawney, and without doubt such language describes the two men. Yet it is also true that 'the department was a product of the intellectual and political climate induced by the Poor Law Minority report'.[112] In so far as LSE came to be associated with the Welfare State (with huge capital letters), the Department of Social Science and Administration was at the heart of it, and the broader humanity of Urwick and of Tawney in particular played a major part in the process.

R. H. Tawney was but one of several new teachers brought into the School by the new department before the war. He came from Manchester, where he had been teaching economic history, to be Director of the Ratan Tata Foundation, though from 1912 he is also listed as an occasional lecturer at LSE. After the war he was made a full lecturer and eventually, in 1931, a professor. Thus 'Tawney remained at "the School" for the rest of his working life, in an atmosphere in which the idea that the state could act as a liberating force persisted across the disciplines'.[113] These are the words of A. H. Halsey, Tawney's intellectual heir and sympathetic biographer, who knows of course that when Tawney spoke of the state he meant something different from the Webbs or even from Hobhouse.

Tawney, then 32 years old, started a little *Commonplace Book* in 1912 which throws an interesting light on the man and his times. He was neither an individualist nor a collectivist but a romantic moralist for whom it was perfectly conceivable that 'England in the middle ages' was 'fairly happy and contented'. Modernity itself is the evil. 'Modern society is sick through the absence of a moral ideal', and since 'modern politics are concerned with the manipulation of forces and interest' it is hard to see how the remedy can be found. Not by constructing Utopias in any case, for they are 'worse than wrong', nor by economics as a science, which 'is just cant'.

When the war broke out, Tawney found the discussion of who began it futile. In his view it merely continued the intrinsic tendencies of modern societies by other means. 'I mean that our whole tendency is to exalt the combative qualities, and to undervalue those of the humble and meek, and that the existing economic organization of society is a perpetual evidence that the world gives its applause to energy, pugnacity, ruthlessness.' 'If we are to end the horrors of war, we must first end the horror of peace,' Tawney concluded. Many would not have agreed but his ethical socialism represented one strand of LSE thinking which has remained present throughout.

Halsey mentions Bronislaw Malinowski in the same context of advocates of ' "positive" freedom'. The young Pole makes his first LSE appearance in the Director's Report for 1912/13 when the ever-generous Martin White made available funds to enable him to lecture on 'Primitive Religions' and on 'Social Differentiation'. At the department itself, Clement Attlee prevailed over Hugh Dalton when a lectureship was advertised in 1912. During war service, his teaching was interrupted but he returned in 1919 and remained until he turned full-time to politics in 1923. Dalton meanwhile held an occasional lectureship and after the war a lectureship; later the Cassel Readership in Economics. Others who joined the staff during these years included Eileen Power in Economic History in 1913, and Morris Ginsberg in Sociology in 1914.

The School in 1913

Student numbers meanwhile remained buoyant, as did the proportion of foreigners among them. Indeed their number was so large—over 200 in all—that in his 1912/13 Report the Director raised 'the question of charging higher fees for foreigners' as 'a means of increasing the School's income'. The subject did not go away. At the time, of course, 'foreigners' were one category, 'Colonials' another, and 'Indians' yet another. Before the war, the School had regularly between thirty and thirty-five students from India, a third of whom were graduates. To look after them, the School had enlisted the services of Sir Theodore Morison, an old India hand though one who made it his business to defend the Muslim minority. This may have been one of the reasons why he was not excessively popular among Indians, the other one being that he 'does not shine as a lecturer, and we do not get very glowing accounts of his efforts in this direction at the London School of Economics'.[114] Still, he remained involved, especially when the Governors appointed a committee in 1914 to draw up a three years' course of study for Indian civil servants. After Reeves's resignation in 1919, Morison was to be on a short list of three possible candidates for the directorship.

The Director's last report before the outbreak of the war appears rather upbeat in mood. In one sense, this was surprising. Not only had Reeves been ill several times and forced to take periods of extended leave but two of his major projects had fallen through. They were both—what else?—projects of physical extension for the School, where living in cramped quarters rapidly became a mode of life. In 1912 W. P. Reeves started an appeal for £25,000 in order to extend the Passmore Edwards building even further. A few thousand pounds were found among governors and friends, and by a legacy; but the nebulous 'wealthy financier' who keeps most appeals going for a while never turned up and the Building Fund was quietly forgotten. Reeves's other plan was to use, and perhaps purchase, Smith Memorial Hall in Portugal Street. The Carnegie Foundation was not prepared to give as much as the Trustees of the W. H. Smith Memorial Institute wanted, so that the first encounter with that company (though remote since the Hall commemorated W. H. Smith but was never used by the company in his name) had a less happy ending than the later one which turned W. H. Smith's warehouse into the Lionel Robbins Building for the Library. In 1914 Reeves had to admit that the scheme he had dreamt of was 'unworkable'.

Yet the Director was right to be upbeat. By 1913 LSE was not only a well-established academic institution of recognized quality but it had also begun to develop its own possibly unique character. This character had, and has to the present day, three main ingredients: the forever unresolved tension between social science and politics, the international composition of students and staff, and the curious intimacy indeed inclusiveness of the institution. Other ingredients could be added. For many decades, the presence of both day and evening students emphasized the borderline of the 'academic' and the 'real' world on which the School existed. This was mirrored in perennial disputes such as those around sociology and social administration, theoretical and professional subjects. Other paradoxes come to mind. The School has always been both a lively and a lonely place. It has combined the unlimited horizons of intellectual flights of fancy with horribly cramped and grubby quarters. It was, and is, very much a part of the great, bustling, unbearable and yet irresistible metropolis, of London.

Being in the centre of a great city has another significance for the School. From its early days, people commented that LSE was more than a university, it was a 'movement'. The word is a little misleading; it seems to suggest that the School was or is the nucleus of a political party, which is clearly not the case. But it is a place where staff and students, even governors and visitors, feel that they are where the action is. Mary Stocks put it well: 'It was the way in which internal vitality reflected external events.'[115] Whether the world spirit has ever visited LSE may be debatable, but the spirit of the times was certainly there and may even have found a tenured position in Houghton Street. The spirit of the times is more than just the *Zeitgeist*, the fashionable intellectual creed of the day; it is the cross-current of ideas which inform whole periods, decades perhaps. At LSE one not only knows which way the winds are blowing but also what changes in climate are probable or likely.

But when all this is said, the core of the London School of Economics is still the mixture of intimacy and tension which holds people of the most diverse origins and opinions together. This is not a cosy or even a comfortable condition. Many over the years could not bear it and drifted away. But those who could bear it found their own selves in the institution of LSE. Many years later they are still pleased when it flourishes and suffer when it goes astray; they are often proud of being associated with it and sometimes ashamed of what it is doing. They can never wholly get it out of their systems. LSE is an inseparable part of their, of our lives.

First World War

On 4 August 1914, the lamps went out all over Europe, and LSE was soon affected by the consequences. Even in the first year many, including Attlee and Lees-Smith, Tawney and Dalton, the porters Wilson and Dodson (whose daughter took over the job for the duration), and of course a large number of students, were called up for war service. Cannan's assistant Theodore Guggenheim had changed his name by deed poll to Gregory in order to serve but was later dismissed because of his German parentage and Reeves fought in vain for permission to let him teach. The routine of the School was affected in other ways. Not surprisingly, the Director reports (in 1914/15) that 'the Social Life of the Students has undoubtedly suffered from the War. There have been, for instance, no dances.' More ominously, the number of Belgians and Russians among the students had risen dramatically. 'All the Belgians and the majority of the Russians were refugees.'

A year later, in 1915/16, the Director's Report contains the laconic paragraph: 'War conditions have made themselves felt in every department of the School's work.' And a little later: 'I fear that there is a great element of uncertainty in our arrangements for next season.' The attempt to 'continue as far as possible as in normal times' was bound to remain a stiff upper-lip attitude in the face of adversity. Student numbers declined from 2,127 in 1913/14 to 1,156 in 1917/18, with the number of women students fairly stable and but a minor decline in railway students. On the other hand, the number of men on the general register went down by 60 per cent between 1913 and 1917. Much of the teaching had to be improvised. Declining student numbers were accompanied by lower income from both government and other sources. A near-epidemic of chicken-pox, mumps, and German measles in 1917 did not help.

Yet for the survivors life went on and plans were made, notably for the development of what came to be called 'commercial education'. On 11 March 1917 Sidney Webb wrote a rather remote—'Dear Director'—and slightly irritated letter to Reeves telling him of LCC plans to expand commercial education 'after the war'. '*If we do not say we are going to provide it*, it will be difficult to prevent some aspiring Polytechnic or Institute from proposing to do it.' Webb proceeded to lay out a plan of teaching in detail. The Director found all this irritating in turn. 'My dear Webb,' he wrote back on 13 March, 'the L.C.C. are queer people in some ways and seem to have left you under the impression that we have not been in communication with them on the subject.' This, however, was untrue. For good measure, Reeves added a memorandum by Professor Sargent which pointed out that many of the courses proposed by Sidney Webb had long been a part of the regular work of the School. All that is needed to expand them 'is a larger staff and more money'.[116]

Even so, the School produced by April 1917 a 'Memorandum by the Director of the London School of Economics on the Development of Higher Commercial Education after the War and the Position of the London School of Economics'. This Memorandum sets out the plan for a two-year course designed 'to turn out a commercially intelligent man, not a mere accurate machine'. This was to be achieved by a first year of general training in economics, geography, history, law, and public administration, followed by a second year of specialized teaching in the five areas of industry, marketing, finance, transport, and insurance. A Higher Commercial Certificate was to be awarded at the end. Once again, the business school seemed to encroach on the research university and many at the School including the Director (and more importantly still, the Secretary, Miss Mactaggart) did not like it.

Among the Governors of the School, the business school was understandably more popular than the research university. Indeed they wanted to be sure that there was nothing second-rate about commercial education, so that they were pleased when in the process of discussions with the LCC and the University Senate the Certificate was transformed into a full degree, the B.Com. On 18 July 1918, thus before the end of the war, the scheme was inaugurated at the Mansion House in the presence of the Lord Mayor. Interest spread far and wide; the Director's Report for 1917/18 even records that the Overseas Trade Minister 'Sir Arthur Steel-Maitland is interesting himself greatly in the matter and hopes to be instrumental in bringing the scheme to the notice of the King and Queen'. More importantly still, funds were forthcoming from public bodies and business. At the end of November 1918, Webb told Reeves that 'I have got a pretty big sum to throw into your City scheme very shortly from Commerce'.[117] The sum, from Sir Ernest Cassel, was to be very big indeed but it was no longer 'your scheme', that is Pember Reeves's, when it arrived.

Sir Arthur Steel-Maitland, MP, had reasons to show an interest in School matters, for by then he had become Chairman of the Court of Governors. Lord Rea died on 5 February 1916. He was an old-fashioned Liberal and true free-trader, whose attacks on Joe Chamberlain make enjoyable reading. His connections with the railways and his position as Junior Lord of the Treasury were useful for the School but he remained an interim chairman. This was emphatically not the case with his successor, Sir Arthur Steel-Maitland, from 1910 to 1935 Conservative MP, and for nineteen years Chairman of the Governors. He helped the School in many ways, beginning with the commerce degree, and presided over the academic miracle of the early Beveridge years.

A Broken Reeves

Meanwhile the Director, William Pember Reeves, was in an increasingly unhappy state, which in the end led to his resignation in April 1919. Reeves's unhappiness had many facets. His health was shaky; at least he felt it that way. In 1916 he failed to shake off a cold and wrote to a friend: 'I am sorry to be a broken reed to you at this time, but the truth is I am really unwell—more unwell than I care to talk about—and it is worrying me a good deal.'[118] While he was recovering, his son Fabian had a flying accident which set the father's condition back. A few months later, in June 1917, Lieutenant Fabian Reeves was missing in France, two months later confirmed as killed in action. The loss 'knocked him to pieces entirely'. Reeves never recovered from the shock, while his wife Maud, the forever active woman who was then working at the Ministry of Food, turned to spiritualist seances to stay in touch with her son.

At the School, relations became strained. Miss Mactaggart remembers that Reeves tried 'to snuff Webb out; he would not let him interfere with the work of the School'. Webb did not like this. Meanwhile Miss Mactaggart had her own problems with the Director. 'He was jealous almost to madness, and vain almost to madness.'[119] Some perceived his policies to be geared almost entirely to making economies. Many found it difficult to talk to him. His irascible temper became proverbial. Also he was blamed for the failure of his plans to attract funds for the extension of the main building and for the purchase of the Smith Memorial Hall. Reeves's biographer Sinclair paints a pathetic picture of the man in his last years at the School: 'Reeves was never quite to recover from Fabian's death. Sick and dispirited, he felt savaged by life. Of his work at the L.S.E., not much remains to be said. In 1917 and 1918, several days a week, always ill, he was to be found wrapped in rugs in his room there. But he did very little.'

Yet this is by no means the whole truth of the man, who was then after all only just 60 years old. At the National Bank he was a much liked chairman, so that he continued in that office for more than a decade yet. Moreover, he 'found his last great love', Greece, and devoted much energy to courting it. With others, he set up the Anglo-Hellenic League (with offices conveniently in the Aldwych). Like others, he took sides with the Nationalist leader Venizelos and against the King as well as all other enemies of the man of whom he became a 'friend and unqualified admirer'.[120] For Reeves, Greece was like another New Zealand, only closer and therefore more real. Or was it a figment of his romantic imagination after all? Did he know what he was defending when he took sides so passionately?

Even Greece brought him into conflict with colleagues at LSE, though most of it happened after his resignation. This was not so much because he incessantly talked about it but for a more academic reason. In his last Director's Report Reeves mentions, rather irrelevantly, that 'King's College has been successful in raising funds to endow a Chair in Modern Greek'. The Koraes Chair was endowed by subscriptions from London's Greek community, which supported Venizelos to a man. Questions were therefore raised immediately by this group when Arnold Toynbee was appointed as the first holder of the chair, and doubt turned to dismay when Toynbee wrote a series of articles in the *Manchester Guardian* exposing Greek cruelty in Asia Minor. The subscribers demanded Toynbee's resignation. Many rallied to his defence, including Harold Laski, who urged 'that universities should not accept endowments under such conditions', R. H. Tawney, who 'was disturbed by the principle involved', and Graham Wallas. Pember Reeves took the opposite line and deplored, on behalf of the Subscribers' Committee, that 'the Chair is occupied by a gentleman who is regarded by Greeks and those interested in Greece as a persistent and mischievous enemy of the Greek race and cause'.[121] In the end, Toynbee had to go. Richard Clogg regards the incident as a sad chapter in the story of 'Politics and the Academy' and has nothing good to say about Reeves's role in it.

This happened in 1922 when William Pember Reeves had long left LSE behind him. His position there became untenable at the beginning of 1919. Since no one else knew how to broach the subject with him, Webb had to step in. Beatrice Webb's diary entry—written on 29 April 1919, several weeks after the event which must have taken place in the very first days of April—has often been quoted. The School was beginning to look up 'after the arid period of the war', but it needed a new Director. Thus, Sidney 'had to undertake the unpleasant task of telling an old friend, W. Pember Reeves, that the time had come for him to resign the directorship'. Reeves did not like the idea and was unimpressed by Webb's story that his appointment was supposed to have been for seven years in the first place. 'It was a painful interview' but

> Sidney was firm and Reeves eventually agreed to resign, remarking, somewhat bitterly, that Sidney 'was ruthless in the pursuit of his causes and allowed no personal considerations, either on his own behalf or on that of his friends, to stand in the way of the success of an institution or a movement he believed in'. Which is of course true!

Beatrice *dixit*.

On 9 April Reeves wrote to a friend in New Zealand that he will 'be giving up the School of Economics very soon now, as the double strain of attending to that and to the National Bank is too much for me'.[122] On 10 April he sent the letter of resignation to the Chairman, Sir Arthur Steel-Maitland. The reason he gave rings true though it might well have led to the opposite conclusion:

> As I have informed you the matter has been in my mind for some time past, but naturally I have been reluctant to give up while the School was still beset by the many difficulties brought upon it by the War.
>
> The extraordinarily rapid recovery, however, of the last five months and the hopeful prospect of a much improved finance have completely changed the position. I feel that I need hesitate no longer.[123]

He would therefore like to be relieved of his duties at the earliest possible moment, such as the end of May.

In writing to friends a slightly different note crept into his statements. His wife's cousin Alexander Carr-Saunders, for example, was concerned that his health might finally have failed Reeves; but Reeves replied that he finds 'the pressure of the work rather greater than I care any longer to face, especially as it is increasing and the Director here will in the near future require to give his entire time to the job'.[124] The Governors accepted the resignation on 24 April with a handsome resolution of appreciation for 'his faithful and successful services during the past eleven years'. Reeves retired to the National Bank, to Greece, and to travels including a trip to his native

New Zealand. He even remained a Governor of LSE though his attendance at meetings left something to be desired.

Reeves also remained a friend of the Webbs. In 1923 he gave moral as well as financial support to Sidney's election campaign, which Webb gratefully acknowledged. 'I have largely drifted out of the things we so long did in common, because of new responsibilities,'[125] Webb added, referring, of course, to LSE. Reeves left in May 1919. Steel-Maitland went on holiday in Scotland after his departure from government. Sir John Cockburn signed the Director's Report for 1918/19. Sidney Webb took charge of the School for one last time, coming in every day to look after its affairs, and searching for a Director who would take it forward to new and even greater heights of achievement.

PART II

A SECOND FOUNDATION

The Beveridge Years

1919–1937

3

AN ACADEMIC MIRACLE

NEW BEGINNINGS

Changing Course

In the summer of 1919 the London School of Economics and Political Science was in rather better shape than the country around it. If we are to believe the historian W. N. Medlicott—and how can we not believe him, given that he was later a professor at the School and wrote his *Contemporary England 1914–1964* after many discussions in his LSE seminar on inter-war history!—the prevailing state of 'peace without tranquillity' left much to be desired. The catalogue of losses and damages incurred by the war was long and ranged from all too many human lives to industrial capacity and social conditions. 'Everything had a run down look after the war', even the dominant élites and the old political parties. Worse still, according to Medlicott, it had 'ceased to be fun to be an Englishman'.[1] But academic life can be counter-cyclical and the social sciences may even benefit from public misery.

William Pember Reeves had been right to point to the 'extraordinarily rapid recovery' of the School after the war. Students came flocking in, many of them ex-servicemen including a significant contingent of Americans, who were sent by the American Educational Mission in England. They were attracted by the evening classes, by the social sciences, and also by the new commerce degree, which gave the School a considerable push forward. The B.Com. may have been controversial among professors but it attracted both applicants and finance. Sir Ernest Cassel's initial benefaction of £150,000 gave the whole University of London a boost which was well deserved in view of the role which the Vice-Chancellor, Sir Sydney Russell-Wells, had played in securing the money. But the School benefited in particular, receiving most of the income from the initial fund; by adding it to the successful Commerce Appeal it was able to put up a new building. Even apart from such munificent gifts, the powers that be, both in London and in the country at large, were inclined to support LSE and did so despite their restricted means. All that was missing therefore was a Director for the

School. Sidney Webb, the founder, had now nearly reached what many would regard as retiring age and was in any case busy with other things, so that he wanted to be rid of his caretaker position as soon as possible.

The search for a new Director is an exercise which concentrates the minds and titillates the nerves of those involved. Names come up and are dropped again and, when one of them accepts at the end of the day, success is also an anticlimax. The first name to be considered briefly in 1919 was that of Lloyd George's private secretary, the Gladstone Professor of Political Theory and Institutions at Oxford, W. G. S. Adams. He had, however, just been elected Warden of All Souls College in Oxford and was therefore not available. It appears that Alexander Carr-Saunders's name was mentioned, but the 33-year-old demographer was then regarded as too young and inexperienced; he had to wait another eighteen years for the opportunity. Sidney Webb kept on 'colloquing' with the Chairman of the Governors though they did not always agree. On 23 April, Webb wrote to his wife that Steel-Maitland 'seems to hesitate about Keynes, whose brilliancy he admits, but of whom he for some reason disapproves as Director'.[2]

Steel-Maitland relented and John Maynard Keynes was asked. He was then 36, the age at which Sidney Webb had founded the School. Keynes had been a Civil Servant twice; in the India Office and later, during the war, in the Treasury. He had made a name for himself at Cambridge and in Bloomsbury. He was now embroiled in the Versailles peace negotiations though about to move to the committed observer's gallery with his *Economic Consequences of the Peace.* Perhaps it is not altogether surprising in view of so many preoccupations that he wrote to his mother on 14 May 1919: 'I've a letter lying unanswered inquiring if I will be a candidate for the directorship of the London School of Economics—pay £1,500 or perhaps more. I shall ask a few questions about it, but have no intention of accepting. I hope Father agrees.'[3] Father, of course, had been a university administrator. It would have been the wrong job for Keynes. For one thing, responsibility for institutions was not his greatest strength; he preferred a more bohemian and idiosyncratic existence. For another thing neither his economics nor his liberalism would have sat easily with the tradition of LSE. He belonged to that other school, identified with Alfred Marshall and also with Gladstone, theoretical and individualist at the same time. LSE represented everything that they did not believe in.

Meanwhile Steel-Maitland went to Italy and a more homely shortlist of three emerged. It included Sir Theodore Morison, who was unsuitable for reasons of age and style; Professor A. L. Bowley, who had long been a key figure for the School's morale, respected and liked at the same time, but was probably not a Director; and Sir William Beveridge. Beatrice Webb wrote of him at the time that 'he has his defects' and she was to be proved right. But the defects were not of the Keynes variety. 'His views are slightly anti-Labour but pro-collectivist.' That apart, 'he is an innovator, not a conventional-minded man'.[4] As important, perhaps, 'there was really no alternative'. For the last time Sidney Webb wrote one of his pressing letters to a possible new Director. On 17 May 1919 'Dear Beveridge' got this note:

> I don't want to hurry you, or to imply anything either way. But you ought to be reminded that time is running on; and it may be found necessary for the Chairman and his colleagues at the School of Economics to come to a decision as between possible candidates. No date is yet fixed, even implicitly; but I am a little anxious lest some one may presently be fixed upon without your name being responsibly considered.[5]

Beveridge was duly impressed. In his autobiography he wrote later that he would have accepted the chance of LSE even without the war; but the fact that he felt out of sorts and without a clear perspective after years of high-pressure war work helped. Thus he accepted the nomination and was confirmed by the University Senate in June. On 1 October 1919 he started his new career as Director of LSE.

The old career had been one of social investigation and Civil Service and William Beveridge had been notably effective in both. When he took on LSE at the age of 40, he had already done more than most do in a lifetime. Born in India as the son of a judge of no particular distinction though with wide non-legal interests, Beveridge had not had an easy childhood. Father was often far away; the elder sister, Laetitia, and the younger brother, Hermann, died early; his mother looked after the sickly boy, and his sister Jeannette became his close friend. At University he blossomed, taking firsts at Balliol in Mathematics, Classics, and 'Lit. Hum.', the humanities degree. His autobiography begins with the undoubtedly truthful yet deeply misleading statement: 'Four years at Oxford left me at twenty-two with no clear idea as to what I should do next.' Beveridge may not have known what job to look for but his commitment to 'power . . . resting on knowledge'[6] (as he later described what he called 'influence') was already evident.

At first, William Beveridge followed his father's advice and reluctantly embarked on preparations for the bar. However, in 1903 he decided to go his own way and accepted appointment as sub-warden of Toynbee Hall. He was probably a little too academic for the liking of some of his peers there, but during the East End years he met many of those involved in the London School of Economics, including the Webbs, with whom he struck up a friendly partnership based on the common commitment to a benevolent state informed by well-trained experts. During the Toynbee Hall years, Beveridge made other friends too, among them Jessy Mair, the wife of his cousin David. She was to be his confidante, collaborator, and companion for the rest of their lives. As an administrator, however, Beveridge was rarely popular. Some may have breathed a sigh of relief even in 1906 when he handed over the sub-wardenship to his friend and future brother-in-law, R. H. Tawney. By that time, Beveridge had already begun a three-year stint as leader writer for the *Morning Post*. This ended when in 1908 he joined the Board of Trade. His Civil Service career there and later, during the war, in the ministries responsible for munitions, for labour, and for food, ended in 1919 with the title of Permanent Secretary and a KCB.

More than usually, however, the bare facts of Sir William's career tell us little about what made him tick. From one point of view this was without doubt the lure of *Power and Influence* (the title of his autobiography). But he sought both in order to bring about informed action to create a better order of society. His work on unemployment had not only helped both the Majority and the Minority Reports of the Poor Law Commission, but also led to the setting up of Labour Exchanges on the German model. (Like so many social reformers of the time, Beveridge was fascinated by Bismarck's social policies and in any case knew and liked Germany, so that the war came as a great shock: the inability to appreciate the political cost of social engineering was the greatest weakness of all collectivists.) Ideas on social insurance were beginning to form in Beveridge's mind though they came to fruition only during the Second World War. Still, wartime administration had strengthened his belief in benevolent state power. And all this was to be based on facts, on firm knowledge of what goes on in the world, on dispassionate investigation, and on the training of agents of change as well as investigators.

What better place than LSE to achieve such objectives? Yet Beveridge did not find the transition altogether easy. He asked for more money and later got £2,000 rather than the £1,500 offered to Keynes. He sought recognition of his own research interests and was appointed a 'lecturer in descriptive economics'. He wrote to his superiors in government that he hoped 'that in fact I am in no sense saying good-bye to public service'.[7] But when the Government wanted to send him to India in 1921 to do some work on tariffs, Webb advised him so strongly against accepting that he had little choice: 'Your position is terribly a "one-man business", and your absence would mean an arrest of development if nothing worse.'[8] At that time, 'development' was well under way. Beveridge was embarked on what may well be the greatest success story of his life, the one in which he gave his best, and left behind a lasting edifice which bore his imprint. Beatrice Webb was the first to call the Beveridge years a 'second foundation' of LSE when Beveridge left in 1937.[9] Twenty-six years later, after Beveridge's death, the then Director Sir Sydney Caine reinvented the phrase in his letter to Mrs Mair's (Lady Beveridge's) son Philip and called Lord Beveridge 'almost a second founder': 'He laid new foundations for us and we who have followed him and those who follow us will always be building on his work.'[10]

The Beveridge 'Impress'

Laying these foundations, both literally and metaphorically, was a process undertaken so quickly and effectively that it left not only later historians but even the second founder himself, a man not normally lost for words, speechless. In 1924 he began his LSE Oration by declaring the task of compressing into one speech the events of two years 'frankly impossible'. It was, he said, 'a time of continued growth and change and proliferation'. When Hayek wrote the history of the School in 1945, he was still overcome by awe in the face of the early Beveridge years. Commenting on the appointments of Sir William Beveridge and the new Secretary, Mrs Jessy Mair, Hayek wrote:

> The era of development of the School which opened with these appointments became one of such rapid change and expansion that it will not be possible to describe it here in the same detail as has been given to its first twenty-four years. Most of these events are, however, still in recent memory and a brief outline will suffice to complete this sketch of the growth of the London School of Economics.[11]

Now that the events are no longer in recent memory, the historian cannot afford the same luxury; on the other hand the distance of three-quarters of the century helps to see more clearly what happened, how it happened, and which features of Beveridge's miracle years led to lasting changes at the School.

In the beginning there was the commerce degree. This turn of events, which brought LSE closer to becoming a business school than any other academic development before or after, remained disputed throughout the ups and downs of the thirty-year history of the degree. Beveridge soon came to realize the fact, though like most people in power he was pleased that the degree had been 'criticized from two mutually inconsistent points of view'. If both extremes are against it, the course must surely be right—or must it? In this case, the critics were in part academics, in part practitioners.

There are some who fear that we may be too technical, and in training for a Commerce degree may lower University ideals. There are others who doubt, because they think that we cannot be practical and technical enough, and that all that is worth learning in Commerce can be learned only in the office and in the world outside.[12]

But the new Director was not going to let anybody take the wind out of his sails; it was just too good an opportunity to miss for someone who wanted to make headway fast.

'Space was the first problem that I met at the School,'[13] Beveridge wrote later with a not unfamiliar Directorial groan. Perhaps the new Director had more reason than others before and after him for this. The School could not have continued its work during Beveridge's first session without use of the famous 'huts' which had been erected on the central island of the Aldwych which is now Bush House. However, plans had already been made for a building scheme that would create both what is now called the Old or Main Building with its entrance on Houghton Street, and generous provision for the Library at the Clare Market end of the new building complex (where the Library was to remain until it moved to the Lionel Robbins Building in 1978). The major part of the money required for the first stage, £75,000 of a projected total of £140,000, had been raised as part of the Commerce Degree Fund, to which the business community had generously subscribed. But what about the rest? A visit by the newly created University Grants Committee in 1920 turned out to be notably successful; the Treasury added £45,000 to cover the building cost. The remainder came from the London County Council and from School reserves. The great plan could go ahead. On 28 May 1920, King George V laid the foundation-stone and the 'hammering sound' began which was to accompany the Beveridge years almost throughout.

There is some dispute as to who first said that LSE is the place on which 'the concrete never sets'. Not uncharacteristically, Beveridge later thought that it was him, or rather, his Secretary and him: 'Janet and I . . . adapted to our needs Christopher North's view of 1829 about His Majesty's dominions: The School of Economics was that part of the University of London on which the concrete never sets.'[14] Most people, however, attribute the remark to the much-loved Eileen Power; even Beveridge seems to have thought so in earlier years.[15] It is thus understandable that Kingsley Martin sounds slightly piqued as he reports his numerous disagreements with the Director and adds: 'I once, and only once, pleased Beveridge. I said that he "ruled over an empire on which the concrete never set".'[16] Whatever tricks their memories may have played on these or other LSE people, everyone remembered the nearly permanent building site on which they lived; and while some may doubt the aesthetic judgement of Beveridge's Wordsworthian belief in '*beauty* born of hammering sound', the School was undoubtedly growing. It actually grew from 57,000 to 134,000 square feet during the Beveridge years. Staff now had rooms, there were even administrative quarters, students had classrooms as well as space for recreation, there were lecture theatres, there was a real library. At least all this was the case once the 'battle of Houghton Street' had been won to which we shall return as we look in greater detail at the achievements of the Beveridge years.

More space was, however, only one of the beneficent effects of the commerce degree. Equally if not more important was the chance to attract new staff, and to improve the position of those already teaching. The core of the Cassel Fund of £150,000 was intended for educational purposes; it led to the endowment of nine University teaching posts as well as annual grants for teaching modern languages and for enabling students to travel abroad. Thus the new degree made possible a significant addition to the School's academic strength and the Director made full use of the opportunity.

There are different views about the condition in which the School found itself at the end of William Pember Reeves's 'reign'. Jose Harris, Beveridge's authoritative biographer, describes it as 'a small and rather obscure institution, providing mainly part-time courses for students living in London'.[17] The picture makes Beveridge's achievements stand out even more starkly, but it also overstates, or rather understates, the truth. Beveridge himself spoke of the School's 'brilliant past, and bright prospects for the future'.[18] Hayek would go along with this judgement. His history is full of praise for the pre-war vitality of LSE and the developments which were only temporarily 'cut short' by the war.[19] This is how the Webbs saw things too; the School had to be reorganized 'after the arid period of the war',[20] but it was already a distinguished institution of international renown. One of the later Sir Ernest Cassel Professors of Law, Lord Chorley, agreed with Beveridge's phrase that the School had to be 'remade' after the war, but not exactly from scratch nor merely by one man. 'Much of what was done would have happened anyway, but without the Beveridge impress and over a longer period of time.'[21] After all, in 1919 the School already had 3,000 students coming from over thirty countries, and fifty-nine teachers, many with great names in their respective fields.

What is true is that, to use Beveridge language with respect to the problem of unemployment, a process of 'de-casualization' was urgently needed. Two-thirds of the students were 'occasional students'; all but three senior members of the academic staff—Professor Bowley in statistics, Professor

Sargent in commerce, Dr Lilian Knowles in economic history—were not in full-time University posts. The Martin White benefaction and the University had begun to remedy this situation, but the Cassel Fund made a major difference. No fewer than eight Cassel appointments were made in the academic year 1919/20: Professor L. R. Dicksee to a Chair in Accountancy, Professor H. C. Gutteridge to the Chair in Commercial Law, Mr T. E. G. Gregory and Dr Hugh Dalton (both already on the School staff) as well as Mr Douglas Knoop to readerships in commerce, and Messrs L. Rodwell Jones, T. A. Joynt, and J. Drummond Smith to lectureships.

This was clearly a major step forward. Looking back at the early years of Beveridge at the School after many decades, another set of names appears even more striking. Around 1920 a number of those arrived at LSE who were to determine more than others the character and appearance of the School for decades to come. In 1919 Sydney Caine registered as a student for the B.Sc.(Econ.). He was followed a year later by Lionel Robbins and also by Arnold Plant, who took both a B.Com. and a B.Sc.(Econ.) as well as finding time to be President of the Students' Union. During the academic year 1919/20 the appointment of Harold Laski to the permanent staff was announced. Bronislaw Malinowski returned in 1921 as an occasional lecturer (which he had first been in 1913/14) and soon advanced to a readership and a chair. R. H. Tawney was one of those whose post now became permanent. Clement Attlee, who had been an Assistant before the war, returned as a Lecturer in 1919. And so one begins to see the contours of the School as later generations knew it.

Attracting these names was not William Beveridge's doing. He had, like all successful leaders, his portion of luck in these early years. Clearly the School at the time was not so puny, parochial, and unknown that it could not attract men—and women—of brilliance and promise. But, as Lord Chorley rightly observed, the new Director hurried things along and there was also what he called the 'Beveridge impress'. Surprising as this may seem to some, the main 'impress' of Beveridge had to do with students. Somehow his peers, the senior academics, never really warmed to him, and such relations are usually reciprocal. But he liked the young, or at least he believed in their education and even in a bit of fun in their lives. Without doubt he was right to emphasize in his own account of his LSE years the fundamental change in the composition of the student body. When he came in 1919 two-thirds were occasional students; when he left in 1937 only one-third. This meant, for Beveridge, that economics had become a form of 'liberal education' both in its general version of the B.Sc.(Econ.) and in the practical one of the B.Com. He got in fact quite carried away by the subject in his opening lecture for the session on 4 October 1920. After the arts and the natural sciences, those disciplines concerned with commerce and with justice must have their say:

> They come by a special channel of their own. They must be imparted not to a few but, if possible, to all the citizens. So we may say that after Prometheus and Epimetheus had given to London, University College and King's College and many other institutions, Hermes (at that time a distinguished member of the London County Council) was commissioned to establish the School of Economics.[22]

If Sidney Webb had been present, he would have squirmed in his seat at being likened to Hermes, the messenger of gods, or rather, the godly messenger and mercurial patron of trade. It is also interesting to note that Beveridge liked to drop the 'and Political Science' when he referred to LSE. However, he clearly believed in the educational value of a certain kind of economics.

He also helped students in numerous other ways. He had not forgotten his Balliol years and wanted some kind of tutorial system at LSE. The invention of Advisers of Studies did not work too well at the School, especially not as a form of supervision exercised by the writing of Oxford-style essays 'on general subjects connected with Economics and Political Science. In doing so, [students] should learn not only Economics, but the essential art of self-expression.'[23] But remnants of the innovation survived the decades, in some departments even as a tutorial system. Over the years undergraduates have often complained that the School is distinctly biased in favour of graduate work and research; Beveridge shared the bias but he also felt a strong responsibility for the needs of undergraduates, which other teachers accepted in theory without being prepared to do much in practice.

The new Director spent much time on and with the Students' Union. He turned it from a voluntary association into a service for all. Whereas some later Directors told Honorary Fellows and selected members of staff after dinner about the latest developments at the School, Beveridge preferred a Students' Union audience. In the Foreword to the Students' Union Handbook of 1922/3, he describes the changes at the School once again in terms evocative of his classics studies. The old School, he says, and means the School before Beveridge, may perhaps be compared 'to the city state of ancient Greece—a city state after the Athenian rather than the Spartan model', as he hastens to add. However, those days are now remote and irrelevant. The School 'has now relatively the dimensions of a modern national state'.[24] It is the task of the Students' Union to carry into this modern state the keenness and vitality of the old *polis*.

Judging from the 1922 list of student societies there is little doubt that the plan succeeded. Dr Hugh Dalton appears as President of the Boxing Club. Mr Sydney Caine organized the Chess Club. Professor Graham Wallas presided over the International Study Circle. The Director, somewhat less plausibly, emerges as President of the Literary Society, but also of the League of Nations Union. R. H. Tawney is listed as Hon. President of the Christian Union and Lilian Knowles of the Economic History Society. The Students' Union itself offered useful opportunities for debate and association, even if one does not entirely follow the hyperbole of its 1922 President Isidore Graul and his exemplary *non sequitur*: 'If "the proper study of mankind is man," the most important part of the School is the Students' Union.'[25]

Beveridge himself was clearly what would nowadays be called a workaholic, so that one detects a slightly grudging undertone in his 1924 Oration statement that 'students have never yet been defined as persons who spend all their time in study'.[26] Yet he did not begrudge others a certain amount of play, as long as they took it seriously, as seriously as he himself had done. Beveridge had been a sickly child himself and at various times of his life he suffered from 'athlete's heart'. Whether this affliction was organic or psychosomatic, he tried to compensate for it by vigorous physical exercise and notably by cold baths, long walks, and badminton. In theory at least, he did believe in *mens sana in corpore sano*. Thus he showed much interest in the athletic activities of students and staff, found a sports ground for them at Alperton in his second year, and a more permanent one at Malden a year later. Another kind of play was offered at Dunford House, Richard Cobden's former home, which the Unwin family offered to the School for summer retreats. The former owners repurchased Dunford House in 1924, but the Director and his Secretary had liked the idea so much that they found alternatives, notably 'Will's cottage' in Avebury, where they often received students and staff for out-of-term periods of hiking, talking, and studying.

The Beaver and Virgil

In his own account of his years at the School, Beveridge somewhat unexpectedly lists another of his early exploits under the heading 'Economists at Play', the adoption of a School coat of arms and motto. When Beveridge came the School had just adopted the colours which it was to keep for three-quarters of a century: black, purple, and gold. The idea of a coat of arms came up a year later, towards the end of 1920, and in the following two years an amusing little story unfolded around it.

At first it was thought that an 'emblem' would be good enough, something that would now be called a logo, for the School. An elaborate proposal compressing all letters of the name of the School into an unpretty ornamental shape was made by Mr Emery Walker but did not find much favour. On 8 April 1921 Beveridge wrote to Sidney Webb for advice, but Sidney replied with a handwritten note: 'This is beyond me . . . I am afraid I am useless on all this.'[27] So a committee was set up, quite a large committee of twelve members, including eight students, with Professor Gutteridge in the chair, which met on 24 May 1921. The committee wanted, as the Director had put it, not a logo but something 'of a more practical nature', incorporating 'the figure of some animal which would be emblematic of the work of the School'. What animal? 'In this connection it was thought that a beaver would be very suitable,' the committee minute states laconically.[28] It also adds that Mrs Mair thought a coat of arms would be too expensive since the College of Heralds would have to grant it, but the students present were eager and promised to find the money.

And so the designers got to work, among them notably Arnold Plant, whose eleventh drawing was finally accepted and after tortuous further changes and much high feeling developed into the official coat of arms of the School as described in the Letters Patent of the College of Arms:

> Sable a Beaver passant Or on a Chief of the Second two closed Books Purpure clasped leaved and decorated Gold.[29]

Why the beaver? Since committee minutes and other authentic data tell us little, we can follow the speculative trail which the Director himself laid. 'I am not quite certain who invented Beaver for us, but I suspect Janet strongly,' he wrote in 1960.[30] The indications are that his suspicion was characteristic but wrong. Nor is there any detectable substance to the theory that the beaver was chosen because of its alliteration with the first part of Beveridge's name. (References to the beaver's 'webbed feet' and the consequent pun on the Webbs seem even more far-fetched.) One student later recorded 'a dim recollection' that the beaver came from the arms of the Hutchinson family,[31] but someone would surely have mentioned the fact in the Gutteridge committee. As so often, the obvious is also the most likely. After all, the Director himself, in his 1922 Oration, found it 'needless to point out the appropriateness to the School of Economics of an animal of such social habits, so constructive, and gifted with such foresight as is the beaver'.[32] It is also 'reputed to be industrious', though clearly much fun was had at the School about that question. Is it not true that 'for five long months in winter the beaver does nothing but sleep and eat and keep warm'? Indeed, could it not be said 'that in fact he never works at all except in September and October when his dam must be built and when the Final Examinations of the University are held'? These of course were soon to be moved to June at the instigation of the Director.

One other beaver story has more significance for the School and its role than for the coat of arms, since it happened much earlier. On 3 November 1912 R. H. Tawney had entered a slightly muddled yet strangely relevant note in his *Commonplace Book*:

> Every community should have a body of bridge-builders, pontifices, a very good name, for the bridge-builder is the real priest. These are the beavers of society, unobtrusive gentle animals, yet with sharp teeth and bright eyes, eyes to see where piles must be driven, what stout timber must be felled. Where the bars to bind and fasten must be set, teeth to cut down obstructions and bite them into place. It is said that the devil builds bridges, and I certainly think that social bridges are not built by men without any devil in them. But he is a good labourer devil, a lubber fiend who does more work than most of the Saints of the Calendar. Never were a gang of bridge-builders needed more than now!

It is not easy to follow the connections between beavers, priests, devils, and bridge-builders, but somewhere in this obscure jumble one approach to the forever unanswered question of the heart and the purpose of LSE may be found. Bridge-builders between theory and practice, whose sharp teeth and bright eyes are put to work rather than to attack and who, despite their fiendish energy, have somewhat saintly pretensions. However, we must not take the beaver too far.

The coat of arms was one thing, but the next was to find a motto. Mrs Mair offered one guinea to a student who would suggest a usable text. Most suggestions, however, came from staff, no fewer than twenty-four—including 'Burrow and Build' and 'Beaveracious'—from the philosopher Dr Wolf. The Director suggested 'Literis et labore'. Sidney Webb suddenly found the exercise no longer beyond him and made two suggestions which would not have done LSE much good at all; 'By Measurement and Publicity' was one and 'Social Service' the other. In the end Professor Cannan prevailed and the words from Virgil's *Georgics* (ii. 490) were adopted in February 1922: *rerum cognoscere causas*, to know, or more accurately, to get to know the causes of things. A good choice.

Every classical drama has its satyr's play at the end. Both the Director and the Secretary have written about the carved wooden beaver presented by four professors in 1925 which contributed much to the general merriment of the School. It was named Felix and, when appointed an honorary student, got its full name Felix Q. Potuit because the verse in Virgil's *Georgics* runs: *felix qui potuit rerum cognoscere causas*. 'Happy is he who can understand the causes of things' was soon rendered as 'Felix who alone understands the causes of things'. 'The School of Economics, between the wars,' Beveridge writes approvingly, 'gave itself to plenty of cheerful nonsense with brains behind it.'[33]

If one surveys the record of the first three years of Beveridge's directorship it is hard to see how anyone could have done more or better. Yet the record is still incomplete. Professor Hayek in his historical sketch rightly puts special emphasis on the publication, in January 1921, of a School journal, *Economica*. In its early years *Economica*, under the editorship of Professors Cannan, Wallas, and Bowley, covered the whole range of the School's subjects. In 1934, as a part of the increasing articulation of the social sciences, it was divided into *Economica (New Series)* and *Politica*, of which only the former survived.

Yet with all this, one important subject—some would say, the most important—has barely been mentioned: the administration of the School. It is, Sidney Webb had written to Beveridge in 1921, a 'one-man business'. One can see what Webb meant and later many would charge Beveridge with having been autocratic. By his own account, however, the new Director tried hard to involve both teachers and Governors more in the running of the School. He activated the largely dormant Professorial Council and gave it a Standing or Office Committee. This was almost the first act of his directorship. Two years later, in 1921, an Appointments Committee and a Library Committee were added. At the same time, three members of the Professorial Council were given seats on the Court of Governors. The Court in its turn formed a standing committee in 1921 which was called for many years the Emergency Committee because its first item of business was the 'emergency' of preparing for the acquisition of the houses on the west side of Houghton Street. Thus the structure of governance began to emerge; it will require a closer inspection because in essence it has lasted to the present day, with the academic Council and its responsibility for appointments and administration on one side, the friendly outsiders' Court to deal with finance and overall development on the other, both intertwined by cross-membership and the Director at the centre of it all.

WILL AND J.

Making Way for a Secretary

However, in the end it was not organization charts which moved the School away from being a 'one-man business'. William Beveridge made no bones about what happened when he reflected later on his early problems:

In this exciting but laborious beginning at the School of Economics, I had one piece of great luck. I found myself able, within three months of starting at the School, to get absolutely first-rate assistance and to lay the foundations of a life-long partnership in work of every kind.[34]

In other words, the 'piece of great luck' had a name, Jessy Mair, and the story needs to be told.

When the new Director arrived, the School had for some time been less a one-man than a one-woman business—Beveridge called it, a 'one woman show'[35]—run, indeed 'ruled' by Miss Christian Mactaggart. She, the Secretary, knew of course all about the place which she had accompanied and often guided through its first twenty-five years, so that she became the natural focus of authority when the old Director, W. P. Reeves, faded away and others were busy with wartime duties. Later in her life, Miss Mactaggart was, as we have seen, quite outspoken about her various bosses, yet it is impossible to tell from her own accounts how she felt about the new Director in 1919. The two sentences which she put to paper sound suspiciously like a diplomatic communiqué:

With a new period of development before it the School was fortunate in the advent of Sir William Beveridge. He raked money in and spent it out lavishly, but by that time persistent over-work had undermined my health and I had a bad nervous breakdown in 1920.[36]

What was cause, what effect? The story from the other side sounded equally straightforward, were it not for a slight taste of crocodile tears:

Continued illness has prevented the Dean from being back at work. . . . The Professorial Council at their meeting on 1st October passed a warm resolution of regret and of sympathy with Miss Mactaggart, and I feel sure that the Council of Management will wish to do so also.[37]

Thus in the Director's Report of 1920 when the severance had taken place, for it was brought about with all the impatience of Beveridge's administrative style. It will be remembered that he arrived on 1 October 1919. Later he wrote that he had 'wanted and expected Miss Mactaggart to go on with me'.[38] But this desire did not last for more than a very few weeks. Soon the Director discovered that at 58 she was older than he had realized, and was perhaps not very well; she spoke of wanting to see the burden of her work reduced. By December Mrs Jessy Mair had left the Ministry of Food, which was then being wound down to be reintegrated in the Board of Trade, and joined LSE as Business Secretary. Miss Mactaggart was initially not displeased to be promoted to the post of Dean. 'More pay, less work, improved status,' though she did not want to be called 'warden', as she wrote to the Chairman, Arthur Steel-Maitland: 'It does so savour of settlements and going round poking a nose into other people's business from the standpoint of a superior person.' She even liked Mrs Mair, who had ostensibly come to help her. 'My assistant and successor is excellent, so far as I can see quite a fit and proper person for the job.'[39] But in fact Miss Mactaggart never found the time to enjoy her new status. In June 1920 she had the breakdown—Beveridge speaks of a 'seizure'—which led to her early retirement. After that she lived for another twenty-three years, much of the time in Italy. 'In fact,' Beveridge wrote with a slightly awkward lack of sensitivity, 'retirement, coupled with kind treatment from the School, gave her a new lease of life.'[40] By January 1921, Jessy Mair was installed as Secretary and Dean.

Much later, when they had already been married for three years, the Director and his Secretary became almost absurdly touchy about their common arrival on the scene. It appears that in an early draft of his 1945 history, Friedrich von Hayek had described Jessy Mair as Beveridge's 'collaborator of long standing'. The draft was sent to the Beveridges, who did not like it at all. He wrote that Hayek 'should say "who had collaborated with him in the Ministry of Munitions and the Ministry of Food"', which Hayek did (except that he used the phrase 'worked with him'). 'Actually this collaboration began only in 1916, three years before we came to the School.'[41] She, in a separate letter, went even further: 'As for myself, my collaboration with him before the Governors invited me to join the School staff was in fact confined to little more than a year when I was his chief secretary in the Ministry of Munitions.'[42]

Companionship and Collaboration

It is not surprising perhaps in view of such mystifications by the protagonists themselves that a huge cloud of rumour and innuendo has come to surround the pair which ran the School for eighteen years. Nor is it easy to get at the truth. In fact the historian finds himself in an unusual dilemma with respect to the Beveridge years. Not only have William and Jessy written much about their years at the School both during their time there and after and thus set the tone of accounts of the period, but being Director and Secretary respectively they have also produced most of the underlying evidence of official papers, minutes, and records. Moreover, Jessy Mair at least has done this quite deliberately and unashamedly with a view to posterity. In a little note 'In Retrospect' written in 1951 she set out her view of recording history:

As I hold the view that history is best compiled by a combination of the contribution made by the contemporary participant, and the scrutiny of any existing documents, I established when I came to the School of Economics a system of records classified and filed, so that the kind of history which I prefer could be written with some hope of accuracy.[43]

Thus wherever one looks one encounters streams and rivulets of information all stemming from one and the same source, and independent evidence is hard to come by. Fortunately, some of it exists and with the help of Jose Harris and one or two others we can try to reconstruct the story.

When William Beveridge and Jessy Mair first met in 1904, the occasion was not a collaborative venture but a family gathering, albeit one of somewhat remote relatives. David Beveridge Mair, Jessy's husband, was the son of a first cousin of William Beveridge's father; in other words the two men shared a great-grandfather whose name was also David and who was a baker in Dunfermline. David Mair was then a former Cambridge mathematics don working for the Civil Service Commission when he was not writing books on the teaching of mathematics. 'He had a rather reserved and hermit-like personality with "an infinite capacity for silence".'[44] One must admire also the capacity of others to remain silent about him. Hugh Dalton on one occasion noted in his diary: 'Dine with the Mairs. I like this mythical husband of hers.'[45] Philip Mair, the son of David and Jessy, has written about his 'unworldly' father in warm yet sad words. He was, even in his own terms, one of life's failures but too decent to take his disappointment out on others. In 1933 he left the family home 'never to return, except for occasional short visits' until his death in 1942.[46]

Jessy Mair was made of altogether tougher material. When the cousins, William and David, first met she was 28 and thus three years older than William. Like her husband, she had studied mathematics, at St Andrews in her native Scotland. In 1904 she was already a mother of two young children, Lucy (born 1901), who was to become a Professor of Anthropology at LSE, and Ethel, who also became known as Marjory (born 1903). Two others were to follow: Philip in 1905, whose little memoir *Shared Enthusiasm* was intended to put the record of the relations between Lord and Lady Beveridge straight, and Betsy or Elspeth (born 1909). Small wonder then that Mrs Mair got a little worried that she might end up as a suburban housewife, or even 'a professional *Hausfrau*'.[47] But she was not made for such meekness. When William Beveridge began to confide in her whatever moved him, she took charge of him and his life, and never let go again. She soon called him 'Will', and he later adopted the 'J.' for her which had actually been 'Jye' in her Scottish family.

Beveridge the man was already the curiously contradictory figure which his biographer Jose Harris understandably found so baffling. Beveridge's mother Annette Beveridge once remarked about the frail prep schoolboy: 'Will is a little skeleton . . . looking about seventy.'[48] Jessy Mair, on the other hand, commented when in 1907 on a climbing holiday Will was mistaken for 'an irresponsible undergraduate' (and his cousin David for a minister of the Scottish Kirk): 'You know you're nothing like grown up.' (Her husband she called in the same context 'the most dogmatically constructed human being I've ever known'.[49]) This was indeed the trouble about Will, that he was 70 and 17 at the same time, but very little in between. He was an old man with a childish, or at least juvenile, streak and a boy with a great deal of the wisdom of the ages. Both parts of him meant that he needed someone to take him by the hand and guide him. For a long time his mother did just that and she continued to hold sway over him even when she was deaf and old and in need of care herself. While he was writing leaders for the *Morning Post* and working on unemployment, Mrs Rose Dunn Gardner, the philanthropic socialite (though emphatically not socialist), became his 'London mother'. But increasingly, Jessy Mair assumed this role. 'It was probably no coincidence that as the influence of Mrs Mair waxed that of Mrs Dunn Gardner waned.'[50]

Not many nice things have been said about William Beveridge the man, let alone about Jessy Mair. It is certainly easier to find venomous comments like those of Beveridge's latter-day research assistant Harold Wilson:

> I found him a devil to work for. . . . All our research work was done in an uncomfortable room we shared above the barn. Early rising was not my forte but Beveridge, after a swim in the coldest water I have ever known, kindly awakened me each morning at seven with a cup of tea. After dressing, without a swim, I put in a stint of two hours' work with him before breakfast, a formidable meal presided over by his cousin and constant companion for many years, Mrs Jessie [*sic!*] Mair. Where Beveridge was difficult to live with, she was almost impossible. I think even he was frightened of her.[51]

This was much later, to be sure, and perhaps half memory and half lore. The comment actually tells two stories; one about Beveridge and the other about his environment. Somewhere behind many scathing remarks about Beveridge there lurks a very English notion if not of niceness then of social acceptability. Keynes knew about this notion and played to it, but Beveridge, even if he knew, could not live up to it. He was not particularly eccentric nor particularly charming; he had no interesting hobbies nor the patience for much small talk; he was neither a homosexual nor a womanizer. Who or what was this man?

As so often with William Beveridge, there are two answers to this question. The first, and in historical terms more important one, is that he was a 'hedgehog' (to use Isaiah Berlin's, or rather Archilochus's language). There is, Berlin argued at the outset of his study of the arch-hedgehog Tolstoy, a 'great chasm' between those humans 'who relate everything to a single central vision', 'a single, universal, organizing principle in terms of which alone all that they are and say has significance' and those others 'who pursue many ends, often unrelated and even contradictory' whose 'thought is scattered or diffused, moving on many levels, seizing upon the essence of a vast variety of experiences and objects'. 'The first kind of intellectual and artistic personality belongs to the hedgehogs, the second to the foxes.'[52]

It is doubtful whether Beveridge was one of the great hedgehogs whom Isaiah Berlin had in mind. (He cites Pascal, Nietzsche, Ibsen, and Proust apart from Tolstoy.) However, if Beveridge was not a great man by the highest standards, he did not fall far short of the measure. He was, in his day, a 'titanic figure',[53] and some of it endured. It is no accident that thirty years after his death, on the occasion of the fiftieth anniversary of the Beveridge Report, numerous articles were written and radio and television programmes produced about him and that he was still controversial. While on several occasions in his life Beveridge was not sure what job to take next, he always knew how he wanted to go about things and what these things were. In sometimes slow and always methodical ways he pursued his objective. He probably wanted to know the causes of things though he did not much care for words like cause, let alone theory. Knowing things, facts as he liked to call them, was enough; facts about unemployment, poverty, the trade cycle which once known would translate almost automatically into influence. Beveridge was also to become the author of the principle that all citizens are entitled to freedom from want as well as the other 'dragons', like ignorance, squalor, illness. The fact that he turned the principle into a blueprint for practical policies made him the father of the modern welfare state.

What Berlin did not say in his study of Tolstoy is that on the whole foxes are more pleasant company than hedgehogs. Jessy Mair knew this and did not hesitate to tell Will so (and thereby strengthen her ascendancy over him), as in a letter of 12 August 1924:

> I have often told you you are a person of a single idea and it obsesses you to the exclusion of so many useful things. It is the reason why you have been called inhuman, and the reason why Gregory Foster and Ernest Barker fear you, but do not like you. It is the reason too why new friendships come easy while permanent ones do not—why in short you are at your best with people who only know you a little.[54]

The description of the hedgehog is nearly perfect but Will cannot have liked it. He probably was not supposed to like it because at the time J. was incensed about his companion. Ernest Barker had written to her that on board the ship which took him and Beveridge to America, Will had shown a much-talked-about interest in an aristocratic lady called Sylvia. Sylvia was then 23 years old and the daughter of the barrister and baronet Sir Richard Paget. Like Beveridge she was on her way, with her mother and family friends, to the British Association meetings at Toronto, where she actually turned up for Beveridge's paper ('It was a shock for a moment to see the Pagets'). J. wrote her full anger off her chest as soon as she heard about all this, though her letters obviously did not reach Beveridge until much later. Still she told Will that this would be 'a most unsuitable combination'. It was, of course, none of her business but nevertheless unbearable to sit at the other end of the world waiting for bad news. 'The whole of Debrett's', she wrote to him on 15 August 1924, 'won't ease the rubs of a lifelong companionship where a real basis is *not*, and it must take some very remarkable common spirit to overcome so great a difference in age and interest.'[55]

Sylvia remained a phantom and soon disappeared from their lives. Still, her brief appearance makes one wonder whether Jose Harris was altogether right to infer from Beveridge's occasional interest in other women that 'Mrs. Mair's emotional hold over him was less powerful than many people assumed'.[56] Jose Harris has compelling evidence that Beveridge was 'desperate to escape from Mrs Mair's clutches, and very anxious to find a wife' during the 1920s and 1930s, but the fact is that he failed in both. J. continued for many years to keep a hold on Will with her remarkable combination of jealousy and detachment. In 1927 she wrote to the Director at length about Eileen Power, with whom everyone had fallen in love. Did she suspect the Director? Of course, J. was anxious 'to see women getting their due'. But did Eileen Power really deserve preferment? 'Finally it is my considered conclusion that Miss Power's main interest is not in the School, which is merely for her a means to a life outside.'[57] Then, some years later, Mrs Neville Rolfe, the Secretary of the British Social Hygiene Council, appeared on the scene, 'the worst of pernicious women'. Everyone hates her and her friend Mrs Adams. 'You aren't at home in that crowd, although they'll fasten on to you like leeches.'[58] Then a younger lady came along and went away again. 'Cut adrift. There is no luxury a young person cherishes, however good and gentle she may be, as throwing crumbs to keep alive the devotion of the patient loving but rejected suitor.'[59]

Clearly J. did not want Will to fall for anyone else, but nor did she ever claim him totally. If anything she felt sorry for him who 'had never had the fulfilment of young love',[60] but there remained a distance, almost a threshold of shame which added to his unhappiness as well as hers. 'I know from you that I have nothing either to gain or to lose, and so very humbly and very gently I am offering my help. I can do only this. If you think evilly of me it is only one thing more. I wish you were happy. J.'[61] That, however, was not to be. Ever since the 33-year-old Beveridge had published his little book *John and Irene*, a collection of quotations about women with an awkward and, to his mother and friends, embarrassing introduction, it was clear for anyone to see that the man would not be happy and most certainly not with a woman.

On the core of the relationship between Will and J., Jose Harris's tactful and plausible attempt to understand answers many of the questions asked at LSE and beyond. Philip Mair, not unnaturally perhaps, did not like José Harris's answers and rallied to the defence of his mother against 'Common Room chatter'. He also, quite fairly, insisted that there is another side to many criticisms of William Beveridge the person. But on the relationship between the two, Jose Harris has shown great understanding. She refers to the obvious closeness of Will and J. and to the many hours of work as well as 'leisure' which they spent together. The quotation marks around leisure simply indicate that what 'free' time the two of them had was mostly spent if not on work then on worklike activity. In ordinary usage of the expression, they did not live together until they got married in 1942 and the contentious 'bathroom' was far too public to be a love nest. (Room 500, on the same floor as the Founders' Room, had been built, in 1928, as a restroom for senior female members of staff, but was soon appropriated by Mrs Mair. Her secretary, Dora Cleather, recalled that she liked to withdraw to it after, or even for lunch which was brought to her there. The room had a divan, and a bathroom and toilet attached.) Beatrice Webb was probably right and certainly well placed to decide that the relationship between Will and J. was 'a platonic relationship'. Indeed, Jose Harris paraphrased from Hugh Dalton's papers his belief 'that Beveridge was in any case incapable of active sexual relationships'. This was the conclusion also of a much more unlikely witness to such matters, Friedrich von Hayek: 'I personally believe that Beveridge was completely incapable of any sexuality.'[62]

There are corners in the lives of his subjects about which even a well-briefed historian can offer no more than guesses. The partnership between Will and J. is one of them. My guess is that their bond was both crucial for their survival and in no relevant sense physical. It was a symbiosis rather than an affair. One actually hesitates to use the word 'partnership'. Beatrice

Webb may have married only Sidney's head but their common *ménage* was crucially one of equals, of two heads and above all two pens. The Webb partnership was symmetrical where the Beveridge one remained distinctly asymmetrical. She was the stronger person, but she needed an object, almost an instrument to exercise her strength and Will served that purpose perfectly. He had his single-minded 'hedgehog' concerns but he needed a strong supporting hand to pursue them successfully.

After Five Years

For LSE all this was only partly good news. The two were clearly content to work together. There is not the slightest reason to doubt William Beveridge's description of the years 1919–37 'as a time of activity and happiness for Janet and myself'.[63] Such contentedness must have conveyed itself to others. Students and junior members of staff have spoken of the Director and the Secretary in glowing terms. It is not altogether clear what Harold Wilson meant when in his 1966 Beveridge Memorial Lecture he said that 'while he was the greatest administrative genius I have ever seen, he was almost certainly the worst administrator'.[64] If he meant that Beveridge was better at propounding assumptions about administration than at practising what he preached, then Jessy Mair certainly provided the other half. But J. was also snobbish and arrogant and uncomprehending when it came to the ways of academics. People minded her style as well as the way in which she stood between them and the Director. She was omnipresent. Marjorie Plant, the long-time deputy librarian of the BLPES, remembered that when the School's motto was sought, 'one former disgruntled member of the School suggested *Immer mehr*' which for him did not mean 'Ever more' but 'Always Mair'.[65] Thus in the end the partnership and Jessy Mair's role in it had much to do with the growing dislike of both, especially by senior members of the School. Beatrice Webb put it in her most vicious terms when she noted that 'Mrs Mair has become a Fury and is in control of his house and workplace'.[66]

However, that was much later in the 1930s. In the early Beveridge years the partnership was seen to work well. It even benefited the School in very practical ways (so William Beveridge thought) when an initially sceptical chairman of the University Grants Committee, Sir William McCormick, discovered that he and the Secretary of the School shared their formative University, St Andrews. 'I do not suggest that the School would not have received a capital grant without this happy accident.'[67] 'Janet and I' is a phrase which appears on many pages of Beveridge's own history of his years at the School, which he describes at the outset as 'an account of what we did together in the School of Economics'.[68] Even after J.'s death, Will found it hard to think and speak of himself without acknowledging her dominant presence.

She in turn had throughout done everything in her considerable power to impress the world with Beveridge's importance and achievements. One slightly painful testimony to this essentially generous intention is a nine-page typewritten paper now found in Beveridge's personal file at LSE which is entitled 'The School and its new Director from June 1919 to June 1925'. In reading it one has to remember that it was written by the Secretary of the School, an officer not a hagiographer or public relations consultant.

The Secretary begins by stressing 'four more or less synchronous events' of 1919: the armistice, government support for ex-servicemen wanting to study, the B.Com., and the appointment of Sir William Beveridge. She then goes on to describe the space problems of the School and the way in which they were solved by the construction of the new building. This was only the beginning, of course; now the battle was on to acquire the remaining houses in Houghton Street which with the support of the University, the LCC, and the Government the Director was about to win.

> This short history indicates some of the qualities of the Director—his vision, his courage, his devotion to the interests of the School, his untiring determination to get for it and to get for it quickly what it needs. It indicates too some of the qualities which make him so valuable on such bodies as the Senate, to which he is appointed by the Crown, at the External Council, at the Academic Council. On these bodies, the lightening [*sic*!] rapidity of his mind, his flair for selecting the important points, and his statesmanlike method of dealing with them, his power of co-operation and conciliation and his moderation and readiness to see and value the other side, all these with his ability in putting his argument make him a force to be reckoned with.

Thus Beveridge got the School recognized as one of the Big Three in London by the University Grants Committee, whose chairman, Sir William McCormick, had rightly commented that in the first Beveridge years 'the School had passed from childhood to manhood'. This passage from adolescence to manhood (Mrs Mair uses the metaphor several times) was soon recognized elsewhere as well, as in America. Here too, the 'leadership of the Director' coupled with 'his prestige as an economist' became a basis for respect and support.

Within the School the Director devoted his tireless energies 'to every side of its interests'. Staff got more money and better facilities for which the Director renounced 'large reception rooms and board rooms'. Students were given space and sports facilities as well as Advisers of Studies. The Library grew rapidly. On the social side, the Director has been 'the moving spirit'. 'No student function is complete without him, certainly no dance!'

And thus, after five years in office, 'there is indeed no part of the School life into which his personality does not penetrate and where he is not at home.'[69] Moreover, while Beveridge changed LSE in his own image, he still made 'important contributions to Economic Science' and taught social administration in lectures and seminars. 'But distinguished administrator, economist and teacher as he is, the qualities which have established him in the affection and the goodwill of his students and his staff are his kindness and his human sympathy.'

J. was clearly rather pleased with this little exercise in 'the kind of history which I prefer' (to use her own words). She sent it to the Chairman of the Governors, Sir Arthur Steel-Maitland, who was then Minister of Labour and probably had other things on his mind. In a short note on 25 May 1925 he told Mrs Mair that he had 'read through the article', 'and I am glad to think of such a good notice appearing of all that he has done'. Fortunately perhaps, only a shortened version of the notice appeared in print, in the *Clare Market Review*. In their next correspondence Will and J. are back to a more formal and appropriate style, 'My dear Secretary', 'My dear Director'.

A SHIP OVER THE HORIZON

Beardsley Ruml and the Memorial

Four years into Sir William Beveridge's directorship, the energy generated by the commerce degree was in some ways spent. The new building had gone up; new students had been recruited; the Cassel professors, readers, and lecturers had become a part of the School. What next? One answer, as always, was space. For the moment, funds to expand the new building into Clare Market were insufficient, while the battle of Houghton Street, that is for the compulsory sale of the remaining small houses on the west side of the street by its owners to the School, was still on. The other, equally if not more important answer had to do with research. The School was established now as a teaching institution in the social sciences but its research facilities were rudimentary. The Library needed expansion. Teachers needed time and space to do their research, as well as research assistance. Graduates had to have a perspective of research scholarships and junior fellowships. Funds for all this, the other half of a modern university as it were, could not be found in the City of London or even the University Grants Committee. They probably could not be found in Britain at all, where the

Oxbridge type of collegiate university was still the dominant model. The only hope lay abroad, in America, and contrary to the illusory hopes of the 'cargo cult' of anthropologists' islanders, the ship did come over the horizon and bring the wares; its name was Rockefeller.

Anthropology apart, the first real person to come across the Atlantic was the recently appointed director of the Laura Spelman Rockefeller Memorial Fund (the LSRM, or 'Memorial' for short), Beardsley Ruml. Ruml was 28 years old when he took up his appointment in 1922 ('he is not quite thirty—but he gives the appearance of being thirty-seven or thirty-eight', one of his sponsors wrote with somewhat spurious precision), and had remarkably clear ideas about what he wanted to do. In fact, he always was an 'ideas man', as Martin and Joan Bulmer have documented so well in their study of 'Beardsley Ruml and the Laura Spelman Rockefeller Memorial'. He had studied psychology at Chicago, where he encountered Professor James Angell, in whose department he held a graduate studentship. He soon followed Angell to the Carnegie Corporation and, when Angell moved on to become President of Yale, Ruml applied for the Rockefeller job. He got it and immediately made his mark by a memorandum proposing concentration of the expenditure from the Memorial on social science research in a few outstanding academic institutions. His memorandum of October 1922 showed a rare combination of insight and sense of purpose. It argued that programmes of social welfare lacked 'that knowledge which the social sciences must provide' and suggested spending some 20 million dollars over the next six years or so for what the Bulmers call 'basic but practically useful social science research'.[70]

Ruml was well aware of the pitfalls of social sciences, their diffuseness, their relatively low level of development, and last but not least their vulnerability to political attack. However, he managed to get his Foundation colleague Lawrence K. Frank to back up his analysis with a more specific report on the state of the social sciences presented in 1923. Frank linked persuasively the low quality of social science work to the 'fluidity' of its institutional base. He also helped make the case for a social science which is detached and relevant at the same time. 'The essence of the situation appears on examination to be not whether a problem is controversial, but rather whether it is studied by men of competence in a spirit of objectivity and thoroughness with freedom of inquiry and expression.'[71] This became one of the principles of the Memorial's strategy and its implementation was helped by the identification of a limited number of institutions worthy of support. During the six years of Ruml's directorship, seven universities between them received more than 50 per cent of the total $21 million for the social sciences: Chicago, Columbia, LSE, Harvard, Minnesota, Vanderbilt, Iowa State (in this sequence of order of magnitude).

Most of the funds went to American institutions. In September 1923, however, Beardsley Ruml came to London to find out about European and above all British social science. It was, then as today, the early autumn conference season so that at LSE Ruml found few teachers at home, though among them Graham Wallas. Wallas immediately wrote to Beveridge, who was then President of the Economic Science and Statistics Section of the British Association, which held its meeting that year in Liverpool (where he had one of his usual rows with Keynes about facts vs. theories). Beveridge in his turn summoned Jessy Mair from her holiday in the Lake District. Before Ruml arrived in Liverpool, he had seen Harold Laski, who had duly impressed on him that most of the Continent of Europe was an 'intellectual desert', but Britain and notably LSE was full of life. He himself (even then forever dropping names) had worked for Churchill and knew therefore 'that advancement of the social studies will have an enormous influence on political affairs'. However, more money was urgently needed; the greatest 'handicap was lack of mechanical, statistical and clerical assistance'.[72] In view of the later debate on the role of American foundations and Laski's own part in it, this early encounter between him and Ruml is worth remembering.

In Liverpool Ruml talked to J. and later to the Director. They got on and struck a relationship which was to last for the next six years. The Bulmers are if anything understating the basis of this relationship when they note: 'The affinities between Beveridge's and Ruml's views of the nature of social science are clear.'[73] Beveridge had no great difficulty persuading Ruml of the requirements of LSE as a centre of excellence in the field. There was the little issue of space, of course, and thus of funds for new buildings; there was the need for research assistance of a variety of kinds; and then there were wider, more long-term projects having to do partly with Beveridge's own research interests and beyond that with the 'natural bases of the social sciences'. From the earliest days of his relationship with the Rockefeller foundations to the somewhat bitter end, Beveridge linked his own quite personal preoccupations with those of the institutions for which he was responsible. 'I hope', he wrote in a letter accompanying the first application by LSE on 16 October 1923, 'the Trustees may be prepared to regard this as a means of relieving my mind from administrative anxieties, and enabling me, among others, to do research.'[74]

On returning home, Ruml needed some advice on the memorandum which he had brought from LSE. He asked 'A.F.', Abraham Flexner, who ran the Rockefeller General Education Board and used it to reconstruct American medical education; later Flexner wrote a comparative study of universities in America, Britain, and Germany based on his Oxford Rhodes lectures in 1928. A.F. was very positive. 'There are very, very few places where the academic and the actual come together,' he wrote on 12 November 1923. He obviously regarded LSE as one of them. 'The School is therefore serving an admirable purpose for the study of current, economic and political problems with a sufficient speculative and philosophical background.' A.F. adds an amusing note, which Harold Laski would not have liked had he seen it, though it actually applied to him as much as to others: 'The fundamentally conservative character of English radicalism is in no small measure due to the scholarship of men of this type [A.F. mentions Beveridge, Wallas, Tawney—R.D.] who are respected in all camps.' For the Trustees, this was sufficient to accept Beardsley Ruml's proposal to make a first grant to LSE. On 31 December 1923, the grant letter was sent allocating to the School $90,000 (£20,500) for 'a fluid research fund for four-and-a-half years' as well as $25,000 (£5,750) for 'additional building facilities'.[75]

'Rockefeller's Baby'

Thus began a story which became so crucial for the history of the School that Jessy Mair later described Beveridge's LSE as 'Rockefeller's baby', no less. The story had, as tends to be the case with all Beveridge's ventures, its own dramatic cadence. The first act was a love affair and lasted throughout the Ruml years at the Memorial. It reached its climax with the large grants of 1925 and 1927 which made Mrs Mair write at one point, in a 'personal' letter 'not for files' (now to be found in the files of the Rockefeller Archive): 'It is rather wonderful to feel that you are weighing our ideals and ambitions in the balance, so far away in New York, and liking our dreams and making them come true.'[76]

The interlude which followed the love affair marked the merger of the Memorial with the Rockefeller Foundation in 1929 and Beardsley Ruml's departure from this particular scene. Ruml moved for three years to the University of Chicago as Dean of Social Studies, then went into business, became chairman of Macy's, later one of Roosevelt's economic advisers, chairman of the New York Federal Reserve Bank, and a respected public figure until his death in 1960. It took Beveridge two or three years to re-establish relations with what was now a much larger and more impersonal organization, though he succeeded. A second phase in the years after 1931 led to the allocation of several hundred thousand dollars to LSE for various purposes. Then came the end of the affair; a protracted divorce foreshadowed by irritations about the Director, about Mrs Mair, about Laski, about social biology, about the Institute of Economic Research, and various other matters. One is tempted to pursue the analogy to a marriage gone stale as one reads the handwritten note by Miss Sydnor Walker, Director of the Social Science division of the Rockefeller Foundation, and generally a friend of applicants from across the Atlantic, in the margin of one of the growing number of LSE requests: 'I have rather an unpleasant impression that our English friends—the LSE, and Chatham House to somewhat less degree—put considerable pressure on us every now and then. There is a sense of "right to command" about them which I don't like.'[77]

That, however, was in 1936, at the beginning of the end of both Beveridge as Director and the Rockefeller connection. When we reach that point, we shall return to the question of what exactly the influence of the Rockefeller foundations on LSE meant. Was it pure philanthropy? Was it, as some have suggested, the establishment of capitalist cultural hegemony over a British institution? Did it help or deflect social studies? But before we get to such questions, a lot more detail about the developments of the School, and also about the LSE–Rockefeller relationship itself is in place.

Fortunately, we know a great deal about this story. At least in retrospect we do, for at the time the Rockefeller Memorial shunned publicity to the point of discouraging grant recipients from making any public announcement. Few could tell in the 1920s what exactly the Foundation was doing. On the other hand, like other American foundations the Memorial and later the Rockefeller Foundation required not only detailed applications but also regular reports. This was, and is, not always appreciated in the rest of the world, though Beveridge and his Secretary knew it and drew great benefit from the care and attention which they gave to both applications and reports. In fact, Beveridge set up a Rockefeller Committee at the School as early as 1924 and included in it some of the most distinguished professors (Wallas, Hobhouse, Sargent, Cannan). The reports to Rockefeller give at least as good a picture of developments at the School as Director's Reports at home, all the way to the final 150-page printed account prepared by Beveridge's successor in February 1938 and entitled *Review of the Activities and Development of the London School of Economics and Political Science during the Period 1923–1937*.

Such formal reports, however informative they are, come to life only by their penumbra of informal letters, office notes, records of conversations. These confirm the familiar experience that all of us find it easier to talk when we are travelling and to strangers who are not part of our normal lives. Thus one internal memo (found by Martin Bulmer in the Rockefeller Archive) refers to the Director tongue in cheek as 'Beverage'. More surprisingly, the Rockefeller representative even found the Director disagreeing profoundly with his Secretary, at least 'when she had left the room'.[78]

Rockefeller money came to the School in three major spurts: in 1923/4, in 1927/8, and in 1932/3. By 1937 it added up to 2 million dollars, or nearly half a million pounds sterling. The School's total annual revenue expenditure during the whole of these fourteen years amounted to just under £1.5 million. Of this total, £168,000 or over 11 per cent was met by Rockefeller subsidies. On the capital side, the picture is even more dramatic; £317,000 of the total capital receipts of £439,000 in the years 1923–37 came from Rockefeller. If one adds revenue and capital for this period, the School found nearly one-quarter of its total income from the Memorial and the Rockefeller Foundation. This makes Jessy Mair's description of LSE as 'Rockefeller's baby' perhaps a little less 'effulgent' (as the recipient of her letter of 26 May 1936 described her style). It even makes her description of the School in 1923 sound like an understatement: 'The infant, although fairly husky, was still in need of Rockefeller nourishment.'[79]

The first instalment of this nourishment had, as we have seen, two ingredients. The fluid research fund of 1924 was extended first by a year and then by another four years in 1928 and 1929 and became invaluable for limited but essential research-related expenditure. It was just the kind of free money which academic institutions need to enable their teachers to visit archives, attend conferences, purchase materials, hire assistance, even buy time for research by paying for replacement teaching. The first building subsidy contributed to the funds already available for the Library extension to Clare Market, the so-called Cobden Library Wing.

Almost as soon as the first allocation was made, the Director, the Secretary, and the Rockefeller Committee got to work on a second, much more ambitious application. This resulted in a Memorandum completed in July 1925 which provided the basis for further applications to the Memorial during the remaining years of its existence. The Memorandum was a curious document. Its main part made, for the first time, the case for completing 'the circle of the social sciences' by adding to the two existing groups of studies—economics, and political science—a third group 'dealing with the natural bases of economics and politics, with the human material and with its physical environment, and forming a bridge between the natural and the social sciences'.[80] This was to become Beveridge's obsession and also a contributing cause for his downfall. Contrary to what all accounts by the Director and his Secretary tell us, the proposal did not meet with immediate, and it never met with enthusiastic, support in New York. The detailed story will have to be told.

But the 1925 Memorandum had a second section dealing with what it called 'urgent needs', as if to say: first we tell you what we regard as most important, and then what we really need. The 'prior needs' at this stage were three: a Chair in Political Economy after the retirement of Cannan; a subject catalogue for the Library (as well as other library support); and funds to acquire the remaining eight houses on Houghton Street. For these, the battle was now really waging. The St Clement's Press, which owned the corner building of Houghton Street and Clare Market, had already bought one of the remaining eight houses with a view to expanding its printing operation. Beveridge had to mobilize all his friends and supporters to prevent this from happening. The LCC included in its 1924/5 General Powers Bill a clause enabling the compulsory acquisition of the houses by LSE. This needed support from the University Senate, the Board of Education, and even Parliament. The Director got all these and more; in the end the owner of the St Clement's Press not only let the School buy the corner building as well but gave it the money to do so.

Beardsley Ruml accepted the 'prior needs' and obliged. On 29 December 1925 three grants were made: $100,000 for the building extension, $40,000 for Library purposes, and $5,000 for each of three years for the Chair in Political Economy, which after an unhappy interval caused by the sudden death of its first incumbent, Allyn Young of Harvard, was to become Lionel Robbins's chair. But Beveridge did not let up. Moreover he now had a hidden second agenda. As a member of the London University Senate, Beveridge had taken a keen interest in acquiring the Duke of Bedford's estate in Bloomsbury in order to move the University offices from South Kensington and create the beginnings of a *cité universitaire*. 'The whole region of Bloomsbury, with the colleges near the Strand, might become a possible academic quarter in London.'[81] In June 1926 Beveridge was elected Vice-Chancellor, which added urgency to his interest. But while much of the funding would come from the Treasury there was no chance of completing the deal without considerable help from New York.

In the course of 1926, the 'natural bases of the social sciences' underwent certain mutations, as once again prior needs were discovered. This time they had to do with international studies. During the Christmas break 1926, Beveridge sailed to the United States to tie up the package. On the way home in January 1927, on board the *Aquitania*, William Beveridge 'reported as usual to [his] mother' on the accomplishments of the journey:

I got aboard here about midnight and we sailed at 4 a.m. There was rather confusion coming aboard, as the single cabin which I had been promised had been occupied meanwhile by a lady, so I have to share with a man from New Zealand. It's a bit cramped but nothing much to complain of.

My coming aboard was cheered by a letter from the Laura Spelman Rockefeller Memorial saying that they had definitely voted the money for which I had asked, viz.:

$700,000, i.e., £145,000 endowment for teaching and research in Social Sciences and International Studies.

$175,000, i.e., £35,000 [corrected in a footnote to £36,250] for building a library development.

This is £180,000 altogether and makes us very rich. I've also a letter in my pocket from Mr Rockefeller which means ultimately saving the whole Bloomsbury site—but this is very secret.

I've also got Professor Allyn Young, and I've prepared the way for Phil[ip Mair] to go to America and earn his living there for a year at least before he settles down in England. So practically I've succeeded in all my objectives.[82]

Which was not to say, to be sure, that Beveridge was now content, especially so far as his growing obsession with the 'natural bases of the social sciences'—which had meanwhile materialized into the desire to find funds for a Chair in Social Biology—was concerned.

From Memorial to Foundation

Spending the Rockefeller money and spending it wisely did not present many problems. On the physical side, the building complex now known as the Old Building emerged in its present shape in the years 1926–9 and 1931–3. This meant that the Library could expand considerably; further lecture halls and rooms for staff were added, and on the new top floor of the building the Founders' Room emerged which gave the Director and his Secretary much aesthetic pleasure and many later generations opportunities for dances and receptions, chamber music, and quiet reading.

Academically, the most important development was the creation of a Department of International Studies and of three chairs in the field; in international law, international history, and international relations. In all these cases, Rockefeller money helped though other funds (Carnegie, Cassel, Montagu Burton) were added. The Chair in Political Economy has already been mentioned. The first major step in 'social biology' turned out to be Professor Malinowski's Chair in Social Anthropology. A New ('Booth') Survey of London Life and Labour was financed. Throughout, of course, there was the fluid and flexible research fund.

Still, Beveridge was not satisfied. He used the interlude during which the

Memorial was merged with the older Rockefeller Foundation to establish relations with the head of the Social Science section (first Dr E. E. Day and later Miss Sydnor Walker) and the European Representatives, located in Paris (Mr Tracy Kittredge and Mr Selskar M. Gunn). Conversations in the winter of 1930/1 concentrate once again on buildings and Library needs. Expansion now turns to plans to purchase the last remaining non-LSE building on Portugal Street, the Smith Memorial building, and the east side of Houghton Street where the St Clement Danes Holborn Estate Grammar School was about to move to Hammersmith and other houses became available. In addition the Library itself needed support. And then there were the academic needs, social biology included.

The Rockefeller Foundation office note of March 1931 on these matters is quite detailed and remarkably friendly.[83] The note was drafted by E. E. Day and his colleagues in the Social Science division. It begins by recounting the story of support for LSE from the Memorial Fund, which is described as an unqualified success, partly because LSE has been able to attract significant support from other sources. 'There is nothing comparable in the field of Social Sciences to the London School, either in England or on the Continent.' The Rockefeller Foundation should therefore remain involved. When the officers describe the purchase of land and the construction of buildings as 'vitally important' for LSE, one can almost hear the Director speaking. The Library is also appreciated. 'In an institution such as the London School of Economics, the library takes the place of the laboratories of institutes devoted to the natural and medical sciences.' Then there is the academic side. The Rockefeller office note never even mentions social biology but accepts the notion of a seven-year grant to strengthen 'the staff of the School with a view to improving facilities for research and postgraduate teaching'.

The School needed $1,750,000 and it was proposed that the Rockefeller Foundation should contribute $710,000. The Trustees were impressed but not exactly bowled over. Europeans have an inclination to turn to American foundations for core funding of their activities without ever thinking of the interests of Americans on the funding bodies. Why should they look after Europe's needs? Why don't the Europeans do it themselves? Judging from the minute of their meeting, this is exactly what made the Rockefeller Trustees wonder and Dr Day had a hard time defending his proposal. In the end he won by using a strange though not altogether absurd argument. Day, the minute records, 'considers the School outstanding in the world. Being outside British tradition, it cannot be expected to obtain a large amount of national support.' One wonders what alternative tradition he had in mind. America? Germany? Or was the School about to set its own tradition, perhaps?

In any case, the School was deeply grateful for the support. Once again, on 9 May 1931, Jessy Mair wrote 'not officially, but just on my own account': 'I feel a personal sense of great friendliness to exist on both sides; a very delightful thing in this work-a-day world.'[84] She was right, the feeling was mutual at that time. Every now and again, Rockefeller people had to defend the School against attacks. In 1928 Laski had written an article in *Harper's Magazine* which annoyed people. The Foundation President Raymond Fosdick consulted his fellow Trustee Dr Wickliff Rose and then wrote to Ruml: 'If a physicist should show the same degree of carelessness as to the facts on which he bases his generalizations,' he 'would not be able to find a publisher. In the field of social relations it seems so easy for two or three swallows to make a summer.'[85] The announcement of the 1931 grant led to a long letter from a gentleman in Bexhill-on-Sea suggesting that 'revolutionaries of every colour are being bred by the thousand at the London School of Economics'.[86] More significantly, the head of Rockefeller's European Office Selskar Gunn wrote to his head office in November 1931 about the 'political flavour' which certain members of the faculty have given LSE.

Dalton and Leigh [i.e., Lees-]Smith, who were defeated in the elections, have returned to the School, and Laski has stirred up a lot of antagonism to the School through his pre-election journalistic articles in the 'Daily Herald'. Mrs Mair informed me confidentially that some of the members of the Board of Trustees of the School are feeling pretty sore about the whole matter, and that it is coming up for discussion at the meeting of the Board.[87]

But the head office, in the person of Dr Day, was not yet prepared to take such attacks very seriously. 'It is easy to exaggerate the partisan character of a school which is sufficiently close to practical affairs so that members of its staff are called from time to time into government service. The "outs" will not hesitate to charge the school under these circumstances with political bias.'[88] In fact, and remarkably, the Rockefeller Foundation was never really worried about the politics, alleged or real, of LSE. Its relaxed attitude both to the involvement of staff and to the public fuss about it showed that it really was outside the British tradition. On the other hand, the Rockefeller Foundation worried a great deal when Malinowski showed himself 'obviously antagonistic to the development of a social biology Department', which was after all the Director's pet project.[89] The Foundation got truly upset when they gained the impression that the academic grant was used merely to support staff whereas the 'and postgraduate teaching' in the allocation letter was conveniently forgotten. But by the time these subjects came to the boil, many other things had gone sour. Their anticipation should not be allowed to spoil the much sweeter taste of the early Beveridge years.

THE SCHOOL AT 35

A Rabbit Warren of Buildings

When LSE was as old as its founder had been at the time of its foundation, it was in fine fettle. Some may regard the title of this chapter, 'An Academic Miracle', as slightly over the top, especially since the miracle was as much physical and financial as academic, but if one looks at the balance sheet of developments between 1919 and 1929 or 1930, the results of change are little short of miraculous. Moreover, these improvements happened before a background of economic and political confusion and, for many, misery. The First World War had left a trail of unemployment and poverty; the mid-1920s boom remained short and was punctuated, even punctured, by the General Strike. Soon the great depression took its grip on the country. The five elections between 1918 and 1931 mark a profound transformation of the party system with the as yet uncertain outcome of the National Government. Yet if one plotted any indicator of the development of LSE during this period, the curve would be almost a line and it would point upward throughout. Despite Attlee and Dalton, Laski and Lees-Smith, and cabinet offices for two Chairmen, Sidney Webb (Lord Passfield) and Sir Arthur Steel-Maitland, it is probably true to say that in no other period of its history was the School as detached from the political world around it as in the 1920s. It was in that sense an academic decade. Thus when the Director gave his annual report to the School on Oration Day in June 1932 he was able to say: 'The economic and political crisis of the world has made till now little difference to the School of Economics and Political Science. The crisis has occupied our minds and darkened our thoughts, as it must darken the thoughts of all who are not frivolous. It has not curtailed or deflected our activities.'

However, 1932 was also the last year in which Beveridge could sound such a confident note without provoking incredulity. The time around the end of the first ten or twelve years of his directorship provides therefore a welcome opportunity for taking stock. In keeping with the 'hammering sound' of that period, where else could such stocktaking begin but with the physical site? It had changed out of recognition since the day on which the heavy, constraining presence of the Passmore Edwards Hall had transformed Clare Market in 1902. But it had also changed in a characteristic manner. The School did not, either then or later, gain an outlet to one of the great thoroughfares around it, Kingsway, the Aldwych, or even to Lincoln's Inn Fields. It remained on the contrary hemmed in and thus physically, as well as in an almost psychological sense, cramped, quite unlike Bush House, or the Royal College of Surgeons, let alone the Law Courts at the end of Clement's Inn Passage. It is thus perhaps understandable that successive Directors tried to find other sites for the School, beginning with Beveridge, who wanted to move it to his never-to-be dream London Left Bank of Bloomsbury; but one can also understand those who were not totally displeased when such plans came to nothing. As the LCC indicated when it helped LSE to expand within, as it were, in 1919, the School belonged where it was, and is, in a congested location which forced it to look for its strength within rather than in another Senate House.

In physical terms the opportunities for expanding within were strictly limited, so that it is not surprising that every step required strenuous new efforts of gaining financial as well as public support. In Beveridge's day, the School first began to take over the west side of Houghton Street (as a result of the 'battle'), later expanded on Clare Market (notably by including the Smith Memorial Hall), then added floors to the existing buildings (such as the Founders' Room), and finally moved across Houghton Street to its east side, where the East Wing went up. After Beveridge an almost twenty-year interval made many people forget the hammering sound, though it came back with a vengeance when the other side of Clare Market was acquired from the St Clement's Press, and later two tall extensions were added on Houghton Street and on Clare Market—the Clare Market and St Clement's Buildings.

The result is not just an architectural mess but a mess in almost every conceivable sense; it has often been described as a rabbit warren. The bridges between buildings connect floors of different levels. While it is possible to walk from one end of LSE to the other—say, from the New Theatre to the Founders' Room and then the Geography Department—without getting wet, most people would prefer a drop of rain to the labyrinth of staircases, in which it is easy to get lost. Some, of course, indeed some distinguished professors, keep their rooms in mysterious corners in which they are left undisturbed because nobody can find them. One room on the east side of Houghton Street could be entered, so people remember, through the roof by lifting a skylight and descending a ladder. There are probably students who have never managed to discover the Graham Wallas Room or the nooks and crannies of the old Library. Yet it remains true that all this is a mess only in *almost* every conceivable sense: it is also LSE as those who love it have always known it, a mysterious mix of aesthetic offensiveness and unending opportunities of discovery.

At the end of Beveridge's first decade, this was not yet quite so true, though even then an intriguing game could have been played of finding the traces of the Passmore Edwards Hall within the new structures. It was true, however, that buildings of a remarkably heavy and undistinguished architectural design had been put up. The East Wing on Houghton Street is perhaps the most extreme example. Judging from the thickness of the walls one is surprised that there is any space inside for rooms. Curiously the later additions, notably those of the 1950s, were as flimsy as the earlier substance was weighty. And of course all this left little open space. The inner courtyard behind the East Wing and the roof of the middle part of the Old Building are about the only opportunities within the confines of the School to breathe air albeit of dubious quality, though the closure of Houghton Street to traffic in 1975 improved things a little.

Some people find it easy to make architects' drawings come to life, others prefer visual images. The stylized picture (preceding p. 429) is an attempt to satisfy the latter if only because they include the author. The picture allows identification of the confines of the School as it was in the 1930s. This was probably the only time in its history when LSE felt fairly comfortable within its constrained space. Developments of student numbers meant that while some lectures were very large and had to be held in the Old Theatre (as it is now called) there was also plenty of space for classes and seminars, and even for recreational purposes. All full-time members of staff as well as some part-time ones now had a room to themselves. The administration occupied a recognizable area. Above all, the Library was at last coming into its own in terms of both reader places and space for the collections.

The BLPES

'Coming into its own' is a deliberate choice of words. Many have described the early School as a library which offers a little space to the rest. In 1925 Beveridge reminded his audience on Oration Day that 'the School at one time in its history was little more than an annexe to the "British Library of Political Science"'. This dates back to the Adelphi and Charlotte Payne-Townshend's mansion, though it was never wholly true. Above all, the Library itself did not have enough space. But with the help of the Rockefeller Foundation and other sources of funding which Sir William Beveridge was able to tap, it gradually became that 'laboratory' of the social sciences which it has remained to the present day. As the Library came into its own it also moved if anything closer to the School. In 1925 the Library Trustees delegated the management of the Library to the School and decided to give it what Beveridge called the 'wider title' of British Library of Political and Economic Science (BLPES).[90]

In its first hundred years, the Library had seven Librarians. The first, Miss Mactaggart's Australian cousin and sparring partner, John McKillop, combined the librarianship with the administration of the School. The second Librarian, B. M. Headicar, appointed in 1910, and in post until 1934, first saw the Library through difficult war years and then presided over its expansion. When his successor was about to be appointed at the end of 1933, the Rockefeller Foundation flexed its muscle, or that of the librarians' profession, which wanted to keep librarianships 'within the trade', and asked the Director what the qualifications of the preferred candidate were. Beveridge replied that Dr W. C. Dickinson had been an able administrator and was also a scholar specializing in Scottish history. 'The School authorities took the view that the post was of such importance that personality and administrative ability were more significant than any special library experience.'[91] Dr Dickinson held the post until 1944, when Geoffrey Woledge succeeded him for twenty-two years. In 1966 Woledge was followed by Derek Clarke, who in turn was succeeded by Chris Hunt in 1985 and by Lynne Brindley in 1992.

The history of the BLPES has yet to be written. Judging by the foretaste provided in a small batch of articles by librarians and the brief historical sketch published by Professor Arthur John in 1971,[92] there is a story to tell. In part it is a story of expansion. Then there is the question of the relationship with the School, never free of tension, determined by the temperament of the main actors as much as the legitimate insistence of the BLPES on its national role. Also, one must wonder whether the analogy to science laboratories suggested by the Rockefeller Foundation is really as apt as it sounds at first. Not all social sciences are bookish, especially not in more recent decades. The Library itself was, of course, at no time wholly 'bookish'. By 1970, when its move to its present abode to what was then Strand House was first considered, the Library had 575,000 bound volumes but over two million 'separate items'. These included the reports and papers which Sidney Webb had from the outset regarded as the raw material of social science. By the 1990s the entire collection of the Library required over 40 kilometres of shelving.

'This growth of the Library's resources has consistently outstripped its capacity to house them adequately.'[93] In the Passmore Edwards Hall a reading-room with forty-eight places had to suffice and books were stored in the gallery. After the completion of the first new building in 1922, three small reading-rooms as well as some space for specialized collections were added. In 1925 the Cobden Wing was built on to the Passmore Edwards Hall, and the cellars underneath were fitted out for reserve stacks. By the early 1930s the Library had got the shape which it kept until the move to the Lionel Robbins Building (as Strand House was renamed) in 1978. The catalogue had been turned into the basis for the London Bibliography of the Social Sciences. Books were stored, many on open access, in special collections as well as a teaching library. There were 550 reader places which by minor extensions were later increased to over 1,000. The British Library of Political and Economic Science had thus become a unique research facility, attracting scholars from all over the world in its own right, and adding greatly to the attractiveness of the School.

A Unique Mix of Students

One reason why the School was less cramped in the early 1930s than at any other time in its history had to do with student numbers. The only line of statistics which while fairly straight did not point upwards during the Beveridge years is that of the overall total of students. In almost every single year of the eighteen years of the Director's reign, the School had just under 3,000 students on its register. There was no plan, let alone *numerus clausus* behind this fact; it just so happened that numbers in various categories added up to about 3,000 each year. What did change were the constituent parts of this total and their change meant that students were distributed more evenly over the facilities of the School and also over the hours of the day.

The fact has been mentioned already that in the years from 1919 to 1937 the relationship between regular and occasional students was reversed from 1 : 2 to 2 : 1. The session 1930/1 marks the midway point in this process when (by a statistical fluke) there were 1,233 regular and 1,233 occasional students to which 471 intercollegiate students have to be added to make up the total of 2,937. Ten years earlier, there had been 978 regular and 1,750 occasional (as well as 173 intercollegiate) students. The process was dramatic and it had other important aspects. The number of higher-degree students had risen from forty-seven in 1920/1 to 190 in 1930/1. It was to rise further to 293 by the time Beveridge left in 1937, whereas from about 1930 onwards first-degree numbers remained roughly stable. Among the first-degree students, those aiming at the B.Com. began to decline in 1926/7, whereas candidates for the B.Sc.(Econ.) continued to increase. The story told by these figures is one of a School which is steadily moving from a multi-purpose institution of higher learning with a strong applied ('business school') element to a university in the prevailing sense of the term.

This is to be sure the truth, but not the whole truth. Before the necessary qualifications are added, one or two other facts are of interest. The proportion of women at LSE was always relatively high—relative, that is, to other universities in the country—but it differed in different student categories. Throughout the Beveridge years, women did not account for more than 25 per cent of all students, though their proportion was significantly higher among regular (about 30 per cent) than among occasional students (15 to 18 per cent), and the difference between the number of women on day and on evening courses was even greater. In fact, more than 40 per cent of all regular day students were women. It is worth noting that among students of the Open University in the 1960s and 1970s the reverse was true, and women dominated the scene. The underlying social changes are evident. If women had jobs at all in the 1920s and 1930s, they were obviously unlikely to go in for further educational qualifications in the evenings.

The number of overseas students at the School was always high; it rarely fell below 20 per cent of the total student body. More surprisingly, during the Beveridge years a majority of these were occasional students. The countries of origin of overseas students generally covered the globe, as they still do at LSE, though during the 1920s and 1930s relatively few came from either Australia and New Zealand or Central and South America. In most years, these two regions of the southern hemisphere together sent fewer students to LSE than Africa, from which about thirty came at any time. This leaves Europe, Asia, and North America. If one were to relate student numbers to the population of countries of origin, Switzerland and Palestine would probably come out on top. In terms of total numbers, India clearly leads the field, followed by China, the United States of America, and Germany. There were also significant numbers of Canadians, Scandinavians, Poles, and South Africans. The analysis of research students produced for the Rockefeller Foundation in 1938 shows China, the United States, India, and Germany ahead of other countries of origin.[94]

What were these students actually doing at LSE? So far as the regular undergraduates and higher-degree students are concerned, the answer is not too difficult. If they were undergraduates, they went to lectures and classes, saw their advisers, wrote essays, and prepared for examinations. The characteristic and perhaps unique feature of LSE was that most lectures and classes were repeated between 5 p.m. and 9 p.m. in the evening. Day students were not supposed to attend the evening classes, and most of them did not. The proportion of regular evening students declined slightly during the Beveridge years from just under 45 per cent to just under 40 per cent but it remained high. Intercollegiate students, mostly in law or geography courses shared with King's College and University College, show a similar pattern.

On the other hand, occasional students differed in many respects from those working for regular degrees. Some were on special courses, like the railway students; 900 of them when Beveridge came though only 300 when he left. The relationship with the railway companies was never wholly repaired; when Sir Arthur Steel-Maitland tried early in his chairmanship to get Lord Claud Hamilton back on board, he failed. Moreover, the role of the railways began to change, and a wider subject of transport studies emerged. Specialized courses were also offered for Treasury officials, LCC teachers and, of course, from 1925 to 1931, the Army Class. Other opportunities for occasional students came and went, such as those for the consular service, for colonial administration, and others which usually betray the influence of a particular Governor and not infrequently that of Sidney Webb. In the early 1920s, 80 per cent of all occasional students attended evening courses, though by the late 1930s this proportion had declined to nearer 60 per cent.

The 'decasualization' of the student population was accompanied and in part made possible by corresponding changes in the teaching faculty.[95] When Beveridge arrived in 1919, only seventeen teachers were technically full-time lecturers, and forty-two worked for the School part-time. Ten years later, the number of full-time teachers had trebled, and by the time Beveridge left it had risen to seventy-nine. The number of part-time teachers rose in certain years, notably between 1929 and 1932, but then declined again to forty-four in 1936/7. Increases in the numbers of full-time teachers in economics (eight in 1923/4, twenty-one in 1936/7) and in political science (from three to seven) were naturally significant, but sociology and anthropology also grew from five to thirteen teachers during the period, and the Law Department emerged from almost nowhere (or rather, one professor) in the early 1920s to four professors, three readers, one lecturer, and two assistants in 1936/7.[96] All this, one remembers, occurred while overall student numbers remained stable at about 3,000, so that the improvement still left the School with an almost unbelievably unfavourable staff–student ratio. This explains both the amount of teaching done by individual faculty members and its method, which was predominantly one of lecturing. Still, improvement there was, from about 1 : 50 to about 1 : 25, and along with other developments this shows that in academic terms the

Beveridge years brought primarily intensive rather than extensive change, the creation of a modern research university of the social sciences.

Financing Improvements

The process cost money, a lot of money, and this is where the Director's success was most spectacular.[97] When he came, the School had an income of just under £25,000 (1918/19); when he left, this had risen to over £135,000 (1936/7). In assessing these figures it is important to remember that during the period in question nominal figures do tell an important story because inflation had not yet become a normal part of life, or of budgeting and accounting practice. If anything, deflation was the order of the day during the 1920s and 1930s. Compared to 1920, when the First World War had driven the cost of living index up to 250 (from 100 in 1913), prices declined to 175 in 1925, 158 in 1930, and 144 in 1935. At the time, there was no 'Tress–Brown' or other index of university prices, and Beveridge made sure that academic salaries (as well as pensions) increased so that some cost inflation took place. Nevertheless, it is safe to assume that the funds brought in were if anything worth even more than their face value and therefore underlined the counter-cyclical development of the School. LSE income had doubled by Beveridge's second year. Three years later, in 1923, it began its steady and rapid climb from £50,000 to over £130,000, a point that was reached in 1932 and then held for the remaining Beveridge years.

Steady student numbers and increasing income clearly meant improvement; Beveridge's first ten years were an unusual period in these terms (see Graph, p. 177). It is worth looking in somewhat greater detail at both the income and the expenditure side of the revenue account. Professor Harold Edey, the distinguished accountancy scholar (and first Pro-Director of the School at the critical time from 1967 to 1970) has helped to put together the relevant data from a variety of sources and thus made it possible to draw a few significant conclusions.[98] LSE at 5 years of age was still a precarious institution, with its income of about £2,800 heavily dependent on fortuitous donations (35 per cent) and the LCC grant (42 per cent). LSE at 10 had a growing fee income (30 per cent) and an increasingly significant grant not just from the LCC but also, through the University of London, from the Treasury (55 per cent) to make up the total of £7,600. LSE at 15 had a total income of £12,000, which after 1912 rose to £17,000 thanks to a larger Treasury allocation; of the total, about 60 per cent came from grants from public bodies and not much less than 40 per cent from fees, so that casual donations and other sources of income had dwindled in

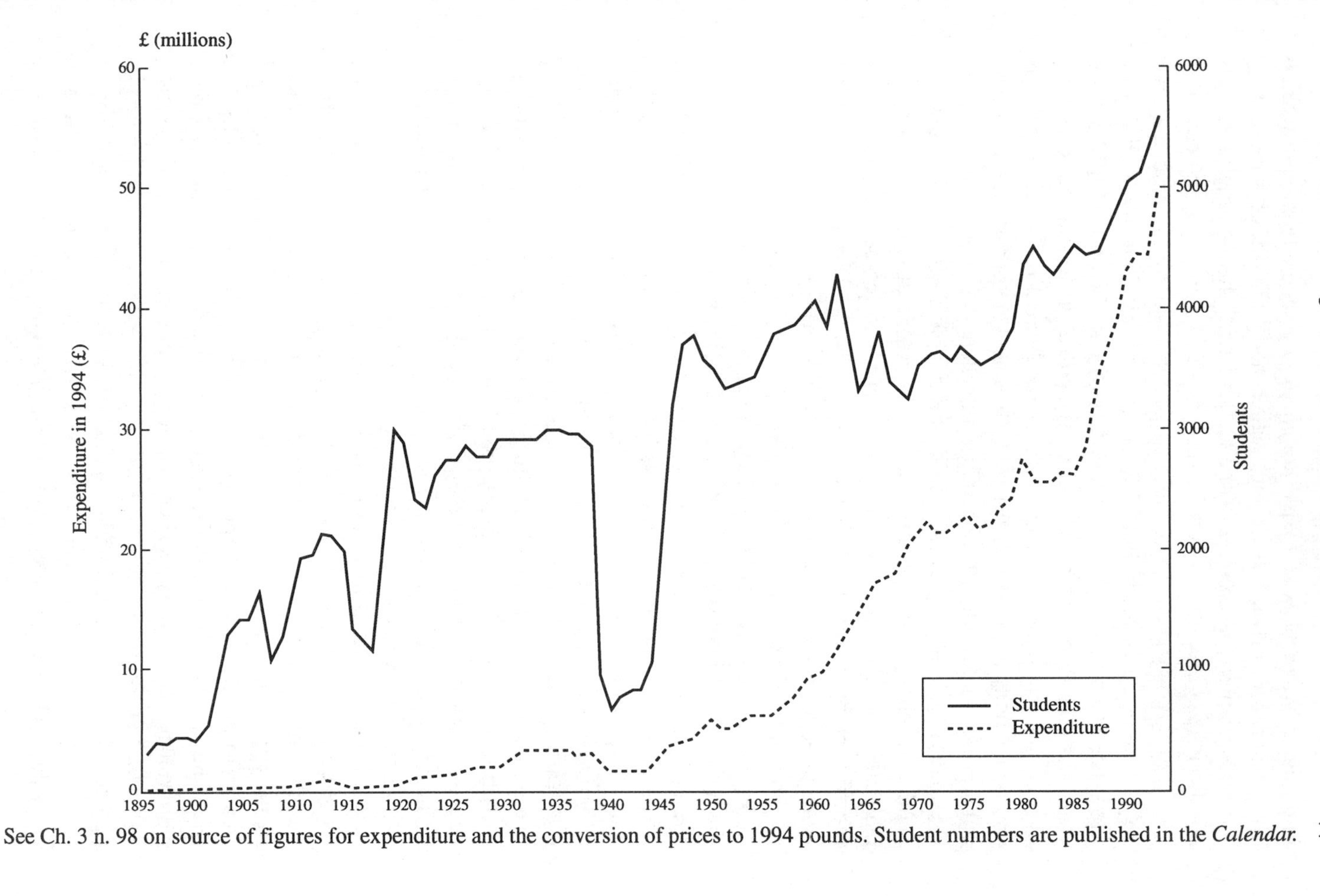

See Ch. 3 n. 98 on source of figures for expenditure and the conversion of prices to 1994 pounds. Student numbers are published in the *Calendar*.

importance. After the war the picture changed. Income from endowments and grants from private trusts, etc., which had been about 1 per cent in 1913, rose rapidly on account of the Cassel Fund, the Commerce Appeal, and from 1924 onwards Rockefeller money. It was just over 20 per cent of a total of £64,000 when the School was 30 years old, and remained at over 20 per cent of a growing income. LSE at 35 had a total income of just under £100,000, which rose rapidly to about £135,000 in the remaining Beveridge years. Its component parts, however, remained fairly stable: apart from over 20 per cent endowment income, fees accounted for some 30 per cent and public body grants for 40 per cent. Income from occasional donations, publications, bank interest, and other miscellaneous sources made up the remainder. This may be called a fairly healthy structure which is not vulnerable to changes in any one factor, though it also demonstrates the dependence of the School on its attractiveness to fee-paying students, its recognition by grant-giving bodies and notably the University Grants Committee as well as the Court of London University, and the providers of endowment funds, such as the Rockefeller foundations.

On the expenditure side, the changes during the first thirty-five years of the life of the School were equally dramatic. In its early years the School spent an inordinate proportion of its income on administration and premises (over 40 per cent). This was inevitable, as anyone who has set up a new institution knows. By 1905 a new expenditure pattern was emerging, although expenditure both for premises and their maintenance varied during years of exceptional expansion. Generally speaking, however, about two-thirds of the School's expenditure went on academic, or at any rate departmental expenditure after 1905. Administrative cost fluctuated between 9 per cent and 12 per cent during the Beveridge years after 1920. Premises accounted for around 12 per cent of total revenue expenditure in every one of the eighteen years of his directorship.

What was, in effect, the School's capital account might as well be called the Rockefeller account, at any rate after 1923. Apart from new buildings, the growing proportion of endowment income reflects the large additions to the capital resources, notably in 1927/8, the year for which the 1938 Report to the Rockefeller Foundation notes an appropriation of £144,000 from the Laura Spelman Rockefeller Memorial for a permanent general endowment. The term 'general endowment' is in some respects misleading; £41,000 of the donation, generous as it was, was also earmarked for certain academic purposes. The point is significant. LSE has never had a general endowment which would guarantee its survival in an extended period of rainy days. Unlike Oxford colleges, or even the University of Oxford—and unlike American universities with which the School might well be compared in other respects—LSE was, and is, dependent on recurrent income essentially from two sources: fees and grants. What expansion there was occurred because special support was found from Charlotte Payne-Townshend for the early Library, from the Cassel Fund and the special appeal for the commerce degree, from a variety of sources for scholarships, from Martin White and Montagu Burton and others for chairs, from the Rockefeller foundations, from public appeals for a multiplicity of purposes. LSE certainly had, over time, its own distinguished court of benefactors. But it never became a financially independent institution. Without the support of its students and the public purse in whatever guise, it could not survive five or even three years.

An Unwritten Constitution

This is a fundamental fact of the life of LSE which has much to do with its recurrent problem, the image of the School in the outside world. LSE had to be acceptable to survive. This was certainly true before the First World War and probably in the years immediately afterwards. Whether it was still true when Beveridge began to go to great and for many unacceptable lengths in asserting the political neutrality of LSE may be open to doubt. Sidney Webb was at all times aware of this fundamental vulnerability of his creation and Beveridge learnt about it soon. Both therefore spent much thought and time on an administrative structure which would combine the needs of external recognition and internal legitimacy. In following the history of the School, we have from time to time alluded to attempts to achieve these purposes, including the early innovations introduced by Beveridge. For those familiar with either the Oxbridge form of collegiate self-government or the civic university structures of autonomy tempered by advice from a council of community representatives, the governance of LSE must however, seem bewildering. In fact it is bewildering and any attempt to describe it systematically soon runs into difficulties.

LSE has no constitution. Its Memorandum and Articles of Association mention the Governors, of course, but say little about their mode of operation except that there shall be an Annual General Meeting (the first within four months of incorporation, 'at such place and time as may be determined by the aforesaid Sidney James Webb') and that the Governors may form committees of not less than three members for certain purposes. Beyond the Articles, the only relevant document is a loose-leaf Staff Manual of uncertain status which describes the committee structure of the School as well as the opening hours of the Library and the pension entitlements of staff. Directors with particularly tidy minds, notably William Beveridge and Sydney Caine, have tried to rationalize the School's administrative structures, but when they discovered that they could not get very far in their attempts, they withdrew to minor adjustments and beyond those the attempt to explain the rationale of what existed already. Even this was not easy, as the only scholar to do systematic research on the School's administration also found when he produced a manuscript which he sent to the new Director on his appointment in 1974 rather than publishing it.[99]

The basic formal fact about the constitution of LSE is that ultimately all power resides in its Court of Governors. The basic substantive fact is that this formal principle is both impractical and unacceptable in terms of academic autonomy. At least these two conclusions are inescapable once the Governors are no longer identical with 'the aforesaid Sidney James Webb', that is since 1911. The formal arrangements are impractical because a body of over fifty, later eighty, and now a hundred members, mostly busy professional people who meet as a Court twice, later three times a year, cannot possibly be the effective government of an institution. The arrangements are unacceptable because they establish a kind of colonial rule over an essentially collegiate institution, the likes of which have over the centuries been jealous of their autonomy as a precondition of academic freedom.

Given the evident conflict between formal arrangements and substantive impossibilities, the locus of power was bound to migrate somewhere other than the full Court of Governors. Again, for a considerable period this need defined the *de facto* role which Sidney Webb played in the School and for a while at least he probably relished it. But such monarchy could not and did not last. The next step was that very considerable powers were delegated to the Director. Beveridge insists (in his 1935 'Reflections on the "School of Economics"', which are almost entirely about administrative arrangements): 'I was not given, when I came, any definition of my powers or duties.' However, when the University of London—like others somewhat puzzled about the School's governance—sent Inspectors to have a look at the arrangements, Beveridge himself wrote to the Senate 'on behalf of the Court of Governors' and produced what may be regarded as a valid constitutional interpretation of lasting significance:

The supreme formal authority in the School as a Company limited by guarantee is, subject to the limits imposed by the Memorandum of Association and by the relation of the School to the University, the Court of Governors of the School, each of whom is also a member of the Company. This authority has, in fact, in practice been delegated by the Governors to the Director for the time being, to whose discretion the Governors have largely left the decision as to what matters he should himself decide and what he should reserve for the Governors. The Governors are satisfied that only by this large measure of delegation and by placing great trust in the successive Directors, could the School have developed as it should do. But it has always been recognised both by the Court of Governors and by successive Directors that anything in the nature of an autocracy is incompatible with the character of a great University institution. In fact, the systematic consultation of the teaching staff has been steadily developed under the present Directorship, the activity of the Professorial Council has been increased, and its practical influence extended through the establishment of the Appointments Committee. The constitution of the Council and the position of the Professors and senior teachers has been fully discussed in the Second Report of the Professorial Council enclosed herewith, and the proposals made in that Report have been approved by the Governors. Closer relations between the teachers and the Governing Body of the School have been established by the appointment of representatives of the Council on the Court and on the Emergency Committee.[100]

The Second Report added little to these principles, save for the growing autonomy of the Appointments Committee. Even so, one would hardly recommend the structure described by the Director on behalf of the Governors to, say, universities emerging from totalitarian rule and seeking genuine autonomy. Beveridge knew this and put his 'administrative genius' to work on the issue; yet his repeated protests that he did not come to LSE to seek power sound less convincing as the years go on. In his later book on the School he positively recommends a structure which gives much power to the Director, and offers at least two strong arguments in support. One is that the collegiate system of Oxford and Cambridge has not exactly been known as a source of innovation. The other is that academics must be allowed to get on with their own work rather than be dragged into administration. 'The function of the administrative staff, as I put it once on behalf of Janet and myself at the School, is not that of getting its own way, but that of setting the academic staff free for what it alone can do.'[101]

Still, three changes were needed in order to implement the Beveridge philosophy of the School's governance: the Court of Governors had to be made more effective; the position of the teachers had to be strengthened; the role of the Director had to be defined. Two of these three needs were met early. The Court, enlarged by nine business representatives in connection with the commerce degree, had become unwieldy. In 1921 an Emergency Committee of five members was created. This was enlarged to ten members in 1925 when two academic representatives were also included. How effective this Emergency Committee was may be open to doubt. The next London University inspection of 1935 not only found that the purpose of the Committee lay merely in 'aiding and advising the

Director in the exercise of his duties', but also that few people know of its existence, and 'no reference to it appears in the annual prospectus'. As a consequence of the 1935 University Report the Emergency Committee was transformed into a Standing Committee and given at first limited, later ever more important powers. In matters of finance and overall policy, it became the effective governing body of the School. Increasing academic representation after 1968 of equal numbers of elected academics as lay members (plus a lay chairman) added legitimacy.

On the academic side, the Professorial Council became the nucleus of increasingly effective representation. From an early point onwards, in formal terms since 1921, two functions of academic self-government were separated. The so-called Professorial Council in fact assembled most (later all) teachers including lecturers (though initially not assistants) and dealt with matters of academic policy. It was the nucleus of what is now called the Academic Board. When it grew too large, it created its own General Purposes Committee (in 1937). Many other committees are based on the Academic Board. For questions relating to appointments, a smaller group called the Appointments Committee was formalized in 1921. Its members were initially elected by the Professorial Council though first by custom and then by rule the Appointments Committee became the actual Professoriate. It too created an executive committee, called the Standing Sub-Committee of the Appointments Committee, in 1963.[102]

Thus a fairly effective Court of Governors and a fairly representative academic committee structure emerged and the two were loosely linked by academic Governors and members of the Standing Committee, but the role of the Director remained crucial. He was not a Governor but Secretary of the Court. But most of its powers had been delegated to him. On the academic side, he was chairman of both the Professorial Council and the Appointments Committee. Even in much later days, elected academics were never more than 'vice-chairmen' of committees of the School, all of which were technically advisory to the Director and chaired by him. If institutions exist in order to protect its members from abuses of power, this is clearly a precarious if not an unacceptable structure. If they are there to allow people to get on with their work and yet make innovation possible, then the arrangements for the governance of LSE are not bad. During the Beveridge years, the School had cause to concern itself with both these consequences of its unwritten constitution.

One footnote which may well be more than that needs to be added. When institutional structures are precarious and offer temptations for abuse, a great deal depends on the behaviour and temperament of individuals. By that are not only meant the autocratic tendencies of people which Beveridge so strenuously and suspiciously denied for himself but more importantly their friends and confidants. Who do they talk to? Who can a Director with the unusual powers of his LSE position confide in? From whom will he take advice? The relationship between Webb and Hewins was critical during the founding years of the School, an asymmetrical relationship but one in which each had his own role to play. Mackinder left the directorship too soon to become an effective autocrat, though he was accused of such inclinations; and Reeves never made the School the object of whatever deeper urges for power and influence he may have felt. What then about Beveridge?

The relationship between Beveridge and Webb was close from the beginning and remained that throughout. It is perhaps in the nature of the two men that it was not particularly cordial, but the mixture of coolness and closeness did not do the School any harm. Webb supported the Director in good days and bad, and Beveridge called on the founder for advice, help, or just a friendly chat. 'Beveridge called here at 5.30—the School is going ahead at nearly all points, and he is very contented.' Sidney Webb supported Beveridge when sticky decisions had to be taken about a Beveridge portrait in 1926 ('It *must* go through now, as Steel-Maitland has circulated a letter to the *Governors*, asking them to subscribe'), or when Kingsley Martin fell out with 'the Beveridge–Mair dictatorship' in 1927 ('There is no need for Kingsley Martin, and the others, to annoy Beveridge further'). In the end, even Beatrice Webb came round to a more charitable view of the Director. On 14 August 1928 she wrote to (by then Lord) Haldane: 'We have been seeing a good deal of Beveridge lately. What an abnormally energetic and adventurous administrator he is.'[103]

However much they saw of each other, none of the evidence of letters and diaries suggests that Beveridge seriously asked for Webb's advice. He kept him informed, complained to him about obstreperous professors, asked him for practical help, but did his own thing. Of course, Webb had long ceased to be Chairman of the Governors. Since the death of Lord Rea in 1916, the Scottish Conservative politician Arthur Steel-Maitland held that post. Webb—who else?—had cajoled him into the job with a strange argument when he wrote to him on 24 June 1916: 'We want to keep the institution in the stream of public life and work—we do *not* want it to become too academic—and yet we do not want it to become the appanage of any one party or school of thought.'[104] Steel-Maitland accepted, was duly elected in July 1916 (at a meeting which eleven of the forty-nine Governors attended, as Beveridge noted even twenty years later), and served the School loyally until his death on 30 March 1935.

Steel-Maitland was a busy man. As Conservative Member of Parliament and of several governments in the 1920s, including a period as Minister of Labour in 1924–9, he had no shortage of a full day's work. His business interests, a variety of public committees, and a wide network of friends did not seem to worry him unduly though they gave his Harley Street specialist occasional pause. Yet Steel-Maitland took his commitment to LSE seriously. For the most part, he was present not just for the Emergency Committee but also for other emergencies. He was an inveterate, and often successful writer of appeal letters. He dealt with complaints or concerns of members of the School with habitual fairness. He found speakers and chaired lectures at the School. He defended LSE against the usual attacks from his own corner of the political spectrum, as when the *Daily Graphic* had made one of the familiar charges against the School's 'socialism'. It was, he wrote to Webb, 'done mainly from the point of view of increasing the circulation of the newspaper in question'.[105] When the Laski affairs of the 1930s excited many, not least the Director, Steel-Maitland kept a cool head.

Yet there is no sign of his getting more deeply involved in the School of which he had actually been an occasional student in 1901/2, let alone of Steel-Maitland becoming the Director's confidant. The Chairman always remained a bit remote. While describing Mrs Mair as 'the sister-in-law of Sir William Beveridge' (in a letter to Lord Gladstone on 3 February 1920[106]) may have been a pardonable early lapse, the archive papers do not suggest that Steel-Maitland was ever particularly close to Beveridge. Beveridge in turn mentions his Chairman exactly once in the 138 pages of his history of the years 1919–37, in a footnote of little relevance. Thus we must look elsewhere if we want to find out in whom Beveridge confided and whom he consulted, and we need not look far. This was precisely where things began to go wrong, for neither his deaf and increasingly frail mother nor his admiring and self-interested Secretary were likely sources of independent advice.

AN ATMOSPHERE OF LEARNING

One Man's Memories

The teachers and students who came to the School in the 1920s certainly heard and remembered the hammering sound of construction work, but they knew little and cared less about the Emergency Committee, the Director's journeys to New York or even Bloomsbury, and the School's budget save perhaps for their own salaries and fees. They had come to learn something about which most of them were not at all clear at first, and many came with some hesitation, at a time of 'confused objectives' in their own lives, even after they had vaguely tried more traditional places and been rejected. But what B. K. Nehru called, not entirely without qualms about the absence of 'distractions', the pervasive 'atmosphere of learning' soon took most of them in.

Of those who have written about LSE in the 1920s, no one has captured the atmosphere of learning more memorably than Lionel Robbins.[107] When his father, in a moment of mixed exasperation and generosity, gave the 22-year-old the money to go to any university of his choice, Robbins lost no time and registered for a B.Sc.(Econ.) at the School in October 1920. The first year of the course included economics by Dr Dalton, money and banking by Dr Gregory, economic history by Professor Lilian Knowles, British constitution by Mr Lees-Smith, logic and scientific method by Dr Wolf, and geography by Dr Hilda Ormsby. Lectures were the main method of instruction; the Director's personal advisers had not yet been introduced. Rereading and memorizing lecture notes helped to pass the intermediate examinations. The final B.Sc.(Econ.) 'was broad rather than deep' and included translations from two foreign languages. Though Robbins knew that he was going to be an economist, he chose 'History of Political Ideas' as his special paper.

In his *Autobiography of an Economist*, Robbins remembers four of his teachers with special affection, two older ones and two near-contemporaries. Edwin Cannan was then still coming down from Oxford every day. Robbins sounds slightly defensive about him because (as he says) Cannan had made a mistake in monetary theory and also advocated the wrong (deflationary) policy after the war. Nevertheless, 'at LSE and within its sphere of contacts, his ascendancy was paramount. We revered him.' It took some time to 'tune in' to the lecturer, who was liable to mumble through his beard, but once that had been achieved, every word he said mattered. His weaknesses were apparent; he was not enamoured with the 'modern' mathematical bent of economists. Moreover, unlike his younger colleague Theodore Gregory, he knew little about the world at large. 'There was something intensely local about his intellectual habits and preoccupations.' Still, 'he was a fine economist: he gave one a sense of the sweep and the power of the subject and its relevance to human happiness.'

Hugh Dalton was not merely much younger—almost a contemporary for Robbins, though in fact eleven years older—but also a very different kind of teacher. He was distinctly 'modern', combining as he did the LSE and the Cambridge traditions. 'Who of those who sat through his lectures can forget that powerful presence, the enormous bald head, the infectious grin and the booming voice?' Robbins is hard on Dalton's later political career, notably as Chancellor of the Exchequer, and also on his character; one can guess that the reference to the 'turbulence of his temper' is not intended to be complimentary. Yet he emphasizes 'the fact that as a teacher [Dalton] always obeyed the rules of academic integrity'. By that Robbins, already inclined to the political right, means that the politician of the left controlled his predilections. Perhaps Dalton's sense of irony was greater than his missionary zeal. 'No one who wishes to understand the history of LSE in those days should neglect the beneficial influence of Hugh Dalton,' writes Robbins, though Dalton's own mind was, judging from his diaries, already as much on his political connections and ambitions as it was on the School. He remained a conscientious teacher until politics finally swallowed him totally in 1935, but he always had a leg in both camps, academia and politics.

Robbins's preferred teacher during the early 1920s was Graham Wallas, like Cannan one of the old guard. 'As a teacher, he surpassed anyone I have ever known.' In one sense, this is strange, for even Robbins cannot remember much of the substance of his teaching. Surely the fact that 'Wallas believed passionately in reason and its power to solve conflict' was not quite enough for 'great inspiration'. But perhaps Wallas was one of those teachers who, in his lectures and even more in seminars, brought out ideas in his students which they came to cherish, and for the emergence of which they liked the catalyst on the podium. Robbins the septuagenarian remembers Wallas as 'the aged seer with his youthful eye and his deep seriousness, discoursing eagerly on the ends of life and the proper mode of conceiving them'.

And then there was Harold Laski, later to be Graham Wallas's successor, now a young lecturer and again a near-contemporary of Robbins. Few men can have been such an endless source of ambivalence in the feelings of students and colleagues alike, and Robbins was no exception. He admired Laski's range of knowledge and 'exceptional powers of presenting this knowledge in an orderly and easily assimilable form'. Moreover, his 'masterly' classes were accompanied by his 'friendly and helpful' ways in private contacts. But on closer inspection he was also 'something of a disappointment'. 'However clever he appeared to be, it seemed to me that there was a slightly *ersatz* quality about much of his thinking.' And then his politics! Unlike Dalton, he was a 'propagandist' and his teaching was therefore 'not conducive to habits of truth and objectivity on the part of his students'. Somehow, Robbins felt, Laski was not, either then or later, quite grown up.

Judged as an adult, he left much to be desired. 'Judged as a very precocious boy with the bewildering mixture of good and bad impulses which usually go with precocity, he had much in him that was positively lovable.'

This is but the first of many encounters with the man who may not, as his recent biographers suggest, have been the founding father of an 'age of Laski' but who was certainly responsible for high passions of love as well as hate at LSE from his arrival in 1920 to his death in 1950. Laski's and Robbins's lives at the School came to be intertwined in numerous ways of which the early 1920s were only the beginning. Robbins found his student years rewarding. 'There was little or no teaching at the School at that time which was not inspired by the spirit of search and discovery'. His friends, with whom he shared 'energetic discussions' in the surrounding tea rooms as well as on Continental holidays, shared his experience. Robbins lists Frederic Benham, Sydney Caine, Jacques Kahane, G. L. Mehta, Arnold Plant, Georg Tugendhat as companions of these early days. When he finally came back in 1929 from a brief stint at Oxford to stay at LSE, he found the School changed and yet similar. The School had got bigger and looked different, but 'the old traditions of friendliness and informality persisted and with them the characteristic atmosphere'.

In later years, Robbins himself was to become the centre of many a controversy. Indeed, he was that already when he was somewhat reluctantly offered what was left of the short-lived Allyn Young chair at the age of 30, though then he not only had Laski's support but his former teacher Hugh Dalton prided himself on having engineered the appointment. Still, lest Lionel Robbins's testimony to the atmosphere of the School in the 1920s be regarded as biased, three other witnesses with very different perspectives shall be heard.

Teachers and Students

One is Kingsley Martin, later for many years editor of the *New Statesman*, the 'wretched editor', as Lionel Robbins referred to him in his *Autobiography*, who in his turn was 'bothered' by 'Professor Robbins and his *laissez-faire* friends' who would not understand the blessings of socialism.[108] When Kingsley Martin arrived from Cambridge in 1924, as a young assistant, and a year later, assistant lecturer in politics, Professor Robbins was still plain Mr Robbins. Martin felt at home immediately. The School, he remembers, was then 'as it always has been, a wonderful home of free discussion, happily mixed race, and genuine learning'. He too had kind memories of Graham Wallas. Of the older teachers, however, R. H. 'Harry' Tawney was for him the most memorable, 'one of the very best men alive, as well as one of the finest minds'. Tawney was always rather unkempt and slightly disorderly, his much-darned untidy suits covered in tobacco ash, his home a clutter of junk from all over the world; he was also liable to give long monologues (to escape 'the boredom of his wife's conversation', Martin thought, his wife being Beveridge's sister Jeannette); but 'his scholarship was impeccable and his personality such that you felt uncomfortable in disagreeing with him'. He was also a socialist of sorts, if of a different sort from Kingsley Martin's 'closest friend', Laski. Martin clearly loved Laski, but his description of the friend cannot escape the familiar ambivalences. 'In appearance you would have put him down as a nineteen-year-old schoolboy.' 'He loved to hold the stage with astonishing anecdotes which at first you believed implicitly but quite soon came to regard as merely samples of Harold showing off.' 'A queue long enough for a cinema waited outside his door in the L.S.E. for the advice he was always ready to give.'

Kingsley Martin found the Common Room in 1924 lively but not harmonious. Everywhere the 'war raged between the Socialists and the advocates of *laissez-faire*'. Well, not quite everywhere perhaps. Martin lists quite a few who were, in his own words, sitting on the fence. There was the beautiful and erudite Eileen Power, soon to be 'carried off' by the historian Michael Postan; there was 'the great Malinowski'; there were Martin's 'friends Morris Ginsberg and Lance Beales', to say nothing of Professor Hobhouse, who 'had a middle view'. In fact, the 'war' was essentially one between economists and political scientists. The former were, in Kingsley Martin's view, so wedded to *laissez-faire* that their brilliant younger colleagues of socialist leanings had to migrate to Cambridge. The political scientists on the other hand, led by Laski, tended the other way, including Martin himself, of course, who got into trouble with the School authorities when he wrote a pamphlet about the General Strike. Beveridge summoned him and told him off in no uncertain terms, which the young lecturer resented. 'I cannot believe that this was a wise way of treating a young member of his staff, vehement and tactless, no doubt, but obviously sincere and troubled about the topics which the L.S.E. was supposed to be troubled about.'

Kingsley Martin left and exchanged his academic career for journalism. In any case it would be quite wrong to give the impression that LSE was only a training ground for future professors. The Hungarian heiress Judith Marffy-Mantuano (later Countess Listowel) came to the School almost by accident in 1926. 'I had never heard the names of the Webbs or any of their successors.' She was lucky; she not only got 'a scholarship three times running' but was also given Eileen Power as a supervisor. 'Without ever scolding me, she brought me gradually to see how narrow my conception of life was.' It was a traditional conception, so much so that Judith Marffy was at first shocked by the many socialists and communists around her, and even by the 'broad-minded liberals' who in her view formed the majority of students. But soon she came to like them, and her own views began to change. 'By the time I took my degree in economic history in June 1929, I knew that I had yet to discover and explore the world; and my attitude towards life shifted from strict Toryism to a friendly Liberalism.'

Apart from Eileen Power, who opened her eyes to the varieties of political life, the need for tolerance, and the road to understanding, Judith Marffy was most impressed by Tawney, Hobhouse, and Laski. Tawney 'looked like a benevolent teddy bear'. While the delivery of his lectures left much to be desired, he had so much to say that he kept his audience attentive by the analysis of 'long deferred results of ideas and actions'. Professor Hobhouse impressed the young lady from Hungary above all by the 'completely detached, balanced and understanding attitude he adopted towards all problems'. Laski 'both thrilled and repelled me'. He attacked everything sacrosanct but was brilliant in doing so, and of course his Sunday teas for students were special occasions.

The soon-to-be Countess of Listowel obviously enjoyed herself. She 'danced miles and miles', and she gave the Conservative Society a certain cachet, as when she organized a collection for distressed Durham miners to the embarrassment of her socialist rivals. But for her the School meant above all the dispassionate search for knowledge. In her autobiography *This I have Seen* she returns to the same features of LSE several times. Her teachers taught her that one had to bury one's prejudices in order to find the causes of things, 'that intolerance was both narrow-minded and uncivilized', and 'they actively discouraged proselytizing'. 'I learnt that statements have to be based on facts, and I was taught how and where to look for them.'[109] Thus for one student at least the School lived up to its promise. Judith Marffy left the School as a curious journalist and embarked on a long and varied life as an author and activist.

Another overseas student who sought a public rather than an academic career, and who succeeded brilliantly, was Braj Kumar 'B.K.' Nehru,[110] the long-serving diplomat from the old Indian Civil Service. When he came to the School in 1929, he already had a science degree from the university of his home town, Allahabad. The School rejected him at first and it was only when he met a senior civil servant who knew Beveridge that the decision was reversed, 'a rather early education in the truth that whom you know is more important than what you know'. B. K. Nehru came and was 'received with the utmost courtesy by Professor Laski, 'who gave me then, and throughout my stay, more personal attention and more affection than my academic or other achievements ever deserved.'

Laski apart, B. K. Nehru felt a little lost. Compared to India, courses were loosely structured and left much time for private study. Textbooks did not exist; one had to read an enormous number of primary sources, books, and articles. 'My most shameful abuse of my new-found freedom was to stay in bed almost till midday.' As a result, B. K. Nehru was one of those day students who attended evening lectures though they should not have done so. Vera Anstey then got hold of the lost young soul and made him write essays as well as read economics. Of course he was drawn into politics. Most students, he found, were left-wing, and Indian students doubly so, not least among them Krishna Menon. However, there were also 'somewhat more rational alumni of LSE' in whose company B. K. Nehru evidently felt happier: Gaganvihari Mehta, R. S. Bhatt, Tony Dias, Tarlok Singh, and others.

The School has always prided itself on its openness to overseas students. It is therefore sobering to read B. K. Nehru's account of the reality of the lives of at least one group among them. Already in 1929, he observes that the location of the School means 'that there is little corporate life after academic hours'. More than that, however, 'the Indian student in England in the thirties was between two worlds'. He no longer accepted British rule as god-given but had not yet acquired the self-confidence of independence. This made Indian students few friends among the natives. Nehru observes that the largest number of his non-Indian friends were Jews. 'They were far from being insular and our common history of oppression formed a natural, though unspoken, bond between us.' David Glass and Samuel Goldman were two of his contemporaries.

But Nehru goes even further. He observes that 'throughout the world foreign students tend to form their own groups for they are all homeless and rootless and friendless. The natives have attachments and interests predating their contact with the newcomers.' Thus when he did not spend his time with Indians, his companions were most likely to be other foreigners, 'Germans, Norwegians, Ethiopians, West Africans, Chinese'. People differ, and some are more sociable than others. Few Americans would have agreed with Nehru but then they had ceased to live 'between two worlds' in 1776. The experience of Oxford and Cambridge is probably different also; life and work are indissolubly intertwined in the old collegiate universities and at least in term-time there is no world outside. At LSE this would be true only for those for whom academic work is the centrepiece of their lives, or perhaps academic work and political activity in their tense relation. LSE can be a total experience but it requires a certain kind of temperament to enjoy it. The School does not necessarily produce well-rounded personalities; it seems to prefer intense, committed, often workaholic scholars, and public figures.

B. K. Nehru, being himself the more classical exemplar of a well-rounded gentleman, could not fail to discover, and enjoy to some extent, the School's 'atmosphere of learning'. His cameos of the teachers of the time are as telling as Lionel Robbins's and more poignant for being less voluble. In fact they begin with the 'young and very handsome Lionel Robbins' himself who, 'tossing back his flowing mane of hair (to the delight of the ladies)' lectured to packed classes in the Old Theatre. Hugh Dalton always punctuated 'his teaching with some anecdote of "when I was in the Foreign Office"' which Lees-Smith, despite having been Postmaster General, did not. Nehru remembers Lauterpacht's lectures on international law for their brilliance, which kindled a lasting enthusiasm. He did not fall for Eileen Power though he puts his immunity down not to her but to the subject of economic history, 'which, I now recognize, I could never really get to grips with owing to my immaturity'. Professor Brogan, on the other hand, stimulated Nehru's lasting interest in the United States, 'about which the British-Indian education of which I was a product had left me with the impression that it had ceased to exist in 1776'. Hayek by contrast 'was so much beyond me that I had to give him up fairly early'. He got to know slightly Malinowski, who made him buy *The Golden Bough*. 'And once or twice I had the thrill of glimpsing the founders of the School, Sidney and Beatrice Webb, whose visits to it had by then become very few and far between.'

Thus B. K. Nehru was far from unhappy. Perhaps he was constitutionally incapable of being unhappy. Yet in later years Nehru did not find it easy to tell what LSE had done for him. He learnt a little economics but remained baffled by political science with its 'unproven assumptions of human nature and wholly speculative theories of social organization'. He sensed that the 'totality of influences' to which he was exposed 'did have a very lasting and profound effect on my personality'. But what effect? He became, he said, a 'historical materialist' believing in the primacy of economics, and he developed a 'total opposition to hereditary privilege'. One cannot help wondering just how likely it is that these were effects of the School. In the end, the ambivalence which B. K. Nehru showed towards Laski may be more telling. When he met Laski after the war and after

Independence, he told him that he had reached the opposite conclusions to those taught by him. But Laski protested that he had not taught conclusions, only how to think. 'That, perhaps, is the greatest claim that a teacher or a school of learning can make for itself.'

The Director's Tenth Anniversary

None of the people mentioned in these pages had much time for the Director, let alone for his Secretary. B. K. Nehru does not mention him again once the connection had got him into the School. Edwin Cannan and Graham Wallas expressed their views privately, though Beatrice Webb lists Wallas among those who 'hated' Mrs Mair from a very early date. Kingsley Martin had his own reasons for being cross with the Director, as we have seen, though his dislike of the 'lonely bachelor [who] relied excessively on the advice and information supplied to him by the school's Secretary, Mrs Mair',[111] went further. But the most devastating judgement of the Director himself came, surprisingly perhaps, from Lionel Robbins,[112] who thought that Beveridge 'in some ways never appreciated the quality or the standing of the senior staff he took over or indeed the tradition of the School as a whole'. According to Robbins, Beveridge was a mediocre scholar ('his ideas of scientific method were primitive in the extreme') and an administrator who could handle things but not people. Robbins's assessment does not fall far short of hatred:

> If he had had a little less self-absorption, a little more scruple, a little less impatience with opposition, a little more common humanity, if there had been some family intimate at hand to encourage his altruism and restrain his ambition, what a much more successful and influential career he might have had. But in fact reverse influences prevailed.

This was more than a little unfair, and even Robbins, as well as Kingsley Martin, knew it. 'His relations with the School at large', Robbins concedes, 'were a mixture of triumph and tragedy,' and Kingsley Martin says himself that he 'must also admit that together [Beveridge] and Mrs Mair built the school from small beginnings to indubitable greatness'.[113] 'Senior staff' were clearly the people with whom the Director got on least well, and since most of those who wrote about his years at the School either were, or were to be members of the senior staff, his image and that of Mrs Mair have become strangely distorted.

Beveridge's ability was clearly not matched by amiability. He commanded respect, not affection, and therefore did not enjoy the loyalty of those who worked for him, with one notable exception. But he knew what he wanted and on the whole also how to get there. His career was by most standards 'more successful and influential' than those of his critics. Even during his first LSE years he had many other 'pre-occupations' and got involved in numerous 'side-shows'. By these, Beveridge himself meant not just 'writing and talking', including, at an early date, broadcasting, but his work on the Royal Commission on the Coal Industry in 1925/6 and the Vice-Chancellorship of the University of London from 1926 to 1928. The election in June 1926 was an unpleasant contest in which Beveridge narrowly beat the sitting University of London MP Sir Ernest Graham-Little by bringing out all the backwoodsmen of the Senate. He then proceeded to concentrate most of his spare energies on wresting the Bloomsbury site from the Duke of Bedford, which, with the help of Rockefeller money, Government, and LCC support, and against a vociferous minority on the University Senate, he managed to do. This meant that the University at last had a prospect of moving from South Kensington to a more central location, once a new building had been erected. Senate House was built long after Beveridge's Vice-Chancellorship; even the foundation-stone was not laid before 1933; but Beveridge remained involved throughout and was above all a member of the team which selected the architect Charles Holden, who was supposed to realize the two dreams of the ever-ambitious Beveridge. One was (as he wrote to his mother in 1927) to find 'some inspired artist in steel and stone and marble—not too much marble—and let him embody it in a group of buildings, of surpassing beauty, which later generations will set beside Westminster'. The other dream was to make this building 'no more than the centre of a University Quarter of London, in which a large and growing proportion of all who lived there should be University teachers or students'.[114] Reality has not judged these dreams kindly. Senate House looks like nothing so much as one of Mussolini's monstrosities; worse still, it is said to have served as the model for George Orwell's dreaded Ministry of Truth. The edifice has undoubtedly not encouraged the prospect of a Latin Quarter in Bloomsbury.

But in its way this too was a story of achievement, though not really comparable to the success of Beveridge as Director of LSE. After ten years, his influence was visible everywhere. Since no one else, except perhaps his Secretary, was going to celebrate the completion of the decade, Beveridge had to do it himself, in his 1929 Oration. He spoke of the School at the end of the war—he meant, at the time of his arrival in October 1919—almost in the terms which Jose Harris adopted, 'with earlier plans for expansion shattered, miserably cramped for means, with a handful of teachers of high distinction, but overworked and under-paid, upon a site which many of our best friends despaired of'. All this was now changed. 'As the decade ends, the School appears to have reached maturity as one of the great colleges of London, with space to grow as large as it need be, with generations of past students pushing steadily up into the responsible work of the world and spreading outwards over all its lands and cities.'[115] With considerable justification, the Director became nearly poetic in his praise for his own achievement and that of those who helped him.

The end of Beveridge's first decade was also a time of changing generations. One after the other those who had formed the early School retired; many died. In 1928, Martin White died, and also Lord Haldane. In 1929 the first President of the School, Lord Rosebery, passed away and also the long-time Vice-Chairman of the Governors, Sir John Cockburn, who was succeeded by (Sir) Josiah Stamp. W. A. S. Hewins, the first Director, died in 1931, and a year later, in 1932, the third Director, William Pember Reeves. Many new appointments to the staff were made in this period, so that Hayek could note in 1945 that 'the great changes, which gave the professoriat of the School very largely the complexion it has at the present day, were concentrated in the few years from 1928 to 1932'.[116] Hayek also noted that quite a few of the new appointees were drawn from the School's own students; one might add that many succeeded the great names of the founding generation. In economics, Professor Gregory's appointment in 1926 was followed by those of Professor Robbins in 1929, and Professor Hayek in 1931. In law, Professor Herbert A. Smith (international law) arrived in 1928; Professor R. S. T. (later Lord) Chorley (commercial and industrial law, succeeding Professor Gutteridge) and Professor Hughes Parry (English law, succeeding Professor Jenks) were appointed in 1930. Morris Ginsberg took the place of L. T. Hobhouse in the Martin White Chair in Sociology in 1929; in the same year, Professor C. A. W. Manning succeeded Professor Noel-Baker in international relations. In 1930 Arnold Plant got his chair; in 1931 Eileen Power, and (somewhat belatedly) R. H. Tawney.

In January 1929 a curious event occurred in that the Webbs had decided to celebrate their seventieth birthday on the same day.[117] Beveridge, forever rationalizing, found the decision 'very sensible', although, of course, only Sidney was born on 13 July 1859 and Beatrice was in fact a year and a half older. The School used the occasion to commission the now famous portrait by Sir William Nicholson, which adorns the Founders' Room and shows the couple in characteristic pose, businesslike rather than loving but also engrossed in the common enterprise of proof-reading. The Webbs had not much enjoyed the 'sittings' (he is clearly standing and the dog does not look as if it could have been a part of the occasions) but came to like the resulting painting. In any case, they were, or rather he was busy with other things. After the election of 30 May, he had been elevated to the peerage and joined Ramsay MacDonald's second government as Lord Passfield and Colonial Secretary. The Webbs never forgot LSE but the School was now far away for them and they increasingly became monuments of a bygone age.

4

SOCIAL SCIENCES COMING OF AGE

NOT JUST A SCHOOL OF ECONOMICS

Defining the Social Sciences

When the founders named their creation London School of Economics and Political Science, neither Fabians nor the numerous correspondents informed of the event at home and abroad raised any objection. Despite Herbert Spencer and Auguste Comte before him, economics for most summed up the study of things social, especially the hard-headed, scientific study. In Scotland, political economy might have sounded more appropriate, but the Webbs did not want to lose the allusion to the Paris École Libre des Sciences Politiques and the Faculty of Political Science at Columbia University, so that political science had to find a place in the name. Once the conjunction of economics and political science was established, no one seemed to feel that anything was missing nor were people surprised when, in its first year, Hewins taught economic history, Mackinder geography, Bowley statistics, and Russell whatever he felt like, at the London School of Economics and Political Science.

The surprise came later, in the 1920s and above all the 1930s, and it was itself a result of the increasing articulation of the social sciences into different disciplines of scholarship. William Beveridge's inaugural lecture of 4 October 1920 on 'Economics as a Liberal Education' began by stating that the author did not want to 'entangle' himself in the attempt to give 'a scientific definition of "Economics"'. He had in mind all subjects represented in the University Faculty of Economics, including '"economics" in the narrower sense', political science, geography, ethnology, and sociology. 'All these make up the circle of the Social Sciences—the study of society and of man as a member of society.' Beveridge then proceeded to position the social sciences in the universe of human knowledge, with the help of his lifelong scientific hero, Thomas Henry Huxley. Like the natural sciences, the social sciences have to observe facts, classify and compose them, then (he thought with Huxley) 'derive general propositions', and 'verify' them by further observation. In this the social sciences still had a long way to go, but they had set out on the journey with justified hope, and even now they had much to contribute to 'the education of every citizen'.[1]

In his farewell lecture nearly seventeen years later, on 24 June 1937, Beveridge returned to the subject. This time he chose the title, 'The Place of the Social Sciences in Human Knowledge'. He referred to his earlier lecture as 'the curate's first sermon' which 'ranges heaven and earth', and promised a more concentrated litany by the 'experienced Bishop'.[2] Concentrated it was, but it was also bitter. (Not surprisingly, while the first lecture was published in *Economica*, the last ended up in *Politica*.) Beveridge used the occasion for an elaborate attack on two objects of his wrath, explicitly on Keynes and economics, and implicitly, at any rate without mentioning the name, on Laski and political science. Laski was one member of the 'professoriate lined up on the platform exposed to public view', and Robbins remembers that the effect of Beveridge's attacks on him and others 'was shattering—and saddening'.[3] To oppose Keynes and Laski, Beveridge set up a notion of social science defined by observation (as against theory) and detachment (as against partisanship). Alas (so he felt), this social science was still a mere prospect. 'The foundation of this School made a new departure in the Social Sciences.' But the social sciences had so far failed to gain public esteem and a recognized place. 'For the Social Sciences that esteem and place are still to win—perhaps in 150 years to come.'

More than one-third of that sesquicentennial has passed at the time at which the present account is written, and we no longer need to fear the wrath of the School's fourth Director in returning to the concept of social science. This has in fact undergone its own transformations, though not all of them warrant more than a passing mention. At the School itself, something strange happened when, in December 1912, the Department of Social Science and Administration was created to accommodate the teaching of social work and related subjects. Was this to be the true home of 'social science'? Nobody commented at the time or even decades later, when a number of authors wrote about the history of the department, on the ambitious choice of name. It may have had something to do with the incorporation of Urwick's School of Sociology, and with the early role of L. T. Hobhouse in leading the new development. However, there was also the use of the term in the 'King's College of Household and Social Science', which came close to what is now called domestic science. In any case, it was never assumed that the Diploma in Social Science had anything to do with Beveridge's 'circle of the Social Sciences'. The Department of Social Science was in fact about social policy; no one ever expected it to represent the core of the social sciences as a theoretical project.

Perhaps the plural, social sciences, is relevant here. Whatever hopes the early advocates of a science of society may have had, it soon became clear that a unified theory of the field would be a vain, and at best a metaphysical hope. (The same is true, of course, in the natural sciences.) Thus successive scholars, Beveridge both in 1919 and in 1937, Alexander Carr-Saunders in an influential report in 1936, William Robson in editing the volume *Man and the Social Sciences* in 1972, O. R. McGregor in writing his contribution on 'The Social Sciences' to a University of London volume in 1990 (to mention but a few LSE-related authors), have used the plural. The Carr-Saunders Report is of interest in this connection. In 1936 Carr-Saunders was still Charles Booth Professor of Social Science (in the singular!) at Liverpool and chairman of a committee of seven distinguished scholars which produced a memorandum 'on the present position and needs of the social sciences with particular reference to population problems'. The committee began with an apparently general definition. 'The social sciences thus understood are concerned with the structure, mechanism and functioning of society.' But it quickly proceeded to differentiate this intention from other social studies, notably economics and political science. These are already well defined, but next to them the social sciences are a 'third field' with their own 'scientific' approach, and their subject is 'population studies', described rather comprehensively as investigation into 'the phenomena of marriage and the family, of social class, of the distribution of population and of social structure and organization in general'.[4]

Other attempts to define the social sciences more narrowly have added to uncertainty. In the 1960s a usage of the concept became widespread which confined it to what some would call the 'soft' social sciences: sociology and political science. The question of whether economics is a social science no longer had an obvious answer. Since the social sciences were not only seen as soft but also as partisan, and for the most part left-of-centre, Keith (later Lord) Joseph, Secretary of State for Education in Margaret Thatcher's first Government, insisted in 1982 that the Social Science Research Council be renamed if it could not be abolished. It became the Economic and Social Research Council as a result. In the non-governmental American Social Science Research Council economics has long occupied a marginal place compared with sociology and politics. The same has come to be true for the *sciences sociales* and *Sozialwissenschaften* on the Continent of Europe.

In the face of such confusion, one may be tempted to withdraw to the two suggestions which William Robson offers without claim to definitional precision. The first is a list. 'The social sciences may be said to comprise economics, politics and public administration, sociology, social administration, law, anthropology, international relations, industrial relations, social psychology, human geography, statistics, demography, management, together with the relevant branches of history.' This reads like the catalogue of LSE departments, though accountants and philosophers may feel left out. Actually, it is quite tempting to define the social sciences 'operationally' as the subjects taught at the London School of Economics and Political Science. William Robson adds one further plausible proposal, proceeding by contradistinction rather than definition: 'The social sciences comprise a group of disciplines which can be distinguished from the physical sciences on the one hand and the traditional arts subjects on the other.'[5] The social sciences can be distinguished from the sciences and the humanities—but how?

The Study of Society, the Study of Nature

The unfolding of the social sciences is so closely bound up with the history of LSE that we cannot let matters rest here. Moreover, some perspectives have changed a century after the foundation of the School. For one thing, we now have a more complex notion of the natural sciences. Taken as a whole, they are neither as unified nor as exact as suggested by the image that guided the self-conscious early social scientists. There lies a vast range of methodological territory between the models of theoretical physics and the working hypotheses of clinical medicine, or epidemiology, between inorganic chemistry and environmental science, between theories of the big bang and the daily weather forecast. Some natural sciences are indistinguishable in method from some social sciences and indeed overlap in subject-matter. Moreover all of them now appreciate that T. H. Huxley's version of it has little to do with science of any description. Science is about theories, their falsification and supersession, and while observation may keep scientists busy much of the time, theories are as likely to emerge from prejudices and idiosyncrasies as from facts. F. J. Fisher put it in his inimitable way when he suggested that 'a hypothesis may come to you at any time', but is more likely to come 'when you are in your bath' or reading something quite remote than 'from data, or from general theory'.[6]

It can nevertheless be argued that the study of society—of man in society, of social action—differs from that of nature in important respects. Complexity may be one of these, though the simplicity of nature is at least debatable. It may be no more than our ignorance, or the memory of the neat symbols, formulae, and models which physics and chemistry teachers purveyed in our schools. Nowadays, economics teachers at least could easily do the same. Two critical differences between the natural and the social sciences have to do with history, and with involvement. Natural history used to be an important subject in its own right; indeed the T. H. Huxley lecture to which Beveridge liked to refer was on 'The Educational Value of the Natural History Sciences'. But in the case of social facts, history actually constitutes them as such, which is why they cannot be reproduced in a laboratory. For all we know, even Keynes's 'general theory' may in the end turn out to have been true only once, that is before it was ever applied; in a Keynesian world, in which his precepts have become almost undisputed practice, we need other theories and policies. This is not to decry theory. We can try to abstract from history and for anyone with a faintly theoretical bent the temptation is at times irresistible, not just in economics; but we pay a price for such abstraction as we come to apply or test our theories. We discover, as it were, that we have been moving in non-Euclidean societies which no one has ever seen.

Involvement has of course a role in the exact sciences as well. Both Einstein ('relativity') and Heisenberg ('indeterminacy') have shown how the observer muddies the waters. We cannot observe without affecting what we see. But again, the involvement of the social scientist runs deeper. Karl Mannheim's *totaler Ideologieverdacht*, the unlimited suspicion of ideological distortion, is not just a definable principle of relativity or indeterminacy. Max Weber, so much more profound than Beveridge in his plea for a value-free social science, suffered from the evident impossibility of living up to his own strictures. At one point in his life, the conflicting pressures nearly destroyed the man. When he gave his famous lectures on 'Science as a Vocation' and 'Politics as a Vocation' (in the winter of 1919/20, thus at just about the time when Beveridge arrived at the School), his student audience in Munich knew that they were listening to a man whose passionate commitment to each of the two noble human activities at once inspired and threatened the other. At times, this was true for LSE as well, which embodies the cost of involvement in the social sciences.

Even so, the hopes for a science of society had been high ever since the Scottish moral philosophers had turned their minds to the subject. Adam Smith discovered that economic processes followed certain laws which transcend the intentions of economic actors and guide them as by an invisible hand. Adam Ferguson and James Millar found that the origins of social equality lay deep in the nature of property and thus of social structure. David Hume concluded that there is no particular reason why the science of politics should not progress, if anything more rapidly than that of nature. But history and involvement caught up with them. Soon, Pierre J.

Proudhon was to declare all property theft and the ideal of equality was modelled in Charles Fourier's utopia. Moreover, the question was asked whether Adam Smith had offered description or prescription. Was his a theory of how things are wherever humans live, or of how they happened to be at his time, or of how they ought to be? Proudhon, Fourier, Saint-Simon, Marx, Comte, Mill, Spencer—what was now the hope for a science of society?

Social research was one of the answers and one close to the hearts of the founders of LSE and their successors. The Webbs and Beveridge shared an almost religious obsession with 'facts' as the purpose of social science. They had not read their Kant, otherwise they would have known that while concepts without facts may be blind, facts without concepts are dumb. They do not speak for themselves, but have to be made to speak by constructs of the mind, by theory. Still, the seminal projects of social research had and have their place in the history of the social sciences, if only because they were never just an accumulation of facts. But only one of these, and a late one at that—late in the sense that by the 1960s the proliferation of social research robbed it of much of its influence—was carried out at LSE: the study of social mobility under the auspices of the International Sociological Association and the guidance of David Glass in the 1950s. Charles Booth had steered clear of the new foundation in the 1890s, and when Beveridge promoted the replication of his survey of life and labour in London by Hubert Llewellyn-Smith and with the help of the Rockefeller Foundation, its effect was swallowed up by the more conjunctural experiences of the 1930s. The path-breaking studies of the Polish peasant in Europe and America and of Negroes and other disadvantaged groups in a large city were conducted at the University of Chicago. In Germany the Verein für Sozialpolitik sponsored the first great empirical investigations of agricultural labour, and later, under the direction of Paul Lazarsfeld, of unemployment. After his forced exile, Lazarsfeld became the guru of social research at Columbia University and produced among other pioneering studies the first systematic analyses of voting. The émigrés of the Frankfurt School who tried but failed to come to LSE were among the initiators of the influential empirical study of the political psychology of fascism, *The Authoritarian Personality*. Somehow the empiricism of Beveridge never matched these classics and after his departure few at LSE were interested in following in his footsteps.

At the other end of the scale, what may be called grand theory also bypassed the School, though perhaps it was less sorely missed. Grand theory after Spencer took several forms. One was the increasingly desperate attempt to integrate the social sciences. Max Weber's *Economy and Society* came as close to success in this venture as any, but then his posthumous work, first published in 1922, is a quarry in which anyone who cares to dig can find material for edifices of theory. Outside Germany, Vilfredo Pareto's *Trattato*, the treatise 'of general sociology' (published in Italian and French in 1916, in English in 1936), influenced economists and sociologists alike, though those who remembered 'Pareto optima' and those who pursued the 'circulation of élites' were rarely the same people. A generation later, Talcott Parsons tried to merge Weber and Pareto as well as Durkheim into one 'social system', but even in his own *Economy and Society* of 1956 he did not manage to break out of the confines of sociology. Parsons spent some time at the School in 1924/5 and remained a good friend (and Honorary Fellow), though he was never a full-time student or teacher at LSE. The other kind of grand theory in the social sciences was of a more sweeping, philosophical, and often historical character. Some of it entered the School through the latter-day disciples of T. H. Green, for example L. T. Hobhouse, but even Hobhouse was a 'positivist' compared with the Hegelians of the Frankfurt School, Max Horkheimer, Herbert Marcuse, Theodor Adorno. When they put out their feelers to LSE in 1934, they were repulsed by a coalition of senior professors, of which more will have to be said.

At the time, the reasons for rejecting them were complex, but one at least had to do with the method of social science (to use the singular for once), and the LSE tradition. Norman MacKenzie has called this tradition 'positivist' and he certainly has a point, considering the Webbs, Beveridge, and facts. One is, however, also reminded of Robert K. Merton's influential concept of 'theories of the middle range', which in strict methodological terms cannot exist but in practice form the bread and butter of social science. There is another way of defining the LSE tradition in the social sciences, which is to ask the quiz question: who did not belong? Robbins thought from a certain point onwards at least that Beveridge was the odd man out, but his view was surely a comment on the person as much as the scholar. Beveridge himself believed, of course, that the theoretical economists should not be there and for preference not exist at all. In terms of Sidney Webb's and Beveridge's aversion to partisanship, Harold Laski clearly did not belong—but what would LSE be without Laski? Some early students wondered about Lilian Knowles—and some later ones about Michael Oakeshott—but theirs was merely the 'wrong' partisanship. Did Friedrich von Hayek belong? And what about the man who is arguably the greatest of them all, Karl Popper?

Diversity of Method

A great university is not and must not be all of one piece. The quiz should therefore remain a game without a winner. Still, 'theories of the middle range' describe a great deal that was going on at the School. Not to lose sight of evidence and be carried away by speculation was a widely accepted mode. Evidence, moreover, was not confined to what one's departmental colleagues accepted as such. Interest in solving problems, not in setting up trade union branches of scholars tilling a particular field, determined the work of most at the School. And whatever people said or wrote, almost everything was tempered by a healthy dose of scepticism. Laski was outrageous, though he made up for it by caring deeply about his students, but otherwise the intellectually outrageous was kept at bay. Some might say, too much so. A puritanical restraint on intellectual extravagance was unmistakable at LSE, and when Beveridge cited Hesiod against Keynes—'The immortal Gods have put sweat on the way to glory. Long is the road and upward sloping and rough at first'[7]—even those who otherwise did not have much time for the departing Director may well have agreed with him.

All this, however, is style as much as method. In the first half-century of LSE, and leaving social research and grand theory aside, three main developments took place in the social sciences. One is, simply, economics. In 1937, when Beveridge gave his farewell lecture, it was no longer possible to use 'economics' to refer to the whole of the social sciences. Economics had begun to become the epitome of the hard, the truly scientific study of a particular set of social facts, of human behaviour in allocating scarce resources to alternative ends, perhaps 'conduct influenced by scarcity' (to cite Robbins). The second development can best be described as history, including what one might call the history of the present. It takes its cue from Max Weber's (and other German hermeneuticists') notion of *Verstehen*, of bringing to bear a multiplicity of methods and a degree of empathy on the understanding of complex realities. Thirdly, prescriptive thinking kept its place within the social sciences, and usually remained associated with some concept of 'policy' or of the 'political', political theory, social policy, even political economy.

Most disciplines as they developed over the years found their place somewhere within the triangle of strict social science, the method of history, and political theory, though the articulation of disciplines meant distinctive moves in one or the other direction. Economics progressively shed its historical past as well as its prescriptive aspirations. Economic history was left with one of these traditions, though more recently, and along with one strand of general history ('cliometrics'), it has moved in the harder scientific direction too. Sociology and political science remained divided between the three methods to the point of becoming almost untenable as disciplines. Demographers had a relatively easy time, though with the advent of historical demography some of the ease was gone.

The departments of LSE reflected all these developments during the inter-war years. Some were, for one reason or another, so important for the history of the School that they will be inspected more closely in subsequent sections of this chapter: statistics, economics, political science, economic history, anthropology, and the poisoned hybrid, social biology. Others have been mentioned already or will come into their own along with later phases of the School's development.

Sociology in the wake of Hobhouse was shaped by three very different men: the wise social philosopher Morris Ginsberg; the archetypical theorist of the middle range T. H. Marshall; and the empiricist David Glass, who most nearly represented 'population studies' as defined by the Carr-Saunders committee in 1936. Geography continued in the synthetic tradition set by Halford Mackinder and L. Rodwell Jones in his succession. Michael Wise has reminded us of Sir Dudley Stamp's view that geographers must range themselves 'with those who see in the study of the earth's riches and the ecology of man a challenge to a fuller understanding that can but lead to the betterment of mankind'.[8] Wise has also paid tribute to the other founders of the Institute of British Geographers, R. Ogilvie Buchanan and S. W. Wooldridge, who were like Stamp associated with the joint LSE/King's College department. As Geography to some extent left the School, the Department of Social Science and Administration returned in 1922 when the Ratan Tata fortunes had succumbed to the slump and to bad labour relations. In her history of the department, Jose Harris observes that 'the misfortunes of the international economy and of capitalist philanthropy . . . turned a dynamic, powerful and independent department into an academic poor relation of the expanding Faculty of Commerce'.[9]

But was not commerce itself, including the B.Com., an 'academic poor relation'? Around economics in the strictest technical sense the School accommodated throughout its history a penumbra of subjects and posts with more specialized, often institutional and applied designations. The Cassel Chairs and Readerships of Commerce, Banking and Finance, and Accountancy are examples. Indeed in the 1920s the strictly technical economists were in the minority. But theirs was the stronger pull; over time, only economics and accounting survived as separate subjects, and in the 1970s all professors in the Department of Economics insisted on the same designation, Professor of Economics. This left the accountants unimpressed and successive professors—W. T. Baxter, H. C. Edey, B. Carsberg—have described with pride the 'revolution in LSE accounting [which] began in the late 1930s' and turned the department into a 'school for accounting professors'.[10] The revolution was, according to Baxter,[11] 'inspired by an outstanding teacher' in economics, Arnold Plant.

In all accounts of his years at the School, Beveridge made a point of emphasizing developments in the teaching of law. Professor Gutteridge, the first incumbent of the Cassel Chair of Commercial and Industrial Law, started a process which by the 1930s had led to an intercollegiate degree in law shared by LSE with King's and University College. In his autobiography, Beveridge describes this process as one further emanation of the fact that LSE 'was always misdescribed in its title', for 'the London School of Economics ought to be a School of all the Social Sciences. My first practical inference from this lay in strengthening greatly its legal side.'[12] It was, however, only in the 1930s and 1940s, influenced by R. S. T. Chorley and then by exiled lawyers from Germany, notably Otto Kahn-Freund, that the School's Law Department acquired its distinctive character as one where lawyers were not ashamed to describe themselves as social scientists.

Bowley and Quantitative Methods

Other departments will be described in due course. Two among them, however, are special in that they straddle all, or nearly all, others. One is Philosophy, or rather, Logic and Scientific Method, taught from 1905 to his retirement by Abraham Wolf, the Spinoza scholar who was also a historian of science and of education. His post was held jointly at LSE and University College, though he was really a University College man, so that for most members of the School he remained a passing presence who came to teach, but who did not really belong. Philosophy became central to LSE only when the joint post had been split in 1945, and the remaining readership filled by the appointment of Karl Popper.

By contrast, the other straddling department had been central to the School from its earliest days. In due course, it had many names, none of them fully descriptive, unless one wants to call it Arthur Bowley's department. The department of facts as figures, perhaps? At 26, Bowley was even younger than Hewins when he was asked to assume responsibility for teaching statistics at LSE in October 1895. Unusually for the School, he had a Cambridge background in mathematics, and even more unusually had encountered the social sciences through Alfred Marshall, who in fact recommended him to Sidney Webb. It took a while for Bowley to get a permanent post as a lecturer and from 1915 Professor of Statistics, but from 1895 to his retirement in 1936, and indeed beyond it until he died in 1957, LSE was his intellectual home if not more. We have already encountered his perceptive and often amusing reminiscences of events at the early School.

Arthur (since 1950 Sir Arthur) Bowley has been described, by his successor R. G. D. Allen, as a 'rather old-fashioned' mathematician but path-breaking early representative of mathematical economics and econometrics. He prepared the ground for measurements of national income, and for economic reporting more generally. He also, and perhaps above all, advanced techniques of sampling which were to become routine for social scientists as well as for opinion pollsters and market researchers later but were still at a very uncertain stage in the 1890s. The wit of his reminiscences notwithstanding, Bowley the teacher and colleague had his flaws. 'Bowley was effective, if rather dour, on committees.'[13] 'He was a diligent teacher, but in his care as an expositor he made little or no concession to the student mind.'[14]

Under the aegis of Bowley and then in succession to him the department evolved into one of the world's leading centres of quantitative methods in the social sciences. R. G. D. (later Sir Roy) Allen had (in Bowley's judgement, expressed in a letter to Beveridge in 1935) 'three sides' to his work, 'more advanced mathematics, statistics, especially on the mathematical side, and applications both to economics and economic theory'.[15] He was one of those members of the growing department who were less concerned with facts as figures than with formal analysis. As this became prevalent in economics, the boundary between the departments was at times imperceptible.

However, when the second chair of statistics at LSE was at last established and filled many years later, in 1949, Sir Maurice Kendall's appointment introduced a different though equally significant range of interests. It extended from computing to demography but above all had a special contribution to make to LSE. His inaugural lecture was one of the great LSE occasions:

The statistician is popularly regarded as a man who knows nothing except facts; and facts, moreover, which are numerical in nature, that is to say, are of a particularly inhuman and specious kind. It is not easy for anyone reared among the humanities or in the practical world of industry and commerce to believe that a person who apparently spends his time merely in counting or measuring what he observes can, by that process alone, acquire any deep insight into what is really happening. That the statistician is a useful person on occasion is now accepted. That he can make great contributions to scientific methodology is becoming realised. But that he has a unified conception of the world at large, that he derives from his science a satisfaction of those desires for a grasp of the fundamentals of things which lie beyond pure reason—in short, that he has a philosophy—may come to you as a surprising claim.[16]

This is rushing far ahead, though not without reason. From its earliest days, LSE statistics provided, curious as this may seem to those who never made their peace with figures, one force for integrating the social sciences. The reason was that those who represented it—from Arthur Bowley and Roy Allen to Maurice Kendall and then Jim Durbin, Claus Moser, Alan Stuart, David Bartholomew—actually liked the social sciences while never abandoning their commitment to what was loosely described as the discipline of statistics. Harold Laski may have had little time for his colleagues Bowley and Allen; we do not know how he reacted to Maurice Kendall's inaugural lecture. Others, like the historians, discovered the uses of numbers much later. But the department was from an early date one of the standard-setters, indeed a backbone of social science as practised at the London School of Economics.

Unity of Intent

Even the briefest of surveys of the departments of LSE in the inter-war period leaves no doubt about the process which we have called the articulation of the social sciences. Specialist disciplines emerged and academic communities around them. These were often reinforced by the professions of those who had studied one or the other specialism. But what about the other side of this process of differentiation—the unity of the social sciences? Was there anything left to hold them together? What would a professor of scientific method have said to first-year students if he had wanted to impress on them that the varied offerings of LSE somehow belong under one roof?

If our imaginary professor had been wise—and how could he, or she, be otherwise?—he would have been cautious in his methodological claims. He knew after all of the variety of methods used by his colleagues. He was aware of the growing self-confidence of the 'analytical' economists, and the self-conscious attempts of sociologists and anthropologists to establish themselves. He could not have been quite certain what exactly Harold Laski's methodology was. In the end, at the level of method rather than grand theory, he would have been left with a few almost trivial statements (though we know today that they are not altogether uncontroversial). There is such a thing as society. The word 'thing' is more than a manner of speaking; there are social facts, *faits sociaux* in Émile Durkheim's sense, which cannot be reduced to others without losing an opportunity of knowledge. For social facts can be studied systematically; they can be gathered and explained, even if there is no royal road to finding out their causes. Social research and 'pure' theory, empathetic understanding and prescriptive argument, will continue to exist side by side and at times vie with each other. Relevant facts themselves are defined differently by different disciplines. This is of practical use but must not lead to any disciplinary imperialism. Social knowledge, whether it is sought for its own sake or for informing practical decisions, often requires awareness of more than one perspective on a problem.

The hypothetical professor is, as one can see, a nice person, but what he has to offer is more in the nature of exhortation than scientific method. Thus we are almost back where we started, with the social sciences in the plural. Almost, but not quite, for there is also the London School of Economics and Political Science itself to consider. This is not meant facetiously, for the School, at any rate in the inter-war years and perhaps until the 1950s and 1960s, is itself a practical definition of the social sciences. They are not a unity; the singular, social science, makes no sense; yet they belong together. The old B.Sc.(Econ.) epitomized this fact. (It also illustrates that from one point of view the B.Com. was an unfortunate aberration, though an overly puritanical approach is as inadvisable in this respect as in others.) It combined a common core with specialization. Lionel Robbins has sung the praises of the degree in terms which are all the more convincing in view of the fact that he also practised what he preached:

> What was . . . important was a sense of the mutual relevance of the various subjects. As I have already explained, the degree of specialization required by the final examination was not such as to preclude a large area of common interest for all of us. The central courses in economics, politics and history were not planned in concert as regards detailed content—the idea of the possibility of such planning is of course a figment of the non-academic imagination—but since they were taken by all, there was no excuse for not realizing the extent to which each was essential to a well-balanced approach to the problems of society. Most, if not quite all, demands for organized cross-fertilization of subjects are nonsense—the sterile chatter of those who have no sense of direction of their own. But in the daily talk of participants of different specialisms in the sort of course which was the B.Sc.(Econ.) in those days, there was an informal process of this kind, the value of which, to me at any rate, it would be difficult to exaggerate.[17]

What is needed, in other words, to keep a sense of the social sciences as a whole, is not one of the familiar nostrums of later days, interdisciplinary research or overarching theory, but the exposure of students to the variety of specialisms, unforced but encouraged, and the hope that somehow this will put each specialism in a relevant context.

One other feature of LSE served to keep the social sciences together under one roof. It has often been said that until the 1960s the School had no departments. 'Indeed the word *department* was taboo,'[18] writes W. T. Baxter, who also adds that pressure of numbers later made this 'happy state' impossible. While technically true, such assertions about the early School should not be misread. Michael Wise makes a telling observation concerning geography after the Second World War: 'At that time, the School was not organized by departments but gave tacit understanding to the desire of the geographers to organize themselves departmentally.'[19] This was no less true for economists and economic historians, political scientists, and sociologists, let alone the Departments of Social Science and of Law. But the absence of a formal departmental structure nevertheless made a difference. Individual teachers counted in their own right and with their own, often varied, interests. Nobody prevented T. H. Marshall from teaching economic history as well as social policy and sociology. Not a few in the early days were what might be called '£50 economists' (or whatever). The story is told by F. J. 'Jack' Fisher, who once asked Tawney how much economics he knew 'and he said "My boy, I was paid £50 to give a course of lectures at Glasgow on Economics and I learnt £50 worth of Economics and that's all I know"'.[20] The price of knowledge rose rapidly in the 1930s, especially in economics, but the Senior Common Room was still small; teachers sat around one big table in the middle for coffee and conversation. While they each had their own discipline, they also met and felt part of a single-faculty school. Moreover, the School maintained the fiction of academic unity not just by the absence of a departmental organization with deans and conveners, but also by its committee structure, including one Appointments Committee with responsibility for all academic staff. In all these ways, LSE remained for a long time, and in traces to the end of its first century, the tangible reality of the social sciences.

One more difficult, even awkward question remains unanswered by methodology and organization: to what extent did the School and its members contribute to the advancement of the social sciences during the Beveridge years? The very existence of LSE and its attractiveness to students from all over the country and the world was of course a major contribution. But what about the scholarly accomplishments of LSE professors? Was there a Weber or Pareto, a Schumpeter or Keynes, a Michels or Parsons among them? Beveridge would no doubt have been unimpressed by this list, as he certainly was by Keynes. But Richard

Greaves tells us—slightly obliquely by putting a statement of Beveridge's in the form of a question—that the Director was unimpressed by LSE luminaries as well: 'Did he not say to me after he had been at the School for seventeen years that no one had produced anything of any value during his time there?'[21] If these words were actually spoken, they betray a very bitter man. Beveridge was bitter in his final years at the School. Still, the question of the signal contributions of LSE teachers to the advancement of their disciplines weighs on the historian. If we look only at the Beveridge years, we rule out of court the comparison with Max Weber's *Economy and Society* and Vilfredo Pareto's *Treatise of General Sociology*, which were both written a few years earlier. It is worth mentioning, though, that no one of quite their stature taught at LSE, nor was there an Émile Durkheim, a George Herbert Mead, or indeed an Alfred Marshall. But let us move on. *The Theory of Economic Development* (Joseph Schumpeter) was written in Bonn and at Harvard, the *General Theory of Employment, Interest and Money* (J. M. Keynes) at Cambridge, the *Sociology of Political Parties* (Roberto Michels) in Heidelberg and Turin, the *Structure of Social Action* (Talcott Parsons) at Harvard. What seminal books emerged from the LSE during the period? Or was the School something different, a unique experience of the social sciences for its students, and a staging post for many of the brightest and most original of its teachers? Before howls of outrage engulf such suggestions, let us turn to some of those who would have a claim to recognition among the great and the good, the authors of *An Essay on the Nature and Significance of Economic Science*, *A Grammar of Politics*, *Religion and the Rise of Capitalism* (or perhaps *The Acquisitive Society*?), *A Scientific Theory of Culture*, as well as—alas!—the editor of *Political Arithmetic* and author of *Genetic Principles in Medicine and Social Science*.

ROBBINS AND ECONOMICS

Cannan and the First Dispute with Cambridge

LSE was not just a school of economics, but it most certainly was a school of economics. In fact, the first reports by the founding Director, W. A. S. Hewins, could be read to suggest that it had become the only school of economics in the country. The English patron saint of the subject at the time, Alfred Marshall in Cambridge, did not like the suggestion. His letters to Hewins from 1899 to 1902 sound a bit disgruntled. Marshall disapproved of the separate listing of 'pure theory' in the LSE Calendar and was uneasy about the strong insistence on economic history. 'The fact is that I am the dull mean man, who holds Economics to be an organic whole.' Above all, he held that Cambridge represented another, earlier, and surely better school of economics. 'But the main point is that Cambridge has an idea of its own which asserts itself in spite of the partially non-Cambridge idiosyncrasies of one or two members of the staff.' It is all very well for Hewins to say kind things about him, Marshall, 'but of course it is for Cambridge that I am jealous'.[22]

A. W. Coats, who first published the early Marshall letters to Hewins, offers a very balanced assessment of their import. 'The two correspondents exemplified the outspokenness and impetuosity of youth and the cautious conservatism of age (Hewins was 34, Marshall 57).' Certainly, Marshall remained in general supportive of the School, as members of the School did of him. 'London and Cambridge', he wrote to Hewins on 19 February 1901, 'have in many respects a closer kinship with one another than with any other economic schools on this side of the Atlantic.'[23] Yet it soon became apparent that London and Cambridge were separated by more than the twenty-three years of the early correspondents. The drama of the intellectual dispute has tempted quite a few authors to use bellicose language. In fact there were two disputes, the first between Alfred Marshall and Edwin Cannan, on economics as perhaps an 'organic whole' but most certainly a technical profession, and a 'simpler economics' addressed to the general public, including evening students. The second dispute was between Lionel Robbins (as well as Friedrich von Hayek) and John Maynard Keynes, and initially had to do with deflation versus demand management as a remedy for the depression, but soon also involved different notions of economics and notably what came to be called macroeconomics. LSE lost on both counts, though it could be argued that each time it prevailed in the long run (about the utility of which Keynes had his own somewhat cynical view) and, more importantly, that it was the livelier, more pluralistic and open centre for teaching and research throughout the four decades of LSE–Cambridge 'warfare'.

Edwin Cannan's place in the early history of the School is unique. Along with Arthur Bowley, Halford Mackinder, and Graham Wallas he belonged to the intellectual founders of the first great school of the social sciences. All four devoted their work largely to LSE and stayed till retirement despite many other temptations. Cannan's daily return journey to Oxford (except during the General Strike of 1926, when he came all the way by bicycle and Beveridge put him up for the night) did not detract from his importance for generations of LSE students, of whom Lionel Robbins was but one grateful member. What made him so different? Why was he regarded as a counterpart to the 'grand old man' in Cambridge (who was actually almost exactly the same age)? Alon Kadish concludes that 'the origins of Cannan's feud with Marshall are unclear'.[24] However, not just the origins of the controversy are shrouded in fog but its substance never became so sharply focused that one could describe it as a great *Methodenstreit*. Gerard Koot has probably got it right when he summarizes his various studies of Marshall, LSE, and historical economics by stating that there certainly was what may be called a 'conflict between inductive and deductive economics'. But:

> In addition to methodology, it revolved around nearly opposite views of the economist as a detached expert or a practical adviser, different estimates of the role of history in social science, diverse judgments on the value of pure theory, and serious disputes on matters of policy. Finally, it was also fueled by personal disputes and competition for scarce academic posts. Most importantly, and from this the methodological debate largely flowed there, was the question of whether economics was a science or an art. The historical economists, as well as nearly all other dissidents, insisted upon the proposition that economic theory *per se* was of less significance than its application. This faith became the ruling principle of economic studies at the early LSE.[25]

The LSE–Cambridge difference, in other words, involved two incompatible notions of economics as a profession. Whenever it turned to specifics, Marshall's defenders could argue that their master had said everything already; 'it's all in Marshall', including (so Keynes claimed) historical economics. Marshall and those around and after him established an orthodoxy, not a very dogmatic and certainly not a narrow one, but one that was intended to dominate the scene, including the British Economic Association, the *Economic Journal*, the filling of posts in universities. A. W. Coats has described the 'sociology' of this process by reference to the frailty of other centres, the needs of the times, but above all the personality of Alfred Marshall, who may have 'led British economics from the rear' but was not short of 'subtle and indirect methods' as well as 'a more direct approach' to achieve his objective.[26] In the end, all others were called 'dissenters'. In fact, they were professionally the more open; no one would ever have said that 'it's all in Cannan', although Cannan too had left barely a subject of economic analysis untouched. Cannan and the LSE left room for different approaches to economics and this remained true right up to the 1930s. 'What characterized LSE in the 1930's was that, despite the holding of firm views, there was a lack of doctrinal commitment, which resulted in an openness to new ideas.'[27]

The man who wrote this, R. H. Coase, had known and heard Cannan but himself taken the B.Com. degree. For his own academic progress,

Arnold Plant was as important as Robbins; and earlier A. J. Sargent mattered as much as Cannan. These were applied economists with a strong institutional interest, who were more likely to lecture on industrial organization than on monetary theory. Their products, however, were academically no less (and in the practical world, for the most part, more) distinguished than those of the analytical economists. In the early years they included, apart from Coase, such men as Ronald Edwards, Arthur Lewis, and Basil Yamey. At the other end of the quality–quantity scale there were the statisticians, Arthur Bowley and in his succession Roy Allen. As we have seen already, the fact that they had decided to turn their brilliant minds to social, and notably economic, subjects was to become instrumental for the emergence of econometrics, the empirical arm of the growing technical subject of economics.

The Master and the Puzzler

When Cannan retired he not only left a gap but it was clear that a new departure was needed. Perhaps Robbins is right 'that the whole organization of teaching of economics [at LSE] was in need of substantial overhaul',[28] but more importantly economics itself had reached a dead end. Harry Johnson, who enjoyed colourful language, observed forty years later that 'the 1920s now appear as a stiflingly dull period in economics', in which little more happened than the adding of 'footnotes and qualifications' to what had come to be a prevailing (Marshallian) consensus.[29] Allyn Young of Harvard seemed to fit the needs of a delicate transition perfectly. His sudden death eighteen months after his arrival at the School forced a more radical departure. Hugh Dalton, who straddled the two generations, describes how both he and Cannan 'believed in creating a deliberate variety of approach and outlook among teachers of economics at the School'. 'Partly with this in mind', Lionel Robbins was appointed to the chair in 1929, though his sponsor Dalton later expressed disappointment about what he thought was the disappearance of variety and its replacement by 'a much more uniform brand of right-wing economics'.[30]

This was Dalton the politician writing and it was also long after the events. In fact young Lionel Robbins had his hands full with the attempt to establish both himself and economics. He had barely published anything and the teaching needs of the School were enormous. However, the 30-year-old already displayed that combination of enthusiasm and measured judgement which was to become his hallmark in many positions of influence. Within three years he managed to write a number of technical articles as well as the influential *Essay on the Nature and Significance of Economic*

Science; transform the teaching, notably of what came to be called 'Economics, Analytical and Descriptive'; create an *esprit de corps* among the School's economists under his leadership to the point at which 'the meetings of the Professors of Economics became recognized as part of the informal constitution of the School';[31] and attract new members of the department, notably, in 1931, to the revived Tooke Chair, Friedrich August von Hayek. Small wonder, then, that soon Robbins had become 'the most influential figure of all'.[32] 'Economics at the LSE was dominated in the 1920s by Cannan and in the 1930s by Robbins.'[33]

Lionel Robbins was a big man in more ways than one. Born in 1898 in a village which had to make way for Heathrow Airport (as he liked to recall in later years), his father had been a market gardener, Liberal activist, and Strict Baptist. His mother, whose 'lively mind animated her beauty', died when he was 11, but his father decided not to send him to a 'minor public school'. Instead, Lionel cycled every day the five miles to Southall County School, where he made friends and enjoyed learning. Still, 'the society amid which my childhood was spent was essentially limited in standing and local in operation. It gave no sense of belonging to anything central or having responsibility in the way which even lesser members of the ruling classes had for what went on at the centre.' These are Lionel's own words, which may surprise those who met the greying Lord Robbins of Clare Market, who to many seemed to epitomize Britain's ruling classes, and not those of a lesser kind, by his suave and civilized manner as well as his superior ways. It is equally remarkable that the great lover and patron of the arts and Chairman of Covent Garden 'never saw a play or an opera until I was over seventeen'. The war drew him out of his somewhat provincial shell, and then university, though of course it was LSE and not Oxford. He never felt at ease with Oxford and 'the curious combination of worldliness and insularity' of much of its conversation.[34]

Perhaps his *Autobiography of an Economist* makes a little too much of a diffidence which in his generation was a useful vehicle for the road to the top. From an early point onwards, Robbins left little doubt in the minds of others that only the top was good enough for him. But the top of what? Friedrich von Hayek developed his own version of 'the hedgehog and the fox' *à la* Archilochos and Isaiah Berlin in a little essay, 'Two Types of Minds', the gist of which sounds even more poignant as he told it to Nadim Shehadi in an (unpublished) interview. 'I distinguish between *the master* of his subject, and what I call *the puzzler*,' Hayek says. The master remembers everything and therefore knows all the answers; the puzzler has no good memory and therefore has to think things out for himself every time he is asked. 'Now that describes very much the relation between me and Robbins.'

But where Isaiah Berlin remained ambivalent because he saw himself as a fox who would perhaps (or so he wistfully thought at certain moments) have liked to be a hedgehog, Hayek had no doubt about which was the better type of mind. Robbins, said the indiscreet octogenarian, for once abandoning his proverbial courtesy, 'was a master of his subject, but he had really very few original ideas'. He was, of course, by far the more brilliant teacher, but 'the questions came very largely from me', and since Hayek always had to think things out afresh, this 'led me sometimes to original ideas'.[35] Such judgements were less than wholly fair; Robbins was rather more fulsome in his praise for Hayek and retained warm memories of their partnership; yet Hayek's assessment contains a grain of truth. Certainly 'in his early years Robbins produced some fine theoretical papers',[36] and his later devotion to the history of economic thought as well as his outstanding role as a teacher left a lasting imprint on the discipline. Moreover, Hayek himself was not beyond criticism. Harry Johnson actually lumped them both together in one devastating comment: 'The failure of the leading figures at the School (Hayek and Robbins) to understand what Keynes was driving at . . . was a major intellectual catastrophe.'[37]

Failure or not, the inter-war economics department of LSE acted as a magnet for bright students from many parts of the world. Some have written about it, like Professor Eveline Burns, who described her conversion from a Geography graduate to an Economics Ph.D. at the 'tremendously exciting place' which the School was in the 1920s.[38] In 1981 the Atlantic Economic Society devoted a panel to 'LSE and its Contributions to Economics', for which Gerard M. Koot, A. W. Coats, and R. H. Coase prepared papers about the inter-war period. The Lebanese scholar Nadim Shehadi has assembled a (hitherto unpublished) remarkable collection of interviews with LSE economists of various orientations—including Hayek and Hicks, J. C. Gilbert and Brinley Thomas, Kaldor and Lance Beales—from which this account has greatly benefited.

A separate conversation by the philosopher W. W. Bartley III with Professor Ludwig Lachman, an early refugee from Nazi Germany who added an M.Sc.(Econ.) to his already formidable qualifications, gives an idea of where economic debate took place at the School: the 'Grand Seminar' chaired by Robbins, Hayek's own seminar, and the 'London Economic Club', in which LSE economists met with those from other colleges and also with Treasury officials. Finally, there was the 'Joint Seminar', a 'late creation' following the publication of Keynes's *General Theory*, 'for the purpose of overcoming what were then the high walls between London and Cambridge'.[39]

Robbins, Hayek, and the Second Dispute with Cambridge

This takes us into the second great war between LSE and Cambridge. Before we get immersed in it, it has to be noted that Robbins and Hayek were involved in two conflicts at the same time, one with Cambridge and one at home, within the School. The latter was by far the more vicious if also less interesting; we have alluded to it already because it had to do with Beveridge. Webb and Cannan (to say nothing of Foxwell, Acworth, even Gregory) were operating on different intellectual wavelengths, and Sidney Webb would have preferred another kind of economics, but he kept his views largely to himself and in any case never interfered with what teachers, once appointed, did at the School. Beveridge, on the other hand, was deeply opposed to the kind of economics represented by Robbins and Hayek. His 'methodology' was confined to a simple either–or: either 'facts' or 'concepts'. 'For Sidney and Beatrice, as for myself, economics was not an analysis of concepts but an inference from facts of society after unsparing examination.'[40]

Robbins's relationship with Beveridge had its ups and downs, or better, perhaps, it had a short up and a long down. Jose Harris tells us that in the 1920s he collaborated closely with Beveridge on the second edition of *Unemployment*, which was of course a characteristically empirical study, and that Robbins probably owed his appointment to the chair at least in part to the warm and friendly relationship established at the time with the Director. Soon after, however, Robbins seems to have changed his mind about the Director's scholarship. Certainly, after Hayek had arrived on the scene the two economics professors did not think of Beveridge as an economist at all. Beveridge in turn had got it into his mind that 'breaking up economics' was one of his main missions.[41]

On several occasions, Beveridge suggested that this curious notion was put to him by Sidney Webb. The evidence for the claim is scant, in fact no more than the transcript of Beveridge's interview with Sidney on 24 November 1932 about his reminiscences, in which Webb made the much quoted statement: 'I was in revolt against one Professor of Economics; I wanted a lot of Professors.' In other words, Webb wanted pluralism rather than any kind of orthodoxy, even, one must suppose, Beveridgian orthodoxy. 'Breaking up' the subject was hardly Webb's language anyway. Beveridge himself later admitted failure. After forty-two years, that is, when he left in 1937, LSE 'had not achieved' the 'purpose of breaking up economics and making it a science in the true sense of that word, and the purpose of making the circle of the Social Sciences complete' (the latter being a reference to the saga of 'social biology').[42]

Thus at home Robbins and Hayek won the battle that had to be fought, but the cost of fighting it was high. It absorbed an undue amount of energy and at times led to painful defeats, as when 'Beveridge's insensate hostility to pure theory'[43] forced John Hicks in 1935 to leave first for Cambridge, where, predictably perhaps, he did not feel happy so that he went on to Manchester. Ultimately, however, the dispute at home was a mere distraction compared to the larger conflict with the Cambridge of Keynes. Many have written about this turbulent episode in the history of economics, participants including Robbins, Hayek, John Hicks, and Nicholas Kaldor, historians like A. W. Coats, Nadim Shehadi, and Emiel Wubben, later economists like Harry Johnson, and, most recently, Brian McCormick.

Brian McCormick's book on *Hayek and the Keynesian Avalanche* contains a useful little table of what he calls the dramatis personae on the LSE side. (At Cambridge, apart from Keynes, the remote Pigou, Joan Robinson, and later some of the 'deserters', fewer names are relevant.) If one expands this table to fourteen leading LSE economists of the time, the most striking fact to emerge is how little exposure to Cambridge they had. Dennis Robertson apart (who in any case taught at LSE only during its wartime years at Cambridge), only Dalton had a Cambridge first degree, and not one had his early teaching experience there. On the other hand, four had Oxford degrees and five had been students at LSE. Two, Hicks and Kaldor, went on to Cambridge, and in both cases the story behind the move was not a happy one. Even in career terms, there were significant differences between the LSE economists and their Cambridge colleagues.

Once again the technical debates between the protagonists reached beyond their apparent substance. A personal element was never absent. Throwaway remarks about the 'muddled' Hayek or the 'suburban economist' Robbins may have been no more than Keynesian *sottises* but they hurt. The Londoners gave as good as they got. Keynes, said Hayek, had 'a great many gifts, but I don't think he was really a great economist'.[44] (Kaldor later returned the compliment on behalf of Keynesians when he commented on Hayek the philosopher: 'That's a field he is much better at than economics.'[45]) Lionel Robbins warned 'the historian of the future, if he wishes to treat of the relations between London and Cambridge during this period . . . that any generalizations that he may wish to make must fit facts of considerable complexity if they are not seriously to misrepresent the situation'.[46] Complex the relations certainly were. Opinion either in

Cambridge or in London was not homogeneous. There were, in the course of the dispute, important transfers of allegiance. Moreover, LSE and Cambridge were working together on common projects, above all the London and Cambridge Economic Service. Yet of the dispute there can be no doubt and, as with the earlier Cannan–Marshall divisions, 'the range of issues involved was broad, including methodology, theory, policy, ideology, the role of the economist in public life, and the proper code of professional conduct'.[47] At the same time the key issue as well as the main difference of view were quite clear in this second dispute between London and Cambridge. It was about the slump of 1929 and the way to combat it, by deflation or by expanding public expenditure.

In July 1930 the Prime Minister, the increasingly beleaguered Ramsay MacDonald, appointed a small 'secret' committee of economists to help him find solutions to the slump. Nicky Kaldor thought even fifty years later that the committee had been so secret that he told his interviewer, Nadim Shehadi, '[you] mustn't write this in your thesis—there was a terrific enmity at that time between Robbins and Keynes'. In fact, Robbins had written about the committee in his autobiography (which Kaldor by his own admission had not read), and now we also have the detailed and characteristically fascinating account by Robert Skidelsky in his biography of Keynes. It is as well to bear in mind from the outset Skidelsky's conclusion that the report by the economists' committee 'is important biographically and intellectually, but left no immediate mark on policy'.[48] In fact it was soon buried by cabinet.

The economists' committee was the subcommittee of a subcommittee within MacDonald's 'economic general staff'. It was a group of five, convened and chaired by Keynes and including the senior Cambridge theorist Arthur Pigou, the government economist Hubert Henderson, the public-sector businessman (and Vice-Chairman of LSE Governors) Josiah Stamp, and the young LSE professor Lionel Robbins. First convened on 11 September 1930, the committee held no fewer than six mostly two-day meetings either in London or at Stamp's place in Shortlands or in Cambridge before the report was initialled six weeks later on 24 October. More precisely, two reports were initialled, to Keynes's and the others' dismay, the second a minority report by Lionel Robbins.

The differences between Keynes and Robbins were essentially two. The first had to do with the causes of the slump. 'Robbins's [view] derived from Austrian economics.'[49] The slump was the 'cure' for the disease created by the post-war failure to adjust the standard of living to reduced circumstances. What was needed now was not just a reduction in real wages but a contraction in public expenditure. Robbins later denied that he had actively advocated deflation, but the effect of his view must surely be described as such. Keynes did not disagree with the need for wage flexibility, but saw high real wages as part of a fall in output and employment (which led to a rise in the share in output of the still employed) and therefore recommended focusing on rising output, if need be by public expenditure. The second difference was more narrowly ideological, though its starting-point lay in the apparently immutable fact that sterling would not be devalued. Keynes and the three others were prepared to recommend revenue tariffs to reduce imports and subsidize exports for a limited period. Despite Keynes's strictures about a 'grin and bear it' attitude which could not be sustained, Robbins stuck to his free-trade last. So did, in the event, a Labour Government in which Dalton was Parliamentary Under-Secretary in the Foreign Office.

Neither Robbins nor Keynes were in their most equanimous mood at the time. They must also have grown a little tired of each other during the almost continuous meetings in the autumn of 1930. Harsh words were said, though they never led to a break in relations. Within weeks, the two worked together again in the committees of the London and Cambridge Economic Service. More seriously, Robbins changed his mind, at least on the first of the two matters under dispute. In his autobiography he wondered aloud how he had got himself into his earlier state of mind in the first place. 'The trouble was intellectual. I had become the slave of theoretical constructions which, if not intrinsically invalid as regards logical consistency, were inappropriate to the total situation which had then developed and which therefore misled my judgment.' Belatedly, in other words, Robbins discovered that Continental theories were inappropriate for the British experience of the great depression. 'I shall always regard this aspect of my dispute with Keynes as the greatest mistake of my professional career.'[50]

In the mean time, another dispute erupted, this time between Hayek and Keynes. It arose over monetary theory. Hayek had published in *Economica* a long and distinctly unfriendly review article of Keynes's *Treatise on Money*. Keynes was clearly stung and reacted in the manner which would have sounded more than strange in countries without gentlemen's clubs or even Oxbridge colleges: 'Hayek has not read my book with that measure of "good will" which an author is entitled to expect of a reader.' Any author? Of all readers? At all times? The dispute went on for years. Its hard core was the question of whether monetary disequilibria and the resulting maladjustments of production require above all a return to 'monetary neutrality', or whether there are desirable interventions to absorb shocks and turn the cycle. 'Nature's cure' versus the State, as it were: here were the first rounds in a bout that was to last half a century.

These rounds went to Keynes. Hayek was never able to deliver a decisive blow against Keynes's interventionism, and perhaps he did not try at the time. He did not even review the *General Theory* when it appeared in 1936. Hayek later claimed that it was impossible to review Keynes because he always changed his mind. Keynes was certainly capable of ducking and weaving through an argument. However, Skidelsky may well be right that 'Hayek did not want to expose himself to another mauling from the Keynesians' either, and above all that there was 'a complete intellectual gulf between the two men' which forbade debate. Instead Keynes resorted to invective, calling Hayek's *Prices and Production* 'one of the most frightful muddles I have ever read', and his later articles 'the wildest farrago of nonsense yet' and 'rubbish', but above all he ceased to read Hayek or to take him seriously. As a result, 'Hayek remained a bystander as the Keynesian Revolution unfolded.'

Like Robbins, Hayek carried on amicable personal relations with Keynes. 'We get on very well in private life,' wrote Keynes in 1933; and Hayek admired in Keynes 'the magnetism of the brilliant conversationalist with his wide range of interests and bewitching voice'.[51] But for LSE the risk was growing that the 'Keynesian avalanche' would engulf its Department of Economics. The School did not know how to react to the evident ascendancy of Cambridge. 'What transpired at the LSE therefore was prevarication, obstruction and filibustering.' Perhaps Brian McCormick's summary is a little overdramatic, if only because it leaves out the two main protagonists, Robbins and Hayek, but it gives one a sense of the lengths to which academic disputes can go:

Kaldor was told by Robbins that he should not lecture on controversial topics. There were rumours that Lerner was refused a seminar room in which to discuss the *General Theory*, although he seems to have avoided the embargo by combining meetings of the executive board of the *Review of Economic Studies* [described by Joan Robinson as 'the children's newspaper'—R.D.] with seminars on Keynes's work. Furthermore Scitovsky has recounted the occasion when students invited Keynes to give a lecture but were refused the use of one of the large lecture theatres. Keynes therefore lectured in a small room on the fourth floor of the School and as a result there was a large audience which listened outside. Scitovsky stood on the staircase of the second floor. In mitigation it must be said that a similar state of affairs occurred at Cambridge. Robertson avoided Hicks when he went for his interview because he feared that the knowledge that he and Hicks were old friends might result in Hicks not getting a lectureship. And when Gregory retired and Robertson was asked to be an assessor for the chair he volunteered to take up the vacant post, although he did not move to London because the School was transferred to Cambridge in 1939. In both institutions differences in research programmes spilled over into personal conflicts.[52]

Between 'differences in research programmes' and 'personal conflicts' there is of course a third dimension of division, politics. The dividing lines were not very clear in those confused years of the late 1920s and early 1930s. Keynes and Robbins had no qualms about advising the same (Labour) Government, and they were probably both vaguely liberal at a time at which Liberals with a capital 'L' had begun their long retreat. But public expenditure and free trade were just the sort of issues which marked the future watershed between left and right. It was in these years that the LSE economists moved to the right and those at Cambridge to the left. Those who physically went to Cambridge (and in some cases elsewhere) from LSE were more likely to be inclined to left-wing views than those who stayed. However open the LSE department was, Robbins 'could make life rather unhappy' for those who took different views, or so Kaldor remembered; indeed he 'weeded out' people 'who were not ready to echo [his] views of economics'.[53] These were views of economic policy above all, and they established the LSE department as one which for many years ran counter to the spirit of the times as well as the dominant mood of the School.

The Cost of Internationalism

In the end, however, it was the research programmes that mattered. Harry Johnson has written of the 'six revolutions' which transformed economics in the 1930s. 'They are the professional revolution, the revolution in welfare economics, the general equilibrium and mathematical economics revolution, the imperfect or monopolistic competition revolution, the Keynesian Revolution, and the empirical revolution.' Younger members of staff at LSE contributed to them all, though in the event the three whom Johnson quotes, Kaldor, Lerner, and Scitovsky, soon left the School. The department as a whole, however, 'can be credited with significant contributions only to the first three'. Johnson associates the first two revolutions with Robbins, the third with Hicks, but also Roy Allen. Two of the other 'revolutions' have to be credited to Cambridge, and the last, the empirical revolution, is 'almost entirely an American creation'.[54]

This is not the whole story of LSE economics, nor is it the end of it. We are after all dealing with the Beveridge years. In these years the foundations were laid not only for those younger scholars who have already been mentioned (and of whom two, John Hicks and Arthur Lewis, as well as the grand old man, Hayek, were to win Nobel prizes in economics) but also for the generation of Roy Harrod, A. W. Phillips, and James Meade (another Nobel laureate) during and after the war. Moreover, during Beveridge's years, and using his own standards, important books were produced, including Robbins's *Nature and Significance*, Hayek's *Prices and Production*, and Hicks's *Value and Capital* (published after his departure but written at LSE). Yet the question remains how it was that the most significant school of economics in the country and beyond managed to miss 'scientific revolutions' twice in its first half-century, and in its chosen specialized subject at that.

Personalities provide a part of the answer, and the deliberate detachment from the mainstream by the founders and by William Beveridge another, but there is a more serious issue. What was Robbins's own analysis when he recanted? 'The trouble was intellectual.' Robbins's errors, Skidelsky argues, 'derived from Austrian economics'. Austrian economics was, of course, present at the School in the shape of Hayek. It was present also by the visits of distinguished economists like Schumpeter and Machlup. Nor did the visitors all come from Austria. The Stockholm School of Economics made its presence felt, indirectly by way of Brinley Thomas, but also by direct encounters with Lindahl, Frisch, Ohlin, Myrdal. The reading lists of LSE courses included books by von Mises and Wicksell, Böhm-Bawerk and Taussig, Walras and Pantaleoni, many of them in the original language. All this came easily to LSE though it contributed to the differences in atmosphere between Cambridge and the School. Cambridge was both less historical and more parochial. 'It was natural therefore for us to seek wider affiliations, both in time and in space.'

But it turned out to be dangerous too. Twice in its early history LSE economics paid a price for its openness to the world. First, it hung on longer than was intellectually useful to the approach of the German historical school. In his *English Historical Economics*, Gerard Koot has argued that invoking the German tradition was designed to undo the practical failures of classical economics from Manchester to Marshall. Whatever the motive, the march of ideas bypassed this intention; it certainly threw economics in Germany itself back by a generation. On the second occasion, LSE economists stuck to theories which were addressed to problems of another time and place. When Robbins rethought his position, he still concluded that his 'Austrian' approach had its use 'as an explanation of a *possible* generation of boom and crisis', 'but, as an explanation of what was going on in the early thirties, I now think it was misleading'.[55] The episode raises big questions about the universal applicability of economic theories, and more, of the real basis of advances in the social sciences.

It also illustrates the melancholy fact of life that strength can turn into weakness (though the reverse is also true, and may even have happened with LSE economics when history was rediscovered and indeed when monetarism, and Hayek, took revenge on Keynes). The internationalism of LSE was, and is, one of its greatest strengths. If it blurred the focus of some economics professors for a short while, it also widened the horizon of hundreds of students and many young members of staff. Harry Johnson, who propounded the theory of the three missed economic revolutions at LSE, never lost sight of what he himself called 'the essential thing': that the School

> is the one centre of economic teaching and research in this country that is genuinely international in its orientation, in the sense that it is not merely an established British university that allows itself the luxury of a few foreign staff-members and students for the sake of variety and balance, but a world university that tries both to keep in touch with whatever of intellectual importance is going on elsewhere in the world, and to admit to its scholarly fellowship students of quality whatever their origin may be.[56]

LASKI AND POLITICAL SCIENCE

Odd Man In

Harold Laski did not take sides in the disputes between his LSE colleagues and the economic gurus of the fens. It would have been difficult for him to do so. Economics was the weakest weapon in his considerable armoury. Moreover, in one way or another he was out of sympathy with all combatants, though in 1930 he was not yet prepared to speak up against some of them. He held his tongue about Ramsay MacDonald's second Government as long as its members sought his advice; there was even talk of a peerage; but deeper down Laski saw his Prime Minister as beholden to the old ruling class, so that he was not entirely surprised by the 'great betrayal' of August 1931. When the Economic Advisory Committee was set up he detailed to the Lord Chancellor, his long-standing acquaintance Sankey, the objections to such committees which he had voiced in the *Grammar of Politics*. In an article in the *New Statesman*, Laski supported Keynes's idea of a revenue tariff (which alienated him even more from his LSE and Labour colleague Dalton), but while he gratefully accepted Keynes's support in the later 'free speech affair' he never fully trusted the Liberal. Keynes for him was a tinkerer whereas he was moving to more and more extreme views. Laski's biographer is probably right in suggesting that 'he failed to examine Keynes's economic ideas in any depth'.[57] With Robbins, on the other hand, Laski had intellectually almost nothing in common, yet in those days at least the two lived in amity, shared their love of books, their dislike of Beveridge, and even edited *Economica* together before *Politica* was born. Neither in 1930 nor later would Robbins have written as unkindly about Laski as Hayek did in 1944, when he held his 'Marxist' colleague partly responsible for *The Road to Serfdom*.

Amity or enmity, it is easy to see that Harold Laski was at the centre of things. He was also at the centre of the School, from almost the beginning of his thirty years as a teacher in 1920 to the bitter end. Being at the centre did not mean, in Laski's case, that he was in charge, as Robbins was, to say nothing of Beveridge; it did not mean either that he was universally liked, as Tawney was and Eileen Power. Nor was he the focus of great academic disputes; there were no such disputes at the time in political science. It meant that he was impossible to ignore. Students adored him; colleagues had strong feelings about him, not excluding envy and hatred; there were few subjects on which he did not express an opinion; his public views at least were often controversial and sometimes embarrassing to his friends, his Party, the School. In short, with Laski there was never a quiet moment even when he was ill (which happened quite often) or in the United States (which also happened on several, extended occasions). It is probably true to say that no one who wrote about LSE after 1920 at any length has failed to mention Laski, which could not be said of anyone else, with the possible exception of Beveridge.

Harold Laski was born into a Manchester Jewish family in 1893. The precocious boy went by way of Manchester Grammar School to New College in Oxford. By the time he arrived there in 1911, he had, to the unforgiving dismay of his family, married (not quite in Gretna Green but in Glasgow) a non-Jewish woman, Frida Kerry, who was eight years older, bore him a daughter called Diana in 1916, and remained his emotional and practical mainstay until his death in 1950, which she survived by twenty-eight years. Frida had kindled his early interest in eugenics, so that Harold at first started on a science degree, but he never got beyond the intermediate examinations. For one thing, he was a very impractical boy. 'Harold's brains were all in his head, and his hands, at that time and all other times, remained totally unskilful.' In fact his hands were Frida. But the imbalance also meant that his ability to write on general subjects was not matched by that 'of grasping any concrete mechanism such as the circulatory system of the frog or the anatomical structure of the earthworm'.[58] Science, biological science at any rate, was not for him.

However, young Laski braved adversity and completed Oxford with a first class degree in history in June 1914. All the time, he was supported by his father on condition that Frida stayed away from him, in Scotland, which so far as the young couple were concerned was more painful than divisive. But after Oxford, what? A brief spell at the *Daily Herald*, writing leaders for George Lansbury, who was then editing the Labour daily, foreshadowed things to come. Laski was clearly a man of the left though at the time that meant Liberalism as much as Socialism. When the war began, Laski was declared physically unfit to serve, and found himself at something of a loss. Thus the invitation to teach at McGill University in Montreal came as relief. The couple left in September 1914, not to return to England for six long years. McGill was a disappointment but it paved the way for Harold Laski's lifelong friendships in America, with his contemporary Felix Frankfurter at Harvard, his transatlantic 'father' Justice Oliver Wendell Holmes, and others in the American political class. In 1916, Laski became an instructor at Harvard, already a political scientist with a special interest in law. All the while, however, he also published articles on contemporary affairs and showed his concern for the downtrodden and oppressed. Few in Laski's generation equal his record as a defender of basic civil and human rights.

The trouble was that whenever he defended a cause he tended to go two steps further than was useful, and even retracting one step left him more vulnerable than his cause. When the Boston police went on strike in 1920 to defend their right of association as a trade union, Laski's public defence of them went far. It certainly embarrassed Harvard, though President Lowell proved a liberal with the courage of his convictions and defended the young instructor's right to speak his mind against hundreds of furious Harvard alumni and potential benefactors. Still, the President must have been secretly pleased when Laski decided to return to his native country. Graham Wallas, who was lecturing at the recently (and with Laski's help) founded New School for Social Research in New York at the time, met and liked Laski—who by articles and letters, and sending out copies of his first book had made sure that he was not forgotten at home anyway—and proposed to the newly appointed Director, William Beveridge, that he be offered a job. Beveridge agreed; Laski came, to a temporary lectureship at first, then, in 1920, to a lectureship, a readership in 1923, and, at the age of 34, to a chair in succession to Graham Wallas from 1926 until his premature death in 1950.

No Science of Politics in England

The Department of Political Science which Laski joined may have carried the other half of the School's name in its title, but compared to the

Economics Department it was small and institutionally almost non-existent; it was essentially a one-man band. A tabulation of its staff corresponding to that of the School's economists and their comings and goings would have made no sense, and that despite the fact that courses were listed in the Calendar under the double heading 'Politics and Public Administration'. In a typical year in the late 1920s, twenty-seven lecture courses and seminars would be announced under this heading. Of these, no fewer than eight would be offered by Professor Laski on subjects ranging from 'Executive and Judiciary Problems' to 'Political Ideas of the Medieval World'. From 1921, K. B. Smellie was there to help Laski, and after Kingsley Martin's departure in 1926, Herman Finer held the third post in Politics. On the public administration side, Sidney Webb held a chair until 1927, though most of the lecturing was done by H. B. Lees-Smith and, from 1926, W. A. Robson. If one looks at the equivalent Calendar section ten years later, very little had changed. Dr Lees-Smith had come back from a stint as an MP and Mr Richard Greaves had appeared on the scene along with Graham Wallas's daughter, May Wallas, but the main burden of teaching in politics was still borne by Professor Laski and Mr Smellie.

The development of public administration at the School deserves a separate mention at a later point in the School's history. Peter Self, who contributed the thoughtful chapter on 'The State versus Man' to the volume *Man and the Social Sciences* at the School, links politics and administration intellectually but observes the backwardness of administrative studies in the Beveridge period. Political science, however, did not have a much better time. While Graham Wallas was one of those who invented the empirical and analytical study of politics in Britain, he had few if any followers on this side of the Atlantic. Perhaps Peter Self is right that, while political scientists are just as keen as others 'to make their discipline genuinely scientific', they have developed 'an inferiority complex because they seem less successful than economists, psychologists, or anthropologists'.[59] But why are they less successful?

Noel Annan, in his 1958 Hobhouse Lecture at the School, detected a peculiarly English problem here. 'We are still trying to produce ore from mines which have for long been worked out, namely the old concepts of state, society, will, rights, consent, obligation; and we have turned our back on the social studies and methods of analysis which alone would restore some value and new meaning to those concepts.' For LSE, this certainly rings true. Harold Laski and his successor, Michael Oakeshott, were much concerned with the 'old concepts'. Modern political science never took a hold at the School, or in most British universities for that matter. A glance at two books published in the early 1960s, one English and entitled *The Nature and Limits of Political Science*, the other American and called *Modern Political Analysis*, makes one wonder not only whether the titles should not have been reversed but also what the two disciplines, 'English' and 'American' political science, have in common.

'American' political science came to have three ingredients, none of which were found to any significant extent at LSE. One is political analysis in Robert Dahl's (or David Truman's, perhaps David Easton's) sense. *Who Gets What, When, How?* was Harold Lasswell's succinct basic question and the much-quoted Dahl offers his equivalent to Robbins's definition of economics for politics as the study of 'any persistent pattern of human relationships that involves, to a significant extent, power, rule, or authority'.[60] Economics is about scarcity and choice, and politics about power and legitimation. One notes that history does not enter. This is true also for the other two elements of 'American' political science. Political survey research has now reached Britain, though even that process started late, and at LSE it did not happen in the Department of Government (as the politics department came to be called from the 1950s). The third element, the economic analysis of politics, though initiated by the Austro-American Joseph Schumpeter, has remained a peculiarly American exercise to the present day.

Why did political science in this sense never really catch on in Britain? Why did Sidney Webb's hopes and Graham Wallas's beginnings not take root at LSE? One reason probably is the enormous strength, at least in numbers, of traditional political philosophy in the old universities, and notably in Oxford. 'An intellectual gulf exists between traditional political philosophy and modern political sociology,'[61] and those happily ensconced on one side were not going to try and build bridges to the other. But LSE? Why did it not break new ground? Was Harold Laski deep down more beholden to the established style than his colleagues in economics? Did the much-cited dapper appearance of the man tell as much about him as the verbal eruptions which carried him away?

Another point must not be forgotten. When it comes to application, modern political science has turned out to be less effective than modern economic science. There may thus be reasons for the 'inferiority complex'. History and values are inextricably woven into the political process. The German cultural historian André Kaiser notes with astonishment that the founders of British political science never conducted a debate about value judgements: 'They were not really interested in basic methodological questions of the discipline. The old themes of moral philosophy were notated as a kind of second voice even in the most positivist programme. Thus political science remained comparatively open for historical or moral-philosophical issues.'[62]

Kaiser supposes that Graham Wallas must have been pained when Laski, in his inaugural lecture of 1926, held the social heritage of history against the 'inherited impulses' of individuals. But this is not said against Laski; on the contrary, Kaiser dedicates his little study in 1993 to the memory of Harold Laski. Professor Laski's inaugural makes such dedication plausible. Even after nearly three-quarters of a century its irresistibly progressing phrases seem to drown all possible objections in sheer aesthetic pleasure:

> My object as the occupant of this chair is not to create a body of disciples who shall go forth to preach the particular and peculiar doctrines I happen to hold. It is rather that the student shall learn the method of testing his own faith against the only solid criterion we know—the experience of mankind. That does not, of course, mean that in the exposition of political philosophy it is one's business to pretend to impartiality. In any case that is impossible; for in the merest selection of material to be considered there is already implied a judgement which reflects, however unconsciously, the inevitable bias that each of us will bring. The teacher's function, as I conceive it, is less to avoid his bias than consciously to assert its presence and to warn his hearers against it; above all to be open-minded about the difficulties it involves and honest in his attempt to meet them. For the greatest thing he can, after all, teach is the lesson of conscious sincerity. More truth is discovered along that road than can be found on any other.[63]

The sceptical reader swallows hard once or twice. 'The experience of mankind'—who says what it is? And honesty and sincerity from the mouth of the 'pathological liar' (as Hayek thought), or at any rate the inveterate inventor of tall stories? Yet Laski was deeply serious when he spoke of open-mindedness and the teacher's function, and moreover lived up to his own precepts. In fact, this is the other and arguably more convincing response to the German *Werturteilsstreit*. In this dispute, which came to a head just before the First World War, the 'socialists of the lectern', for the most part Imperial German 'socialists', unashamedly defended the right to proclaim values *ex cathedra* in academic lecture theatres, whereas the other side, in the person of Max Weber, advocated a selfless, indeed self-effacing attempt at 'objectivity', a puritanical control of passions and values. The first position is unacceptable, and the second unachievable; combining explicit partiality with openness for doubt, debate, and difference thus makes a great deal of sense.

Max Weber's doubts in Laski's position would have arisen from the fact that the university teacher has an authority which it is easy and tempting to abuse. Laski in particular had such authority, since he combined position with flair. Laski the teacher influenced at least two generations of students, British, Indian, American, and others, many of whom became the standard bearers of the social democratic age. Some, like Judith Marffy-Listowel and B. K. Nehru, whose judgements we have heard, or Laski's colleague Lilian Knowles, who was also one of his steadfast defenders, did not adopt all or any of Laski's views. Even those who were influenced by his arguments did not follow him all the way to his sometimes Marxist conclusions. But this merely shows that Laski, the man of strong and at times extreme views, did not abuse his authority as a teacher.

The more difficult question is, what remains of Harold Laski's political science. We now have a handful of biographies of the man, two of them published in the 1990s. Tributes written in India, at the Laski Institute in Ahmadabad, betray the lasting gratitude of Indians to the man who stood up for their independence early, who propounded a view of the modern state which seemed relevant, and who taught many of those who led the new country in its early stages. Of the British biographies, the earliest, by Kingsley Martin, still furnishes the most real and endearing portrait of Laski the person. Granville Eastwood has assembled much material to tell us how Laski was seen by others, in his various roles as teacher, political theorist, politician, and friend. Michael Newman has tried to salvage Laski's reputation as a political scientist, notably from the elaborate attack by Herbert Deane. Isaac Kramnick and Barry Sheerman have produced what is likely to remain the most comprehensive and in some ways authoritative biography of the man, though in the course of their labours they perhaps got a little carried away in claiming that Laski's 'life and career stand monumentally astride the first half of the twentieth century', or that Laski is 'one of the twentieth century's principal public intellectuals'.[64]

None of his biographers claim that any of Laski's numerous books can be described as a lasting classic. To be sure, his contributions to the history of political thought will be read for a long time to come; they show the teacher and his phenomenal knowledge (including that of the collector of rare books) at his best. Recent readers have also studied with benefit Laski on sovereignty, or on American democracy. But even the *Grammar of Politics* would not make it to the hypothetical list of great books written at LSE during the Beveridge years. Michael Newman has weighed the strengths and weaknesses of the '*magnum opus*' very fairly. He sympathizes with Laski's 'doctrine of rights', which anticipates the notion of 'social rights' or 'positive freedom' and sees in it 'a *moral* justification for democratic socialism', but he admits that, even if one accepts the objective, the theory has few original features and the book itself is more a collection of arguments on specific topical subjects than a treatise of the foundations of politics.

Laski's books, including the *Grammar* in its various editions from 1924 to 1938, are in fact a running commentary on his times and the author's intellectual movement with or more often against them. Gradually since the mid-1920s (as Newman endeavours to show), or suddenly in 1931 (as Deane argued), or in fits (as Kramnick suggests), Laski moved from his early guild socialism through social democratic gradualism to a version of Marxism. First he advocated a civil society American-style, then a benevolent, well-administered state, and in the end a revolutionary, even violent breakup of the alliance between economic interest and political power that prevented change. All the while, however, Laski wanted to be involved, not just as a professor of politics but as a political professor, as teacher and journalist, author and campaigner, friend at one and the same time of needy students and mighty political leaders. Thus he talked revolution but enjoyed living here and now. He occasionally spoke of violence but always practised kindness.

A Memorable Teacher

To say that Laski's 'knowledge of economics was limited'[65] is both true and beside the point. There is a sense in which Laski did not want to know any economics. Economics is the science of more or less, but Laski was interested in all or nothing. This is more than a matter of temperament. Economics has to do with provisions which can be scarce or abundant but are certainly measured on a scale of gradations; economic choices are never absolute alternatives. Recent American theories of democracy have tried to apply this approach to the political process using votes as the yardstick equivalent to dollars or pounds. For Laski this was an alien way of thinking. It is no accident that he turned to law as a favoured alternative subject and sought a doctrine of rights as the theoretical foundation of politics. His interest was always in entitlements which people have or do not have rather than in gradations and quantification. As a person, Laski enjoyed the political game, and the friendship of its actors, but as a political scientist he got carried away by abstract and absolute reasoning. In fact, using his later colleague Karl Mannheim's typology of political styles, Laski was the epitome of the 'abstract' progressive who as a rule loses out when confronted with more reality-conscious conservatives.

It is easy to see why two biographies of Laski were written more than forty years after his death, and why many others remember the man with affection or sometimes with dismay. He was a person of several layers, not all of which fitted neatly together but which added up to a most unusual figure. The top layer was that of the dapper, chain-smoking, alert, well-informed, and opinionated intellectual of the inter-war years. Had it not been for his quintessential Englishness even in his love of America, he would have seemed equally at home in Berlin or Paris. Talking was for him the medium of life, so much so that he allowed his words to run ahead of his thoughts and his knowledge, which often amused his audiences, who soon found out about this weakness, though on two or three occasions such weakness got him into deep water, and worse. The inflammatory speech to the wives of the striking policemen of Boston in 1919 was, as we shall soon see, only a foretaste of things to come.

Laski the intellectual also saw himself as a politician, which was more unfortunate because he had no real sense of power, how to get it and what to do with it. When he exercised influence, as he did on several occasions from the General Strike of 1926 through the relationship with American politicians including Franklin Roosevelt to the chairmanship of the Labour Party in 1945, he left few traces. There is no Laski equivalent to Keynes's new economic policy or Beveridge's Reports. Laski's political career, for what it was, had more to do with being there and knowing people than with achieving particular ends. Could it be that a growing sense of such futility made him use more and more extravagant language?

The next layer is that of Laski the scholar, which we have seen shows a mixed picture. But the closer we get to the inner person the brighter the colours become in which he must be painted. He was an intensely loyal man for whom treason or even betrayal were thoroughly alien. His loyalty extended not just to people but also to a small number of institutions. America was one, the Labour Party another, and then there was the London School of Economics, to which he formed a deep and lasting attachment. For the most part, this was mutual, but over the years the School and its Directors have not always honoured the profound commitment of the School's more difficult members. Harold Laski had a story to tell in this regard, as did, in his later years, Lionel Robbins. There were others.

Then we get to the layer of Harold Laski the teacher. Testimonies to his teaching are legion and deeply moving. Laski was a memorable lecturer, unrivalled even in a School which was not short of public speakers. His Socratic ways with individual students whom he helped discover unexpected talents have been described by many. He made everyone feel important. And he cared—for the intellectual development of young people, but also for their next hot meal. Many a young visitor to Laski's office found an unexpected pound note in his or her pocket when they left. The student tributes in the special issue of the *Clare Market Review* after his death are moving and suffice in themselves to testify to a life well lived.

I was only one of a very large number of people to whom Professor Laski brought great personal happiness—apart from the intellectual joy he gave to everyone he taught. His generosity in terms of human happiness meant far more than can ever be recorded. For myself, I realize with deep gratitude that most of the things which I now hold to be good in life were made possible for me by Harold Laski.[66]

Of Laski the most private man we know pleasingly little. For that we must thank Frida, and Harold too. There was no vanity about their relationship. It was sustaining and real.

All this needs to be said before the shadows of other and later events are allowed to temper the picture. Laski was almost the epitome of the academic teacher. But advancing political science is not the achievement for which he is remembered. In later days even students were heard to say: 'At lunchtime I went to Laski for fun, and at five to Smellie to learn something.'[67] Laski did not establish the subject of political science at LSE. Since his successor did not do so either, a question mark remains. But Laski instilled the excitement of a school of (economics and) political science, its forever unresolved tension with the outside world, into the minds and hearts of more than one generation of students. He defined, as it were, the inimitable LSE experience.

TAWNEY AND ECONOMIC HISTORY

Inventing the Subject

The more technical economics became, the more distinctive did the economists at the School appear. 'In a sense the economists were a group by themselves,' said Hayek later, and even added that the Economics Department 'was something quite isolated'.[68] One can hardly call it marginal; after all, 'the economists sort of regarded themselves as having the key to the universe'; but since 'they hadn't' (this is Lance Beales speaking[69]), the claim made them not only separate but for some slightly suspicious. Politics on the other hand was, at least as long as Laski was there, central, though almost more as an experience not to be missed than as a developing discipline of the social sciences. Leaving public administration on one side, it was not so much a discipline or a department as a one-man show. The subject which had the strengths of both economics and political science and the weaknesses of neither was economic history. There are not enough organs in the body to allocate one to every department of the School and in metaorganics we have used up the heart, Laski's political science, and the mind, the economics of Robbins and Hayek, already. But it would not be wrong to describe economic history as the soul of LSE in its first fifty years. This had much to do with those who taught the subject, and above all with R. H. 'Harry' Tawney.

There is also a more substantive reason why economic history was central. Michael Postan has explored it in the little gem of analysis entitled 'Time and Change' which he contributed to the volume edited by Robson on *Man and the Social Sciences*. Postan mocks and applauds at the same time the 'imperfect insularity' of English social science. What went on elsewhere was both noted and not taken very seriously. This meant that Britain did not fall lock, stock and barrel for the historicism of Continental and notably German social science before the First World War, nor did it swing all the way to the other extreme when the mood of the intellectual 'stock exchange' turned. 'The lessons of history had been overproduced; their market value fell.' LSE certainly began with a strong emphasis on history, but when 'the anti-historical reaction on history itself' occurred, it remained for the most part non-violent. History 'peacefully occupied and then equally peacefully evacuated the field of L.S.E. economics', though Malinowski's fight against historistic anthropology (as practised above all by the 'diffusion theorists' at University College London) was not quite so pacific.

Postan is looking for 'a compromise between the time-oriented and time-allergic thought'. Economic theorists may think that the more empirical and historical disciplines are there to provide a testing ground for their hypotheses, but they are mistaken. For theories to be usable even as such they have to be 'reincorporated into the totality of the situation from which they have been abstracted. This is the work which the more broadly based social scientists, and above all the economic and social historians, can do best.' What is more, the second generation of LSE historians did exactly that. Postan mentions Eileen Power, who became his wife in 1937 and died suddenly at the age of 51 in 1940, and more particularly, Tawney. Far from being split between the economic historian and the social moralist, Tawney made both fruitful for each other.

> His historical work correspondingly centred on the problems which his social tracts laid bare: above all, the origin, the limits and the consequences of the market economy and the unlimited profit motive which actuated it. It was as problem-oriented a history as there ever was, and it reached to propositions and conclusions as general as social science could ever hope to establish.[70]

One does not have to accept every word of Postan's sweeping analysis—or even the singular, social science—to be stimulated by it. The first generation of LSE economic historians were the ones who did not distinguish between economics and economic history. Actually, a separate department did not come into being for some years, and even then it was called Department of History, although economic history was a special subject of the B.Sc.(Econ.) from the introduction of the degree itself. In so far as an identifiably separate subject existed during this period, its name was Lilian Knowles. The formidable yet (even for her students and colleagues of different political inclinations) lovable Tory lady had been (as Lilian Tomn) one of the first LSE research students in 1896 and later an occasional lecturer, before she returned in 1903 as a lecturer, later reader, and in 1921 Professor of Economic History. Along with George Unwin at Manchester, she was the first holder of such a chair in the country.

Lilian Knowles published notably her joint research with Cunningham on commercial history. However, few would disagree with N. B. Harte: 'She was destined to become a great teacher of a new but established subject, rather than a pioneer in its development.'[71] Eileen Power did initiate certain developments, yet much the same would later be said of her, and also of Lance Beales. Indeed even Tawney—like his colleagues in other departments, Robbins and Laski—is sometimes praised at least as warmly for his teaching as for his published work. F. J. 'Jack' Fisher, an assistant in economic history from 1930 and later successor to Power and Ashton in the chair, had a characteristically caustic comment on the suggestion that 'the leading people at the School were exciting teachers': 'Yes, I think the outside world said that they were charlatans. I think you have to be a bit of a charlatan to be an exciting teacher. But they were exciting teachers.'[72] A charlatan pretends to be what he isn't, and to pass as knowlege what he invented on the spur of the moment. Laski certainly did that from time to time, though Eileen Power wrote out every word of her brilliant lectures (which led Lance Beales to complain that 'her thought got stereotyped'[73]) and Lionel Robbins describes how 'the finest lecturer of his time', Graham Wallas, 'wrote his lectures out word by word, learnt them by heart and practised them in front of a looking-glass'[74] (though Robbins decided not to imitate such teaching by mirrors).

Jack Fisher was not just a past master of the caustic putdown. When interviewed in 1978 by his successor Theo Barker about his memories of the School in the 1920s, he made the slightly facetious yet important comment that 'all the ideas hadn't been used up then'. 'Economic History teaching in 1920 was marvellous. There was very little written. You could make it up as you went along. [Charlatans after all?—R.D.] In fact anything you said would come as a revelation from on high to people who had read even less than you.'[75] Lance Beales made a related point to O. R. McGregor by way of 'justification of my habit of not writing'. 'I was more interested in the material. I felt that I had a subject to which I owed a duty.' And the subject was new, so that he 'used to use original material even for lectures'[76] and suggested research topics to others. In some ways, economic history, like other social sciences, had to be invented in the 1920s. There were classics, to be sure, great works of the past, and much teaching was concerned with them, but by themselves the classics did not establish the discipline. Nor was it good enough just to write books on the nature and significance of particular social sciences. Teaching was crucial for at least two reasons. It could afford a certain experimental approach appropriate in this exploratory phase, and it was designed to create adherents, a profession, even what came to be called a scientific community.

Notable Women

In the case of LSE, one other aspect of teaching is worth mentioning. Whatever the method, the School was arguably unrivalled in teaching considerable numbers effectively. Despite Beveridge's endeavours, LSE had not become an Oxbridge where tutors spend long hours on individual students and the weekly essay dominates the experience. But it was not a Harvard or Columbia either—let alone a Sorbonne or a Heidelberg—where great men display their knowledge and others, graduate assistants or even hired crammers, pick up the pieces. At LSE the great men and women were the teachers, and at best they were both models of public speaking and caring mentors of their students. They also knew what they were talking about; in that sense they were anything but charlatans. Curiously, or so it might seem in the 'publish or perish' age half a century later, they preferred the living audience of the lecture room to the abstract public of buyers and readers of books, let alone academic journals. The deepest mistake of Beveridge's suggestion that during his years 'no one had produced anything of any value' at the School was that he seemed to attach value only to great books when what was uniquely 'produced' at LSE was students who were imbued with the excitement of knowledge in the social sciences and its uses in the outside world.

Few were more effective as teachers at LSE than Eileen Power. When Lilian Knowles died in 1926 Power was already a reader at the School, where she too had been a research student before the war. In 1931 she became Professor of Economic History. Her talents as well as her interests were multifarious. A principal instigator of the Economic History Society in 1926, she helped found the *Economic History Review* a year later and these were but two of her lasting achievements in the field. She did write books, including a very big book on *Medieval English Nunneries*. 'I think myself,' said her Cambridge colleague John Clapham somewhat unkindly in her obituary, 'that it is just a little too big.' It included, he added even more unkindly, though in this case not towards her, 'bits of what I should call Ph.D. timber'.[77] No one ever commented adversely on her teaching. 'She lectured widely and brilliantly, continuing the nineteenth-century tradition of the public lecture as intellectual and political forum.'[78] Of Lilian Knowles it was said that she left behind 'little intellectual progeny';[79] Eileen Power, on the contrary, taught both future economic historians and many others who remembered her teaching for the rest of their lives.

They remembered also the radiant woman, the pleasure she took in fine clothes, the delights of her salon in Mecklenburgh Square, the temperament of a professor who listed in *Who's Who* 'travel and dancing' as her favourite pastimes. Eileen Power was the epitome of the fox, not the hedgehog, which is why she could not bring herself to leave London and apply for Clapham's Cambridge chair after his retirement in 1938. (Eileen's husband Postan applied and was appointed.)

> I do find the L.S.E. a much more stimulating place to work in and London a more congenial place to live in than Cambridge. I like people to be all different kinds—I like dining with H. G. Wells one night, and a friend from the Foreign Office another, and a publisher a third and a professor a fourth; and I like seeing all the people who pass through London and putting some of them up in my prophet's chamber.

Maxine Berg, who found this letter in Eileen Power's papers, resents the fact that 'the response of many of her male contemporaries was to feminize and thus to trivialize her achievements'. Yet it would be a sad world if the only correct response to the richness and radiance of this most feminine feminist were to reduce it to one dimension of achievement. After all, Berg herself comments on Power's 'elegant learning' and even 'feminine charisma'. The more interesting question is why economic history—contrary to both economics and political science—seemed to attract outstanding women. Maxine Berg points to an even more specific connection, the 'direct route through Girton College and the Cambridge Historical Tripos to research at the L.S.E.',[80] which Knowles and Power took as well as more than a dozen others, though not Vera Anstey, who had come to the School from Bedford College. Why economic history? Is it a 'soft' subject compared to the 'harder' disciplines of economics and politics, and therefore a niche for women in a man's world? Is it a multidimensional subject with which professionalized men find it more difficult to cope than more broadly educated women? Or is the importance of women in early British economic history merely one of those accidents of intellectual history which one should not try to burden with unnecessary meaning?

The Saintly Scholar

In any case, women were not alone in economic history. Eileen Power's closest partner and friend at the School—as well as neighbour in Bloomsbury—was for many years R. H. Tawney. One hesitates to call the humble and often self-deprecating man (whose published lectures usually start with an apology which is more than a *captatio benevolentiae*) a towering figure; yet he was. Laski's recent biographers like to quote Max Beloff's article which refers to the first half of the century as 'the age of Laski', but there would be at least as much justification for speaking of 'Tawney's century' (which F. J. Fisher did with quite different intent). For Britain, no author represents the social-democratic century more fully and impressively than R. H. Tawney. Moreover, he achieved this not as a politician nor even a political theorist, but as 'a scholar, a saint and a social reformer'.[81] Others have taken up Beatrice Webb's description of Tawney as a saint. He was certainly 'one of the great spirits of our time'.[82] His colleagues, however, have a point: 'Tawney was the most deceptively simple of this generation of economic historians; he has been linked with Namier as one of the two greatest British historians of the century.'[83]

Like William Beveridge—and more to the point, like Beveridge's sister Jeannette, whom he married in 1909—Richard Henry Tawney was born in India, the son of a senior professional man who served in the Imperial colony. However, Tawney's father was an educationist rather than a lawyer, the Principal of Presidency College in Calcutta, and Tawney himself never took a serious interest in the country in which he was born; his later Asian connection was with China rather than India. Born a year after Beveridge in 1880, Tawney straddled the generation of the early teachers at the School, Edwin Cannan and Graham Wallas, and that of those who were about to dominate LSE when he himself became a professor in the early 1930s. He had a traditional education, at Rugby and Balliol College, Oxford, though the second-class degree with which he left showed that he had other things on his mind as well as an academic career. 'And so, with the education of a gentleman, the curiosity of a historian and the instincts of a democrat, he moved out into the world.'[84]

The world meant soon, and for many years to come, the world of working people who were striving to be their best selves. The phrase is deliberately chosen. Tawney was unimpressed by those who simply wanted to better themselves in terms of career and material wealth. His concern was with people's ability to live a decent life, a good life even. In his 1919 Fabian pamphlet which was later turned into the classic, *The Acquisitive Society*, he argued that industrialism and the concomitant individualism had led people astray in that many had lost a sense of 'function', of common purpose. 'If society is to be healthy, men must regard themselves not as the owners of rights, but as trustees for the discharge of functions and the instruments of a social purpose.' Lest this be misunderstood, he added ten years later in his *Equality* lectures that no one was going to be forced to be happy. Contrary to so many others, Tawney remained unimpressed by the totalitarian temptations of the 1930s. He never lost sight of those rights which he called 'primary, essential and fundamental', basic civil rights which rule out the attempt to impose a notion of positive freedom. Thus his socialism never made him forget the open society. 'A right to the pursuit of happiness is not identical with the right to attain it.' And even when it comes to equality, 'the important thing . . . is not that it should be completely attained, but that it should be sincerely sought.'[85]

Such quotations, all quotations perhaps, fail to recreate the deeply religious, self-contained, modest, but also caring, active, and outgoing figure of the man of whom his colleagues said on the occasion of his eightieth birthday that 'he is, indeed, the embodiment of his own principles'. The statement can be turned on its head, as J. R. Williams, R. H. Titmuss, and F. J. Fisher did in the second half of the same sentence: 'The severest criticism of *Equality* as a social theory is that it would be easier to realize in practice if all men were Tawneys'.[86] A. H. Halsey similarly described Tawney's utopia as 'a society of sober and strenuous Tawneys'.[87] Tawney's concern with people's chance to be their best selves led him first to Toynbee Hall, where he was resident during Beveridge's sub-wardenship from 1903 to 1906; he was to return to Toynbee Hall for extended periods in 1908 and 1913. The two years at the University of Glasgow in 1906–8 exposed him to academic life and to economics. But he left both, the first to engage actively in workers' education (in Rochdale, Wrexham, Chesterfield) and the latter for economic history. In 1913 he was put in charge of the new Ratan Tata Foundation at LSE, under the chairmanship of L. T. Hobhouse. The war saw him as a non-commissioned officer in France, where he got severely wounded. After a brief spell at his old college in Oxford, he began his LSE career as a Reader in Economic History in 1920. In 1931 he was made a professor.

Tawney's activities outside the School were at least as intensive and varied as those of Laski. He failed four times to get into Parliament, but remained close to generations of Labour leaders (though not to Ramsay MacDonald). He continued his association with the Workers' Education Association, in which he held office from 1905 to 1947. He served on the Sankey Commission, which recommended the nationalization of the coal industry. He published articles and made speeches. Others, notably Ross Terrill and A. H. Halsey, have written about his ethical socialism and its wide appeal. He was, as Ross Terrill put it, 'a socialist for all seasons',[88] or in Chelly Halsey's words, 'he is perhaps the only man who can be saluted by Fabians, Marxists, guild socialists, trade unionists, co-operators and Christian Socialists alike'.[89] Is this why Tawney, unlike Laski, never became a centre of controversy at LSE? Or did his wife, Jeannette Beveridge, protect him from the Director's wrath?

She probably did not. Many of Tawney's friends regarded her as a liability rather than an asset, and not just because she did so much of the talking in their untidy London household and the positive slum of a cottage in the Cotswolds. 'She was intelligent, she was kind, but she was a scatterbrain.'[90] Tawney wrote *The Acquisitive Society*, Kingsley Martin quipped, and Jeannette illustrated what it meant. She certainly saw to it that the Tawneys never got prosperous. There was more to her, to be sure. She got involved in many good causes. She was loyal and supportive to her husband. When she died in 1958 after nearly fifty years of a childless marriage, it was evident to his friends that for him too the end was near.

If his wife did not shield him from Beveridge's wrath, who or what did? Perhaps the most significant difference between LSE's two leading socialists of the inter-war years is that by contrast with Laski, Tawney was all of one piece. Postan made the point that there was no split between Tawney's historical works and his social tracts. Ross Terrill builds his intellectual biography of Tawney on the foundation of this thesis. 'Tawney the historian, political analyst, and socialist, were all one; a persistent, smouldering idealist, consciously responding to situations presented by the vices and virtues of Victorian industrialism.'[91] However, Tawney's historical patch, as it were, was not the Victorian age, but England before the Civil War, the late sixteenth and early seventeenth centuries. From his first book on *The Agrarian Problem in the Sixteenth Century* (1912) to the last one on Lionel Cranfield, *Business and Politics under James I* (1958), he wrote and lectured widely on the subject. The period from 1540 to 1640 is of course what F. J. Fisher had in mind with his contribution to the *Essays in the Economic and Social History of Tudor and Stuart England* on 'Tawney's Century'.

Tawney's most accessible and widely read contribution to economic history is the 1926 book, *Religion and the Rise of Capitalism*. Characteristically, it is based on a series of lectures given in 1922. Contrary to Max Weber's famous essay on *The Protestant Ethic and the Spirit of Capitalism*, it does not propound a thesis. In fact, in a long footnote to his book, Tawney objects to Weber's thesis, and even to the attempt to introduce anything so simple as a 'thesis' into history at all. Weber overlooks capitalist developments in Catholic regions of Europe, simplifies Calvinism for his purposes, reduces the capitalist spirit wrongly to pecuniary gain. 'Both "the capitalist spirit" and "Protestant ethics", therefore, were a good deal more complex than Weber seems to imply.' Tawney's own book is full of the complexity of interrelations between religion and early capitalism, and does not even hint at any causality.

Max Weber, of course, had asked himself a theoretical question: why is it that at a certain point in European history considerable numbers of people, a class even, started to save rather than consume? What motivated such 'deferred gratification' (as American social psychologists would say later)? Testing God's benevolence was one (Calvinist) answer; adhering to the moral demands of a vocation another (Lutheran) one. Tawney did not have much time for this kind of approach. 'The concern of the sociologist, as I understand his work, is primarily with the general. . . . The concern of the historian begins with the particular'. It does not end there, to be sure, but even when it turns to the general, it is about testing applicability and respecting the sequence of change. Tawney was even more impatient with methodology, which he compared to the succession of curtain-raisers in Chinese drama where the spectator 'discovers that the performance is over at the moment when he hoped that it was about to begin'. By contrast, economic history for Tawney is, like all history, *l'histoire intégrale* which overcomes all specialisms. In substance, it is 'the study, not of a series of past events, but of the life of society', so that the paradox is true 'that all history is the history of the present'. In style, it has an almost literary commitment to the richness of language which reflects the richness of life. 'It is permissible to hope that science and art are not finally irreconcilable.'[92]

Tawney's inaugural lecture as Professor of Economic History at LSE, from which these quotations are taken, was given on 12 October 1932. Sidney Webb, Lord Passfield by then, was in the chair. The occasion, and the combination, showed the original LSE at its best. Not surprisingly, Sidney Webb and Harry Tawney got on well. Had it not been for an unfortunate altercation with Margaret Cole, who first asked Tawney to do the job and then proceeded to edit a volume on the subject herself, we might now have a great biography of Sidney Webb. In the event, we have Tawney's two lectures, 'The Webbs and their Work' (1945)—the precise title later chosen by Margaret Cole for her book—and 'The Webbs in Perspective' (1952), as well as many testimonies of Sidney Webb's appreciation of Tawney's help in drafting reports and pamphlets and Beatrice's assurance that the scholar and saint 'is loved and respected by all who know him'.

Most importantly, Webb and Tawney shared an attitude to the vexing question of social science and values which was unlike Laski's. Both were partisan, and their party was the same as that of Laski, of course, but their commitment to scholarship went much deeper. Tawney's biographers are probably right to suggest that he chose his subjects of study guided by certain values; even the 'value-free' purist could not object. Nor is there any objection to giving special emphasis to the non-acquisitive dimensions of social life, or for that matter to examples of equality. But the medium Tawney chose—the one which Webb had in mind in his early statements about the avoidance of bias in research—was that of scholarly research and writing. He did not separate partisanship and scholarship, thus thrusting one into the political arena and the other into outer space, but united them. For that reason, he had no need to say in his inaugural lecture that he was not proselytizing; nobody ever suspected him. If there is such a thing as committed scholarship which accepts all the strictures of objectivity while leaving little doubt about the author's preferences, Tawney is the example. Or is example the wrong word? Does his economic history like his socialist society require Tawney to be viable?

The LSE Department of History in any case had its Tawney. Like political science, and unlike economics, the Department of History found the shape in the early 1920s which it kept for the following fifteen years or more. Mr Tawney became Professor Tawney and Dr Power Professor Power but they still taught 'Economic History from 1485' and 'The Growth of English Industry' respectively. For a seminar on the 'Economic and Social History of Tudor England' they actually joined forces. It is striking in any case how many courses were offered jointly by at least two teachers in a department in which people obviously got on well: 'Economic Development of the Overseas Dominions' by Mr Beales and Mrs Anstey, 'Industry and Trade in the Later Middle Ages' by Dr Power and Mr Postan, in later years 'Economic History since 1815' by Professors Power and Tawney, Mr Beales and Mr Durbin, and soon Mr Fisher as well. Occasionally, outsiders were allowed in. Mr T. H. Marshall taught economic history before he became Professor of Sociology. Professor Laski offered a course of twenty lectures on 'Constitutions of the Great Powers'. For many years, the other half of history, political history or, as it came to be called at the School, International History, remained in the care of a lecturer, L. G. Robinson (from 1930 reader) and a part-time professor, Arnold Toynbee, who soon left for the Royal Institute of International Affairs; but from 1932 the Stevenson Chair, first held by C. K. (later Sir Charles) Webster, became the nucleus of a separate department.

MALINOWSKI AND ANTHROPOLOGY

Anthropologists All

Robbins, Laski, and Tawney were not only names to conjure with at the LSE of the 1920s and the 1930s, but most students, certainly most research students and young assistants, had actually encountered them. They were real figures on the School scene. Bronislaw Malinowski was also a name to conjure with, but when Kingsley Martin says that 'the great Malinowski sat on the fence'[93] (which separated socialists and anti-socialists) one wonders whether he actually knew much about him. B. K. Nehru finds it necessary to record that he 'never attended any of Professor Malinowski's lectures' because he was after all a great name.[94] Robbins does not tell us anything about Malinowski beyond listing him as a professor. Between Laski and Malinowski there was little love lost. 'Laski never warmed to the great anthropologist Malinowski, who seems to have actively disliked him.'[95] So far as Tawney is concerned, he regarded anthropology as a kind of equivalent in space to what economic history does in time and, in his inaugural, mentions 'the admirable book of Dr Raymond Firth on the economic life of the Maori' but not Malinowski. A. H. Halsey suggests that Malinowski shared Tawney's 'belief in the possibility and desirability of "positive" freedom',[96] but the evidence from Malinowski's posthumous *Freedom and Civilization* (edited by his second wife and widow Valetta) is at best circumstantial.

More surprisingly, Beveridge has little to say about 'the great Malinowski', although at one time he played a key role in one of the two objectives—or were they obsessions?—to which he repeatedly referred. One was, it will be remembered, to 'break up' economics, and the other, 'making the circle of the Social Sciences complete'. The Director's ideas on the subject were quite precise. There was economics and there was the study of political and social relations.

To complete the circle of the social sciences, a third group of studies is required, dealing with the natural bases of economics and politics, with the human material

> and with its physical environment, and forming a bridge between the natural and the social sciences. On the side of human material, there should be included here such subjects as Anthropology, 'Social Biology' (genetics, population, vital statistics, heredity, eugenics and dysgenics), Physiology so far as it bears on problems of fatigue and nutrition, Economic Psychology, and Public Health. On the side of physical environment come Geography in its widest sense as a study of natural resources, Agriculture and Meteorology.[97]

The 1925 submission to the Rockefeller Foundation from which this is quoted had been agreed by both the Professorial Council and the Emergency Committee. (The submission did, of course, also include the 'prior needs' which for many, not least in New York, defined more real priorities.) The practical proposal for completing the circle of the social sciences was to create four chairs, in anthropology, social biology, economic psychology, and public health.

In the event, only one chair was created immediately, in anthropology, and Bronislaw Malinowski, already a reader at the School, was appointed to it in 1927. It is not altogether clear what the Director had in mind when he included anthropology in his third basket. From the context one must assume that he was thinking of physical anthropology as much as social anthropology. But by the mid-1920s the School already had a distinguished tradition in the field. E. A. Westermarck had been lecturing in anthropology since 1904 and was still a part-time professor (until 1930). His fellow Martin White Professor L. T. Hobhouse was as much an anthropologist as a sociologist at least so far as the data used for his evolutionary theories were concerned. With his student and successor Morris Ginsberg he had published *The Material Culture and Social Institutions of the Simpler Peoples* (1915). Since 1913, moreover, Professor Seligman had been a (part-time) Professor of Ethnology lecturing on general ethnology as well as the pagan tribes of the Anglo-Egyptian Sudan.

Malinowski grew out of this tradition. Westermarck and Seligman were his teachers and mentors. He liked and admired Westermarck, with whom he had much in common. On the other hand, 'Seligman and Malinowski were temperamentally very different'. The difference was not just one of temperament but one of concepts of anthropology too. 'Seligman essentially was one of the believers in the association of all branches of the discipline, physical anthropology, archaeology, cultural anthropology'. Raymond Firth adds that 'he didn't touch the "social stuff", as he said'.[98] Malinowski, on the other hand, though he owed intellectual guidance and practical help to Seligman, was all about the 'social stuff'. When he became a reader in 1923, he insisted on the post being described as one in 'social anthropology', which distanced him not only from the cultural anthropologists at University College but also from the 'natural bases of social science'. The tension between the 'antiquarian' Seligman and the 'scientific' Malinowski (whose epithets these are) had to be contained by the Director at times.

Born in Cracow in 1884, Malinowski belonged, like Tawney and Beveridge, to the middle generation between the founders and the new leaders of LSE. That, however, was all he had in common with Tawney and Beveridge, or with anyone else who has been mentioned so far in this story. He was born in the Austro-Hungarian Empire, became a Polish citizen after the war, British in 1931, and shortly before his death in 1942 considered American citizenship. At the School, people liked to think that he was of aristocratic origin, but then it is never easy to place foreigners and Malinowski made a point of looking distinguished. His father had been Professor of Slavonic Languages at Cracow and both his parents came (as Malinowski's daughter Helena put it) from a class without equivalent elsewhere, gentry perhaps, 'but certainly not aristocracy'.[99] When Malinowski's father died in 1898, the family fell on hard times; moreover Bronio (as he was called by family and friends) was a sickly child, close to blindness for a period, in need of care throughout. One wonders how the man, plagued by illness throughout his life, was able to spend long periods on fieldwork in New Guinea, until one reads his diary. 'I gave myself an injection of arsenic, after sterilizing the syringe in the kitchen.'[100] Arsenic and exercise alternated, and the question remains how much of his illness was real and how much hypochondria.

Bronio Malinowski studied mathematics and physics at Cracow with considerable success. Yet soon after he had received the Imperial Prize for outstanding academic achievement in 1908—his dissertation had been on the physicist–philosopher Mach—he began his rootless, wandering life, often drawn and almost always accompanied by a woman. In Leipzig, he studied *Völkerpsychologie* under Wilhelm Wundt, in London anthropology with Westermarck; he had come to join his lady friend, the pianist Annie Brunton. Between 1910 and 1914 he divided his life between England and Poland, including a year as 'special lecturer' at LSE, but he managed to get out of Europe, to Australia, just in time as the war broke out. Australia became the starting-point for the short initial trip to New Guinea and then the two long fieldwork expeditions to the Trobriand Islands, as well as for an engagement to one and marriage with another woman. After the 'penniless Pole' Malinowski and the 'English Miss' Elsie Masson had got married in 1919 they spent several years partly in London but partly first in the

Canary Islands and later in their home in the Dolomites above Bolzano. To say that Malinowski commuted from there to London would be going too far, but he led what his daughter called 'a divided existence' even after the appointment to a chair in 1927. The Malinowskis' house in Primrose Hill was a home only for a short while after 1929. His ailing wife moved to Austria, where she died in 1935; he travelled and wrote and taught and enjoyed his students, especially the women among them. Some he visited in Africa, where they were doing their own fieldwork. In 1938 he took a sabbatical to go to America, where he married again, and at the outbreak of the war became a Professor at Yale University. In 1942 he died of a heart attack at the same early age as his father, 58 years old.

Lest this be considered a slightly disrespectful sketch of the life of a great man, it should be added that its purpose is solely to show how different Malinowski was from his colleagues at LSE. He loved England on the whole ('I say on the whole,' his daughter remarked of his similar love of America, 'because he went through so many changes in his attitudes to countries and places and peoples'[101]) but he was very un-English. Firth called him not only 'a very obvious foreigner' but 'a somewhat disturbing foreigner'.[102] Malinowski loathed Mussolini, and even more so the Nazis, and he never toyed with communism; yet he was as near to an unpolitical being as was possible at LSE. When he became a celebrity, it was more for the explicit sexuality of his ethnography than for any public pronouncement. He was certainly no saint; he wanted to live and let live rather than reform the world; he was egocentric rather than altruistic—in short, he was no Tawney, let alone a Webb.

Fieldwork and Theory

But he was a great teacher, perhaps that above all, and the founder of a discipline. 'Social anthropology began in the Trobriand Islands in 1914.'[103] Malinowski's pupil Edmund Leach overstated and understated the effect of his teacher at the same time. Others had gathered data before, and the whole point of Malinowski's anthropology was not just to gather data. He sought that elusive union between fact and theory which no other discipline of the social sciences has pursued with greater intensity than anthropology after Malinowski. His ethnography became a model for research as well as for the formation of young anthropologists. Few before him had spent two years living among 'natives', learning their language, observing their customs, and describing it all in meticulous detail. Not even the indiscretions of the diary ('I see the life of the natives as utterly devoid of interest or importance') can detract from the significance of the example. Raymond

Firth, Malinowski's student, junior colleague, successor, biographer, and appraiser, thought that too much was made by a later generation of the necessity for fieldwork; but it certainly became the hallmark of anthropology not to allow any theorizing without previous immersion in some alien culture.

On the other hand, Malinowski knew from the outset that observation without theory is dumb. He put it, not surprisingly, in more positivist language: 'Only laws and generalizations are scientific facts.'[104] Fieldwork needs to be informed by interpretation in the light of general rules. Malinowski was groping for such rules throughout his life, and his most extensive attempt to assemble the results of such groping was published after his death. Two concepts are central for Malinowski's theoretical approach: 'function' and 'culture'. He saw himself as the inventor of functionalism and he struggled with a theory of culture. The former meant above all that artefacts as well as customs and activities serve an identifiable purpose within a whole; the latter, culture, is intended to define the whole, though more often than not Malinowski approaches this objective by classifying, for example, 'Types of Institution' and 'Forms of Integration'. In the end we learn that culture 'is an integral in which the various elements are interdependent'.[105]

Are then the *Argonauts of the Western Pacific* and *The Scientific Theory of Culture* candidates for the list of Great Books produced at LSE? Not if we are to believe one of Malinowski's earliest research students, E. Evans-Pritchard. In his scathing 'Notes and Comments' on Malinowski he calls the *Argonauts*

> a failure, for [Malinowski] offers us no sociological interpretation of it of any sort. Why is this? Malinowski had no idea of abstract analysis, and consequently of structure. . . . All he tells us could easily have gone into 50 pages rather than into over 500 pages. In a sense it is a piece of book making on the model of a sociological novel, for example by Zola.

Not surprisingly then, Evans-Pritchard thinks little of Malinowski's *Scientific Theory*. 'It is a good example of the morass of verbiage and triviality into which the effort to give an appearance of being natural-scientific can lead. Malinowski was in any case a futile thinker.'[106]

Such comments are above all illustrations of a polemical style uniquely characteristic of anthropologists among all the social sciences. It was by no means alien to Malinowski himself, so that one is not surprised that his qualities 'made for him "unfriends" as well as friends'. These are Raymond Firth's words, which matter not only because Firth came to represent LSE anthropology both during and after Malinowski's time and belonged to the 'new' generation of Robbins, Plant, T. H. Marshall, and others, but also because Firth almost alone among his anthropological colleagues maintained a reasoning and civilized style of discourse throughout. For Firth, Malinowski's ethnography stands as a model for later generations. Malinowski also was the first to show conclusively 'how fact was meaningless without theory and how each could gain in significance by being consciously brought into relation'.[107] Of Malinowski's theory itself, Firth remained unconvinced. Without Radcliffe-Brown's concept of structure, function remained a tautological notion.

Few if any other anthropologists have given rise to such an extensive and sustained debate of their work as Malinowski. Books and articles trying to solve the riddles left by his intellectual personality continue to appear even fifty years after his death. Perhaps Ernest Gellner (in a lecture in 1990) got closest to what made the man tick whom he described as the 'William the Conqueror of anthropology'.[108] (Frazer of *The Golden Bough* was 'King Harold'.) He tried to combine two traditions of knowledge which were also two personal experiences. One was the communal and slightly romantic notion of culture which took Malinowski back to Zakopane and the Gorals in the Tatra mountains, in other words to the roots of which he was so reluctant to speak. The other tradition found its early expression in Malinowski's dissertation on Mach, who was not only an arch-positivist but also a classifier rather than a dynamic theorist. The two traditions do not go well together; their conflict is a part of the explosive personality of Malinowski; yet it defined and set in motion the new discipline of social anthropology.

Once again it is the teacher rather than the author to whom we have to turn to explain the extraordinary effect of Bronio Malinowski. Even Evans-Pritchard had to recognize that 'he taught most of the other social anthropologists who subsequently held chairs in Great Britain and the Dominions'.[109] Some of them held more exalted office, like Jomo Kenyatta, the 'Father of Kenya', who was one of many overseas students in Malinowski's seminar. Edmund Leach confirms what others felt: 'During his lifetime Malinowski's main academic influence was through his teaching. . . . He was a dynamically powerful personality, a "charismatic leader" who aroused intense emotional feelings of love and hostility among all those with whom he became closely associated.'[110] Raymond Firth agrees: 'In the academic field, quite apart from his writings, it was his contribution as a teacher that was important.' Firth also describes it in detail. Malinowski 'could use the rostrum brilliantly, on formal occasions, but what he really liked was the seminar'.[111] There he sat in his (actually Westermarck's) armchair, listened, asked questions, commented caustically or seriously, probed and guided students to the real problem, until in the end he pulled the various threads together. This not only happened while the teacher sat in an armchair. Malinowski talked, and taught, constantly. Students 'learned to discuss their theses on bus-tops or dodging the market-barrows down Holborn side-streets'. Audrey Richards adds: '[They] might be irritated by his intolerance, or inspired by his enthusiasm. They were never bored.'[112]

Small wonder, then, that a 'strong Malinowski clan'[113] formed around him. It included many women, among them Audrey Richards, who became a friend of the family but just shrank from marrying Malinowski after Elsie's death; she was then a lecturer at the School. Edith Clarke, Lucy Mair, Hortense Powdermaker, Camilla Wedgwood, Monica Wilson were all students in Malinowski's seminar. Like their male colleagues Raymond Firth and Edward Evans-Pritchard, Edmund Leach and Meyer Fortes, they went out to universities elsewhere in the country and abroad, though a few stayed. Isaac Schapera, who had been a Ph.D. student and lecturer in Malinowski's early years, left for South Africa, and Raymond Firth was appointed to his readership until he succeeded Malinowski himself in the chair. Lucy Mair took over the course in Colonial Studies which had become a part of the Department of Anthropology. It was designed to educate colonial administrators and had the strong support of both Malinowski and Beveridge as well as, more importantly, Lord Passfield (Sidney Webb) during his period as Colonial Secretary.

The Department of Anthropology remained quite small, though perhaps compact is the better word. It was also remarkably self-contained. This had much to do with its somewhat strained relations with the immediate neighbour, sociology. While anthropologists were busy inventing 'a new social science', no less (Maurice Bloch used the words in his interview with Raymond Firth), sociologists remained stuck in their ways. Even the normally kind Firth observes that Morris Ginsberg had 'missed the boat' of new developments and now 'became a bit jealous', which did not help relations. T. H. Marshall 'was a sensitive interpreter, whom I respected very much, but he wasn't a fieldworker'.[114] The new social science was not slow to develop its own somewhat hermetic ethos. The sociologists responded in kind. When a B.Sc. in Sociology was proposed in 1936, Anthropology was totally left out of the syllabus. Malinowski was incensed and wrote officially to the Director and privately to 'Dear Jessy'. 'It is not my personal opinions which matter but the plain fact that anthropology as we are teaching it at the School is a training in the only really empirical approach to sociology.'[115]

The emphasis on 'direct observation of living communities' may have pleased the Director but may also have been the reason why anthropological theory failed to inspire others outside its disciplinary confines. True, Talcott Parsons attended Malinowski's seminars a few times, and some later anthropologists like S. F. Nadel and M. Gluckman had an impact on other social sciences, but for that they were all but rejected by their anthropological colleagues. Anthropology developed its own culture, as it were. *Homo oeconomicus* was already flourishing; *homo sociologicus* had yet to be born (though this was unlikely to happen in Morris Ginsberg's department), but what Edmund Leach called 'Malinowskian man' had come into being at LSE in the 1920s. The creature was bent on reciprocity though not necessarily symmetry in relations with others. When Leach speaks of a world in which 'each individual is constantly seeking to operate the prevailing social conventions so as to maximize his private satisfactions',[116] one is bound to think of the inventor of 'Malinowskian man' as much as his invention.

HOGBEN AND SOCIAL BIOLOGY

The Natural Bases of Social Science

Long before Seligman's early retirement in 1932, anthropology at LSE had become emphatically a social science. One disgruntled student later described it even as a 'no-links-between-biology-and-culture' science.[117] The Director clearly thought so too. However pleased he may have been about Malinowski's emphasis on observation and fieldwork, he had not succeeded in using anthropology to 'complete the circle of the social sciences'. Since there clearly were not going to be chairs in economic psychology or public health, Beveridge's hope now rested on the establishment of the discipline and even the Department of Social Biology. The saga is worth telling. It raises real questions on the boundaries of social science, and more particularly on the London School of Economics and Political Science. It also marks the beginning of a phase in which things began to go wrong for Sir William Beveridge.

The saga, at least in its unambiguously documented part, begins with the LSE Memorandum to the Rockefeller Memorial of 16 July 1925. The Memorandum was Beveridge's work. So how did 'Social Biology' get into it? This is one of the points at which we pay for the fact that those who made history also wrote it, and occasionally allowed their memory to be deflected by interest. A predecessor of social biology, 'various aspects of the problem of population (including eugenic and other questions)' came up in the very first conversations between Beveridge, Jessy Mair, and Beardsley Ruml of the Laura Spelman Rockefeller Memorial Fund.[118] Lady Beveridge thought in 1952 that she remembered Beardsley Ruml putting special emphasis on the idea.

He took the line, however, that [a chair in economics] should be contingent on the creation of a chair designed to bring biological aspects of mankind to bear on his [*sic*!] place as an economic entity in human relations. His insistence on this condition had to be accepted as the prelude to the development of other less traditional developments.[119]

Beveridge's own published account is more cautious. Ruml 'was looking for something big and new. The Natural Bases of the Social Sciences, the theme in effect of my first Address at the School, on "Economics as a Liberal Education" began to be bruited.'[120]

In his inaugural lecture, Beveridge had in fact paid homage to his academic hero, the biologist T. H. Huxley, whose almost Comtean hierarchy of the sciences he had cited. Mathematics is the most perfect but also the least complex science. Biology is much more complex and therefore less perfect. 'But there is a higher division of science still, which considers living beings as aggregates . . . I mean the science of Society or Sociology.'[121] Beveridge quickly substituted 'Economics' for 'Sociology' and gave it the benefit of late birth and therefore permissible imperfection. However, he insisted also that 'Economics' would be a worthwhile scientific endeavour only if it followed the methods of the older sciences as he saw them, and kept the boundaries to them, and notably to biology, open.

It would be wrong to charge Beveridge with simple reductionism. He did not seek biological explanations for social facts; moreover, and with significant consequences for the saga of social biology, he opposed the political project of eugenics. In his early contribution to the Sociological Society (in 1906) he had even rejected demographic explanations of unemployment and instead referred to the 'nature of our industrial system'. He pursued the subject further in his lecture on 'Population and Unemployment' at the British Association meeting of 1923 from which he was called away to meet Beardsley Ruml. On the other hand, Beveridge was so obsessed with what he regarded as the scientific method of fact-gathering and induction that he feared the science of man would go astray if its advanced social departments were not chained to the older and more perfect natural sciences, from economics to biology. Thus the idea of 'bio-economics' or at least social biology was very much on his mind—and also, so he thought, on that of Beatrice Webb, whose 'desire for biology as a preliminary to economic studies' he cites—when Beardsley Ruml said what all foundation officials have said ever since: he wanted to support 'something new', if possible something unheard-of, which would enthuse Trustees who were not happy about spending their precious funds on projects which simply provided institutions with more of the same.

It is thus not surprising that Mr Ruml suggested, when he saw the first draft of the 1925 Memorandum, that the new departures should take precedence over the 'prior needs' of building and research funds in the School's submission. Beveridge later made much of this reversal of the two parts of the text. When he had to write a paper in 1935 purely for internal consumption to justify his actions he went almost as far as Jessy/Janet did later:

> It was made clear, however, by Mr. Ruml that while he was prepared to consider grants for such purposes [as the 'prior needs'] he felt that these purposes should come second, and that the prospect of getting help for any purpose at all from the Trustees of the Memorial depended considerably upon our indicating that we proposed developments of some new kind. He was in fact particularly interested in extension to the natural bases of the Social Sciences.[122]

In the published account of his LSE years, the Director chose more ambiguous language: 'Thinking over the emphasis which Ruml had laid on our developing the Natural Bases of Social Science, I reversed the order on July 16th [1925].'[123] When he came to describe the formal process of adopting the Memorandum, ambiguity operated the other way round. In the book of 1960, Beveridge said without ifs and buts that he was 'instructed by our Emergency Committee' to send the Memorandum as amended, whereas according to the internal paper of 1935 he merely 'reported my discussions with Mr. Ruml' to the Emergency Committee, which at that time 'did not keep formal minutes, but from subsequent action it is clear that my application to Mr. Ruml was made on their authority'.[124]

Such details matter because later evidence suggests strongly that the idea of social biology was Beveridge's alone, and that he used the alleged interest of the Memorial (which was primarily one in 'developments of some new kind') in order to keep his colleagues at the School in line, whereas he put pressure on the Memorial by arguing that School committees were strongly in favour. When the game fell apart in the early 1930s, this did not help either the Director or the School in its dealings with what by then had become the Rockefeller Foundation. Even now, in 1926, the Memorial decided in the first place to make funds available for buildings and the fluid research fund, the prior needs. Only a year later, in 1927, was the endowment grant given, and by then new prior needs, notably in international law and international studies generally, had emerged.

But the Natural Bases did not evaporate. From May to December 1926, a paper on the subject underwent a number of changes. It set out what the School wanted in the field and was unanimously agreed by the Professorial Council on 1 December 1926. The changes introduced by members of the School in what was initially Beveridge's paper are not without interest. For instance, Beveridge had written 'Economics' throughout, but Malinowski insisted on adding 'Social Science'. Beveridge had suggested that one could either look for natural scientists who then study the social sciences, or vice versa; 'for various reasons the former alternative is likely in practice to be preferable'. In the final version, at Laski's and Hobhouse's insistence, the statement had been replaced by what amounted to the opposite: 'The latter course has been followed in the teaching of Sociology at the School of Economics. While continuing and even extending this method, it is desired now to initiate and develop the former method as well.' A nice piece of committee work by the political scientist and the sociologist!

Throughout, people had trouble defining social biology, which perhaps, given that those who had a hand in it were all social scientists, is not surprising. There was much talk of the 'borderland' between natural and social sciences and the need to know the 'territory on each side'. But how was one to describe the territory on the other side? Everyone was agreed that the 'subject may be defined broadly as the application of Biology to human Society'. But what does it mean? Beveridge had a go: 'it would cover such topics as Instinct in Man, Inherited and Acquired Characteristics, Quantity and Quality of Populations, Racial and Economic Tests of Fitness.' The Professorial Council did better and prevailed: 'it would cover such topics as variation and heredity in man, selective immunity, relative importance of environmental factors in social structure and changes, questions of race and class in relation to hereditary endowment, economic and biological tests of fitness.' This was, as it were, the academic part of the resolution. When it came to practicalities, the professors saw to it that 'extending existing methods' remained firmly in the foreground. Sociology gained a little more; social anthropology got all it wanted (including the rapid transmogrification of a 'genetic anthropologist' into an undefined research fellowship); there were firm commitments to social psychology and criminology; the Geography Department with its own 'borderland' received an honourable mention. In the end, little seemed to be left for social biology. 'The subject probably does not involve the setting up of a biological laboratory.' In the long run a chair would be needed as well as research staff, but since 'it is possible that no suitable candidate for a Chair would present himself if it were established', the idea should be suspended for the time being and two temporary research fellowships for trained biologists who want to study economics and social science should be created instead. The whole plan was experimental in any case, and 'to allow of experiment there should be considerable liberty in postponing the establishment of any teaching posts or holding them in suspense'.[125]

Enter Lancelot Hogben

The Laura Spelman Rockefeller Memorial Fund did not seem to mind. Two other features of the debate are noteworthy. One is that the School clearly did not like the Director's proposal. Polite as people are, at least most of the time and when they meet in committees, the professors did not want to reject the idea outright, but they moulded and bent it until they thought that it suited their own interests. The other point is that the debate was curiously abstract. No particular set of unresolved scientific problems required the presence of a biologist, nor was there a person in the Director's, or anybody else's, mind whom the School wanted to attract. At one point before the paper was finally adopted, Hobhouse suggested the name of Alexander Carr-Saunders to the Director—'Now C.S.'s views may differ a good deal from yours or from mine, but I do think he is competent in this subject'[126]—but despite some support for it among the professors the idea was not pursued. Initially no name was seriously discussed at all, since not even the subject could be clearly defined. As a result, this became precisely the kind of pointless debate which is liable to generate ill feeling, and turn into ideological warfare. The issue was the Director's view of science. The economists took no part in the debate, the sociologists and anthropologists bent it to their advantage, but the issue did not go away.

For one thing, Beveridge was nothing if not persistent in pursuing his goals. When the Rockefeller grant was made in 1927, he agreed to postpone the establishment of a Chair of Social Biology 'till enquiries showed that suitable candidates were available' but set aside the necessary funds and actually began to make enquiries. Eight years later he had to defend himself against the accusation of having diverted funds from other purposes, which was clearly unjustified. People may have been surprised that Beveridge kept on looking for a social biologist but he was entitled to do so. Letters were sent out in many directions, both by the Director and by Professor Seligman, who apart from the Professor of Scientific Method, Wolf, was the only one to take an active interest. In 1927 it was still suggested that a young trained biologist should come to study economics (and social science) first, but unsurprisingly those asked saw little attraction in the prospect. The chair re-emerged. In October 1929 it was formally established. Despite a last-minute attempt by the Director to call it a chair in 'Bio-Economics', Social Biology prevailed.

After the establishment of the chair and its advertisement, things moved quickly. A Board of Advisers met; it soon disposed of the ten applications (which included one by Solly, later Lord, Zuckerman) and turned to other names. Beveridge was travelling, but the Secretary attended the meetings and kept him posted. 'Oh dear, how I loved talking to the two of them,' the scientists Professor Watson and A. V. Hill, she wrote on 26 November 1929.

What joy they had in their work. What dears they are. How pure, pure science is. How free from worldliness, how perfect a brotherhood. Can you wonder why I have always wanted that spirit in this place. How I by nature and by training sheer off from the mind which is bent on this or that propaganda, this or that social reform, this or that religion. The cool detachment of the dogged search after what is true: that is where the soul reaches its completion, believe me. That is the real university spirit and only that.[127]

Jessy had, of course, been a scientist while Beveridge slipped from astronomy to biology and further to economics. Did he share her almost erotic if asexual attraction to 'pure, pure science'? Was he as unhappy as she must have been among all those worldly reformers? In any case, the short list was soon reduced to two—J. B. S. Haldane and Lancelot Hogben—and, when the former dropped out, Beveridge's interest focused on the latter.

Hogben was then Professor of Zoology at Cape Town. Born at Southsea in the year of the foundation of the School (as he himself liked to stress), 1895, he had moved about a great deal in his young career. He had a degree from Cambridge, then worked at Birkbeck and at Imperial College in London, moving on to posts in the Universities of Edinburgh (1922), McGill in Montreal (1925), and Cape Town (1927). By late 1929, when Beveridge got in touch with him, he had developed quite a reputation, two reputations to be exact. One was in his subject, where his fellow researcher, the marine biologist Carl Frederick Pantin, who had just become a Fellow of Trinity College, Cambridge, described him as 'one of the most brilliant (perhaps *the* most brilliant) of the younger biologists', adding temptingly though perhaps with somewhat insufficient knowledge: 'There are few with so wide a knowledge of Biology, and such a unique philosophical outlook and there are, I think, none at the present time who combine these with a deep interest in sociology, except Hogben.'[128] Professor Watson of University College was less fulsome but praised his range, teaching skills, and 'intellectual honesty'.[129] The other reputation was more ominous though curiously it did not deter Beveridge. Professor Pantin knew Hogben well. 'He is a rather strange person and many people do not get on well with him. . . . His advanced social views (he is rather red) have rubbed many of the old school of Zoologist the wrong way. This is unfortunate for him as a zoologist as it has made it difficult for him to return to this country to a zoological chair.'

Beveridge naturally proceeded to make further enquiries. On 24 October 1929 he wrote to Arnold Plant, who was then also at Cape Town. In his letter he used yet another definition of social biology, thus demonstrating the continuing elusiveness of the subject. 'Broadly this means, I suppose, population, inheritance, race and sex distinctions, eugenics and dysgenics, etc. etc.' And a little later: 'As to what the holder of the post exactly would do, it is a little difficult to be precise; he is really going to do whatever he makes of it', though for the first time the notion of a 'research chair' was mentioned and also the presumption that the incumbent 'would naturally want to have the use of a laboratory'.[130] Beveridge's questions to Plant were the obvious ones: what sort of person is Hogben and would he come to LSE? When Plant replied on 15 November, he had done his homework. Yes, Hogben wanted to come very much. He had even given Plant an outline of what he would wish to teach, notably in relation to giving genetics its proper place in the study of human society, neither overemphasizing nor neglecting its importance. But Plant also added that while Hogben's 'outstanding ability' compensated for much, he had caused 'disturbances and annoyances'. 'I have been, and still am, one of his closest friends here, but there are long periods during which we avoid each other, and I can never be sure whether I love him or loathe him more.'[131]

Beveridge was not put off, not even by the foretaste of other difficulties which the prospective colleague himself raised. He did not want to apply for the chair; he had to be asked. He would bring a small but expensive research staff, including his wife, Dr Enid Charles; the necessary money had to be found from the Medical Research Council. He needed laboratory space; the old grammar school on the east side of Houghton Street had just been purchased and offered opportunities. His research had increasingly moved to the study of animal behaviour and genetic questions connected with it; he not only needed space for his animals but it was doubtful whether this research had any social applications. And his 'rather red' politics? Beveridge asked him outright, and on 27 January 1930 Hogben sent him a long confessional which demonstrates both the political passions and the political naïvety of the scientist. At Cambridge he had been a socialist. His 'ethical prejudices' had not changed but his 'views about politics have become more and more vague'. His business in life was scientific investigation, and 'it is impossible to arrive at the truth about any question—social ones especially—unless one keeps aloof from activities and interests in which ethical values play a prominent part'. He was not a member of any party but would vote for one which supports 'the formation of a Ministry of Scientific Research'. This would probably be Labour, though he would not support any party 'that was not pledged to reduction of armaments'.[132]

Ants and Antiquarians

On 27 February 1930 the University Senate offered Lancelot Hogben the Chair in Social Biology tenable at LSE with the title of 'Research Professor' and membership of the Faculties of Science and of Economics. Hogben accepted, and in writing to the Director remembered his early and continuing friends at the School: 'Perhaps you will convey my kind regards to Mrs. Maer [*sic*!], Professor Bowley and Harold Laski.'[133] On 19 October he arrived in London. Less than a week later he gave his inaugural lecture on 'The Foundations of Social Biology'. The chair was taken by H. G. Wells, who had terrible things to say about a social science whose 'generalisations still float loose and away from its observations', and wonderful things about the 'stirring' developments in biology. Now the two can be combined. 'Upon this basis Economic Science will be built anew.' Professor Hogben will do it because he is 'no mere specialist' and also an 'adventurer'. 'We all welcome him most heartily.'[134]

At least one person in the audience thought that Hogben's inaugural lecture was 'brilliant' though 'Mrs. Maer' was probably as pleased by the impression that 'there had never before come to the School so distinguished an audience'.[135] She meant an audience of scientists, of course. The lecture had many of the qualities of the classic inaugural; it introduced the scholarly personality of the new teacher, put him in the context of the history of his discipline, and sketched out his scholarly programme. Moreover, Hogben had chosen a fairly accommodating style, citing Beveridge and Bowley with approval, and even holding out some hope for 'political economy' and 'social anthropology'. However, the lecture was also confusing. In so far as it had a point, this was: do not jump to conclusions before you have done more research, especially when it comes to the relative importance of biological and social factors! In the process he almost seems to say: stop history and go back to the laboratory! 'One of the greatest dangers is an undue haste prompted by enthusiasm for legislative applications of half-assimilated knowledge.' But what research? At one stage, Hogben says that 'the biologist, as a biologist, is primarily concerned with those general characteristics which ants and antiquarians have in common'; at another stage, and having obviously enjoyed his own play on words, he not only charges philosophers of evolution (Hobhouse?) with 'the habit of making puns' but also insists that biology is more than genetics and when it comes to the central nervous system the whole environment enters the realm of (social) biology. He offers neat criticisms of Carr-Saunders's explanation of the distribution of Protestants and Catholics in Europe by 'innate characteristics' or of the assumption that 'the I.Q. necessarily measured the inborn capacity of an individual'—but what exactly is the scientific status of his own elaborate historical explanation of the emergence of Christian denominations, or of his declared preference for 'behaviourism'? The confused and confusing lecture is held together by the repeated insistence on an ethically neutral science of investigation and of inference, and perhaps by the plea for supporting the new science of social biology:

> Let me ask you to cherish it tenderly while it is sending forth its first green shoots. If, on the other hand, you are inclined to regard this new departure with suspicion, I have the assurance that I shall encounter in a school with a unique tradition of free inquiry the same cheerful tolerance for my misguided optimism as I hope I shall extend to your legitimate distrust.[136]

Well said, but alas it was not to be! The fault for such failure lay by no means on one side alone, unless one believes that the Director should never have introduced the contentious subject into a school of the social sciences. Some of those who later claimed to have this belief had after all supported the original Professorial Council decision and the appointment of Hogben. Over the three years following his arrival, Hogben set up the Research Department of Social Biology with a growing number of 'antiquarians' as well as 'ants'. The 'antiquarians' were his research staff. Foremost among them was Hogben's wife, the mother of their four children, Dr Enid Charles. She was a demographer of note, and also an active social reformer and feminist. Harold Laski wrote the Foreword to her book on *The Practice of Birth Control*: 'The movement for the scientific control of population is only part of a wider movement for making mankind the master of its fate.' Enid Charles the feminist and Lancelot Hogben the idiosyncratic scientist 'found a degree of happiness and companionship with each other at a critical time in their lives',[137] but in the end went their separate ways. Among Hogben's other assistants were Dr Louis Herrman, Pearl Moshinsky, an LSE B.Sc.(Econ.) in Sociology, and most importantly the demographer Dr Robert René Kuczynski, one of the early political refugees from Germany.

The 'ants' raised more difficult questions. Few subjects have produced greater LSE myths than Professor Hogben's experimental animals. As the myths expanded, the animals grew in size until in the end they became screaming apes in their Houghton Street cages. In fact, the *Xenopus laevis* was Hogben's favourite victim, the South African clawed toad whose pituitary gland helped him understand the reproductive process (and develop the 'Hogben pregnancy test') as well as changes in colour pigments, which were one of his subjects. Hogben used other animals in Houghton Street; the School's application to the University Grants Committee of 1933 refers to 'about 150 rabbits, 300 rats and guinea-pigs, and 1,000 amphibia' or toads, but the latter were a particular concern of the porters, and not just of them. Ted Brown remembered the 'enormous great African toads' who after experiments were put in an ether chamber and then in the dustbin. But they were not always dead. 'One day they had these toads jumping all over the road. When they put them in the dustcart they jumped out. They weren't dead, they had recovered!' One can understand Mr Brown's feelings. 'We were glad when they went, because the place smelled awful.'[138]

Lancelot Hogben was a strange condiment for the London School of Economics and Political Science, and that not just because of *Xenopus laevis.* Even as a scientist, he was a man of several faces. His own most technical research, which earned him the Fellowship of the Royal Society in 1936 at the age of only 41, had little to do with the School's interests or even with social biology in any conceivable sense. One wonders what the Rockefeller Foundation made of it when 'The Chromatic Function in the Lower Vertebrates: A Study in the Analysis of Behaviour' was listed as one of the publications resulting from their LSE grant. Hogben's second scientific face also had little to do with the School, for he was a great popularizer of science. Bertrand Russell encouraged him to publish his *Mathematics for the Million*, which earned Hogben handsome royalties. Harold Laski, claiming somewhat incredibly that he was 'soberly measuring [his] words', called it 'one of the indispensable works of popularization our generation has produced'.[139] Then, just as he was leaving LSE in 1938, Hogben completed *Science for the Citizen*, subtitled 'A Self-Educator Based on the Social Background of Scientific Discovery'. The huge compendium, or so the author claims in a preface entitled 'Author's Confessions', was written on the weekly train journeys from London to Devon, where the Hogbens had bought a cottage for the family soon after their arrival. 'If the Southern Railway had not provided third-class Pullman cars to Exeter, this book would not have been written.' It was dedicated to Harold J. Laski and praised Beveridge for 'the enlightened view that some of the students [at LSE] might benefit from a general introduction to natural knowledge'.[140]

Demography

The third face of Hogben the scientist had to do with demography, the one field in which he left a lasting imprint on LSE. Hogben was not a demographer, though he had done research on fertility and the reproductive process, but population was one of his lifelong interests, and he did much to promote population studies. His interest in 'ethically neutral science' made him oppose the tradition of eugenics. C. P. Blacker, then Secretary of the Eugenics Society, regarded Hogben's inaugural (which he had attended with Sir Bernard Mallett, the President of the Society) as 'a formidable attack on eugenics—biting, amusing and very pleasing to his audience'.[141] For reasons which are not altogether easy to understand, Pauline Mazumdar, in her highly informative study of the Eugenics Society, called it an 'attack from the left'. 'In Britain, the explicit use of environment as a means of attacking eugenics began in 1931 [1930?—R.D.] with the work of Lancelot Hogben.'[142] Had not on the contrary eugenics been a part of the left hopes of social reform? What about young Laski and his Frida, for example, to say nothing about Shaw and Wells? Hogben's argument against eugenics underlined the recurrent thesis of his inaugural lecture. Eugenics was propaganda based on insufficient knowledge. If politics entered his attitude at all, it was the aversion to sorting people by alleged biological characteristics which had led him to cause 'annoyance' in Cape Town when he opposed the beginnings of apartheid.

Mazumdar called Hogben a 'reductionist' though her own detailed analysis of his work does not bear out the title. If there was a reductionist among serious scholars at the time, it was Carr-Saunders, whom Hogben had not spared in his lecture. Judging from Hogben's autobiography, Carr-Saunders was not in his good books anyway. 'It used to puzzle me how anyone so facelessly devoid of charm and with so mediocre intellectual equipment attained such eminence.'[143] Still, this did not prevent Hogben from maintaining close relations with Carr-Saunders, whom in correspondence at least he addressed by his student nickname 'Carcinius',[144] even after he had become Director, and from working with him to advance population studies. The eminent group which in 1936 produced the Report on the Present Position and Needs of the Social Sciences to which reference has been made already, included, apart from Colin Clark, G. D. H. Cole, P. Sargent Florence, J. Marschak, and E. A. G. Robinson, both Carr-Saunders and Hogben. There is no record of any disagreement between them.

The dismissal of the simpler assumptions of eugenics cleared the decks for establishing the discipline of demography. Eugene Grebenik has described the process by which 'the purely biological approach' was 'replaced by one which was more socially oriented'. The appointment of Hogben 'had a profound effect on the development of population studies in Britain'.[145] Hogben brought or attracted such students of the subject as his wife, Kuczynski, and soon David Glass (who had actually taken his BSc.(Econ.) in 1932 in another subject in the 'borderland' between science and social science, geography). Hogben focused attention on critical subjects like 'an incipient cessation of population growth in Britain'. His Memorandum on a Coordinated Programme of Research into the Population Problem of 1936 mapped the field. The vitriolic introduction apart, his testament and balance sheet of the LSE years, the book *Political Arithmetic*, assembles contributions on fertility, mortality, family structure as well as the subject of ability and educational opportunity, which was to occupy social researchers at LSE in the 1940s and 1950s.

Even Blacker in retrospect expressed his 'opinion that Professor Hogben's criticisms have been useful to the [Eugenics] society'.[146] Views were changing, partly as a result of a more critical scholarly debate and the development of the social sciences, and partly in response to the horrible abuses of the notion of eugenics even in the early years of the Nazi regime in Germany. When Carr-Saunders suggested an organizational framework for investigating population matters systematically, the Eugenics Society set up, in 1936, the Population Investigation Committee (PIC). Carr-Saunders became its chairman, David Glass its research secretary. In 1948 Glass succeeded Carr-Saunders in the chairmanship. By then, the PIC had become closely associated with the School.

Sans Taste, sans Everything

'Unfortunately, the impact made by Hogben on the London School of Economics was not great.' Grebenik surmises that this was due to there having been 'little personal contact between him and his colleagues in other departments' and moreover to the fact that 'whatever personal qualities he possessed, tact and diplomacy were not his strong points'.[147] This is putting it mildly. If one reads Hogben's (unpublished) autobiography, especially in the unexpurgated version, one must conclude that it was his personality rather than any great methodological dispute which led to the bitter taste of the entire saga of social biology and to its unhappy end. The autobiography, written by the 79-year-old little more than a year before his death, begins with illustrations of the 'taste' of second childishness. 'I came of poor but intellectually dishonest parents.' Father was a Methodist missionary for seamen in Portsmouth before the family moved to London. And mother? 'Her French fluency is the only conceivable attraction a woman so sexually frigid could have had to so procreative a male of the human species.' No more of that ilk. The son had a somewhat mixed-up childhood but managed to get a state scholarship from Tottenham County School to Trinity College, Cambridge. The young zoologist took his examinations quickly, which was as well because as a Fabian and pacifist he had a difficult war, including three months in Wormwood Scrubs prison. Hence his insistence that the party of his choice had to be 'pledged to a reduction of armaments'. At the end of the war, he met his wife 'Charlie' Charles, herself a political agitator until four children and scientific research put an end to that, and the family began its wanderings.

Depending on one's humour, Hogben's chapter on LSE is a document to relish or to cause disgust. The School in 1919 (he wrote) had 'little prestige in academic circles'; it remained 'an institution dedicated to the propagation of a political creed'; above all, a 'brick and mortar curtain' isolated it 'from contact with the natural sciences of other colleges of the University'. Beveridge changed all that, albeit 'with at times scant approval from political activists of the academic staff'. Thus he, Hogben, had misgivings about 'the public image of the School as an academic institution'. But he liked Beveridge spontaneously and trusted him; he got on well with the Webbs; and he was also 'inveigled' by Laski probably to counteract the 'truculently anti-socialist' bias of the Eugenics Society. In any case, while his 'sympathies were more with Beveridge than with Laski, when there was friction between them over some of the political publicity which the latter invited', he 'retained a warm regard for Harold as a person'. In fact, many of his friends were Communists, and he himself was impressed by the 'stupendous effort' made by the Soviet Union 'to mobilize scientific discovery for greater productivity'. 'Many of the younger scientists in Britain and in France regarded the restrictions of one-party rule as a small price to pay for it.'

Elsewhere in the School, he found it harder to make friends. 'Apart from Wolf and my colleagues in the department of Economic History I recall few with whom I had much in common.' By the colleagues in economic history he did not mean Tawney, though he did once refer to his 'wholesome wisdom',[148] nor did he have in mind Eileen Power, 'an exceptionally charming woman, though intellectually far less gifted than her second-in-command', but thought above all of this 'second-in-command', Michael Postan. Hogben's venom was reserved for economics and the School's economists. 'I regarded, and still regard, as a mental exercise on all fours with astrology, economics as taught by Hayek and Robbins.' Actually, this was one of his milder comments about the 'Hayek–Robbins circus', 'the last stronghold of the most ultra-individualist metaphysical nonsense masquerading as economic science west of Vienna'. Hogben positively detested 'the Viennese invaders and their apostate English allies'.[149] (Only Hugh Gaitskell was given worse marks by the neutral scientist as 'a woolly-minded, emotionally shallow, self-satisfied and pompous careerist'.) By contrast, Laski and Malinowski got a somewhat surprising pat on the back. 'Both Laski and Malinowski had some appreciation of the difference between authentic science and argumentation.' This was soon half retracted, though, for Laski was one of those responsible for the 'disputatious attitude to learning' at the School and 'Malinowski's functionalism was a plea to leave the native content with leprosy, malaria, protein deficiency, disease and illiteracy'. Hogben's scorn was not confined to his colleagues. 'I cannot say that I took to my heart students I encountered in the L.S.E..' They were all in his view more disputatious than scientific.

Who then remains? Beveridge, 'the least ostentatious of men', 'endearingly informal host', and of course believer in a neutral science of facts. The immediate colleagues obviously get good marks: Enid, Louis Herrman, R. R. Kuczynski, J. L. Gray, David Glass. Then, more curiously, not just the Webbs but the 'enduring friends' Kingsley Martin and Barbara Wootton. And all this added up to an LSE with 'an atmosphere reminiscent of a Liberal, Fabian or Young Conservative summer school'. Or does it add up? Was the fact that nothing quite added up perhaps the deepest reason for the ultimately destructive role played by Hogben during his LSE years?

The rest of the story is quickly told. Hogben spent in all just over six years at LSE. Throughout this period, he and his department remained an alien body in the School. Even the rather laconic page on social biology in the Calendar ('Professor Lancelot Hogben . . . will welcome postgraduate students who wish to undertake realistic enquiries') seems strangely out of place. Some may have resented the privileges of a research professor, others the idiosyncrasies of this particular research professor; most did not like the toads and the rabbits; almost all wondered what exactly this very visible and audible and smelly department was contributing to the main purposes of the School. As early as 1931, S. M. Gunn of the Rockefeller Foundation notes about a conversation with Malinowski: 'M. obviously antagonistic to the development of Social Biology Department, but does not show any animus against Hogben.'[150] By 1935 such noises had become so general that the Foundation decided that it was time to terminate support for the department. A University of London Report on the School in the same year had concluded that social biology as practised by Hogben did not belong and had to be discontinued.

Still, Beveridge, and Hogben, hung on. For a while Hogben even got 'much more optimistic' about developing social biology at the School, and Beveridge was delighted when Hogben sent him his book, *Genetic Principles in Medicine and Social Science.* Is this one of the great books of the Beveridge years? Certainly, Beveridge had a very high opinion of Hogben. In 1936, he wrote to the Chairman of the Governors, Sir Josiah Stamp: 'He is certainly the most distinguished member of our staff and a brilliant mind which it would be a disaster to lose.'[151] Strong words, but by then things had become desperate. Hogben's initial resolve to brazen out the growing antagonism to him and his work weakened when, in 1936, the possibility of a chair in Aberdeen—a chair in biology with no 'social' about it—was mooted. From the correspondence at the time it appears that the Director was surprised when Hogben finally decided to leave for Aberdeen, but on 11 February 1937 his resignation was accepted and he departed.

Once More: The Social Sciences

Even then Beveridge wanted to fill the chair again. For him, it had become a matter of pride and *amour propre*. It is amazing how little he sensed what others thought, or else how stubborn he was in his beliefs. Shortly before Hogben's resignation, on 8 February 1937, John Van Sickle at the New York office of the Rockefeller Foundation had written to Tracy Kittredge, the Paris Representative, encouraging him to let it be known 'in conversations with Robbins, Hayek, *et al.*' that the School would get no additional funds for anything unless it discontinued social biology. 'Sir William and Mrs. Mair have been repeatedly and frankly informed of our views, but have apparently kept this knowledge from the general staff.'[152] Still in February 1937, four distinguished members of the professoriate, Ginsberg, Laski, Plant, and Robbins, laid a Memorandum before the Emergency Committee which did not mince words. (Two of the authors, Laski and Plant, had been long-time supporters of Hogben.) It may be that one day biology has something to offer to the social sciences. 'Our analysis of the six years' experience of this Chair does not lead us to believe that this time has yet come.' The only lasting results of Hogben's work are in areas not specifically biological, like demography. A readership for Dr Kuczynski is therefore justified, as is the continued employment of J. L. Gray and P. Moshinsky for 'further studies in realistic sociology', but 'in our view, the continuance of a Chair of Social Biology on the present lines would, at this juncture, be a serious error of judgment.'[153]

The views of the four professors prevailed. A deeply resentful Hogben went north. His letters to London during the subsequent period make painful reading. The Director also departed, with a sense of sadness and failure. He himself had been converted to the methods of science at Charterhouse and Oxford. 'But my dear old School of Economics, after I had spent more years there than at Charterhouse and Oxford together, would have nothing of my Science of Society as learned from Huxley and other men of science.'[154]

Leaving personalities as well as professional jealousies on one side, the question remains: what went wrong? Beveridge and Hogben both believed that the practitioners of an imperfect and immature social science were unable to appreciate the contribution of the natural sciences to their fields of knowledge. But in stating their case, Beveridge and Hogben were, I think, confusing two quite different matters. One is that, in exploring the causes of things scientific, enquiry must never be allowed to become hide-bound by disciplinary boundaries. This applies across the board and includes boundaries between so-called social and so-called natural sciences. Hogben is quite right to mock a condition in which 'political science, economics, and sociology are entitled to arrive at incompatible conclusions so long as each refrains from examining the credentials of the others',[155] and the same is true for genetics and sociology, or geology and geography. When departmental trade unionism takes over from open enquiry, all is lost. The inflationary use of Thomas Kuhn's notion of 'scientific communities' has favoured this temptation not just in the social sciences.

This means that there are numerous problems which are typically dealt with by social scientists but require for their solution an input of biological knowledge. The recent *Social Science Encyclopedia* contains a number of relevant entries including 'Ecology', 'Evolution', 'Genetics and Behaviour', and even 'Sociobiology'. (It does not mention Lancelot Hogben.) From this point of view, it is actually not difficult to agree with Donald MacRae's suggestion (in 1972) that 'perhaps the L.S.E. should reinvest in the chair of social biology we lost when Hogben left us' in order to pursue 'ethological studies' and add 'inter-species comparisons of behaviour' to 'inter-societal' ones.[156]

However, the emotional twist which made the Hogben controversy—which was in fact a Beveridge controversy throughout—so explosive had little to do with defining scientific problems irrespective of disciplinary boundaries. It is rather the result of a particular view of the logic of scientific discovery shared by Beveridge and Hogben but deeply flawed and above all not applicable to the existing and perhaps inevitable plurality of the social sciences. In what he himself called his LSE 'Memorial Volume', *Political Arithmetic*, Hogben went so far as to start with a 'Publishers' Note' to inform the readers: 'Any profits arising from the sales will be devoted to publication of other *factual* social studies.' The emphasis is Hogben's, of course. Not a word is said about the great path-breakers of science who were predicting facts rather than observing them, and whose predictions, if proved false by experiment or observation, gave rise to new and better theories. Here lies the reason why neither Hogben nor Beveridge qualify as great or even important scientists. They reined themselves in when it came to theory and always found a reason why conclusions could not be reached. As a result they always ran the risk of drowning in a sea of facts. Beveridge's work on the history of prices is an example.

So far as the social sciences are concerned, Beveridge–Hogben's problem is a different one. One might call it the inability to live with complexity. This has something to do with Mrs Mair's brotherhood of unworldly pure scientists. Social scientists clearly are not that, at least for the most part they are not. Their methods are variegated, and many are contaminated by history, involvement, and value preferences. Models, theories of the middle range, and social research coexist with hermeneutics (*Verstehen*) and normative thinking. The common elements of a scientific method for all the social sciences must therefore be sought in the realm of style as much as content. Like all disciplines of scholarship, the social sciences are about finding out the causes of things, discovering truth. They too are characterized by the interplay of general claims, theory, and specific observations, research. But above all they must proceed in ways which are not hermetic. Model-building as well as reasoned argument, systematic fact-gathering as well as historical understanding have to remain open for critical scrutiny. Nobody must be allowed to turn his Trobriand Island into a fortress. In that sense the style of the social sciences is necessarily 'disputatious' (as Hogben called it). It has to be one of critical debate, though the rules of such discourse raise important issues of methodology themselves.

One of the conditions of effective critical discourse in the social sciences is less a rule than a description of context. Perhaps an institute of zoology can drift off into a world of its own, separate from other institutes, not even part of a university. For political science or economic history, anthropology or even economics this would be deadly. Social sciences have to be exposed to the variety of approaches in their entire range if they are to flourish. They have to remain together to prevent ossification and error, to which the Institutes of the former Soviet Academy of Sciences succumbed not least for this reason. Not only do the social sciences have to remain together but departmental boundaries should be conveniences rather than barriers. In all these respects, LSE in the 1920s and 1930s came as near to perfection as we are likely to see.

5

DISCONTENTS, DISTRACTIONS, AND A LONG FAREWELL

A QUESTION OF ACADEMIC FREEDOM?

Into the 1930s

In 1931, the political climate took a decided turn for the worse. The 'Great Betrayal', Ramsay MacDonald's switch from a minority Labour Government to a National Government against a large section of his own party, did not come out of the blue. Its immediate *raison d'être* was to persuade international banks to come to the rescue by introducing savage budget cuts, including a reduction in unemployment benefits, for which MacDonald had only the slenderest majority in his cabinet and none at all in the Parliamentary Labour Party. But the great depression had already taken a heavy toll on the Labour vote, which made it virtually impossible for the party to govern in any case. Unemployment stood at two million people. Business did not seem able to get out of the slump. Government was all but bankrupt. One of MacDonald's own ministers, the former Chancellor of the Duchy of Lancaster, Sir Oswald Mosley, had set up the New Party, which soon turned into the 'Blackshirts' of the British Union of Fascists. Others within the Labour Party, including Harold Laski, were moved by their despair of gradualism and democratic reform to embrace not just Marxism but its Soviet variant.

The point about these developments was that no one with any antennae for the public sphere, let alone the London School of Economics, could remain unaffected by them. The 1930s (and 1940s) were one of those deeply unhappy periods of history in which politics encroached on individual lives and institutions virtually all the time. Gone were the days in which, apart from occasional forays by individuals into parliamentary politics, the School could concentrate its energies on developing the social sciences and teaching them to students who were quite prepared to learn about the causes of things before they went out to transform them. Now, even those who refused to be drawn in had taken a stance. An anonymous editorial writer in the Michaelmas 1933 issue of the *Clare Market Review* contrasts the politically minded 'intelligentsia' and the seemingly non-political 'athletariat' and adds the disturbing innuendo: 'The athletic union which refuses to think about even conservative politics, is still with us; nor can I find any real justification for the rumour that it is interesting itself in Fascism.' Does he mean to suggest that the rumours may after all have some justification? The 'intelligentsia'—or perhaps just the author of the editorial—is also torn and confused by the twin attacks on Fabianism by 'Marx and Robbins'. Since Marx is not likely to be tolerated in the University of London, there is the risk that 'the School will become that barren polytechnic of big business which is already threatened as its fate'.[1]

Such musings may be inchoate but they have a serious core. The 1930s were to be the decade of the great temptations of totalitarianism. In the face of new economic and social problems, reformism and with it democracy came under pressure. It is unfortunate though perhaps not altogether surprising that some of the Fabian advocates of gradualism yielded to the temptation. The ageing Webbs travelled to Stalin's Soviet Union in 1932, and again in 1934, and produced several editions of their mammoth *Soviet Communism: A New Civilization?* From the second edition onwards, no question mark at the end of the title, though Beveridge was not alone in thinking that 'the title should be nearly all question marks'![2] But, as Norman and Jeanne MacKenzie put it, the Webbs came to regard the Stalin regime as 'applied Fabianism—the old Webb notion of a threefold state of citizens, consumers and producers united by a moral creed and efficient organization'.[3] There is no hint of the horrors of what was after all one of the three or four unspeakable public crimes of the century, the 'harvest of sorrow' following collectivization and described so vividly by Robert Conquest. The Webbs' 'detached and philosophical interpretation' of the Moscow Trials (written, to make matters worse, after the second visit) can only be described as sickening. Laski was often a little unbalanced, so that one is not particularly surprised to read in his own account of Moscow experiences statements like: 'Russia is a land of hope. The masses have no doubt that the sacrifices of today will be justified by the achievements of tomorrow.'[4] But the more sober Webbs, also, took the line which was to become so fashionable among fellow-travellers of Stalin and Hitler that you cannot make an omelette without breaking eggs, meaning human beings: 'We suggest that [the trial proceedings and sentences] are the inevitable aftermath of any long-drawn-out revolutionary struggle that ends in a successful seizure of power.'[5]

Few remained entirely immune from the temptations of totalitarianism

The Beveridge years begin with change

14. *The Sphere* depicts Mrs Janet Mair, the Secretary (1919–38) and Sir William Beveridge (1879–1963), fourth Director (1919–37) at work 1922

15. Aldwych and the Bush House site, *c.*1919, showing the huts, sections of which were used by the School to ease its space shortage

16. King George V and Queen Mary arriving in Houghton Street to lay the foundation stone for the LSE 'Old' Building, 28 May 1920

17. Lunch-hour dances remained an LSE institution until the 1960s. Here (*centre*) in 1920 Sydney Caine (Director 1957–67) with his future wife Muriel Harris

18. Centre of the School's social life. Smartly attired students and staff in the new refectory, 1922.

After the First World War some of the great teachers arrived

19. Lionel (later Lord) Robbins (student 1920, lecturer 1925), during the LSE 'evacuation' to Cambridge, 1940

20. Harold Laski (lecturer 1920), in his room at the School in 1948

21. **Scholars and teachers of the School's heroic age**

(*clockwise from top left*) Bronislaw Malinowski; L. T. Hobhouse (*l*) with A. L. Bowley; Eileen Power giving one of her BBC broadcasts; R. H. Tawney (*r*), pictured here with M. M. Postan in 1959; Lancelot Hogben

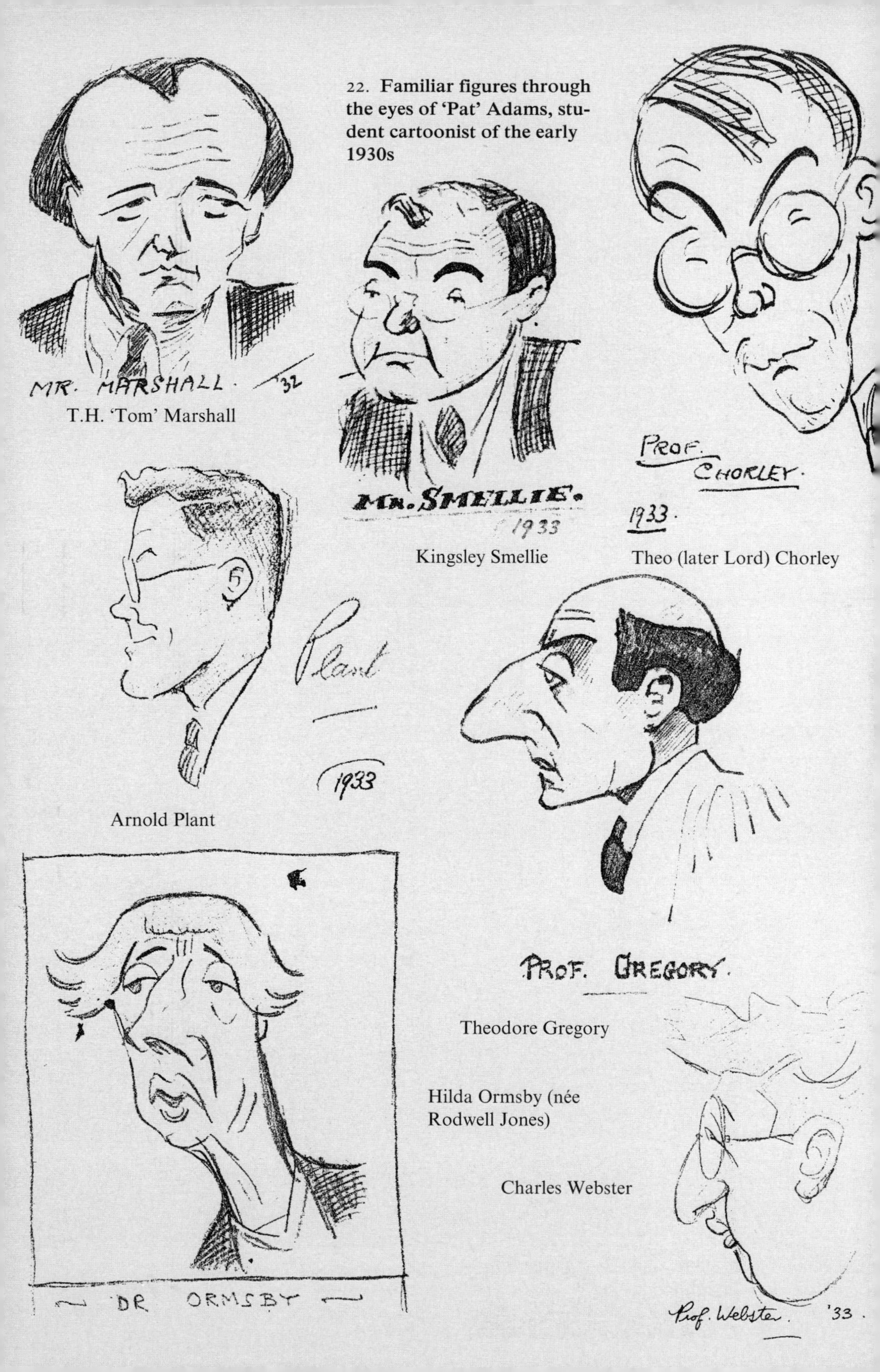

22. Familiar figures through the eyes of 'Pat' Adams, student cartoonist of the early 1930s

T.H. 'Tom' Marshall

Kingsley Smellie

Theo (later Lord) Chorley

Arnold Plant

Theodore Gregory

Hilda Ormsby (née Rodwell Jones)

Charles Webster

The porters were always the gate-keepers of the School

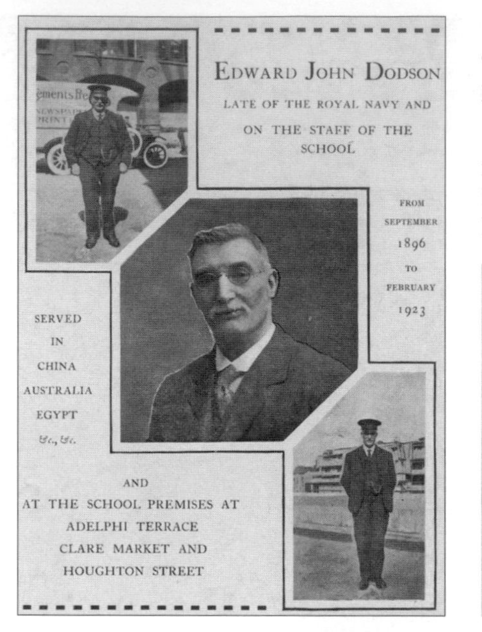

23. Two head porters who served for sixty-six years between them: E. J. Dodson (*left*) and George Panormo (*below*) (1912–51).

24. Off-duty, *c.*1951

25. Full order in the Founders Room, 1981

26. **Forever changing, forever the same...**

(*clockwise from top left*) Houghton Street from the new Bowley Roof Garden, 1936; the original Three Tuns public house in the 1930s; the familiar main entrance after the war; late 1950s, showing the St Clements Press Building before conversion for LSE and, *r*, the former Holborn Estate Charity Office; the Three Tuns building; Clements Inn Passage, first home of the Economists' Bookshop 1947; Aldwych, 30 June 1944.

27. Around 1940: six major figures spanning eighty years of LSE history

The Director Alexander Carr-Saunders (1937–57) with (*l*) the then Secretary Walter Adams (later Director 1967–74) in Cambridge, 1940

Sir William Beveridge and Mrs Janet Mair *c.*1940

Sidney and Beatrice Webb at Passfield Corner, 1942

in the 1930s. William Beveridge was one. But, alas! He had his own travails which led him, in his later autobiography, to give the chapter on this period of his life the heading 'Fading of Dreams'. By this he meant, to be sure, in the first place the dreams of the age. 'In the nineteen-thirties one dream after another that had amused mankind since fighting ended in 1918 faded and vanished.' The much hoped-for world of prosperity, justice, and liberty was no more. Then, immediately following a reference to the Abyssinian war conducted by 'Hitler's jackal Mussolini', he finds it appropriate to say: 'In the same years I came to realise that the hope with which I had come to the School of Economics . . . was as far from realization as it had ever been.'[6] Economics had still not become an inductive science.

In fairness it has to be recorded that ever since the completion of his first ten LSE years, Beveridge had explored 'ways of leaving, not the School of Economics, which I had come to love, but the position of Director in it'. Beveridge is quite specific on this intention. 'When Allyn Young died in the spring of 1929—just after my mother's death, just before my father's—I explored with Steel-Maitland and Stamp the possibility of my succeeding Young as Professor of Political Economy at the School and of finding a suitable successor to me as Director.'[7] Steel-Maitland and Stamp, the Chairman and Vice-Chairman of the Governors, were not convinced, and the Webbs, whom the Director consulted also, positively dissuaded him. Perhaps they were wrong. After ten years in an important office one is liable to make mistakes. The initial enthusiasm is gone, yet routine is a bad substitute. The habit of governing leads to brashness where sensitivity is needed. Moreover, people have seen it all and begin to wonder about alternatives. It is more than a little doubtful whether Beveridge would have been an appropriate Professor of Political Economy but evident that after a brilliant decade as Director he became increasingly accident-prone and vulnerable.

Article 28

Most of the mistakes which Beveridge made, largely in his last five years as Director, had to do with his notion of pure (social) science and its institutional requirements. This was true, of course, for the saga of social biology, but also for a sequence of incidents at the inevitably tense boundary between LSE and political reality. This boundary had worried Beveridge from his earliest days as Director. It had led him to almost total political abstemiousness, as a curious phrase in his own account of the LSE years reveals: 'Laborious search for political activity in all my documents of this time, including the written conversations with my mother where nothing was concealed, has brought to light four examples only for mention here as verging on politics.'[8] As one looks at the examples, which include a fleeting (Beveridge says, 'shadowy') connection with a 'Liberal Summer School', one is not surprised that he has forgotten them. William Beveridge was devoid of political passion.

This was not true for most others at LSE, of course. So how could he rein them in? The first systematic attempt to do so was made when, in February 1923, the Director got the Governors to set up a Committee on Partisan Propaganda.[9] At the time there were, of course, a number of teachers who aspired to, or had achieved, seats in Parliament, though not on one side of the House only, if one thinks of Halford Mackinder as well as H. B. Lees-Smith, Clement Attlee, and Hugh Dalton. The remit of the Committee was phrased in neutral terms; it was to consider 'what steps could be taken to dissipate the view held in certain quarters that the School is being used for partisan propaganda'. In fact the old bogeyman had reappeared, this time in the shape of the lay Governor Mr Wilson Potter, who had found his City friends reluctant to accept LSE graduates 'because they had the impression that the teaching given at the School was of a socialistic tendency'. He himself did not regard this as correct, of course, but something had to be done to dissipate the view held by so many. Should there perhaps be 'a regulation prohibiting teachers in the School from taking part in active political propaganda'?

Sidney Webb and the representatives of the professoriate, notably Mr Lees-Smith and Professor Hobhouse, did not think so. They argued forcefully that 'any interest or vitality in lectures given or books written' would be lost if teachers had to detach themselves from the real world. In any case, nearly three decades of similar accusations had not prevented the phenomenal success of LSE. The recommendations made by the Committee reaffirmed therefore the terms of Article 28 of the School's Articles of Association, and concentrated on the appointments process and the academic integrity of teachers. All that remained of the wider issue was the exhortation 'that when speaking or acting in their personal capacities outside the School, teachers should take every possible precaution to make it clear that they speak for themselves and should avoid emphasising their association with the School'.

In subsequent boundary disputes, the interpretation of Article 28 of the 'Memorandum and Articles of Association' of 1901 remained the central issue:

> No religious, political or economic test or qualification shall be made a condition for or disqualify from receiving any of the benefits of the Corporation, or holding

any office therein; and no member of the Corporation, or professor, lecturer or other officer thereof, shall be under any disability or disadvantage by reason only of any opinions that he may hold or promulgate on any subject whatsoever.

Academic freedom indeed, since the 'Corporation' in question is that of a university! Yet something seems to have irked the Director about Article 28. 'After ten years I began to think that something should be done to bring home to everybody concerned the impartial character of the School. After twelve years, I felt certain of this and I wrote a long memorandum putting the problem.' But why? The three reasons which Beveridge himself gave later sound strangely remote, indeed irrelevant. First, changes in political structure had brought to the fore the Labour Party, which was supported by prominent teachers at the School. The second reason was Laski's appointment in 1920—more than ten years before!—and the consequent 'deepening [of] our red colour in many eyes'.[10] Thirdly, by defeating the University MP, Sir Ernest Graham-Little, in the contest for the Vice-Chancellorship in 1926, he himself had made an enemy who was always ready to attack the School.

It would have made more sense for Beveridge to refer to the saga of social biology, not, of course, for Hogben's politics, but because of the conflicts about pure science and propaganda. More specifically still, Selskar Gunn of the Rockefeller Foundation had noted the unhappy 'political flavour' at the School after the 1931 election, in which Dalton and Lees-Smith lost their seats and which had been accompanied by a bevy of controversial Laski articles.[11] But then Beveridge got itchy whenever he encountered political activity at the School. In October 1932, the Director explained his views in some detail in his address to new students. He spoke of the practical life and that of science. 'The essence of science is to say nothing until one is absolutely sure.'[12] In practical life, on the contrary, decisions have to be taken. It is 'like a journey'; if one comes to a fork in the road and the signpost is not legible, one has to take one or the other direction anyway. The scientist, by contrast, 'has to put the writing on the signposts or at least say what should be put there'. Scientists may thus be called 'indecisive', but advocates of practical reforms are 'unscientific'. The two cannot easily be combined, as Beveridge argued by using a surprising metaphor. 'A University teacher of Social Science should be content with one mistress. He cannot at one and the same time with advantage be a scientist and a practising politician looking for votes, any more than he can with propriety make love to two women at once or marry two wives.'[13] It takes little guesswork to give the bigamist a familiar name.

The careful reader will have noted the words 'with propriety'. What is proper can be written into rules and protected by sanctions. When the Director addressed the students in the autumn of 1932 he had actually completed the tortuous process of doing just that. His initial proposal of December 1931 had passed the Emergency Committee of lay governors easily, but in the Professorial Council 'there followed a riot of discussion, with amendments and counter-amendments, lasting nearly five months'.[14] It was not before May 1932 therefore that the Resolution was finally adopted which stipulates in paragraph one:

That while members of the staff of the School of Economics should, in the full sense secured to them by Article 28 of the Memorandum and Articles of Association of the School, be free from regulation or censure by the Governors of the School in respect of their writings or public speeches, they should regard it as a personal duty to preserve in such writings or speeches a proper regard for the reputation of the School as an academic centre of scientific teaching and research.

The four remaining paragraphs of the Resolution are more technical and refer to the rights and duties of members of the staff who intend to be, are, or have been Members of Parliament. The statement, on the other hand, which the Professorial Council forced the Director to add by way of authoritative interpretation was anything but technical. It not only emphasized that having pursued a political career would not disadvantage present or potential teachers in any way, but also elegantly pulled the carpet of sanctions from under the edifice of the Director's new rules: 'On Resolution 1 it was the intention of the [Emergency] Committee that the Director would be expected to make a "light intervention" in cases of infringement.'

With the rules and their interpretation in place, when and how were they going to be applied? The School did not have to wait long. In the space of one academic year, 1933/4, three problems arose which left lasting scars on a number of people and inflicted irredeemable damage on William Beveridge, who might so easily have become the most popular as well as the most successful Director of the School: the case of the *Student Vanguard*, the Laski affair, and the story of the Frankfurt Library. Only one, the Laski affair, is strictly an Article 28 matter, but all three touch on politics, the reputation of the School, and the difficulties which a Director committed only to pure science encounters in dealing with possible infringements 'lightly'.

At the end of the academic year 1933/4 the Director knew that things had not gone his way. It was in fact his *annus horribilis*. In his Report of 21 June 1934, he made what he himself later called, an 'oddly solemn reference'[15] to the troubles of the preceding year, which he contrasted with the unusual absence of any disturbance by building activity:

When the history of our School comes to be written, it will have to be recorded that this session of unexampled physical peace in the School has been a session of acute psychological disturbance. The future historian may be able to explain this paradox himself, or he may call in aid his colleagues from a neighbouring field—economics, psychology, politics or some other study. The historian will at least be able to place our experiences in due perspective: he will see them, I think, largely as a sign of youth in the School. For all its thirty-nine years, the School of Economics is in many ways like the ship of the story on her maiden voyage, making first experience of stresses and problems long familiar to her older sisters, improvising machinery to meet new situations, still needing to find herself and the sense of unity that only time can bring. The School is still as it was thirty-nine years ago, a place of immense potentialities not yet realised, a place of new difficult studies on whose successful pursuit, whether in our School or elsewhere, depends perhaps the peace of mankind, depends perhaps, I am almost tempted to say, the survival of civilisation.

Beveridge himself has not helped the future historian. In his autobiography the incidents of 1933/4 hardly figure at all, and in his own account of his School years they are alluded to with that mixture of bewilderment and bathos which the 1934 quotation reveals. Others have since done better, including the biographers of Beveridge and Laski. The story of the Frankfurt Institute and Library started the 'acute psychological disturbance'; but it lasted longer and is also closely bound up with academic assistance for German academic exiles, so that we shall deal with it in the subsequent section. The case of the *Student Vanguard* and the Laski affair had in any case wider ramifications within the School.

The Case of the Student Vanguard

The Students' Union correspondent of the *Clare Market Review* remembers hearing in early 1934, in the second week of Lent Term, 'the distant rumblings of the approaching thunder'. The thunderer was none other than the Director, surprisingly perhaps in view of his close and good relations with LSE students in general and the Students' Union in particular. One lingering subject of tension had to do with the eternal source of student revolt, refectory prices and arrangements. But on 15 January 1934, in a letter to the President of the Students' Union, Frank Straus Meyer, Beveridge added a more specific and contentious issue. In response to a request from the Chairman of the Marxist Society, H. J. Simons, Beveridge ruled that School rooms would not be available for a course on Marxism planned by the Society. The Union, at a full meeting, expressed its regret at such restrictions, and in the following weeks an increasingly acrimonious correspondence between the Director and the President of the Union ensued. Mr Meyer insisted that it was for the Union and not the Director to decide what student societies did. Beveridge not only disagreed but lost his temper, as in his letter of 23 January 1934:

Dear Meyer
Thanks for your letter of yesterday. Do please be sensible. When you have come off your high horse and are better able to see the difference between your position in the School and mine, I shall be delighted to consider further any proposals affecting the welfare of the Union.[16]

Personal chemistry apart—Beveridge clearly did not like the 'odious American Jew',[17] as Beatrice Webb described Frank Meyer in her Diary—the Director persuaded himself that the conflict was about the question of who runs the School, no less. In his highly emotional 'Note on the Present Discontents' written later that spring, Beveridge describes as the 'immediate cause of the present troubles', 'the assertion by the Students' Union of a dyarchy, leading to anarchy, in the management of the School and its buildings'.[18] Dyarchy, however, the monarch—the autarch perhaps?—was not going to have.

The Director proceeded to devise a Code for the Union and a Memorandum for student societies establishing clearly that he was the only authority when it came to notices, meetings, and the allocation of rooms. All the time he dealt with Mr Meyer in a notably hostile manner, until in the end the Union rallied to the defence of its President and resolved 'that this Union regrets that the Director has treated the question of the Code as a personal one between himself and the President'.[19]

In this unhappy atmosphere, the storm broke on 27 February. While Meyer wrote a long letter about Union autonomy, the Director had discovered a new and to him much more serious infringement. The February issue of the left-wing newsletter, the *Student Vanguard*, had carried an article about alleged incidents of spying on Indian students by Oxford and Cambridge tutors. An editorial note had been added at the end of the article:

The above statement, coming from a reliable source, adds considerable weight to a thesis we have been led to hold, namely, that a well-organized system exists for spying on colonial students. It is not restricted to Oxford and Cambridge: in the London School of Economics a retired Indian policeman fulfils the same function.[20]

Beveridge immediately banned the sale of the paper on School premises. Meyer, Simons, and other members of the Union saw this as a part of the Director's campaign to restrict their activities and defied the ban. The process was set in motion which led to the expulsion of F. S. Meyer and H. J. Simons from LSE.

The 'retired Indian policeman' might not have noticed the libellous statement, but his colleague Harold Laski, who never failed to notice anything, had shown him a copy, which not surprisingly led him to consult his solicitor. Professor John Coatman was one of Beveridge's somewhat anomalous acquisitions. Born in 1889, he had indeed been a member of the police in India after he had passed the (Indian) Civil Service examinations in 1910. After the war, he had taken an Oxford degree but returned to India as Director of Public Information. When Beveridge, through his friend Stephen Tallents, then Chairman of the Empire Marketing Board, managed to attract funds for a Chair in Imperial Economic Relations, the Appointments Committee acquiesced and accepted John Coatman as the beneficiary. In 1934 the funds ran out and the research professorship was discontinued. Without any link to the *Student Vanguard* case, Coatman left LSE and joined the BBC for the remaining two decades of his working life, leaving little trace at the School—except that he was the supervisor of H. J. Simons and one of the other students initially implicated, Daphne Trevor.

That 'the effect of [the Director's action] on the School was electric', as 'Argus' wrote in the *Clare Market Review*, is an understatement. The subsequent events add up to one of those stories which read very differently in School Committee minutes from the accounts of either victims or bystanders.[21] Fortunately, the bare facts are not in dispute. On 27 February Professor Coatman brought the libellous *Student Vanguard* to the attention of the Director, who immediately drafted and posted a notice stipulating that 'no papers or periodicals may be sold, distributed, or advertised in any part of the School premises unless the sanction of the Director has been obtained in advance'. The Director told Meyer and Simons that this ruling applied to the *Student Vanguard.* On 28 February Meyer wrote to the Director to complain about such 'censorship', which he regarded as 'another addition to the series of restrictions on freedom of speech and assembly which you have been attempting to impose upon the students of the School', and asked Beveridge to reconsider. Six students then proceeded to defy the ruling and sell the *Vanguard* on School premises: Meyer, a 25-year-old American Ph.D. student in Anthropology with Princeton and Oxford (Balliol) degrees, and also a part-time research assistant to Raymond Firth; H. J. Simons, a South African Ph.D. student in Colonial Administration supervised by Professors Coatman and Malinowski, who had been in the South African Public Service; Phyllis Freeman, a Canadian Ph.D. student in Economic History and holder of a Daughters of the Empire fellowship; Daphne Trevor, another South African working for a

Ph.D. under Professor Coatman; and two undergraduates, one, A. M. Miller, supervised by Professor Laski, the other, G. E. Marmont, by Professor Power. They were reported by the Academic Assistant for Student Societies, and the Head Porter. The Director sent for them and decided in the case of five, but not in that of Miss Trevor, to 'exclude them from the School until further notice pending a definite decision upon their case'.

Several statements given in the following days make it clear that the students regarded the incident as part of a larger campaign by the Director, who, on the contrary, tried to treat the *Vanguard* case in isolation. On 2 March the six students sent a letter of apology to the Director, recognizing their error of judgement and disregard of the Director's authority, and emphasizing that their motive had not been that of 'furthering the publicity of the paragraph [on 'spying'] in *The Student Vanguard*'. 'We therefore sincerely apologise for our offence and venture to hope that you will accept this expression of our profound regret.' On 5 March the Emergency Committee held a special meeting attended by the Director, the Secretary, and seven Governors, including Lord Passfield (Sidney Webb), Professors Ginsberg and Rodwell Jones, and Mr Lloyd. The Director reported the facts, including the apology by the six, their individual statements of regret, and the readiness of the *Student Vanguard* to comply with the demands by Professor Coatman's solicitors.

The Emergency Committee then decided that in the case of the two undergraduates and Daphne Trevor, 'having regard to their youth and the fact that they acted under the influence of elders who alone had seen the Director . . . it would be possible to accept from them a full apology and undertaking for good behaviour in the future'. For Phyllis Freeman, the same conclusion was eventually reached, though with some reluctance. On the other hand, Meyer and Simons were 'required to leave the School'. On the following day, the Union, in a unanimous resolution, 'express[ed] its grave concern and deplore[d] the extreme and unprecedented severity of the decision and the measures taken against the President and a member of the Executive of the Union'. A deputation was sent to the Director to appeal for clemency. On 13 March, a petition signed by 400 students supported this request. Mr Simons's South African teacher, Professor Brookes, sent a plea for his former student.

However, the Emergency Committee did not relent, at least not entirely. On 22 March it resolved 'that the decision for expulsion stands, but in the case of Simons, if after the expiration of a term he presents himself for re-admission, it shall be within the authority of the Director in light of the situation then existing to consider his re-admission'.[22] At this point what had been a School matter, albeit one which aroused considerable emotions, became a public scandal. Throughout April 1934, petitions were addressed to the Director and to the Court of Governors, letters were written to newspapers, meetings were held to expose the injustice of expelling students who had apologized for what was after all not a capital offence. Some of these representations came from Students' Unions, but there were others. The Dean of Canterbury and the Canon of Westminster joined, along with Victor Gollancz, J. B. Priestley, and two dozen others, an appeal to the Governors by cultural figures. Prominent Labour peers, MPs, and intellectuals, including Lansbury, Attlee, and G. D. H. Cole, wrote an open letter. Student and civil liberties organizations at home and abroad pleaded and appealed. More established personalities expressed themselves in their own way but to the same effect. One wrote: 'Of course boys will be boys and I am wondering whether it is not possible that some lesser punishment than that which would bring a complete blight on their career might not meet the case.'[23]

However, all this was to no avail, perhaps because the School itself, its Governors and teachers as well as many students, remained largely silent, acquiescent if not supportive of the Director. Simons was, after a term's absence, readmitted and obtained his Ph.D. in 1936. Meyer, on the other hand, remained not only expelled but had to leave the country. The Secretary of the School had written to the Home Secretary to that effect, and when last-minute attempts to have him reinstated at LSE failed, Meyer had to leave Britain within a matter of days. He went to the University of Chicago, where Professor Malinowski had helped him find a place to complete his doctorate in anthropology. After that the man whom Beveridge described even in his LSE book as 'a very red politician from America'[24] turned to a lifetime of activity on the extreme right, which George Nash has described in his book on *The Conservative Intellectual Movement in America.* Meyer worked for the *National Review*, became active in the Conservative Party of New York, wrote books about *The Moulding of Communists* and *The Conservative Mainstream*, the two strangely intertwined experiences of his life, and 'became famous for his crackling, uncompromising creed, often expounded over the telephone long after midnight to conservatives all over the land'. Before his death in 1972, the 'odious American Jew', described by Nash as an 'intense, cheerfully argumentative, chess-playing, zip-code-hating metaphysician of the Right',[25] even joined some of his prominent conservative friends by converting to the Roman Catholic faith.

Meanwhile at LSE, the Emergency Committee had to deal with the rules for students which Beveridge regarded as 'much more important' than the incident which gave rise to them. They arose from his angry paper about 'the present discontents'. The Students' Union, he said there, 'has fallen under the control of people who are not students in the ordinary sense at all, but grown men and women, some with axes to grind, some perhaps with no more than a desire for easy leadership'. Graduates aged 25 'not students in the ordinary sense at all'—the School had clearly come a long way since its beginnings. Against their political doings, as well as 'dyarchy', the Director tried to introduce a gamut of rules about Union accommodation, student societies, charitable collections, and meetings, and penalties to go with them all. A fine 'not exceeding £5' remained as a penalty for disobeying 'all rules made and instructions given by the Director of the School or under his authority' but otherwise the Director's powers were formally limited by the creation of a Board of Discipline and the strengthening of the role of the Emergency Committee.[26]

So far as the other side of the story is concerned, it would be too easy to cite students. They were naturally upset and aggrieved by the Director's wrath. But others also began to wonder about Beveridge. Where was the 'light intervention' which the Professorial Council had demanded as a matter of principle? Sir Arthur Steel-Maitland, the Chairman, tried to calm Beveridge down, but ultimately let him have his way. In May 1934, E. E. 'Rufus' Day and John Van Sickle of the Rockefeller Foundation came to see Sir William Beveridge and Mrs Mair, and found them much agitated about 'the dismissal of a certain Myer [*sic*!], a New York Jew, and President of the Student Union'. What would they have done in the Director's place? Mr Day 'questioned the wisdom in general of expulsion on the grounds of insubordination'. Sir William did not like such 'criticism of his handling of the Myer case. The discussion at times became a bit acrimonious. Sir William revealed unexpected sensitiveness. At one point he said to E.E.D., "For God's sake, don't go around England advocating the rights of the students as against the authorities of institutions." However, the evening closed most amicably over Scotch highballs.'[27]

The Laski Affair

Some authors suspect that there is a connection between the *Student Vanguard* case and the Laski affair, which overlapped both in time and in theme. Kramnick and Sheerman suggest, for example, that Beveridge attributed the 'outburst of student protest' in 1934 'in part, to Laski's influence'. In a general sense, this may well be true, but Jose Harris is quite right to observe that 'Beveridge's actions on this occasion [i.e. the expulsion of Meyer and Simons] seem to have commanded widespread support among governors, teachers and students.' Obviously, Laski was involved, but there is no evidence of his taking sides for either the President of the Students' Union or the Chairman of the Marxist Society; on the contrary, it appears that it was Laski who persuaded them to write the letter of apology because he believed that 'the completeness with which they were in the wrong was beyond discussion'. By the time the case of the *Student Vanguard* was settled, however, Laski was already involved in his own serious and again highly personalized dispute with the Director whom he had supported for so long.[28]

There is no precise beginning of the Laski affair. Lingering unease about Professor Laski's public utterances had informed the Professorial Council interpretation of Article 28 in 1932 and continued throughout the next two years. It came to a head in the days of February 1934 in which Meyer and his friends were suspended. Beveridge told Laski that his articles in the *Daily Herald* were doing harm to the reputation of the School. Apparently, Laski's short but irreverent portraits of contemporary figures raised particular comment, though of course his opinions on political events too must have sounded increasingly extreme to a more and more conservative public and political class. For a few weeks nothing happened; but in April the Director took the matter up again. The two men had a meeting which left both, and certainly Laski, deeply disturbed. On 18 April the Director wrote to Laski telling him that he was acting, by his *Daily Herald* articles, in violation of what Beveridge kept on calling 'the 1931 Resolution of the Professorial Council'. (In 1931 Beveridge had tried to push through the Resolution reminding teachers of 'proper regard for the reputation of the School' in their public utterances, but it had taken him many months and important concessions before he finally succeeded in May 1932; a fact which he had conveniently forgotten.) Laski replied in a sad little letter that he now wanted 'the Emergency Committee to rule upon the differences between us about my articles in the *Daily Herald*':

> In the light of this, I do not think I ought to comment on your letter beyond saying that I, too, found our talk of the other night very painful. I had not realized until then how profound had become your personal antagonism to me, and it came upon me with all the force of an unexpected blow.[29]

The Emergency Committee was not at all eager to pick up this new bone of contention. The Chairman, Sir Arthur Steel-Maitland, tried to calm the waters. Laski had written to him and explained that writing the 'Pen Portraits' took about two hours a week of his spare time, so that there was no question of his neglecting his professorial duties, and in any case he always had the welfare of the School at heart. Moreover, Steel-Maitland and others wanted to wait 'until a decent interval had elapsed after the Meyer affair'.[30] This turned out to be unfortunate. Many eyes and ears were now on Laski. When he accepted an invitation to Moscow and gave a series of lectures there at the end of June, he could not expect his words to go unnoticed. And they did not go unnoticed.

On 6 July the *Daily Telegraph* published a leading article attacking Laski's 'pestilent talk of class war' in his third Moscow lecture. Laski had sounded a theme which was to prove his undoing exactly eleven years later. He argued that, if radical changes did not happen soon in Britain, there might well be violence and revolution. This was bad enough in 1934. On 11 July, ugly questions were asked in Parliament, as by Captain P. MacDonald: Does LSE get any Government money? And 'is it not a fact that Professor Laski is an avowed Communist, as well as of alien origin?' The Vice-Chancellor of London University reacted by writing—Kingsley Martin suggests 'as the result of sheer panic' after an intervention of the Visitor of the University, King George V—letters to several newspapers dissociating the University from Laski's Moscow lecture, which would 'doubtless form the subject of an inquiry by an appropriate body'. On 14 July the affair got another, even more serious if ultimately self-defeating twist by the University MP, Sir Ernest Graham-Little, writing to the *Daily Telegraph* demanding 'disciplinary correction' by LSE and adding that 'the London School of Economics, where Professor Laski functions, has long been regarded as a hotbed of Communist teaching, and such action by the governing body is consequently unlikely'. As if insult was not enough, Sir Ernest heaped injury on top of it: 'In the absence of a spontaneous condemnation by the School of Economics of this regrettable outburst by one of its teachers, the Court [i.e. the committee allocating resources in the University] of the University might conceivably take action by reducing the allocation it makes to the London School of Economics.'

The position of University MPs was obviously anomalous, and nowhere more so than in London. They were elected by former graduates, members of Convocation, and in the case of London many of these were external students who had never set foot in a London college. Sir Ernest, as Chairman not only of the London Graduates' Association, but above all of the University's External Council, held a controlling position which made it impossible even for the Conservative Party to put up another candidate. Moreover, Harry Henry has shown, in an article in the *University of London Union Magazine* in February 1937, that the election was anything but secret, with tellers ticking off the names of voters on a list as they received the ballot papers. Still, the MP had clearly gone too far. A lively debate ensued in the columns of daily and weekly papers. Not just George Bernard Shaw, or Clement Attlee and Stafford Cripps, but, in a letter to *The Times*, five LSE Professors—Chorley, Gregory, Power, Tawney, and Webster—rallied to the defence of Laski and of academic freedom. Captain MacDonald, MP, was made to apologize for his error about Laski's 'alien origin'. The Chancellor of the Exchequer made it clear that he saw no reason to take any action on the financial side. Finally, John Maynard Keynes, in an impressive letter to the *New Statesman* (21 July 1934), made the crucial points. It was, he said, 'well established in England, as distinguished from Moscow or Berlin, both that a professor is entitled to the unfettered expression of his opinions and that no one but himself has any responsibility in the matter'. The suggestion that financial pressure should be exerted on LSE was 'monstrous'. Sir Ernest Graham-Little was 'obviously unfit to represent a University in Parliament'. If this kind of inquisition into opinions would not stop, civilization itself was threatened.[31]

For the Emergency Committee of the School, however, this was the beginning rather than the end of the story. While the public debate was waging, on 19 July 1934, Beveridge produced a Memorandum in which he made three points. First, Laski had, by his regular columns for the *Daily Herald*, violated the 1928 regulations concerning outside work by teachers, and a salary cut equivalent to his outside earnings should therefore be considered. Secondly, Laski had contravened 'the 1931 [*sic*!] Resolution of the Professorial Council' and done damage to the reputation and interests of the School. What, one wonders, did the Director mean when he remarked in this context: 'A claim by University teachers to say on every occasion just what comes into their heads (particularly on their own subjects) will not be admitted by the public and will not, in the long run, save academic freedom for the world'? Thirdly, the Emergency Committee may wish to take action against Sir Ernest Graham-Little's letter both publicly and with the University authorities.

Between July and October the Emergency Committee had the issue on its agenda at every meeting. Indeed the question of Sir Ernest Graham-Little's attack on the School recurred well into 1936 and led to several internal as well as public statements about the true nature of LSE. The School has 100 teachers (a letter to the Vice-Chancellor recalled) but none of them 'so far as is known is a Communist, and certainly none of them gives Communist teaching in the School'. On the contrary (it was added perhaps somewhat unnecessarily), the teachers in Economics and Commerce 'are critical not only of Communism, but of nearly all other forms of state intervention in the economic sphere'. This and other communications to the University led the Vice-Chancellor to inform LSE on 5 October of the conclusion of the relevant committees that 'no further action was called for'.[32]

Nearer home, more painful actions were taken. The Emergency Committee decided almost immediately, on 19 July 1934, that Professor Laski was guilty on the two counts of the Beveridge Memorandum relating to him. His agreement to contribute regular articles to a national newspaper 'should be submitted for consent of the School', and he had, unwittingly perhaps, done harm to the reputation of the School. On 25 July, Harold Laski, who had been sent a copy of the minutes of the Emergency Committee, replied with a dignified letter to the Director. In it he stated that he 'will write to the "Daily Herald" to explain that I must cease to contribute a weekly article to it as from the end of the vacation'. He would obviously continue to make 'occasional speeches on political topics' but regards himself 'as bound, in making such speeches, by the general sense of the resolutions passed by the Professorial Council in 1931 [*sic*!]'. Beyond that, he wants the Emergency Committee 'to be aware of my anxiety, in the fourteen years I have served the School, to add, as best I can, to its reputation as a centre of teaching and research'.[33]

By the autumn of 1934 the affair was, in a technical sense, closed. In practice, to be sure, little changed. Laski continued to make public and controversial pronouncements. The School continued to be subjected to attacks for its alleged political bias. But deep scars were left not just on Laski but on many of those involved.

University Teaching and Political Activity

What then does the 'future historian' have to say about the 'acute psychological disturbance' of Beveridge's *annus horribilis*? The first relevant comment was made right in the middle of the disturbance, on 12 March 1934, by someone who despite her often misguided enthusiasms never lost her capacity for acute analysis, Beatrice Webb. In her *Diary* she records a visit by a worried Beveridge on the previous evening.

> The trouble at the School is, I fear, destined to become worse during the next decade. Political and economic studies, carried on in London, one of the hubs of the political and financial world, by an assembly of some three thousand students of all races and professions—undergraduates and post-graduates, youths and maidens, and men and women in the prime of life—under the direction of a large and miscellaneous staff of professors and assistant professors, is bound to develop heated antagonism of creed and class. Beveridge cites Laski, with his Labour Party

journalism and his close association with political personages, as the centre of the mischief. Laski denounces Robbins and his group of fanatical individualists for their resolute refusal to study the economic facts of today or to permit any deviation from abstract reasoning from disputed premises. And all this fanaticism will grow worse in the coming duplex struggle for power in country after country—first in the sphere of politics, between parliamentary libertarian democracy and the totalitarian state, and secondly in the sphere of economics, between capitalism and Communism.[34]

In other words, the conjunction of time and place, of the particular time of the early 1930s, and the particular place called London School of Economics and Political Science, was bound to be explosive.

Secondly, Beveridge was surely right in suspecting a 'psychological' element in any explanation of the 1934 discontents, though he failed to see that his own 'psychology' was a big part of the issue. 'Beveridge in 1934 felt himself beleaguered on several fronts—domestic, political, intellectual—and the behaviour of Laski was perhaps the last straw in a period of intense personal strain.' Jose Harris adds that the Director's objections were to an 'emotive and flamboyant'[35] style as much as to the views held and propounded. The extent to which Beveridge personalized conflicts suggests that he was worried about his own position, his authority, his ability to assert himself as Director and perhaps even as a man. Since he must have known that his victories—Meyer's expulsion, Laski's retraction—were at best pyrrhic, it is not surprising that the year left him depressed and searching for alternative lives.

A third point is perhaps less an explanation than a lesson from the affairs of 1934. The notion which comes up time and again in accounts of these affairs is that of 'academic freedom'. Beveridge himself used it in his 1935 'Reflections on the "School of Economics"': 'There has been no interference of any kind at any time with academic freedom in the School.' The five LSE Professors who had defended Laski in their letter to *The Times* in 1934 called the public attacks 'a menace to academic freedom'. Laski expressed doubt whether 'we had more than a bare majority in favour of academic freedom' since the Governors' 'only real concern was endowments'. Laski's biographer Michael Newman takes this view a big step further and surmises: 'It was surely inevitable that [Laski] would see some parallels between his own position and that of German academics.' Jose Harris's account of the discontents of 1933/4 is entitled 'Social Science and Academic Freedom'.[36]

All this is understandable and in some respects justified, yet distinctions have to be made. Academic freedom in the strict sense is the freedom of teachers to teach what they believe to be true, and the freedom of students to learn without dogma or hindrance—in an institution of higher learning. If academic freedom is not guaranteed, universities rapidly decline into second-rate centres of indoctrination of which this century has seen far too many. Academic freedom is not a luxury, or even just a basic right in a civilized society, but a necessary condition of the search for truth and the education of citizens. Thus understood, academic freedom was never in any doubt at LSE, nor would it have been if the 'left' had purveyed its views as unashamedly as the 'right' often did. At all times, LSE students were exposed to a variety of views in a climate which encouraged questioning authorities rather than acquiescing to any kind of political correctness. In its core activity, teaching and research, the School was and is a model of academic freedom.

This is not to deny the reality of threats. Some of them came from without. Sir Ernest Graham-Little's interventions were flagrant attacks on academic freedom. Some of the City Governors at times got close to similar encroachments. The Rockefeller Foundation, on the other hand, never did. And throughout its history, the School's academic councils have defended academic freedom jealously and effectively. They have been almost invariably supported by the leading members of the Court of Governors, notably their Chairmen and Vice-Chairmen.

But of course the issue of the Laski affair was not what professors were saying in the classroom or the lecture theatre, but in the market-place, in the *Daily Herald*, and in Moscow. This may seem a fine distinction but it is important. Some may have wondered whether every word Laski said—and every place which he picked to say it—was perfectly chosen but even his more extreme critics did not intend to deny a British citizen his freedom of opinion. This has nothing directly to do with *academic* freedom. Professors do not have any particular right to say outrageous things in public, nor do they have a particular claim to wisdom in political matters.

Here we enter delicate territory, to be sure, especially for the London School of Economics. As we survey it two pitfalls emerge, and both have given the School much cause for anguish over the years. One has to do with the repercussions for the institution of actions by professors and students who are exercising their civil rather than academic liberties; the other arises from the expectation that a school of higher learning should provide a platform for the expression of opinions as well as knowledge. Strictly speaking, neither of these issues affects the core of what is meant by academic freedom. Professors making outrageous speeches in Moscow or anywhere, and students preventing an unwelcome politician from stating his or her case, are not applying or violating principles of *academic* freedom. But it would be vain to deny that both impinge on the ability of the institution to conduct its academic affairs in freedom. The need exists therefore to find rules of conduct in keeping with the values and interests of a university.

Such rules are not easy to come by. The conduct of members of the institution outside their strictly academic responsibilities is as much a matter of taste as of regulation. Who decides when a professor or a student has done unacceptable harm to the institution? Sidney Webb resigned as Chairman when the repercussions of his 'railway speech' dawned on him. Harold Laski was indignant when the Director rebuked him for his *Daily Herald* articles. Both were probably right. But throughout its history LSE was more vulnerable to attack on academic grounds than others. Even the famous resolution of its Union not to fight for King and country has not led the interested public to regard the University of Oxford as unpatriotic, whereas a few red banners in Houghton Street confirm widespread prejudices that the School is 'socialistic'. Some mixture of sensitivity on the part of members and generosity by the institution is probably the right approach. Neither was always present during the later Beveridge years.

Similarly when it comes to outsiders using the School as a platform to expound their views, the School is much in the public eye. It is debatable whether politicians can actually claim a right to be heard in an academic institution. However, it is a bad sign if a school of the social sciences is not prepared to listen to diverse views. Giving speakers a rough ride is one thing, attacking them physically or preventing them from speaking another. It could be argued that LSE should if anything be prepared to listen to a wider range of views than the general public may find acceptable. In any case, every effort should be made by those in charge to maintain a climate of open debate.

The style of these comments is carefully chosen. We are not—to repeat the crucial point—talking about academic freedom and the compelling need to preserve it against all attacks. We are, however, talking about a school of the social sciences in the centre of London. Max Weber's prescriptions for a value-free social science, and Sidney Webb's version of it as an LSE of dispassionate study and research, are too saintly for the real world, and certainly for that at Clare Market. Here, accepted rules of engagement and a firm but relaxed approach to transgressions are as important as saintliness.

The School has not found such an approach easy to take. Perhaps the attacks which accompanied its life from the earliest days hurt too much. Most if not all major mistakes in the history of LSE have been made by the authorities leaning over backwards to prove that the School was not what many seemed to think it was. Only the neutral, for the most part politically Conservative chairmen remained calm in the midst of the storms; men like Sir Arthur Steel-Maitland, Lord Stamp, Sir Otto Niemeyer, and Lord Bridges. For those closer to the coal-face, heavy intervention and an exaggerated emphasis on the presence of partisanship of the other side, from Lilian Knowles to Michael Oakeshott, was frequently the line of defence. Perhaps this was the price LSE had to pay for the excitement and vibrant creativity of its location astride the boundary between theory and practice, scholarship and politics. It may even have strengthened the School to fight the battle against public prejudice in its peculiar way. However, it was also painful. Frank Meyer and Harold Laski were not victims of an attack on academic freedom, nor even of the wrath of a beleaguered Director, but of the almost constitutional vulnerability of a great institution. Meyer did not much care, but Laski did, and as a result he epitomized the School's weakness as well as its strength.

IN DEFENCE OF FREE LEARNING

The Academic Assistance Council

Not everything the Director took up in that fateful year 1933/4 went sour on him. On the contrary, his decision to put the considerable weight of his personality and position behind the cause of helping exiled Central European academics was, in Lionel Robbins's words, 'one of Beveridge's great moments—his finest hour I would say'. Robbins is referring in particular to a meeting in Vienna at the end of March 1933. Hitler had seized power in Germany on 30 January. One of the first acts of legislation by the Nazi-led coalition was the innocuously called 'Restoration of Civil Service Act'. It was in fact a pseudo-legal encouragement for removing 'undesirable' people with civil service status—and German professors were among those who had this status—despite their assured tenure. A first wave of sackings, affecting those with a high political profile, many of them Jews, took place immediately. Beveridge's finest hour followed the darkest hour of German universities which offered no resistance.

At the end of March, Beveridge was in Vienna to meet some of his collaborators on an international history of prices and wages project. Lionel Robbins and his wife Iris were also in Vienna, though for other reasons. One day when Robbins had arranged to spend the evening with the much-admired senior figure of the Austrian school of economists, Ludwig von Mises, he and Beveridge ran into each other and spent an hour together 'in one of the Vienna cafés',[37] as Beveridge remembers not implausibly, though Robbins tells us that it was in fact the Hotel Bristol and before Robbins and Mises went out for dinner. When Mises joined Beveridge and Robbins, he brought

> an evening paper carrying the shocking news of the first academic dismissals by the Nazis—[Moritz] Bonn, [Karl] Mannheim, [Hermann] Kantorowicz and others. Was it not possible, he asked, to make some provision in Britain for the relief of such victims, of which the names mentioned were only the beginning of what, he assured us, was obviously to be an extensive persecution.

Since there is no dispute about the facts, it is as well to follow the story as told by Lionel Robbins who did not ordinarily have many good words to say about William Beveridge:

> All his best instincts, his sympathy with the unfortunate, his sense of civilized values, his administrative vision and inventiveness, were quickened by the question. Slumped in a chair, with his great head characteristically cupped in his fists, thinking aloud, he then and there outlined the basic plan of what became the famous Academic Assistance Council—later the Society for the Protection of Science and Learning—to which hundreds of *émigrés* even now living in the English-speaking world owe the preservation of their careers and, in some cases, probably their lives.[38]

Other encounters in Vienna, notably with the physicist Leo Szilard, who was staying in the same hotel, confirmed his resolve. On returning to London, Beveridge lost no time. He started, quite properly, at home, that is at the School. At its meeting on 3 May 1933 the Professorial Council, on a proposal by the Director, set up an Academic Freedom Committee consisting of Professor Chorley, Dr Dalton, Mr Judges (a lecturer in economic history), Professor Laski, Professor Power, Professor Robbins, and Professor Webster, with the Director.

The Committee met the following day, and again on 8 May. On 17 May it reported back to the Professorial Council with two important proposals. One was that exiles of 'international reputation'[39] should be invited as teachers to the School, notably for graduate teaching. The other proposal, based on calculations by Professor Robbins, was that all members of staff should be urged to give 1 per cent (lecturers), 2 per cent (readers), or 3 per cent (professors) of their salaries annually in order to collect at least £1,000 a year for an Academic Assistance Fund. The proposals were accepted and endorsed by the Governors. Three years later Professor Chorley was able to report that in the intervening period a total of £2,850 had in fact been collected. 'With these monies [the Academic Freedom Committee] has been able to afford relief to a substantial number of distinguished refugees and young people of outstanding promise, most of whom have now been satisfactorily placed in permanent positions in different parts of the world.'[40]

LSE discontinued its 'tax' in June 1936, though the Academic Assistance Fund continued in operation, managed by Lionel Robbins, for another three years. The School was able to reduce its activity because of parallel national developments set in motion by Beveridge in the same early May days of 1933 in which the School Committee was created. Beveridge went to Cambridge and talked to George Trevelyan, Frederick G. Hopkins, and Lord Rutherford. Hopkins was then President of the Royal Society. He drafted a statement for signature by the great and the good of science. On 24 May 1933 the Academic Assistance Council was launched, with Lord Rutherford as President and Sir William Beveridge, and Professor C. S. Gibson, FRS, as Hon. Secretaries. The next morning the first cheques came in, from John Maynard Keynes and Michael Sadler. By 1 June, £10,000 had been raised.[41]

The story of the Academic Assistance Council has been told by several authors, including Beveridge himself in one of his retrospectives, the book written in 1959: *A Defence of Free Learning*. The most moving account is that of the 'Conversations with Tess [Dr Esther] Simpson', who accompanied the Academic Assistance Council from the earliest days in the Royal Society offices in Burlington House as Assistant Secretary, later Secretary of the successor Society for the Protection of Science and Learning (so named in 1936) and the Society for Visiting Scientists until she formally retired in 1978; the book is called *Refugee Scholars*. From Norman Bentwich's *Rescue and Achievement of Refugee Scholars* to the recent *International Biographical Dictionary of Central European Émigrés*, the extraordinary success of the Council and the Society have been documented. In all, 2,600 scholars had been helped by 1958, initially mostly from Germany and Central Europe generally, later also from Spain and Portugal, and from the Soviet Union.

It soon became clear that the Academic Assistance Council needed someone who knew the life of scholarship from inside. On 1 July 1933, Beveridge reports: 'We lured Walter Adams from safe pensionable lecturing at University College, London, to be our full-time Secretary—to work harder for the same salary without pension or prospects.'[42] Adams was then a 27-year-old history don who had specialized in history of science. He had also shown his mettle by trying to set up a Residential Club where 'coloured students' could meet 'Europeans' and overcome the isolation of individual lodgings by 'group life'. The very first thing he did after his appointment was to go on a trip to Germany, visit universities, and talk to many of those who were threatened. He was instrumental in getting the Warburg Library to London, and later in extending the activities of the Council to other countries. Many have paid tribute to his work as Secretary of the Academic Assistance Council. 'I found him ideal to work with', Dr Simpson wrote, 'and the refugee scholars were devoted to him.'[43] Even in her nineties, the indefatigable Esther Simpson remembered the 'idealism', the devotion, and the appetite for work of Walter Adams. When he left in 1938 to become Secretary of the London School of Economics in succession to Jessy Mair, a group of refugee scholars commissioned a portrait of him by Sir William Rothenstein.

The Academic Assistance Council and LSE worked closely together. In a book by the German Society for Exile Studies, Gottfried Niedhart described the 'English years' of the historian of the labour movement and biographer of Frederick Engels, Gustav Mayer. Summarily dismissed by the Rector of the University of Berlin at the age of 62, he came to London to look for a future. He met Laski and Tawney and Sidney Webb, who were all welcoming and supportive, but for getting started in London, Adams was critical. He was helpful, Mayer said later, 'beyond all praise'; 'he has done everything possible for me'.[44] LSE gave Mayer an Honorary Research Fellowship and a room; the Academic Freedom Committee and the Assistance Council contributed a little money. When he finally decided to emigrate, the School made him a Special Lecturer so that he could get an exit visa and a residence permit. The financial guarantee was provided by the Society for the Protection of Science and Learning. Mayer was able to work for the International Institute for Social History in London, surviving by occasional awards and help by the School and by Harold Laski. He died in 1948.

This is but one of many stories, tinged with sadness, for Mayer remained a lonely man cut off from his roots; at the same time one creditable to those who tried to alleviate the burden of an unbearable fate. In their response to the true threat to academic freedom in Germany and elsewhere in Europe, the School and its Director showed their moral fibre. They acted quickly and appropriately. When representatives of the Rockefeller Foundation came to see Beveridge a year later, on 22 May 1934, they seemed surprised: 'Sir William displayed an extraordinary bitterness toward the existing regime in Germany; he thought it might be pardoned in the eyes of the Lord but not in the eyes of man.'[45] On 13 October 1935 Beveridge gave that dual sentiment full expression in his sermon at Carrs Lane Church in Birmingham. In fact, he made the point of active liberty long before most:

What is happening in Germany today is a challenge to us in Britain above all. It is a challenge not to be taken up by protests. Protests butter no parsnips. It is a challenge given by deeds, not words—and must be met not by words, but by deeds. It is a challenge given by deeds of hate; it should be met by deeds of charity.[46]

The Story of the Frankfurt Institute

Not all Beveridge's initiatives in helping refugee scholars went as well as the setting up of the Academic Freedom Committee and the Academic Assistance Council. In the middle of May 1933, when the plight of German scholars in exile was uppermost in his mind, the Director had the first of several conversations with Friedrich Pollock, then Assistant Director of the Frankfurt Institut für Sozialforschung. Pollock was concerned about the Institute and above all its library which, while legally the property of a Swiss holding company, was still in Frankfurt and therefore clearly threatened. The 'authorities in charge of the Institut', Beveridge noted after his conversations, were prepared to give the library to LSE 'with a view to establishing also in London either the headquarters of the Institut or a branch where members of the Institut could continue to use the library'. The Director was, perhaps naturally, impressed and got in touch with officials at the Home Office and the Ministry of Labour about necessary permits before he responded to Pollock on 1 June 1933: 'I write to say that the School will be happy to accept this valuable gift, and thus make certain of keeping the library together and available for the purposes of social and economic research.' He added, nobly, that if the Institute decided to settle elsewhere, he would regard it 'as an obligation of honour' to return the library.[47] The Emergency Committee endorsed the Director's actions.

Enter Lionel Robbins. One day during this period, 'I had strolled into the Director's room for a talk about some minor departmental matter when he informed me with great satisfaction that he had virtually completed an agreement with a certain research organization whereby, in return for the deposit with us of its valuable library, it would be given accommodation at the School and authority to use it as its headquarters.' Robbins was horrified. This 'certain research organization' was 'well known as a stronghold of Marxism in Germany', and the Director had not even consulted teachers at the School. Going ahead with the plan 'might involve the introduction of something quite alien to the principles on which the School had been founded; and it certainly ran the danger of appearing to confirm all the ignorant misrepresentations to which we were always subject'. Beveridge was annoyed, though when Pollock insisted on finalizing the deal with undue haste, he was also puzzled. Webb was brought in and seemed to support the Robbins position. Other refugees in London were asked about the Institute and confirmed Robbins's judgement. The deal did not come to pass. Indeed Beveridge (according to Robbins) recanted. 'I cannot imagine what possessed me,' he is supposed to have said.[48] Many years later, Hayek, in response to questions by Nadim Shehadi, was still full of venom about the Frankfurt Institute and its possible move to LSE:

> The London School of Economics would have become a Communist institution in effect. . . . Some of them are dreadful creatures. I had my doubts about Adorno. I forget all the other names. Mr. Marcuse. . . . It's the kind of Marxism which I dislike the most. It's a combination of Marxism and Freudianism. I am equally opposed to both of the sources, and in its combined form I find it particularly repulsive.[49]

Jose Harris regards the story of the Frankfurt Library along with the Laski affair as 'the most serious'[50] of the incidents of Beveridge's dreadful year 1933/4. The judgement is arguable in so far as one aspect of the story is Beveridge's 'autocracy', the lack of consultation with academics. But so far as the Institut für Sozialforschung is concerned it was all a storm in a teacup. Max Horkheimer, the real head of the Institute, was not a man to leave his own destiny, or comfort for that matter, in the hands of others. Two years before the Nazis came to power, he had transferred the endowment of the Frankfurt Institute to Switzerland. By the time Beveridge entered this particular scene, the Institute had been incorporated in Geneva, and Horkheimer himself as well as almost all other senior members were settled there. Two worries remained. One was the library of then 60,000 volumes which the directors had not been able to move. The other was, in the case of the ever-cautious Horkheimer, concern that Switzerland might not be altogether safe.

The latter concern led first to some diversification, with small branch offices in Paris and a little later at Le Play House, the seat of the Sociological Society in London, and soon to the move to the United States, where the Frankfurt School found a new home in New York before it went on to Los Angeles. There remained the library, which had been confiscated as a reservoir for Communist propaganda before Horkheimer managed to move it. Clearly, Horkheimer was using the School, and his own administrative dogsbody Pollock, who was devoted to Horkheimer but never taken seriously as a scholar by his master, to get it out. Horkheimer, Adorno, Marcuse, and Fromm were hardly Communists. They were what was called in Weimar Germany *Salonbolschewisten*, who were always more prepared to give up their political beliefs than the good life. Even their sympathetic historian Martin Jay remarks: 'And so, the Institut's members may have been relentless in their hostility towards the capitalist system, but they never abandoned the lifestyle of the *haute bourgeoisie*.' Horkheimer was a very wealthy man, and a great schemer, too. He later played a problematic role in controlling funds and even visas for immigration to the United States. The desperate plight of his one-time friend Walter Benjamin, who failed to get to America in time, has recently been revealed. Beveridge must have suspected some of this, and in any case realized that the Institute was not totally dependent on his goodwill. On one occasion, he wrote to his friend at the Ministry of Labour that 'there is, I understand, some division of opinion among those in charge of the Institut' as to where they should go. But in essence Beveridge showed the commendable if in this case not entirely appropriate naïvety of someone who wanted to help.[51]

Unfortunately, this naïvety was bound up with the 'obligation of honour' which he felt not only to return the Frankfurt Library once he had got it, but to get it out of Germany in the first place. Over the next two years Beveridge used all legal and political means at his disposal to try and get the library out. Sir Josiah Stamp even talked to the German Economics Minister Hjalmar Schacht about it. Worse still, on 14 February 1935, the Emergency Committee 'agreed that Sir Arthur Steel-Maitland should consult Lord Lothian with a view to exploring the possibility of a direct private approach to Herr Hitler'. When the Foreign Office told the Director that nothing more could be done, lawyers were instructed to pursue the matter. But Beveridge would not discuss it with anyone, not even with the Secretary, Jessy Mair. In her desperation she wrote to him on 16 March 1935 imploring him to drop the matter. 'We are not legally bound to go on.' Pollock is using the School for litigation which he or other Institute members could not pursue. Pollock is probably not a 'gentleman and a man of honour'. Everybody is against the deal. 'You have embarked on this in a spirit everyone would respect. But no one would respect your obstinacy in trying to go on with it.' The Emergency Committee persuaded the Director to seek release from earlier promises, and a cable was sent to Pollock (who was by then in America) accordingly. 'Dr. Pollock immediately replied by cable, later confirmed by letter, to the effect that he released the School from the obligation of honour.' It took another year, until 16 July 1936, for the episode to be finally closed when the School's solicitor informed Mrs Mair that 'there is clearly nothing more to be done'.[52]

Jose Harris thinks that the incident was 'highly significant in three respects'.[53] It shows that Beveridge's 'impartiality' was not confined to one political direction; it demonstrates the depth of ideological divisions within the School; and it is an example of undemocratic governance. These points are well taken but in the history of the School and, contrary to the case of the *Student Vanguard* and the Laski affair, the story of the Frankfurt Institute remained an episode, telling but without consequence.

Exiles at LSE

One feature of the defence of free learning by Beveridge and others in 1933 which strikes the historian as curious is the repeated insistence that the initiatives taken were not only, or even primarily, about Jews. The Report by the School's Academic Freedom Committee of 17 May 1933 spoke of assistance 'as a measure of defence for academic and scientific freedom, and not specifically as a Jewish question. The threatened teachers included many who had no Jewish connection at all and many others who were no longer members of the Jewish Community.' The launching statement of the Academic Assistance Council similarly emphasized that 'many who have suffered or are threatened have no Jewish connection'. In the course of the preparations, the Royal Society in particular took an unequivocal line. 'They were strongly of opinion that no signatory of the Appeal, and particularly that neither of the honorary secretaries should be of Jewish origin.' In the event, at least one signatory, Samuel Alexander, was Jewish, but (Beveridge writes) 'his twenty-eight years as Professor of Philosophy and the loving esteem in which he was held by all who knew him, to say nothing of his academic honours and Order of Merit, would have made objection to him absurd'.[54]

Whatever interpretation one wants to put on such statements, many of the academic exiles who came to the School were Jews, and while they may have been 'no longer members of the Jewish Community' at home, Hitler's actions could not fail to remind them of their roots. The School took in three groups of exiles in the 1930s: those who passed through on their way to other, more permanent positions; those who maintained a loose though useful association; and those who became central figures at LSE. The first of these were probably the largest group. Some of them stayed for a few weeks, some for a year or more. They included men like Franz Neumann, the labour lawyer who turned his law into political science under the supervision of Harold Laski and Karl Mannheim before he went to the USA in 1936; the social historian Ernest Kohn-Bramstedt, who later joined the BBC and then the wartime intelligence services; the ex-communist Social Democratic organizer and activist Richard Löwenthal, who came as a research fellow for a year before he returned to political activity and journalism, though after the war he became an influential Professor of Politics and International Relations in Berlin. Gustav Mayer, who has already been mentioned, was one of many examples of the second category.

By far the most influential category of exiles consisted, however, of those who stayed and became a part of LSE. Jessy Mair wrote to the Director, who had not been able to attend the occasion, about the first lecture by the political economist Moritz Bonn on 20 October 1933: 'Bonn had a great welcome at his inaugural address. Robbins made an admirable introduction, and students packed the theatre, standing patiently to hear it all. It was indeed both remarkable and moving.'[55] Bonn combined practical experience with an academic career and had a long-standing interest in Britain. He became the *Wandering Scholar* of the title of his autobiography, for after his retirement from the School in 1938 he spent five years in various North American universities.

The most unlikely and remarkable group of exiles was that of lawyers. It was unlikely because by moving from one legal tradition to another, their knowledge and experience became all but useless. One young legal scholar put it nicely: 'Their more fortunate brethren, the doctors and scientists, found to their benefit that human bodies and the integral calculus are the same the world over, even in Britain. Whigs in the Strand and conveyancers in Lincolns Inn portended otherwise for the law.'[56] The legal exiles were also remarkable because it is probably not too much to say that they were responsible as much as anyone for reconciling the Law Department with its social science environment. Their personal need became the Department's gain. Hermann Kantorowicz went to the New School for Social Research before he came to LSE as research director in law with a special interest in jurisprudence and sociology of law. Hermann Mannheim, who had been a professor of criminal law and judge in the criminal division of courts in Berlin, turned to criminology, which he taught at LSE for twenty years, from 1935 to 1955. Hersch Lauterpacht had degrees from Vienna and London and taught international law. He was later joined by Georg Schwarzenberger, who began as secretary of the London Institute of World Affairs, and added a Ph.D. to his Tübingen Dr.jur., before he started teaching at LSE in 1945.

Possibly the most influential of them all was Otto (later Sir Otto) Kahn-Freund. Like Ernst Fraenkel and Franz Neumann, he belonged to the generation of labour lawyers who grew up around that new branch of law in the Weimar Republic. Their choice of subject set them apart; they were true Weimar Republicans, for the most part Social Democrats, and naturally interested in the social sciences. Kahn-Freund's early book on the 'social ideal of the Reich Labour Court' is a classic in socio-legal analysis. But contrary to Fraenkel and Neumann, who felt compelled by the culture change to turn to political science, Kahn-Freund stuck to law, became a barrister and QC, and in some sense invented labour law for England. He taught at the School from 1936 to 1964, after which he accepted a chair at Oxford, and he left many traces of academic achievement, like the *Modern Law Review*, and of warm appreciation by his students and colleagues.

Other departments did not attract quite such a gaggle of stars but they also benefited. R. R. Kuczynski came by way of social biology but stayed as a statistician. Then there was the representative of the other Frankfurt, as it were, the Hungarian-born sociologist Karl Mannheim, who had been a Professor at Frankfurt without being associated with the Horkheimer Institut für Sozialforschung; Frankfurt had its own history of academic infighting. Mannheim's fame rested on his 1929 book *Ideology and Utopia*, which laid the foundations for a 'sociology of knowledge' but also caught the relativist mood of the times by casting doubt on all claims to objectivity. The suspicion of ideological distortion (Mannheim argued) is total. At the School, Mannheim began with a graduate teaching fellowship financed from Rockefeller money. Despite Malinowski's strong support, he failed to attract funds for a major research project on regimes and their élites, perhaps because with Malinowski he felt 'that the whole of the sociological program of the London School of Economics needs reconstruction'. Morris Ginsberg did not like this much and there arose what Carr-Saunders later called 'the Mannheim–Ginsberg problem'.[57] Social analysis based on research versus social philosophy in the Hobhouse tradition was the issue. Beveridge tried to get Mannheim a place in the social biology group, which made little sense; he did eventually find him a readership; but in 1945 Karl Mannheim accepted the professorship at the University of London Institute of Education. By then he had become a much-read analyst of the contemporary scene (*Diagnosis of Our Time*) and an advocate of 'democratic planning'. Two years after his departure from LSE he died, at the age of 54.

Others came later, after the aggravation of persecution in Germany in 1938, and the annexation of Austria. The latter caused Karl Popper to leave, perhaps the greatest of all exiles to come to the School, though the long detour by way of Christchurch, New Zealand, meant that he did not arrive until 1945. By then a younger generation of exiles was about to appear on the scene: those who had left as children, mostly with their parents, and now made their mark on the life of the School. Some of them, notably (Sir) John Burgh, (Professor) Ernest Gellner, (Sir) Claus Moser, even made Esther Simpson's 'roll of honour' of her most successful 'children'. However, a glance at a later School Calendar, say in 1972, reminds one of many other names: Professor Hilde Himmelweit, W. H. N. Hotopf, and A. N. 'Bram' Oppenheim in Psychology; Professor (now Lord) Peter Bauer, Lucien Foldes, and Kurt Klappholz in Economics; Professors John Hajnal and Frank Land in Statistics; George Grün and Walter Stern in History; Professor Michael Zander in Law. Some of these second-generation exiles had already gone through the firm yet caring hands of the long-time postgraduate officer of the School, Anne Bohm, herself an exile straddling the two generations.

There were other administrators like Ilse Boas. The Library gained a most extraordinary man in the person of Eduard Rosenbaum. He was 47 when he came, in 1934, with a grant from the Academic Assistance Council. By then he had acquired a doctorate of philosophy in Kiel; taught workers' education courses; worked in an import–export company in Hamburg and with the Hamburg Chamber of Commerce, for a year as its librarian; helped the German war effort as a civil servant, which was recognized by a civilian order of merit; assisted the German Peace Delegation to Versailles; taught at the universities of Kiel and Hamburg; and edited an economic newsletter. He must have been the most polymath acquisitions officer in the history of the BLPES, where he served from 1935 to 1952, while still writing articles for learned as well as popular journals, and also an eminently readable history of the Warburg Bank.

Thus LSE alone benefited immeasurably from the great exodus of German and Central European scholars and intellectuals. If one adds other places of destination, Oxford and Cambridge, the New School and Columbia and Harvard, and the many outstanding natural scientists who had to leave, it becomes clear why German academic and intellectual life would never be the same again. At the School, the new teachers of the first and the second generation of exiles meant a further broadening of range, and an injection of new energy. It is a comment on LSE that those who came were made to feel at home, and that those who received them on the whole felt at ease with the newcomers. The Director deserves much of the credit. In the worst of his eighteen years at LSE he did one of his best and most consequential deeds.

AT LSE IN THE 1930S

Remembering the Student Days

In writing history it is tempting to recall great events and their protagonists, but for more ordinary humans these are often remote and a disturbance rather than an enrichment of their lives. This was certainly true at

LSE in the 1930s. Betty Scharf, an undergraduate B.Sc.(Econ.) student in the mid-1930s (and later a distinguished teacher and Governor of the School), remembers: 'The institution was such that what went on high up was not known to humble folk like us.'[58] The distance was felt by many. 'On my peregrinations back and forth,' Aubrey Jones wrote later, 'I sometimes caught glimpses of the Great.' He means 'the Director, the bird-like Beveridge, and the Secretary, the over-powering Mrs. Mair'. His own dealings with the administration, if any, were, however, with the 'ever kind and ever helpful Miss Evans'.

All students remember at least some of the Great, though almost invariably—and with the possible exception of Laski—from a distance: a social distance, or a physical distance. The former became manifest in those formidable boards which allocated scholarships or grants, and which left one student at least with the 'impression of having entered a Sultan's petitioning chamber; the Sultan (Beveridge) sat in the centre, flanked on either side by his advisers, myself deferentially distant. There was a vague pinkness about it all; it may have been the decor; or it may have been the gowns, worn unusually to impress.'[59]

For the most part, however, the distance was as much physical as social. All remember the lectures and some of the great lecturers. Many students, whatever their subject, heard Lionel Robbins and felt that he 'was a mountain' (N. Seear). Hayek gave them more trouble, especially in the early years; 'his accent in English was thick and his thought appeared tangled' (A. Jones). Morris Ginsberg was another widely appreciated figure; 'his lecturing method was not as spectacular as that of Professor Laski, but what he had to say was always relevant'. Many also liked T. H. Marshall, 'who had one of the most beautiful speaking voices I have ever heard' (E. Townsend). Eileen Power fascinated her audience by her 'art of speaking' and otherwise. Tawney presented more difficulties; he 'had a very low voice and one had to sit in the front row in order to be able to hear him' (A. K. Dasgupta). Arnold Plant, Raymond Firth, Otto Kahn-Freund all had their admirers. And then there was Laski, of course, 'a spell-binder' (E. Townsend) whom no one missed, though some thought he was above all 'a great showman' (N. Seear).[60]

Much depended on the subject which people were studying. In the 1930s there were a growing number of women who took up social science courses. They felt at times on the fringe rather than at the centre of the School. In part this was no doubt due to what one might call the class structure of academic courses and disciplines, in which social science ranked roughly where its often poor and downtrodden clients rank socially. But equally importantly, social science students spent as much time outside the School as inside. Many lived in settlements, those 'hives of splendid "do-goodery"' which were far away from 'theory as propounded at LSE' (E. Townsend). Moreover, the course involved extended periods of practical work away from the School. Thus, 'your centre of interest and activity wasn't really focused in LSE' (N. Seear). Favourite teachers mentioned by all are Edith Eckhard and 'Minnie' (M. Louise) Haskins. Edith Eckhard was for over three decades, from 1919 to 1952, the heart and soul of the Social Science Department, and for many years, although only a lecturer and 'senior tutor', also its *de facto* head. A quiet person, she was a devoted teacher, always available, always helpful. The *Times* obituarist says that she was 'loved and admired by the changing colleagues' and that 'there are hundreds of social workers throughout this country and overseas who hold her in affectionate memory for all she gave them in their student days'. 'Minnie' Haskins, a tutor and lecturer in the Department from 1919 to 1944, was remembered for her warmth as a person, and also her poetry. Neither, however, had much of a following outside Social Science. Indeed, Eileen Younghusband, who was not just known but was famous elsewhere, was at times resented in her own department for that reason. The one hundred social science students, rising to 170 in the 1930s, were living in a world of their own.[61]

It was a very different experience to be an economics student, and more so, an economics postgraduate. Not only did the economists regard themselves as the centre of the School, if not the universe, but their own core activity was very far from practical work; it was 'Lionel Robbins' great seminar' in Economic Theory. A. K. Dasgupta has described the way it worked. Lionel Robbins would allocate papers on subjects topical because of some important publication—like imperfect competition in the year after the books on the subject by Joan Robinson and Edward Chamberlin had appeared—to about eight research students. They included in Dasgupta's time Ursula Hicks, Marian Bowley, and R. H. Coase. Teachers like John Hicks and Kaldor and sometimes Hayek and Plant would also be present. The papers would be read, then 'cyclostyled' and discussed in a kind of second-reading debate a week later. Over tea, groups would be formed to discuss a particular paper further in considerable detail. Sometimes one paper remained on the seminar agenda for several weeks, with Robbins in the Socratic role. 'So he would give an impression as if he was an outsider and the writer of the paper was the authority.' In the end, Robbins would turn to the teachers. 'Hicks, do you have anything to say?' Sometimes he did, and sometimes he did not.[62]

Other departmental stories could be recounted. What they show is that even in the 1930s there was not just one LSE experience. Nor was this merely a matter of the department or course which had brought people to the School. National stories could be told. Dasgupta, like B. K. Nehru before him, says without resentment but as a matter of fact: 'In those days there was not that kind of mixing between the students coming from outside England, so that I cannot claim that I had very intimate association with my colleagues, although we were members of the same seminar.' Rudolf Edler was one of those who embarked on the adventure of adding English legal training to his German law degree. He liked Krishna Menon, was amazed at Meyer's expulsion ('where are the limits of liberty?'), and observed: 'All the good will of friends found there for life could not obliterate the difference in age of brawn and bone.' The other side of the picture, as it were, confirms its colours. Nancy (later Baroness) Seear had come from Cambridge and 'found LSE scruffy in a great many ways'; she had her own circle of friends and therefore 'didn't get to know people who were here'.[63]

There were other cleavages. Some students were poor, very poor. It hurts to read Nell McGregor's account of how she worked her way out of a Manchester working-class family in the middle of the depression to LSE. Her grant of £40 just about covered basic cost, though not fares, lunches, let alone stationery and books. When she fell ill, her grant was at risk; in the end she got her degree on tea and buns and baked potatoes and not much else; but when she declined to pursue teachers' training (for which the grant had initially been given) she was in trouble. 'A revolutionary, I hoped that the revolution was imminent.' When it failed to materialize, she had to repay much of the money. Elizabeth Townsend remembered her own student life as frugal, but instead of £40 after fees she had £100 for full board plus £5 a week for fares and lunches and £80 'dress allowance' from her parents.[64]

Nell McGregor was by no means alone. 'Yours in disgust' from Argyll, an angry octogenarian wrote to *The Times* in 1993 to attack today's 'whinging students' and recounted how for three years he had 'walked from the City to Aldwych every night, stopping on the way for a 6d Welsh rabbit [*sic*!] at Joe Lyons', then attending lectures and working in the Library before he walked to Waterloo for the 'workman's train' and in Surbiton another three miles to get home. He was an evening student, and they had a hard time anyway.[65] Roland Bird remembered 'four years of drudgery and delight spent at LSE' in the early 1930s. Every night in winter and summer they would be at it, 'snatching a meal off a Scotch egg in five minutes flat between lectures and searching the library for a missing volume of the *Economic Journal* that contained the key to all knowledge'.

There was some fun on occasion, but my main memory is of sounds of every unharmonious kind—the mechanical screams from St. Clements Press, the concrete mixer never long absent from Houghton Street even in the depths of the depression, the six-piece jazz band that used to entertain the queues outside the Stoll, the rehearsals of the 'Pirates of Penzance' that went on in the theatre throughout one vacation.

And yet that was obviously not all. 'The real joy was to be able to share in a school of ideas without any hint that we who came in the evening from the City, from industry and from public service were in any way less eligible than the day students.' It was not the hope of advancement or even the search for knowledge that kept him and others going 'but the encouragement and devotion of the staff'.[66]

With all these varieties of experience, a whole lot of students have hardly been alluded to: the woman undergraduate, the wealthy (and the not-so-wealthy) American, the union activist, the 'specialist' in Colonial Administration, or in the Army Class, the badminton champion. There were others in the 1930s, not types but names, as it were: two who were to become Chairmen of the Court of Governors later, Morris Finer (LL B 1939) and Huw Wheldon (B.Sc.(Econ.) 1938). Huw Wheldon shared a room for a while with David Rockefeller (General Course 1938), though John F. Kennedy, who wanted to follow his brother Joseph's example (General Course 1933), only enrolled (General Course 1935), then fell ill and went home.[67] The Nobel Laureate-to-be Arthur Lewis was at LSE then (B.Com. 1937, Ph.D. 1940), and the man who was to become the conscience of the School and guardian of its tradition in critical times, John Griffith, arrived as a Law student in 1937.

Was there anything that brought the diverse individuals and groups at the School together? Betty Scharf describes the Graduands' Dinner in 1939,[68] where she and Douglas Allen (later Lord Croham) made the speeches on behalf of the 'class of '39', but that, of course, is the farewell occasion. The Director's address to new students may belong in the same category, and perhaps Oration Day. The Students' Union attracted some but most certainly not all, nor did the sports ground in New Malden, or the lunch-hour concerts, or the weekends at Avebury. The School in the 1930s was a remarkably diverse place. And yet, as one reflects on the memories of those who were there at the time, certain features stand out. LSE was not the kind of carefree place where young men and women could live out their adolescent confusions without paying too high a price for the results. Compared to the old undergraduate universities it was serious. Real life was never far away. The attempts by the Director and the Secretary to introduce an element of Oxbridge lifestyle into LSE—the Avebury weekends, the emphasis on athletics—sound slightly forced and probably passed the majority of students by. Universities have often been described in terms of the distinctive space which they create for their members, a 'pedagogical province', an 'ivory tower'. None of these descriptions apply to the School, where on the contrary truly academic pursuits always involved a battle to keep the noise of the outside world out, for a few hours in the evening at least, and perhaps for a few terms of frequently interrupted study. LSE was therefore more serious but also more seriously cherished by its students even if they were desperately poor or felt that their 'delight' was almost outweighed by 'drudgery'. In this way the School produced a particular frame of mind; as one student wrote about the 1930s half a century later: 'The closed mind was alien to everything about LSE 50 years ago.'[69]

The Gatekeepers

In an institution of the variety as well as the apparent hierarchy which LSE displayed in the 1930s—and to a large extent also before and after—there are bound to be some who link the diverse and disparate parts and for that reason alone command special respect and affection. They keep the gates open between all those separate gardens, those of the 'humble' and the 'high-ups' as well as those of the social science diploma students and the economics postgraduates, the Indians and the Americans, the athletes and the intellectuals. Some of these gatekeepers are simply unusual individuals. Harold Laski was one of them. Others are administrators, though they must be accessible by virtue of their 'intermediate' position and at the same time open for the concerns of students and teachers. Miss Mactaggart had played that part in the early days of the School, partly by temperament and partly because she was, the more 'difficult' McKillop apart, the only one around in that kind of administrative position. In the 1930s and later, Eve Evans played a similar role for many.

When Eve Evans finally retired in 1954, Professor David Hughes Parry expressed the hope that a later chronicler would devote to Eve a 'more cordial and laudatory paragraph' than Friedrich von Hayek had done in his 1945 History. This she certainly deserves. She came to the School in 1920, on the recommendation of Beveridge who had known her as one of the temporary clerks with the Civil Service Commission who was now becoming redundant. She was then 26 and charged with organizing the hitherto virtually non-existent registry. This she did in her characteristic 'unflappable' way, efficient and friendly at once. She had a classics degree from

Royal Holloway College, but from the moment she was appointed Registrar in 1921 'she lived for the School which absorbed all her energy' (Kathleen Lewis).[70]

Many have praised the talent with which she 'kept [the School's] operations smooth, calm and efficient' as Registrar, in difficult times, as Acting Secretary from 1941, and Secretary in succession to Walter Adams from 1945 to 1954. But it is her role as a gatekeeper which made her special. Perhaps it actually helped that in the hierarchical order of things at LSE she could not be, as Registrar, a member of the Senior Common Room. 'Her consideration for all those with whom she came in contact' was rightly praised by the Director, Sir Alexander Carr-Saunders, when she retired. Kathleen Lewis stresses 'her unfailing interest in students' which 'gave her a remarkable memory for them, even after many years, and it was a great pleasure to her to follow their subsequent careers'. Sir David Hughes Parry goes a step further and recalls how students went to her when they needed guidance, but staff too said to students in trouble, 'Go and see Miss Evans'. 'In doing so they knew that she had all the understanding and sympathy of a friend and none of the arrogance of the person in power.'[71] Perhaps Lionel Robbins summed up best what many felt in his obituary for Eve Evans in *The Times* of 16 August 1971:

> But Eve's position in this lively community [of LSE] was much more than that of an efficient administrator. With a natural elegance of manner enhanced by her height and distinction of appearance, she was a warm human being, intensely loyal to the institution to which she belonged and devoted to a wide circle of friends. She was a shrewd judge of character, not unaware of human weakness and absurdity, yet tolerant of idiosyncrasy and abounding in sympathy for anxiety and distress. Juniors and seniors alike regarded her as a sister who understood their problems and shared their hopes and fears. Lucky the institution which, at a critical stage of its development, could command such dedication and such qualities of mind and character.

Eve Evans was an unusual individual in a position which others might have interpreted quite differently, an *ad personam* gatekeeper, as it were. There is, however, one group or category of people who have played this role throughout the history of the School: the porters. It is almost unfair to single out names—and those of head porters at that—because all of them contributed much to hold the complicated, in more ways than one disparate institution together. However, the head porters were themselves a very unusual collection of individuals, and a handful of them, exactly five, span the time of the first eight Directors: Dodson, Wilson, Panormo, Brown, and Kearey.

Edward John Dodson came first. He has figured in this story before as one who began a tradition of keeping the real School on the straight and narrow by showing the self-confidence not to join any group or take sides, and thus commanding universal trust. When he was first employed on 6 October 1896 the Director tried to define his duties 'as Porter and General Attendant here [at 10 Adelphi Terrace], to clean basement, steps & windows, attend to lift, carry up coals & parcels, go errands, and make yourself generally useful'. The latter became his primary task and accomplishment. He had a career in the Navy behind him but now his loyalty to the School was total. Indeed it became 'infectious', as 'E.T.R.' noted in the *Clare Market Review* when Dodson retired in 1923. 'He never forgot a person or a name', which is one of the characteristics of gatekeepers. Dodson commanded the entrance to the School. Those who came for the first time met him and were guided to their destination; those who came every day got a friendly greeting; those who came back 'would again be greeted by name by Dodson as of yore, as an old friend, and he would make them feel as if it were only yesterday that they were at the School'.[72]

When Dodson was hired by Hewins he was firmly told that 'wages will be 25/- (twenty-five shillings) per week, without quarters or perquisites, hours probably will [be] 8 to 8 (but this must be settled later)'. When he retired over twenty-six years later in 1923 his weekly wage had risen to £4 per week, but there were still no perquisites. In fact, the Governors thought that the wage bill of £1,105 per year for seven porters was too high. However, Beveridge not only disabused them but also argued successfully that Dodson, who had become 'one of [the School's] institutions, universally liked and respected' should be given an *ex gratia* pension of 35*s*. per week. More than that, the Director persuaded the Governors to introduce a general pension scheme for porters. Dodson left a sick man, but was clearly much appreciated.[73]

Next came Walter Wilson, who began as junior porter under Dodson in 1905, and later served as porter, Head Porter, and Clerk of Works until his retirement in 1957. His tasks too were as varied as those of porters have remained to the present day.

> In his early days at the School it is said that the Secretary, Miss Mactaggart, used to arrive regularly at 10 a.m. and often used to run her handkerchief along the rail of the staircase on her way to her room. It was one of Wilson's tasks when he was warned of her approach to race up to the top of the staircase and slide down with a duster on the rails so that the Secretary could not find any dust to complain of.

On the occasion of his fiftieth anniversary at the School, Eve Evans paid tribute to him. 'In the course of fifty years, Mr. Wilson has met emergencies with the cheerful alacrity and unfailing resourcefulness which characterise him.' He too had the gatekeeper's virtues, a 'tenacious memory', an 'affectionate interest' in students, and that all-important 'heartwarming welcome' to those who somewhat hesitantly return to the School, 'with the right name put unhesitatingly to the right face and the highlights of the visitor's career recalled!'.[74]

Then came George Panormo, made famous by the conversation with him which Mrs Mair published in the *Epic of Clare Market*. Actually it was not much of a conversation, more a sequence of prompted answers.

LADY B. When did you first come to the School, Panormo?

PANORMO. I first came temporarily in 1911; then I came on to the permanent staff in 1912, when the Hon. Pember Reeves was Director. Mr. Headicar was the Librarian—and Miss Mactaggart was the Secretary. We did not see lots of the Director in those days—it was Miss Mactaggart.

LADY B. You had a colleague called Dodson. I believe he was here when you came?

PANORMO. He was the Head Porter; he had been in the Navy. Yes, he took me on.

LADY B. I think he came in the very first year at Clare Market, when LSE first came from Adelphi Terrace. When I came in 1919 I found that the people who really ran the School were Dodson and Miss Mactaggart. And very well they did it.

PANORMO. Yes, I agree.

All this changed, of course, when Beveridge took over the reins, got money in America, and put up new buildings. Many years later, long after his retirement, Panormo had a recorded conversation with his successor as Head Porter, Ted Brown, and the School's Information Officer, Shirley Chapman. Then he was a little more outspoken, talked about the rows between the 'red-hot Conservative' Dodson and the Labour man Wilson, and about Mrs Mair, who used him for all kinds of things, 'Panormo this, Panormo that'. Once when Beveridge had sailed to America, Panormo had to go to the shipping company to check the passenger list in order 'to see who was travelling with him'. His job did not lack variety, standing in for a student at Lees-Smith's 'mock trials', going down to Tilbury with Hayek's Home Office papers to make sure that the economics professor was allowed into the country, collecting gowns from Ede & Ravenscroft for Oration Day—and taking the names of students at the main entrance. This is how he, like his predecessors, stored names in his memory. Panormo came in 1912 and retired in 1951, after which he lived another thirty-three years to the age of 98.[75]

Mrs Mair appreciated the porters, if perhaps with a slight tinge of condescension, and it appears that they appreciated her in turn, as her sad reminiscence suggests: 'When I left the School on the last day of my service there after I had made my public farewell at the Commemoration Ceremony, they were the only members of the School body corporate to bid me a personal goodbye.' Thus ends Mrs Mair's—by then Lady Beveridge's—piece on 'The Jolly Goodfellowship Porters', in which she praises the 'door-keepers' in her own way:

> It is often said that it is in the minds of such a group standing at the entrance as stood the door-keepers of old that the full knowledge of the work and personalities of those who cross their threshold is stored. There they are, detached but deeply interested observers, while humanity in all its manifestations passes before their observant eyes. There can be few institutions where the united nations, and many others, exhibit their particular traits and peculiarities more frankly and more free from inhibitions than in the vestibule, the common rooms and the lecture halls of LSE. What could not Wilson and Panormo, with their companions, tell us if they chose about the world as they have seen it there.

Lady Beveridge adds: 'How I wish they would choose to do so.'[76] Not everyone will agree. Discretion is part and parcel of the loyalty of gate-keepers. The porters exercised it throughout. The recorded conversation with Panormo and Brown in 1974 adds touches of colour to what we know but is never unkind or sensationalist. The same is true for the long interview which Edward ('Ted') Brown gave to Professor Michael Wise in 1989. Brown was a lift boy, porter, Head Porter, Clerk of Works, and House Manager at the School from 1927 to 1976. He was succeeded by L. F. Kearey, who also had a long School record from his arrival as a porter in 1935 to his retirement in 1980. None of them served in order to get into the newspapers or even to write memoirs; they served because they came to like the School and its people.

Part of the loyalty and the discretion of porters is their impartiality. At the School this was often tested. Perhaps Mrs Mair tested it more than she knew by using porters for personal errands. Frank Meyer certainly tested it during the *Student Vanguard* affair. Ted Brown tells an interesting 1968 story. He was standing and watching the students build a barricade when one of the plain-clothes detectives around who wanted to stop them asked him to hold his attaché case. One of the students saw this and immediately decided that Ted Brown was one of them and turned against him.

The porters were loyal to the School, impartial in its disputes, and discreet with their store of memories, but living to a ripe old age, as many of them did, they tended to become a little nostalgic. Ted Brown, like some others, found the School scruffy, impersonal, unfriendly—but only the School of his retirement days. Before, and notably in the 1930s, it was very different. 'It was almost, in the old days, like a stately house. Everything was elegant, very clean, very smart, very efficient. And everybody so polite.'[77] True or not true, Lady Beveridge would have loved to hear such praise from one whose affection has added a golden glow to his memories.

Reading the Director's Reports

Beveridge's annual reports in his last seven years at the School give as good a picture as any of what LSE looked like and how it changed, seen from on high, of course, rather than from a student or gatekeeper perspective. In 1931, the Director was very upbeat. He spoke of the flowing tide of regular students, whose numbers were now up to 1,233. He recorded the inaugural lectures given, or about to be given, by new professors: Coatman (Imperial Economic Relations), Manning (International Relations), Chorley (Commercial Law), Parry (English Law), Ginsberg (Sociology), Hogben (Social Biology), Plucknett (English Legal History), Plant (Commerce). The building of 'a strong law faculty' is underlined. Good news on finance as a result of increases in the Treasury and University Court grants and large Rockefeller monies are duly reported, as are the athletic achievements of students and the plans for major research projects (in this order). But alas! 'this account misses—misses inevitably—the reality of the School'.

> The School which I have spoken of in terms of statistics and finance and buildings, of research projects and publications, is, in reality, some hundreds of individual lives, filled with individual hopes and ambitions, with desire to know and desire to serve, with plannings and achievements, with dearly-bought success, and unprofitable or profitable failure.

In 1932, Beveridge notes that 'the economic and political crisis of the world has made till now little difference to the School of Economics and Political Science'. Student numbers are up. Teachers have been promoted (Tawney, Power, Beales, Benham) or newly appointed (Hayek in Economics, Webster in International History, Paish in Commerce). Mr L. G. Robinson has become the first Dean of Post-Graduate Studies. New financial resources make the next phase of construction in Houghton Street possible, and notably the extension of the Library into the area at the corner of Clare Market and Houghton Street which used to house the St Clement's Press. New grants include the establishment of Leverhulme scholarships. Universities are 'more than learning from teachers and books and passing examinations'; once again achievements in football, rowing, hockey, and badminton are duly recognized. The death of two former Directors, W. A. S. Hewins and W. Pember Reeves, is mourned.

'Individually and collectively we in the School have anxious times ahead, but we have also the material for much cheerfulness.'

One statement in the 1932 Report gives pause for a reminder concerning the real basis of it all. About the financial position of the School the Director says that it is possible 'to sum up that position in a sentence. In respect of capital resources and liabilities our position is satisfactory, or at least better than we could have expected a year ago; on the side of recurrent income and expenditure our position is worse and hazardous.' It is true that the capital receipts of £41,946 in 1931/2 and £61,998 in the subsequent year were two of the three largest in the first fifty years of the School; after 1933 they dropped to about one-tenth of those sums per year. On the other hand, School recurrent expenditure rose from £98,000 in 1929/30 to £115,000 in 1930/1 and on to £137,000 in 1931/2, a level at which it remained for the rest of the Beveridge years. Moreover, much of the increase was accounted for by salaries and thus truly recurrent. On the income side, only the Rockefeller grant for postgraduate teaching and research, and 'Donations and Subscriptions' showed a similar increase, though the Treasury grant—from 1930/1 paid through the University Court—did go up by some £4,000 in the year, which led the Director to his understandable remark about the hazardousness of the School's income position. In subsequent years this did not change to any significant extent.

In 1933, the 'anxious times' foreseen by the Director had arrived. 'The School, like the world, has been filled with tumult and confusion.' In characteristic style, Beveridge confounded the universal and the parochial. The School's tumult was caused by new building activity. Again, a moment's pause to take stock is in place. By 1933 what is now known as the Old Building of LSE was complete in all its complexity. The Library had a satisfactory home; offices and seminar rooms were provided, as well as the Old Theatre; common rooms, refectory, and the Founders' Room were but some of the general facilities. On the other, east side of Houghton Street the rear of what is now the East Wing complex had been completed and nearly the entire property acquired for School use. The East Wing proper was added on the site in 1937/8. Once that was completed, space for lecture rooms and also for the administration had doubled since Beveridge's arrival, that for 'social rooms' trebled, for 'teachers' rooms' quadrupled, and for 'library and seminar rooms' quintupled. There was also three times as much general circulation space at the School.

In 1933 the Director concentrated, however, on reporting other matters. Student numbers were now stable, though their composition had changed. Whereas fewer students had come from Europe and Asia, 'imports from

North America have nearly doubled', from 59 to 113. This went hand in hand with an expansion of postgraduate studies. Another new initiative, the course in Business Administration, had shown success. Financial anxieties had been relieved to some extent by assurances that the University Court would not cut the School's Treasury grant, and also by a certain amount of 'hedging' with Rockefeller dollars. 'The circumstances of the time and the increasing number of students bring to the fore the problem of their subsequent careers.' To help students, the post of Appointments Officer had been created and filled by Brigadier E. de L. Young. Then there was the problem of German exiles. In assisting them the School had shown the kind of fellowship which it needed, but did not yet have, more generally. 'Till it is accomplished, the School, though it may be built in steel and stone, in colour and line, is not yet built in reality.'

The period between 1933 and 1934 was, of course, Beveridge's *annus horribilis*. An unusually long Report records above all positive developments. The 'lull in the building operations' means that the School has now found its physical shape for a while. Student numbers have risen again, especially those from Germany and from China. The arrival of the first exiles from Germany has added lustre and strength to the academic staff. The Director reports the split of *Economica* and first appearance of *Politica*, the publication of important volumes of the *New Survey of London Life and Labour* and the 'Annual Digest of International Law'. The discontinuation of the (Coatman) Chair in Imperial Economic Relations is regretted. The Students' Union has experienced 'disturbances' (the name of Frank Meyer is not mentioned), but in any case, the Director states, real student life takes place in the 'fifty clubs and societies', including 'twenty for different forms of athletics from Rugby football to Folk-dancing', nineteen for special interests like music or India, seven political societies. The 1934 Report ends with Beveridge's invitation to the future historian to explain his woes to which this account has tried to give a response.

The 1934/5 Report begins with a reference to the death of Sir Arthur Steel-Maitland, for nineteen years Chairman of the Court of Governors. Steel-Maitland had been a steadying if somewhat remote presence at the School throughout these long years. He had also shielded LSE from many an attack by those of his own political persuasion. Beveridge and the School had much reason to be grateful to him. His successor, Sir Josiah Stamp, was of a different ilk. For one thing, his was to be one of the great LSE families. His younger brother, (later Sir) Dudley Stamp, had been appointed the Cassel Reader in Economic Geography at the School in 1926 and remained as a Professor at LSE until his retirement in 1958; Josiah's son Maxwell Stamp was to be an influential Governor in later years. Josiah Stamp himself was born in 1880 and had begun life, after the scantiest of private schooling, as a tax inspector and civil servant. His early success in this career encouraged him to take an external degree at LSE, in which he did so well that he was invited to do research there from 1912 to 1916. Armed with a D.Sc.(Econ.), he embarked on a career partly as an economist and partly as a businessman. After his death in an air raid in 1941, Beveridge found it necessary to comment on 'the conflict between scholarship and affairs'[78] in his life; in fact he was one of the great and much sought-after advisers of governments, businessmen, and even LSE. A Governor since 1924, Vice-Chairman since 1930, he was the born successor to Steel-Maitland in 1935. In 1938 he was created Baron Stamp. As Chairman he gave the School the full benefit of his intellectual vigour and personal charm. In 1940, a year before his death, he resigned the office because (so Jessy Mair wrote to *The Times* on 24 April 1941) 'he felt that such a tenure should not be permanent'.

In 1935, the Director used his annual Report not to speak about the future but largely for a kind of retrospective. He lists with pride his own achievements, such as the rise in total income from £17,800 to £130,000, or the building growth 'almost beyond reckoning'. Recent changes are few by comparison, though they include the acquisition of Smith Memorial Hall in Portugal Street, and the transfer of language teaching for commerce students from King's College. The fact that the Rockefeller Foundation has given notice that a change in policy 'may make impossible further help such as we have received in the past', provides an opportunity for another look back. Other issues, retirements, student societies, etc., are mentioned, but history dominates the Report, 'partly because this session has in some ways had less history than most'. The Director is in mellow mood. He senses in the School 'a stronger feeling of unity in spirit and interest' than ever before; indeed, 'in this, as perhaps in other ways, I have a sense that this uneventful session may prove a turning point in our road'.

Turning-point to what? The 1936 Report is the most businesslike of all. There are now 1,446 regular students; the number of postgraduates has risen dramatically to 276. Evening students remain buoyant; they want lectures to begin not earlier than 6 p.m. There are more women. Intercollegiate work has also expanded. The turnaround in the composition of LSE students has been mentioned before. Of the steady 3,000 or so students at the School, two-thirds were now regular or intercollegiate students and only one-third occasionals. The number of higher-degree students had risen throughout the 1930s; there were more overseas students too, especially in the immediate aftermath of Hitler's seizure of power. 'Occasional' railway students mixed with 'regular' candidates for the B.Sc.(Econ.), the LL B., the B.Com., the smaller business administration course, or the various diplomas, notably in social science. Evening students amounted still to 40 per cent of all regular and occasional students. What these figures confirm above all is the extraordinary diversity of the student body, which is by no means one of artificial statistical categories. There was, from the beginnings of the School history, a unique LSE experience, but it also had variations for the different groups, courses, and corners of the institution.

The Director has other facts to report in 1936. The new Librarian, Dr Dickinson, 'has proved himself a master in the productive use of scarce means'; he had been an administrator before. The Royal Statistical Society has moved from Adelphi Terrace to the Smith Memorial Hall. The athletes are no longer doing quite so well; they have continued 'the tradition, begun last year, of losing in the final of as many Inter-collegiate competitions as possible'. Two of the longest-serving teachers, Professors Bowley and Sargent, have retired. Unusually, no peroration of a general nature concludes the Report.

In fact, the real turn of events was long foreshadowed before it was documented in the Director's 1936/7 Report. 'To-day's ceremony is the last at which I shall appear as Director,' Beveridge told an unsurprised audience on Oration Day, 24 June 1937. Two new developments are noted, certain changes in the School's constitutional arrangements, and the beginning of the construction of the East Wing on Houghton Street. Otherwise, the Director contrives to convey an air of decline: fewer students, major staff losses (including Hogben) and no new appointments, the end of Rockefeller money. But not all news is melancholy. 'The Students' Union has had a year of harmonious activity.' And, of course, both the men's and the women's Badminton Cup (Beveridge's preferred sport) had been won. Beyond that, 'one cycle in the history of the School is near its end.'

THE SECOND FOUNDER RETREATS

'Schema' for Economic Research

Hard as Sir William Beveridge and Mrs Mair tried to write as well as make the history of their LSE years, the gaps between what they said and what actually happened got so wide in their later years at the School that questions were bound to arise. One of these concerns the Director's attempts to find for himself another position. We have seen that Beveridge made the first tentative move to this end in 1929. In the following years he missed no opportunity to tell those around him that his real vocation was research rather than administration, though from time to time he also flirted with a return to the Civil Service. By 1934 such ideas had crystallized in the attempt to create an institution for 'realistic economic research' in London.

From Beveridge's point of view, this meant that an institute should be set up at LSE, a School within the School almost, with him as the head and Mrs Mair as the Secretary, while someone else was to take over the Directorship of LSE. For his own benefit as well as a small number of others, he set out his intentions in what he called, 'in emulation of H. G. Wells's engaging hero Kipps', 'Schema'. He was soon going to be 56 ('Schema' was written in the autumn of 1934) and did not think that he could give his best as Director any more. The School's numerous problems 'must be dealt with by the Director, but do not call for any special qualities of W.H.B.'. On the other hand, 'scientific work in the field of Economics applied particularly to social problems'[79] does call for these special qualities. He should therefore become a Research Professor at the School.

By November 1934, 'Schema' had been turned into 'Project A', a proposal sent to the Rockefeller Foundation as a highly confidential document, though with supporting letters from Steel-Maitland and Stamp. 'Sir William intimated', the head of the European Office and Assistant Director of the Rockefeller Foundation, Dr Tracy Kittredge, noted after a long conversation with him on 12 November 1934, 'that at the present time there are only three persons, aside from himself, who are familiar with his plans, these three being the Chairman of the Board of the School, Sir Arthur Steel-Maitland, the Vice-Chairman, Sir Josiah Stamp, and Mrs Mair'.[80] In other words, the Director left his colleagues at the School in the dark about his attempt to compete with them on their own territory.

'Project A' is a four-page paper which begins by describing some of the research activities of the School. 'But there has been no senior member of the staff with interest in and effective leisure either for Applied Economics in general or for that particular section of it which may be described most conveniently as Social Economics.' To fill this gap, as well as to meet his personal desires, Beveridge proposes to direct this research. In fact he could serve the School and the purposes which the Rockefeller Foundation had supported 'better in the way that I now propose than by continuing as Director'. He asked for a five-year Research Professorship without administrative or teaching duties in order to create a 'laboratory of Social Economics'. Four subjects are briefly introduced: the (continuation of the) history of prices and wages; unemployment; government intervention in the economic sphere; population. He, Beveridge, would, of course, 'work in co-operation', notably with Professors Malinowski, Robbins, and Hogben (none of whom, however, knew anything about 'Project A').

Contrary to the imaginings of their latter-day critics, people at the Rockefeller Foundation were stunned. They were also embarrassed. Beveridge was in a great hurry and therefore wanted to send Mrs Mair to New York as a special messenger. Cables were needed to dissuade her from boarding the *Majestic*; Beveridge was told that no decision could be taken before April 1935 in any case. The Memorandum on 'Project A' thus sailed unaccompanied except for a somewhat pathetic letter by Beveridge emphasizing the urgency and the secrecy of the plan. Tracy Kittredge's letter, which went off on the same day as Beveridge's, was overtly supportive but mentioned that 'there has been in recent years increasing dissatisfaction in London with the administration of the School'. 'The time has probably come when a younger man is needed for the general direction of the School's future development.' Kittredge added: 'I fully recognize that Mrs Mair is not the best possible ambassador that Beveridge could have chosen to present his proposal to you.' However, the proposal of 'Project A' would 'permit him to retire gracefully from the directorship at this time'. Was that why the Chairman and the Vice-Chairman of LSE's Governors had supported it?

In any case, it was not to be. Soon after the *Majestic* had arrived Miss Sydnor Walker pointed, in a letter to Mr Kittredge, to a whole string of doubts and difficulties. One of these turned out to be especially relevant. 'The fact that a known group would deprecate having such a program take the place of an independent institute of economic research points to the necessity of careful consideration of the results likely to follow the establishment of Sir William's program in the London School of Economics.'[81] This was a reference to what came to be known as the Halley Stewart project of a National Institute of Economic Research. Curiously, its main promoter was Sir Josiah Stamp, who knew that the City—and charities like the Halley Stewart Trust—were more likely to respond to an independent initiative than to the LSE.

The tangled situation is set out with admirable clarity in a long letter to Tracy Kittredge written by Professor Noel F. Hall, then Professor of Political Economy at University College London, on 13 December 1934. Hall had discussed the matter with Stamp and Robbins, with Carr-Saunders, with the initiators of the National Institute plan, Henry Clay, economic adviser to the Bank of England, and with Hubert Henderson,

Secretary of the Government's Economic Advisory Panel, as well as their supporters Israel Sieff and C. I. C. Bosanquet. Three projects were in fact floating about London at the time; Hall called them the 'Institute', the 'School Institute', and Robbins's modest proposal of an 'archive' of economic statistics, the word 'archive' being an allusion to the term used by German Verein für Sozialpolitik for its institute as well as its journal; Hall therefore called it the 'Archiv'. 'Everyone, including Stamp, agrees that the Institute is a more desirable project than the School Institute,' because it could get funding which the School would not attract. However, Robbins's Archiv was compatible with the Institute.[82]

Hall then proceeded to look at strands of interest and opinion within LSE. Beveridge wanted a professorship, even the Chair in Economic Statistics shortly to be vacated by Bowley. The Governors 'clearly could not contemplate for one moment having Beveridge in the School as a Professor'. Then there were two other issues: the position of applied economics at LSE, and that of Hogben and social biology. To start with the latter, Hall felt 'that the first practical step in getting back to realities in the whole of this business is the liquidation of social biology at the School of Economics'. So far as applied economics was concerned, an able successor to Bowley could surely be found. Moreover, he knew, though many others had not yet noticed, that Professor Robbins took a growing interest in applied economics. He should be given the £2,000 a year he needed for his Archiv.

This leaves Beveridge and the Institute. Given Beveridge's 'most intense yearning to carry out some descriptive research before he is too old', 'the best solution would appear to be to accelerate the development of the Institute of Economic Research which Clay, Henderson and myself have been working on, to enlist Beveridge's interest in it, and to invite him to become Director of it'. He would do the job well, 'providing that [the Institute] is housed at least one mile from Houghton Street'. He, Noel Hall, had no personal interest in the Institute.

In the event, things worked out more or less as Hall suggested, though with two important exceptions. One was the time it took to achieve the aims. The Department of Social Biology was finally 'liquidated' three years after Hall's letter. The School Institute was rejected by the Rockefeller Foundation almost immediately, or rather, the officers of the Foundation agreed at the end of December 1934 that 'for the present, we shall plan to do nothing in regard to Sir William's proposal'.[83] Yet it took another three years for the National Institute to be set up. 'The establishment of the institute was delayed by problems concerning Sir William Beveridge's activities and the London School of Economics.' Kit Jones's statement in her 'Brief History of the National Institute of Economic and Social Research' is an understatement.[84] Many delicate meetings had to be arranged between the spring of 1935 and the summer of 1936, when the prospectus for the National Institute was agreed. At one of them, which took place at the Rockefeller Foundation offices in Paris on 14 November 1935, Robbins, Hayek, Gregory, and Plant jointly as well as severally expressed their dismay about Beveridge and his ways. They concluded that as long as Hogben was at the School there would be no funds for 'realistic research'. Moreover, if Beveridge is to become Director of the National Institute, it must be clear that there is 'no provision for Mrs Mair on its staff'.[85]

But Beveridge did not become Director. The second important variation of Hall's 1934 proposals concerned their author himself. 'Relations became strained and delay ensued while alternative names to that of Beveridge for the Directorship were explored.' In the end, Professor Noel Hall was chosen and persuaded to become Director, Lord Stamp President, and Beveridge Chairman of the Council of Management. Beveridge attended once and then faded away to Oxford and to other things.

In his memoir of the early years of the National Institute, Austin Robinson expressed himself 'puzzled as to what its functions were intended to be'. More than that, he thought 'that the founding fathers were equally confused'.[86] That, however, may be a Cambridge perspective from which it did not seem too difficult, for example, to choose the name, Department of Applied Economics. Elsewhere, the desire to anchor economic theory in empirical research, including the economic study of social problems—the word 'Social' in the title of the National Institute was actually Beveridge's legacy—was strong. Robinson may well be right when he goes on to say that this desire was in the end satisfied by unhappy accident rather than design, that is by the war and the role of economists in 'economic warfare'. At LSE, however, the problem of applied economics had already led to confusing organizational decisions, the commerce degree, and the Department of Business Administration; and it remained unresolved.

An Alliance Crumbles

For Beveridge, the complete failure of 'Project A' was clearly devastating. When the Social Science Director of the Rockefeller Foundation, Beveridge's old ally Edmund Day, wrote to him on 8 May 1935 to explain the decisions of the Trustees, he began the letter with an awkward attempt at a lighter touch: 'Now that "the smoke of battle has cleared". . .' When it finally did, not just the Director was wounded, but the School as well.

After extended discussion, the Trustees decided to terminate as soon as practicable the general program in the social sciences initiated by the Laura Spelman Rockefeller Memorial in 1923–24, and supported since the reorganization of the Boards in 1929 by the Foundation. The officers of the Foundation are now under instruction to work out the details of early withdrawal of Foundation support from the various undertakings which have been receiving current support under the former program.

To some extent, this 'withdrawal' resulted from a wider shift in Foundation policy, away from comprehensive 'programmes' and towards more specific 'projects', but the personal element was none the less unmistakable. In defeat, Beveridge showed himself to be the gentleman he was. On 16 May 1935 he wrote to the Foundation President, Max Mason:

I think I need only say now that nothing can impair my appreciation of the help which has been accorded by the Foundation in the past, both to the School and to the University of London, or of the personal courtesies and kindness from yourself and all others connected with the Foundation with which this help has been accompanied.[87]

This sounds final, even terminal. In fact, contacts with the Director and other members of the School remained cordial during the remaining two years of Beveridge's directorship. Old grants had to be administered; a number of new applications were made (though that for an archive in economic statistics was unsuccessful). There were even signs of a revitalization of the relationship when Carr-Saunders arrived on the scene. The lengthy printed *Review of the Activities and Development of the School* submitted to the Foundation in February 1938 contained a detailed programme of future plans for which the new Director hoped to enlist the support of the Rockefeller Foundation.

But while the Foundation seemed to waver once or twice, in the end its determination held. The alliance between LSE and the Rockefeller Foundations had some of the traits of a marriage with mixed fortunes. The courtship of the early years after 1923 had been followed, after the interlude in which the Memorial was absorbed by the Rockefeller Foundation, by the consummation in the form of the huge grants to LSE in 1931/2. But the period from early 1935 onwards must be described as a protracted process of divorce which kept both sides preoccupied and unhappy. In the end, the alliance crumbled quite rapidly.

The influence of the Rockefeller subventions on the progress of LSE had, of course, been considerable. Let us remember the facts: 25 per cent of the total expenditure of the School between 1923 and 1937 was financed by Rockefeller grants. But it will be remembered also that nearly two-thirds of these grants went into the capital account to pay for buildings and library acquisitions. Of the revenue account, only 11 per cent were met by Rockefeller subsidies. 'Rockefeller's baby' thus had other sources of nourishment as well, but the influence of the Foundation was undoubtedly great.

The word 'influence' is chosen deliberately. In fact, the new Director himself, Sir Alexander Carr-Saunders, used it when he prepared the School's Report to the Foundation. Under the heading 'The Influence of the Subventions', the Report states: 'It will be seen from the above [figures] that the great increase of the activities of the School which has been recorded in the historical section, is due in very large measure to the subventions of the Foundation.'[88] Language is important here because in recent years an amusing little debate has developed about the Rockefeller Foundation, LSE, and the hegemony of capitalist values.

One of the protagonists of this debate is Donald Fisher, who in a Berkeley Ph.D. dissertation of 1977, several articles since, and a book on Rockefeller philanthropy and the development of the social sciences in 1993, has argued in Marxian and Gramscian terms. For him the class basis and hegemonial intent of Rockefeller philanthropy is evident. Beardsley Ruml, in 1923 a 28-year-old new recruit from the University of Chicago at the head of the Laura Spelman Rockefeller Memorial, is according to Fisher one of the main agents of these sinister forces, and LSE one of his chosen victims. Two substantive developments demonstrate the effect of the conspiracy: the insistence on allegedly impartial empirical social research, and the replacement of diffusionism by functionalism in anthropology. Both contribute to strengthening the status quo and thereby serve the interests of the ruling capitalist class.

When Fisher first propounded his theories they were not altogether new. As early as 1930, Harold Laski, one of the first supporters and beneficiaries of Rockefeller philanthropy, had contributed an article on 'Foundations, Universities, and Research' to a volume called *Dangers of Obedience*. Clearly, such dangers did not threaten Laski. Actually, his piece reads like a vicious attack on his Director, notably the insistence on 'concrete facts', fieldwork, and co-operative research, except that on this occasion Laski holds foundations responsible. His 'Marxism' is of a more traditional kind: 'The man who pays the piper knows perfectly well that he can call the tune.' Laski also knows how this happens in this particular case: 'Trustees look to university presidents to pick the professors likely to attract endowments from the foundations; university presidents look for professors who can produce the kind of research in which the foundations are interested; professors search for healthy young graduates who can provide the basis for the ultimate generalizations.'[89]

How is one to read this? Perhaps: Sidney Webb picks Beveridge who gets money from the Rockefeller Memorial, appoints Laski to a chair... No, this cannot be entirely right. Martin Bulmer, who with Joan Bulmer has done much detailed research on the Laura Spelman Rockefeller Memorial, tried to counter Fisher's theories by pointing to the role and independence of foundation officials, the autonomous developments of the social sciences, and the fact that both the Memorial and the Foundation supported very little substantive research and very few posts at LSE. He might have added that of all the Rockefeller moneys to LSE the economists, who presumably most nearly fitted the image of defenders of capitalism, got least. But Fisher was not to be deflected by facts. Bulmer, he countered, 'lacks a clear theoretical base'. 'It goes without saying that we view the world very differently.' They had simply 'different definitions of what counts as evidence', so that he felt free to repeat his own conclusions.[90]

More recently, Salma Ahmad, herself the author of a dissertation with much useful material about the Department of Social Biology and other aspects of the development of LSE, tried to mediate between the protagonists. 'I suggest that while Fisher and Bulmer both offer interesting insights into the development of social science during the inter-war period under Rockefeller influence, neither author has been able to offer a totally convincing account of the foundations' political, social and intellectual role in society.'[91] One cannot help wondering what a 'totally convincing' account would have to look like. In some ways Beveridge, Robbins, in fact all Directors and Chairmen of the Governors of LSE might well have been delighted to read the attacks on their hegemonic activity. For once, the School is not described as part of a Marxist plot to subvert the existing order, but as the agent of capitalism paid to stabilize it. But beyond that, the pseudo-theoretical game leaves no one the wiser.

The plain fact is that half a dozen influences on LSE were more important than the Rockefeller Foundation: the Webbs for example, Beveridge and 'his Janet' with their natural-science obsession, Laski in his own way, but also (if one wants a vested-interest analysis) the City members of the Court, railway barons, Chamber of Commerce members, and all who wanted the School to concentrate more on commerce and business administration and less on functionalism or even 'realistic economics'. The Rockefeller Foundation never once supported this line of development. In fact, American foundation officials are often indistinguishable from the social-science community (and move back and forth between academia and foundations), and foundation trustees like to register their own preferences or antipathies but meet intermittently and cannot really control either the preparation or the implementation of strategic decisions. The first tenet of social science tells us that social facts, including foundations, have intended as well as unintended consequences, and the present account provides some useful material for reflection on the latter; but whatever the alliance between the Rockefeller foundations and LSE was, it was not an alliance between a hegemon and an unsuspecting victim.

Let us leave this little distraction, then, and look at the reasons for the divorce between Rockefeller and the School. In part, they had to do with changes in policy by the Foundation. In 1934 the Rockefeller Foundation had decided to cease large-scale institutional grants in the social sciences. Its officers and trustees had also grown wary of brick-and-mortar subsidies. Beveridge's 1934 request was, of course, not a large-scale grant. Indeed he made the point that he was asking for small money, £5,000 a year at the most and perhaps only £2,500. Little did he realize that this added to the discomfort of his erstwhile friends. For eleven years the relationship had been one of common but not vested interest; now one side wanted something personal which cast a shadow of doubt over the whole alliance.

A number of niggling little matters marked the beginning of the end. The Foundation officials did not like being taken for granted by their English partners. On one occasion in early 1934 they complained that the School had used a windfall from devaluation gains for purposes that were not agreed. Increasingly, they had misgivings about Mrs Mair and her role. These were not helped when at one meeting in London she did as usual all the talking but as soon as she left the room, Beveridge was 'very frank in saying that had they planned more wisely from the beginning, funds could have been saved'.[92] A running dispute had accompanied the funding of research because the Foundation felt that Beveridge had systematically ignored the second half of the designation, 'and postgraduate instruction'.

Then there were the two big issues of social biology and of the Institute of Economic Research. When teachers at the School discovered that contrary to the Director's assurances Foundation officials neither supported nor liked the Hogben exercise, they were surprised, to put it mildly; but the Foundation officials were even more surprised by the reaction of the teachers. Thus in 1935 Beveridge's two-faced game was suddenly up. The story of the Institute was not altogether dissimilar. More and more distinguished professors of the School, including Hayek and Robbins, Gregory and

Malinowski, spoke quite openly and very critically about the Director, his Secretary, and the problems of LSE.

It is hardly surprising therefore that by 1935 if not earlier the Rockefeller Foundation had come to the conclusion that the School was in a mess. In their view it was disorganized, unhappy, and lacking a sense of direction. Sir William's prestige in the country was high, but did he still have 'the freshness and vigour of mind' to do new things? Whatever plans are proposed, there is always 'the unsolved question of Mrs. Mair's future position'. And the Governors? 'We are not convinced that Sir Josiah Stamp and Sir Arthur Steel-Maitland have any solution for these questions.' From the perspective of the Rockefeller Foundation, the School was drifting. 'They seem to be asking that we proceed in the hope that a way will open up.'[93] One can hardly blame the officials of the Rockefeller Foundation for losing interest in the School. The alliance had begun as a conjunction of a clear sense of institutional direction in London and the search for worthy social-science partners in New York. By 1937 the School had virtually lost its sense of direction; new developments happened elsewhere, as with the National Institute of Economic and Social Research; and the Foundation had grown rather weary of everyone at LSE complaining about everyone else, and all about the Director and his Secretary.

A Working Constitution for LSE

It was perhaps only a matter of time for the various incidents which punctuated Beveridge's last years at the School to converge on a constitutional debate about the role of the Director. Beveridge did not mind constitutional discussions. He had begun his time at the School with a report on its governance leading to certain changes in July 1921, and returned to the subject at regular intervals, notably in order to interpret Article 28 and define the status of teachers' outside work and political activities. By 1934, however, the question was no longer how the Director defined the position of everyone else at the School but how the School could circumscribe the position of the Director. Pressure to that effect came from several directions. Lingering doubts remained about both the case of the *Student Vanguard* and the Laski affair. The Governors, and especially Sir Josiah Stamp, urged the Director to study the constitution of other colleges of the University of London carefully. The University conducted one of its 'visitations under Statute 114' and sent an Inspection Committee under the Warden of All Souls, Professor Adams, to look at the ways in which academic opinion was recognized in the decision-making process of the School.

Academics at LSE, both senior and junior, pressed for more formalized procedures.

Thus a Constitution Committee of the Professorial Council was set up in June 1934. It consisted initially of the Director, the Secretary, Professor Laski, Mr Marshall, Professor Parry, Professor Plant, Professor Robbins, Professor Smith, and Professor Webster. In 1935, representatives of the lecturers, Mr Beales, Mr Robinson, and Mr Judges, were added. For the first year, the meetings of the Committee remained rather dilatory. The minutes of 12 December 1934 record that 'discussion was general', though one 'fruitful line of investigation' was to split the Professorial Council into an Academic Board and a professoriate. In fact this line of investigation bore no fruit at the time, though the nomenclature re-emerged when the constitution of the School became an issue again a quarter-century later. At times the early discussions of the Constitution Committee took on an almost frivolous air. Someone, probably the Director, devised a mock examination paper:

7. If you propose any change in the present arrangements, how do you propose that the Director of the future should deal with

. . . .

(b) such an offer as that of Mr. Lawe's Committee to provide funds for a semi-independent Department of Business Administration?'

Even this little game tells us something about the underlying issues.

In 1935 the work of the Constitution Sub-Committee intensified. The University Inspectors had reported, and the Director tried to regain the initiative by circulating his 'Reflections on the "School of Economics"'. We have encountered this rather discursive, not to say disjointed, paper before. It raised all sorts of issues. Attacks on the Economics Department prompted a spirited reply by Lionel Robbins which, by relying 'on the method recommended by the Director, namely, observation and measurement',[94] demonstrated how much 'applied economics' was in fact taught at the School. The suggestion that the name of the School should be changed because it was neither a 'school' nor confined to economics and political science, was quietly dropped by the Committee. On one important matter, the Director confessed to having changed his mind, that is the relative role of academics and outside governors in running the institution:

I would still rather think of myself as a senior colleague among my fellows in the School, with special functions of general initiative and co-ordination (as each of them has a special function) than as delegate of the Governors. But, for a variety of reasons, I am not sure that to-day I should put my view as to the government of

University institutions from within and not from without quite as strongly as I should have put it a dozen years ago.[95]

Quite possibly, these were the very reasons which led the Director's academic colleagues to put the case for government from within in no uncertain terms. The Revised Draft Report agreed by the Constitution Committee in November 1936 is as explicit as one is likely to find such a text in that age of respect for authority. 'We do not feel,' the Committee conclude, 'that the present machinery affords sufficient scope for continuous survey and initiative regarding academic policy as a whole'.

The School has grown to its present position largely as the result of the initiative of particular individuals, in particular as a result of the zealous care of the present Director. . . But we submit that the time has now come when, if friction between different parts of the School is to be avoided, and the reserves at its disposal are to be used in the best way, some machinery for more systematic organization of initiative is desirable.

Further critical comments abound. 'We do not think that it is untrue to say that decisions are sometimes taken without consultation.' When there is consultation, it is often with small *ad hoc* groups chosen by the Director. 'The power of initiative is in a sense arbitrary.' A way has to be found to 'eliminate the feelings of uncertainty and apprehension with regard to procedure which in the past have sometimes impeded dispassionate consideration of complex issues'.[96]

What then is to be done? As often happens, proposals for reform do not quite match the drama of analysis. The proposal, adopted on 13 May 1937 by the Court of Governors, was that a General Purposes Committee of the Professorial Council be set up, composed of the Director and fifteen members of the Council (but, with a notable absence, not the Secretary). Four of these were to represent non-professorial staff, whereas five should, by rotation, be senior professors and 'heads of department'. The Committee was to be 'an organ for the formative discussion of academic policy', including the development of new branches of teaching, the establishment of posts, the development of research work, methods of teaching, and related issues.

While the constitution was under review, a number of other matters were formalized. Most notable among them was the renaming and clearer definition of the Emergency Committee. The University Inspectors had observed that not only its name but even its existence was obscure for most. From May 1937 it was called Standing Committee. Its minutes, hitherto written but not circulated or agreed, would be available to members. It would have certain clear powers, including the suspension of students and staff. Above all: 'The decision of the Standing Committee on any matter will be binding on the Director.'

Not everything was formalized, however. The Constitution Committee could not make up its mind whether the School in fact had, or should have in future, Departments and Heads of Department. The ambiguous position, by which departments existed to all intents and purposes but were not explicitly recognized, continued. There was also considerable debate over the need for a formal Charter. The notion was favoured by the Director, but narrowly rejected, on a vote of 5 : 4, by the Constitution Committee. One result was that a curious pamphlet could be produced; it combines historical description with excerpts from Committee minutes and the formal decisions of 13 May 1937, and is called *The Working Constitution and Practice of the London School of Economics and Political Science.* Another result was that the powers of the Director remained considerable, and remain so to the present day. The Court of Governors and its Standing Committee are in formal terms the supreme authority. But, as Governors and Professorial Council agreed in 1925, 'this authority has, in fact, in practice been delegated by the Governors to the Director for the time being, to whose discretion the Governors have largely left the decision as to what matters he should himself decide and what he should reserve for the Governors'.[97] Everything else is convention, though in 1937 it was clear enough in which direction such conventions were pointing.

The End

The last years of William Beveridge's directorship of LSE were full of incident, even drama. We have told half a dozen stories about this period, one of student as well as professorial politics, one of German and other Central European immigrants, one of methodological idiosyncrasies, one of financial heights and depths, one of constitutional changes, one of normal life at LSE. All these stories involve a Director whose role was critical and often highly visible. They also show one who had lost the sure touch of his early years. Somehow everything seemed to turn sour on him, even his active concern for those persecuted by the Nazis in Germany.

There is a story within the story. The Director had wanted to leave office ever since 1929. He was no longer happy with the School, and increasingly the School ceased to be happy with him. At some points, notably in 1935, persistent rumours had him depart almost immediately. Tracy Kittredge reported to the headquarters of the Rockefeller Foundation 'an increasing amount of discussion in England as to a possible change in the direction of the School' before the end of 1935. But where would Sir William go? The answer was vague, too vague to be real: 'To some other position which will enable him to concentrate his future activities on the special studies of social insurance, unemployment and demographic problems, which constitute his chief fields of intellectual interest'. An invitation to rejoin the Civil Service with a brief to set in train preparation for rationing and other economic plans in case of war was, in Beveridge's own words, 'blocked by the Treasury', which was not prepared to offer him equivalent pension rights.[98]

By the beginning of 1937, it had become clear that Beveridge's days at the School were numbered. He was not pushed out—there would have been no statutory procedure for doing so—but his position had become very uncomfortable indeed. In a letter of 28 February 1937 Beveridge told his 'London Mother' Rose Dunn Gardner of prospects 'which seem certain to take me away from the School of Economics quite soon'. For one thing, the National Institute of Economic and Social Research was now about to appoint a Director. For another thing, and 'this is *horribly* secret', 'the possibility [has arisen yesterday] of my going back to Oxford—in a way which I believe I should like most of all'.[99] The Senior Fellow of University College had asked him whether he would allow his name to go forward for the Mastership. He said 'yes', and was duly elected at the end of March. On 30 March he wrote to the Chairman of the Governors resigning his LSE post by the end of September. His Oxford election was publicly announced on 4 April. The Governors met on 7 April, accepted the resignation, and set up a Selection Committee which by 3 May 1937 had concluded to propose Professor Alexander Carr-Saunders for the directorship of the School.

In June, Sydnor Walker noted after a conversation with Beveridge and Mrs Mair that 'Sir William was in excellent spirits and seemed to have only a few regrets of a personal nature in leaving the School. I think that he is quite delighted with the change in occupation.'[100] Sadly for him, the School was delighted too. The warmth with which its senior professors, notably Laski, Robbins, Chorley, and Plant, welcomed the new Director was by the same token a comment on his departing predecessor.

Moreover, the departure did not go without a hitch. This had to do with Jessy Mair. As Beveridge was seriously pondering his resignation in 1936, he tried to persuade the School to extend Jessy Mair's contract beyond the normal retirement age for women which she would have reached in 1937. The School, or at any rate the professoriate, was in uproar at this proposal. Beveridge turned to the Chairman, Sir Josiah Stamp, for help, but to no avail. Mrs Mair professed to be heartbroken; Beveridge was outraged. In the end the Webbs had to be asked to intervene. In the process of bilateral and multilateral conversations, a 'compromise' was worked out. Mrs Mair was offered an extension of her contract by two years, of which she would spend the first, 1937/8, in office, and the second, 1938/9, on leave. After that she would be entitled to a pension of £500 per annum.

When the Emergency Committee had finally taken this decision, everybody acquiesced, though Beatrice Webb noted in her diary: 'If all parties to this unseemly business are now satisfied with the outcome, we are amply repaid, even if B.'s soreness does not wear off—as I think it will.' Somewhat curiously, Beatrice Webb added at this point: 'It certainly did so with W. P. Reeves, who had stated to a friend, with some resentment, after an interview with Sidney in 1919, that "Webb had kindly but firmly ordered him to resign the directorship".'[101] Had Webb asked Beveridge to resign? Probably not, for when the Mair settlement was reached, Beveridge had already been elected to the Mastership of University College, Oxford. Moreover, in his history of the School, Beveridge writes that 'Sidney and Beatrice had been most unwilling for me to leave the School'.[102] 'Mrs Mair must go' was the order of the day in early 1937. When she finally left LSE in 1938, the new Director found generous words for 'her many and great services'. 'In particular I would mention her care for student welfare and her active association with the whole range of student activities.'[103]

Accounts of Beveridge's last months at the School vary. Some who were there at the time have described the scene of the Director sitting all by himself at the professorial table in hall, with no one wanting to be next to him. Sydnor Walker of the Rockefeller Foundation reported, on the other hand, that 'his final appearance on Commencement Day, where he delivered the Oration, was a great occasion; in fact there were a series of entertainments and farewells at the time I was in London, and everyone was doing him honor.'[104] So far as the Oration was concerned, she may not have realized the full significance of Beveridge's scathing attack on the senior academics present on the platform. His Oration, as Beveridge himself later wrote, 'had two themes, neither of a kind to be welcomed by most of my academic audience'. One was, it will be remembered, the need for a natural science of society, and the other, the demand for total political detachment on the part of social scientists. 'Both keys must be used' to win the social sciences respect, which perhaps would take 150 years, like the prehistory of the modern natural sciences. 'Having exploded this bomb in Houghton Street, I went off to Oxford in September 1937, to see if there was anything that at the age of fifty-eight I still could do to shorten that second period of 150 years.'[105]

Beveridge's World

In fact, as we now know, Beveridge's reign as Master of 'Univ' was soon interrupted by events outside his control. However, these events found him ready to make the major contribution for which he is above all remembered. 'Beveridge did not really have a calling for a life of scholarship.'[106] Jose Harris's judgement may seem a little harsh, but it is certainly true that his deliberate forays into scholarship were invariably deflected by more practical demands. Actually, he had to wait longer than others to be invited to help the war effort. 'His' former junior lecturers like Attlee and Dalton undoubtedly knew his weaknesses as well as his strengths, and Churchill, whom Beveridge saw with largely unrequited admiration, may have had his own doubts about the near-total conversion of the 'realistic economist' to central planning. In July 1940 Beveridge was at last asked to help mobilize the civilian labour market as chairman of the Manpower Requirements Committee of the Production Council. G. D. H. Cole and Harold Wilson helped him in that job. In December 1940 he was appointed Under-Secretary in the Ministry of Labour. But he never got on well with Ernest Bevin, so that he was shunted sideways to the innocuous-sounding Inter-Departmental Committee on Social Security and Allied Services, which was supposed to tidy up a rather fragmented set of arrangements. The Report by the Committee was to become William Beveridge's greatest triumph.

The Committee set up in June 1941 consisted of civil servants and initially did not engage Beveridge's full attention. But soon he caught fire. With the help of the secretary, Norman Chester, later to be Warden of Nuffield College, and a number of others who were to rise to positions of distinction, Beveridge set to work and produced the historic Beveridge Report. It proposed to slay the five giants: Want, Sickness, Squalor, Idleness through unemployment, Ignorance. However, its main thrust lay in the simple principle that everyone should as of right be entitled to a certain basic income, to which everyone should also contribute by a system of national insurance. The success of the Report after its publication on 1 December 1942 was spectacular, and lasting. Over fifty years later, numerous celebratory lectures and articles met with a wide response. Beveridge had become and was to remain the national and international hero of the welfare state.

If 1933/4 was Beveridge's *annus horribilis*, 1942/3 became his *annus mirabilis*. Two weeks after the publication of the Beveridge Report, Sir William and Mrs Mair got married, her husband having died during the previous summer, and Jessy turned finally into Janet Beveridge. A triumphal journey through the United States ensued. But then the lucky streak broke. The Report became a political football and was disembowelled in the process. Beveridge was not offered another government appointment; *Full Employment in a Free Society* was a privately sponsored report. When he was elected (unopposed under the wartime agreement between parties) the Liberal MP for Berwick-upon-Tweed in 1944, University College demanded his resignation. In 1945, and despite being one of the most popular men in the country, he lost the seat, and his elevation to the House of Lords a year later offered little consolation. He was not exactly a good constituency MP, and perhaps the Nonconformist rugged Liberals of Berwick did not take to his bureaucratic social liberalism. In subsequent years Beveridge chaired a variety of commissions and continued his research, but he became a bitter and lonely man. In 1959 Janet died. He followed her on 16 March 1963. They are buried side by side in Northumberland.

Jose Harris makes many points about the personality of Beveridge, but two are particularly important in the context of the history of LSE. One is 'the highly collectivist character of Beveridge's political ideas' during the war years. His Liberal Party inclination notwithstanding, 'he envisaged a more "socialistic" organization of society which would be achieved, not through working-class pressure, but through an enlightened, bureaucratic, public-spirited elite'. One understands why his relationship with the Webbs remained so close. Jessy Mair's advice to Beveridge as he began to write his Report goes right to the heart of the story of LSE; she urged him to concentrate on 'plotting the future as a gradual millennium'. The Gradual Millenium could almost have provided an alternative title for this history of the School.

The other point made by Harris over and over again is that Beveridge was so much better at thinking up schemes than at implementing them. In fact, Jose Harris suspects that Ernest Bevin knew this and therefore did everything in his power to remove Beveridge from line positions in the Civil Service. 'He had acquired a reputation as a brilliant inventor of model schemes, but was imperious and unrealistic in translating them into action.'[107] Beveridge had no sense of politics in both the good and the bad sense of the word. He thought that, once an idea had been born, it would also be able to walk and nothing must stand in its way. Persuasion, patience, empathy into the motives of others were not his strong points, and the lack of them proved his undoing not just as a government servant. His greatest strength lay in being a hands-on chairman of an advisory committee. Actually, at times he showed awareness of his limitations, as at the end of his autobiography, when he spoke of 'the Power which I had never enjoyed and . . . the Influence which I had seldom lacked'.[108] Contrary to his wife's insistence, he was therefore neither 'the compleat civil servant'[109] nor even 'the compleat Director of LSE'.

In 1954, Janet Beveridge published a strange book entitled *Beveridge and his Plan*. The book is strange because it is in fact a biography of the man, childhood, adolescence and all, but one which interprets his life as one long preparation for the Beveridge Plan. LSE hardly figures at all, not in the index, and only in two or three asides in the text. Yet it is no exaggeration to say that William Beveridge has two extraordinary achievements to his credit. One is the Plan, that is the principle of universal benefits as a statutory right backed by universal contributions. This is not only one of the central ideas of the century, but also the basis of the social-democratic consensus which gradually emerged in the developed world after the Second World War. The other Beveridge achievement, however, is LSE, that is the firm and lasting establishment of a school of the social sciences which in teaching and research has become a world centre.

Looking back over nearly a century there can be little doubt that Beveridge was the greatest Director of LSE. He found the wherewithal to give the School a firm financial and physical foundation. He transformed a flimsy semi-academic institution geared to part-time teaching and learning and certain special courses into a modern research university with a strong undergraduate component. He attracted notable teachers and scholars to cover the whole range of the social sciences, and indeed to explore and at times transcend the boundaries to the sciences and the arts. He created constitutional arrangements which have not been changed significantly since his departure. In short, he put the School on the map, in London, Britain, and the world, among scholars, public servants, and business people. He was indeed the second founder of LSE.

Much, perhaps too much, is made of his flawed personality. 'Beveridge was outwardly opinionated and extrovert, inwardly troubled by doubt and self-reproach.'[110] Jose Harris is undoubtedly right. And yet, would an untroubled, self-confident man have achieved the same successes against many odds? Contemporaries may wonder why a man who seemed to care so deeply about the poor and the needy was not more likeable as a person. For one thing, some liked him, notably younger colleagues and students. Even thirty years after his death Peter Bayley remembered with affection the 'benign Beveridge' of later years. 'He retained a sort of boyishness and excitement, and his charm and tact could be consummate.'[111] His problem was that he never managed to bridge the gap between abstract schemes and human sentiments. Paul Addison, in an Oxford lecture to celebrate the fiftieth anniversary of the Beveridge Report, claimed that Beveridge 'was not interested in alleviating the fate of the poor but in social policy'. On the same occasion, Lord Longford told the anecdote about Evelyn Waugh and Beveridge. Waugh asked Beveridge how he got pleasure in life. 'I get mine by leaving the world a better place than I found it.' Waugh was of a different ilk: 'I get mine by spreading alarm and despondency and I suspect I get a lot more of it.'[112] Shortly before his death, Beveridge remarked to Peter Bayley, who took him home from a dinner at University College: 'Oh dear, I really haven't been human enough. I must try to be more human.'[113]

Clearly Beveridge held strange views about social science, and used dubious methods in exercising authority. These and his somewhat abstract ways made him fierce enemies. But even apart from the fact that others were charmed by him, he was widely respected. His reputation in the City and Whitehall as in the world at large reflected well on the London School of Economics. Beveridge made dreadful mistakes, not least by not leaving when he had the good sense to think of doing so, but his mistakes are far outweighed by his achievements.

His Secretary and companion Jessy Mair has to be seen in this context. Clearly, Beveridge was a fragile man, not so much in physical terms as in the frailty of his basic vitality. He needed support, but he was not made to garner it by persuasion and friendship. Whatever else one may say about Jessy Mair, she kept her man going. Without her he would have been lost and ineffectual. She did this in often objectionable ways, which included a dangerous adulation of his achievements. Also, in her heart of hearts she probably did not believe in the project of LSE whereas he did, whether 150 years were needed to complete it or not. Yet she cared, about Beveridge in the first place, but also about students and about staff, including not least the porters. The fact is that the School could not have had Beveridge as its greatest Director without her.

In a sense it does not matter whether Beveridge was happy at the School, nor whether the School was happy with him. It is fashionable these days (or are such fashions overtaken by new ones already?) to play down the role of individuals in history. Yet some institutions would not be as they are—some might even not exist at all—if it were not for the presence of the right person at the right time. Beveridge at LSE was just that. Alexander Carr-Saunders, not a man of unduly many words, recognized this in his first Director's Report in 1938. He pointed to the facts, many of them quite appropriately statistical, which 'illustrate the immense growth of the School in size and complexity' during Beveridge's directorship and then added:

It may fairly be said that in his time the School was reborn. Future generations will attribute to Sir William Beveridge its definitive establishment as one of the major colleges of the University of London. The debt of the School to him is immense, and I am glad to record that the first act of the Court of Governors during my term of office was to elect him as one of the small but distinguished group of honorary governors.

PART III

STRENGTH IN ADVERSITY

The War and Reconstruction, Expansion
and Troubled Times

1937–1995

6

A SAFE PAIR OF HANDS

NEW THEMES AND A NEW STYLE

Carr-Saunders Arrives

With Beveridge's departure, the heroic age of the London School of Economics and Political Science came to an end. The act of creation had been the work of Sidney Webb and W. A. S. Hewins, helped by Fabian and Oxford colleagues, friends in high places, and the commitment of early students eager to improve themselves through an education in the social sciences, but also by a time which sought new answers after the exhaustion of the Gladstonian universe. Creation is one thing, survival and firm establishment another. The process of establishing the School was William Beveridge's lasting achievement, with the help of Jessy Mair, the Rockefeller Foundation, and the growing number of those who made LSE their intellectual and professional home. The dynamism of the social sciences in the face of bewildering social, economic, and political trends gave the success of this era its flavour.

By 1937, the work of creation and establishment had been done. From now on, LSE would be faced with different challenges originating partly in the vagaries of the times, and partly in persistent pressures for growth. The School had to prove its mettle, and it did. The decades which followed Beveridge's resignation were thus full of incident—the wartime evacuation to Cambridge; the two waves of post-war university expansion; 1968; and from the 1970s an environment increasingly hostile to higher education in general and the social sciences in particular. Throughout this period, the School tried to preserve its uniqueness, and for the most part it succeeded. Yet inevitably, in the wake of the 1944 ('Butler') Education Act and the 1963 ('Robbins') Report on Higher Education, LSE took on features of a normal university. The themes of normalization and defence of identity are less dramatic, less grand than the earlier ones of creation and establishment. Whereas in the first forty years a confident School changed and at times shaped its environment, subsequent decades frequently saw the School reacting to challenges from outside. The story which has to be told is therefore one of how the interwoven strands which make up the uniqueness of LSE fared in the face of the challenges of the time: the social sciences; the forever tense relationship between scholarship and politics, dispassionate enquiry and passionately held values; the commitment to both London and the world; the needs of diverse groups of students; the ability to provide not just technical training but critical faculties and an intellectual home; the deep belief in change by enquiry and knowledge.

The School was fortunate in finding a safe pair of hands to direct it through the transition from its heroic age to normality, and more particularly through the upheavals of the war and reconstruction. Alexander Carr-Saunders was 51 years old when he took up the directorship in October 1937, much older than his predecessors (other than Reeves, who also came at 51), for Hewins had been 29, Mackinder 42, and Beveridge 41 when they were appointed. He stayed in office for nineteen years, longer than any other Director in the first century of LSE. He was also the first not to go on to another job but to retirement when he left the School.

Carr-Saunders's character helped to bring about the steadying influence which he exerted on the troubled institution which he inherited, and which was much needed in the turbulent times which would soon engulf it. The new Director combined reticence with firmness. Soft-spoken, not given to great flights of imagination, he yet had a clear sense of direction, and a knack for preparing the ground before events occurred which would have overwhelmed others. Beatrice Webb noted with her usual candour that she doubted 'whether he is as able, either intellectually or administratively, as Beveridge; far less power of initiative and fulfilment'. Still, he had other advantages, such as 'more judgement and far better manners'.[1] Sir Hector Hetherington, the Principal of Glasgow University, put this rather more kindly when he recommended Alexander Carr-Saunders to the Selection Committee:

> He is rather reserved,—at least in the expression of his feelings. But he is admirable on Committees, practical, clear-headed and judicious, very even-tempered, and magnanimous. He doesn't inspire; but he encourages. People would like and trust him; and he would stay the course. If you could get him, you needn't be afraid to leave the School in his hands.[2] .

This was exactly what the School wanted to hear. The Selection Committee first met on 15 April 1937. It was a strange committee, consisting only of lay Governors. In the absence of the Chairman, Sir Josiah Stamp, and despite the presence of Lord Passfield and the Vice-Chairman, D. O. Malcolm, the outgoing Director became its dominant member. Several names were considered, among them the Oxford philosopher H. A. Smith, who was soon to be elected Warden of New College, and the political economist Alexander Loveday, who was then working in the League of Nations Secretariat. Although the senior professors had agreed not to put up an internal candidate, Lionel Robbins's name remained in contention at least as a backstop. However, all this soon turned out to be unnecessary. By 3 May Beveridge was able to inform Carr-Saunders that his name would be put forward to the Governors on 13 May. 'I am sure that I needn't tell you how much pleasure it gives me personally to write this letter. There is no one whom I would sooner or as soon have seen in my place and I shall have special pleasure in doing everything to ensure you a fair start.'[3]

More importantly, Beveridge's main adversaries, the senior professors, when told of the news, were unanimous in expressing their pleasure. Several of them wrote to Carr-Saunders.[4] Harold Laski: 'We are eager for a man of distinction who understood the academic tradition and would sympathize with its implications. . . . It is a great satisfaction to me to know that you have been willing to cast in your lot with us.' Lionel Robbins: 'Those of us who are old students of the School and knew the great men who founded it, have a very jealous care for its repute and its traditions; and I think I can say without reserve that we shall all feel when the news is known, that its repute and its traditions will be worthily upheld by the decision which the Governors have taken.' R. S. T. ('Theo') Chorley: 'If I may end on a personal note I may say that it will be a great pleasure to me to work under one who is an old mountaineer and one whose wide and liberal outlook in connection with social problems have always appealed to me.' Arnold Plant: 'You can count upon a great welcome and the most loyal support.' On 13 May, Carr-Saunders was duly elected, though not without certain contractual reservations which betrayed the lingering memories of his two predecessors: he was appointed for one year at a time, under no circumstances renewable after the age of 60.

Carr-Saunders was well prepared for the job. At the time of his appointment he had been Charles Booth Professor of Social Science in the University of Liverpool for fourteen years. One of five children of a well-to-do father in what would nowadays be called financial services, Alexander Carr-Saunders had read natural sciences at Magdalen College, Oxford. His first-class degree gained him a research studentship at the Marine Biological Laboratory in Naples. From Italy, he returned to Oxford to teach zoology. In 1910 he went on to London to join Karl Pearson in his genetic research. During this period he got involved in the Eugenics Society and, as sub-warden, in Toynbee Hall, and began to move from the natural to the social sciences. The experience of war—largely in the Royal Army Service

Corps—may have helped this transition. In London before and after the war, he also met Harold (and Frida) Laski, Beveridge, and others, so that he was more than a dark horse when his name was brought up for the directorship in 1919, though he was young at the time and entirely unproven as an administrator.

In the event, he turned to genetic research, and then to population studies. His 1922 book on *The Population Problem* established him as a scholar whose empirical research effectively straddled the disciplines which Beveridge had so strenuously tried to bring together, but did so easily and naturally. Five years later, in his *Social Structure of England and Wales*, Carr-Saunders (with his Liverpool colleague D. Caradog Jones) used census statistics as well as social surveys and his considerable analytical powers to present an instructive, if non-theoretical, account of society and economy. The same can be said for his 1933 book on *The Professions*. He later regretted (so his son Edmund, a medical practitioner himself, reports) his early excursions into the theory, not to say the ideology of eugenics. In due course he came to represent and to some extent create the English tradition of population studies. Above all he was everything that Beveridge had ever dreamt of, a natural social scientist as well as a non-ideological scholar—except that he lacked the aggressiveness, the need to prove that he was different, which had marred his predecessor's way.

Carr-Saunders knew the School well when he came and was already linked with it in many ways. He even had a family connection with the Reeveses; he and Maud Reeves shared grandparents. When he and his wife visited the Webbs a few weeks before he took up his post, Beatrice 'entertained the pleasant little lady for a suitable spell'. Carr-Saunders had married late, in 1929, and Teresa bore him three children, Edmund, Flora, and Nicholas. Her interests were very different from his, farming rather than books and music, and she played little part in the life of the School though she was much liked by its members, especially during the Cambridge years. That evening at the Webbs, Beatrice and Teresa withdrew to their rooms early, but Carr-Saunders and Webb stayed up until after midnight, 'Sidney telling him the story of the School and all its many difficulties with the four successive Directors'. Beatrice did not think that Carr-Saunders would stay in office very long, certainly no more than ten years, 'which is all to the good. The directorship of the School is a post in which the holder gets quickly "worn out".'[5]

It did not take the new Director long to put his stamp on the School. Tracy Kittredge, head of the Paris office of the Rockefeller Foundation, and a great friend of LSE, noted as early as 16 November:

There can be no question but that Carr Saunders has made an exceedingly satisfactory beginning as director. In the first month he has completely won the confidence of all groups in the Faculty of the School. His only difficulty seems to have been with the Secretary, Mrs. Mair, who apparently is feeling keenly the fact that she is no longer treated as a co-director.[6]

And again, a few months later, on 10 February 1938: 'Carr-Saunders is conscientiously endeavouring to introduce more democratic methods of administration.' Nor was this merely a brief honeymoon, as an office note in December 1938 shows: 'Under the leadership of Carr-Saunders a new spirit has developed in the School. The Faculty are cooperating to a greater extent than might have been thought possible two or three years ago.'[7] It is too bad therefore that all attempts by the new Director, and even the support of Mr Kittredge, failed to move the Rockefeller Foundation to renew its interest in LSE at that time.

Between Peace and War

The job to be done at the School was not just one of style, much as style mattered. When Sidney Webb came back from the Governors' meeting on 30 September 1937 at which Carr-Saunders took over from Beveridge, he told his wife: 'All was promising: accounts satisfactory, building progressing, students increased and peace reigning. Even Mrs. Mair showed herself as highly efficient and gracious in manner.'[8] Not many had quite such a rosy impression. Beveridge himself had alerted his successor, in a conversation on 2 June 1937, to seven major issues: the East Wing on Houghton Street, for which money was available, had to be built; social biology had to be saved; the Business Administration Department needed attention; a new Secretary had to be found; there was a surplus in the accounts; student numbers were declining along with the quality of students; the B.Com. was too specialized and had to be reformed.

Carr-Saunders listened, and quietly decided his own priorities. He had never thought much of social biology, laboratories and all, so he closed the department and replaced Hogben's post by a readership in demography for Dr Kuczynski. It is perhaps contrary to his democratic ways to say that 'he' did it, but in this matter the whole School was on his side in any case.

A new Secretary clearly had to be appointed. Who was it to be after the experience of the last eighteen years? The decision to appoint Walter Adams met with widespread approval. His record with the Academic Assistance Council, later the Society for the Protection of Science and Learning, was promising. He was not a domineering administrator but a quiet, effective, and above all a fair man. Perhaps he did not make much of an impact in the two and a half years after his arrival on 1 August 1938 and before his departure for war duties. William Pickles, in 'When a Director Was Secretary', an article written in 1974, found remarkably little to say about Adams, though one statement suggests that the new Secretary was the right man at the right time: 'He walked boldly into a problem situation in 1938 and helped to make it no longer a problem.'

For the rest, Alexander Carr-Saunders used the *Review* for the Rockefeller Foundation of February 1938 to state his policies. The historical part of this *Review* was drafted by Mrs Mair, but the new Director wrote part III on 'Future Development'. There he listed four areas of concern: staff, briefly; the Library, in some detail; buildings, very briefly; and research at considerable length. The statistician in him led him to begin with a remark about money. 'The School depends for its income roughly as to three-sevenths upon parliamentary and local authorities' grants, as to two-sevenths upon fees, as to one-seventh upon endowments and as to the remaining seventh upon donations and subscriptions.' One-seventh, or over 14 per cent, is a significant proportion; without major shifts in the other sources, the Rockefeller contributions would be sorely missed if they were really to dry up.

On staff the Director states clearly that the School does not intend 'to enlarge the circumference of its studies'. But there are gaps, and it is highly desirable to provide for 'studies which form links between existing specialisms'. Population studies is a good example for both. Economics and sociology should join forces to study modern social structure. The philosophy and methodology of the social sciences should be developed.

The Library was close to the new Director's heart and he sees it 'faced with grave problems of growth'. If it continues to take in 12,000 volumes a year, it will run out of space in four years' time. For 'a national and even an international library for research in the Social Sciences' it is seriously underfunded both on the staff and the acquisitions side.

So far as buildings are concerned, the only immediate project is that on the east side of Houghton Street. In Carr-Saunders's first year, the wreckers pulled down the old houses and construction began. It was completed before the beginning of the war. Contrary to Beveridge's assurances, however, the School had to borrow £25,000 to complete the East Wing, which meant, as the Director put it in his 1937/8 Report, that for some years to come 'there can be no plans for further expansion which any increase in the scope of our activities would make desirable'.

Then research. Carr-Saunders made a case for funding what he liked to call a 'better research organization', an infrastructure which frees scholars for the work at which they are best. The case was that much stronger by not overstating the comparison with the natural sciences. 'The comparison must not be pushed too far, but it is true to say that in both cases organisation, costing money, is required in order that research workers should not have to spend over-much time in the collection and analysis of material.' The Director then proceeded to list important research themes of existing departments, but once again concentrated on adding 'projects arranged by groups of departments'. Demography, studies of the national income, of trade, and of economic and social behaviour in Greater London; the Institute of Business Records; the Land Utilization Survey of Britain; a survey of international social organizations; studies of the impact of law on industrial and economic development, of the influence of law on changes in the structure of the family; a bureau of local government research; the study of comparative social institutions, and the sociological study of ethics, are some of the subjects mentioned. 'These subjects are not likely to obtain immediate support from routine university finances.' They do not lend themselves to undergraduate teaching. 'Yet it is clear that in indirect ways there is here that which would have an important bearing on the solution of many practical problems at present confronting the world.'

Except for one, war. When the Report to the Rockefeller Foundation was written, the dark shadows of war seemed still remote. They were not mentioned either in the *Review* or in the more formal request for further support from Rockefeller, or indeed in the Director's first Annual Report. The Director's application to the Rockefeller Foundation in December 1938 began with what in later years would be called a mission statement and a justly proud one at that:

> The London School of Economics is the only institution in Great Britain of which the undergraduate teaching is exclusively devoted to the social sciences. It is also the chief centre of postgraduate work and of research in this field in Great Britain and the size of its staff and the number of its students makes possible a range of specialisation which is more intensive than in other places. The School is young as well as unique, and its rapid growth has only been made possible by support from the [Rockefeller] Foundation which has been accompanied by generous help from British sources both governmental and private.[9]

The scope of studies, the composition of the student body, the distinctive Library are then described. A section is devoted to 'the external influence of the School'. 'The location of the School in London has enabled the members of the staff, in addition to fulfilling their duties of teaching and supervising research and of prosecuting their own studies, to render many public services.' Nor is such influence confined to Britain. The School has had a major influence on the social sciences world-wide, and has in turn benefited from its international contacts. The School has grown rapidly and now defines by its chosen subjects the scope of the social sciences. 'But this does not mean that it has reached a stationary condition.' It needs to develop its research with the help of external funds.

Carr-Saunders was the first Director to find words to describe what was special about LSE, and to unite the School around the resulting self-image. The reason was that he was from the beginning at one with the School. 'He absolutely loved it,' his son Edmund said in conversation. 'He thought he had found what he wanted to do.' But the effect of such love was of necessity more internal than external. The Rockefeller representatives liked what they saw but did not change their position. A few grants, including one to use the surplus of earlier building allocations in order to acquire the remaining parts of the Houghton Street site, continued through 1938, but the Foundation was not prepared to reopen the question of research support. Tracy Kittredge tried ('a very good case has been made') but his superiors did not respond. 'It appears unlikely that there will be any early action by the Foundation that would provide for general support of the London School.'[10] In November 1939, when war had broken out and the School had moved to Cambridge, the Foundation made available an Emergency Fund of £12,500 for research that could otherwise not be sustained. A pang of conscience perhaps? In any case, for the time being the last word on the subject.

The fact is that the more difficult Beveridge, always at odds with his professors, at times gentlemanly but rarely nice, idiosyncratic in his preferences for people, methods, and styles, was also more effective when it came to cajoling people into giving money. On the other hand, Carr-Saunders was precisely the man to see the School through the difficult years ahead, and when, after the war, the question of money became essentially one of public funds, he may well have been more effective than his predecessor would have been.

The Director's Report on the year 1937/8 was also Carr-Saunders's only normal report for many years. He had given four members of the junior staff representation on the Professorial Council. A few changes in staff had to be reported, notably the appointment of Professor Condliffe as Professor of Commerce with special reference to International Trade. A reciprocal arrangement for students had been reached with the École Libre des Sciences Politiques. The Students' Union, not to be outpaced by the School, had produced a new draft constitution. It was also promised its own premises in what was still the separate Three Tuns pub. Lord Passfield had resigned from his remaining offices as Chairman of the Library Committee and member of the Standing Committee. The Chairman and Vice-Chairman of the Court of Governors had both been honoured by the King and were now Lord Stamp and Sir Dougal Malcolm.

In 1937/8, total student numbers remained at 3,000. First-degree candidates were up, higher-degree and diploma students down. The number of overseas students, notably from Europe, had increased. All this began to change in 1938/9, when overall numbers declined for the first time since the early 1920s. 'Last September everyone was preoccupied with the crisis, and since that date many potential students have no doubt found that the leisure time, which they might have devoted to taking classes, has been consumed by various war preparations.' The Director's Report on the Work of the School in 1938/9, given on Oration Day, 22 June 1939, is full of allusions to the threatening clouds. 'Universities, more than any other institutions, require an atmosphere of peace to carry out their proper functions. Rumours of war are distracting, and preparations for war may draw students away.'

Soon the rumours became realities. In fact, the Court of Governors agreed at its meeting a few days after Oration Day, on 6 July 1939, to pursue an arrangement by which, in the event of war, the Office of Works would take over the whole of the Houghton Street buildings. Carr-Saunders had not remained inactive in preparing for a difficult future. He foresaw the need for evacuation and made sure that the School would not be divided. Government plans saw one-half of LSE in Glasgow and the other in Aberdeen. But the Director had made arrangements with Cambridge where as early as 25 July 1939 the Governing Body of Peterhouse College minuted its decision: 'The proposal of the University Committee that the College should house the London School of Economics in the event of war was approved. It was agreed, subject to any general agreement between Colleges, to charge £3.3s a week for board residence.'[11] Others were less fortunate or perhaps had less foresight. University College London, for example, was split several ways and moved all over the country, to St Catharine's in Cambridge, to Aberystwyth, to Bangor, Cardiff, Swansea, Sheffield, and Southampton. Keeping the School intact during the war is Alexander Carr-Saunders's greatest though by no means his only achievement. It had a price. Friedrich von Hayek was right to note in his history of the School's first fifty years: 'The first two years of the new Directorship, though nominally still years of peace, were already overshadowed by the threat of imminent war. Though many plans were made and some even brought into operation, the outbreak of war cut short all new developments.'[12]

A HAPPY WAR IN CAMBRIDGE

Peterhouse

At first sight, Cambridge was a strange choice for LSE. Even on closer reflection, one wonders about the chemistry of the two academic institutions. Six years after the move, when the Director was made an Honorary Fellow of Peterhouse, Alexander Carr-Saunders thanked a well-wisher in tellingly comparative terms:

> As a result of the war I find myself more strongly linked with Cambridge than with Oxford. Until 1939 I hardly knew anything about Cambridge. I may not understand Cambridge very well yet but the personalities of the place are better known to me than those of Oxford.[13]

Carr-Saunders was far from alone in finding Cambridge rather unfamiliar. Six of the ten Directors of the first century of LSE had an Oxford past, and the only one who came from Cambridge, I. G. Patel, was there during the war and thus exposed as much to LSE as to the ancient university itself. Two Directors went on to become heads of Oxford colleges. Most of the senior professors, Cannan and Wallas, Robbins and Laski, Tawney and Hobhouse, had Oxford degrees. Only one of the famous names, Lancelot Hogben, had—not surprisingly—graduated from Cambridge. The differences lead us deeply into the mythology of Britain's ancient universities. Still, it can safely be said that LSE economics was established against the prevailing Cambridge orthodoxy of Alfred Marshall. Both he and, later, Maynard Keynes had complicated relations with their colleagues in Clare Market. Moreover, Oxford had an affinity to political power and those who sought or had it which was not as pronounced in Cambridge but shared by the School of Economics and Political Science. Quite generally, Oxford may be called a more worldly place, whereas Cambridge is more bound up with the scientific community. These are sensitive issues, and it may be best to hide behind the man who has thought hardest about them, Lord (Roy) Jenkins. He wondered how Oxford and Cambridge managed to stay at the top of the European League in the nineteenth century in which so many universities at home and abroad began to compete.

> The short answer is that Cambridge did it through mathematics and Oxford through religion, and that mathematics being on the whole a more serious subject than religion (at least as pursued by the Oxford liturgical disputes of the second quarter of the nineteenth century), there stemmed from this a certain Cavalier/Roundhead bifurcation which had not really occurred in the seventeenth

century but which in the nineteenth century sent Oxford in a more metaphysical, frivolous and worldly direction and Cambridge on a more enquiring, serious and austere course.[14]

In other words, LSE had something in common with each of them, and as so often before and after, opted for the 'enquiring, serious and austere' rather than the 'worldly' option when challenged. Thus Cambridge became the home of the School for six important years, and clearly the more inward-looking atmosphere of the small university town had its own advantages. It may even have helped preserve the Clare Market tradition as in a cocoon unthreatened by its environment.

However, events moved far too quickly to allow time for such musings. On 1 September 1939, the Ministry of Works laid claim to the Clare Market/Houghton Street site. It was to be used for the Ministry of Economic Warfare which, incidentally, was then headed by the LSE man Hugh Dalton. The School immediately proceeded to a triple move. Administrative and lending-library material was taken to Cambridge, where the new headquarters of the School were set up in the building next to the Master's Lodge opposite the main gate of Peterhouse, called New Court at the time, and later The Hostel. The remaining Library was kept in Houghton Street but concentrated in more confined space. The Director discovered that there was a continuing interest by evening students in taking part-time degree courses and rooms were rented in Canterbury Hall, a Church Assembly building in Cartwright Gardens near Euston Station, to offer lectures and classes.

The School quickly found its place in Cambridge. The Director, the Secretary, and Mr L. G. Robinson, who lectured on international history and was also Dean of the Graduate School, were given High Table privileges at Peterhouse; these were later extended to other members of staff. Students were found, for the most part, private lodgings. Grove Lodge, a large house with a garden close to Peterhouse, was rented to house a student lending library as well as common rooms and facilities for lectures and classes. In all, 620 students registered at LSE in 1939/40 in Cambridge, to which 359 evening students in London have to be added. Forty-six teachers, fewer than half the staff of the preceding year, had to do a lot of commuting to teach both groups and, in many cases, discharge other war duties as well.

At the time the Director and most of his colleagues believed that their Cambridge sojourn would be but a brief interruption of normal LSE life. This belief was strengthened not just by the illusions of the 'phoney war' but by the fact that the Ministry of Economic Warfare actually moved out of the LSE buildings on 16 March 1940. Mr Wilson, the head porter, and others left behind or sent to investigate the condition of the site, reported severe dilapidation and above all the sorry state of the Library. Its reading rooms had been used by the Ministry. The School immediately set about restoring the Library and prepared to take over at least the East Block on Houghton Street for use in October 1940. Extended discussions in Cambridge about the location of the School in the autumn of 1940 found the staff split. 'Unanimity was not to be expected when so much depended upon unpredictable events,' the Director reported when he wrote about the Session 1940/1, though he also said: 'It was clear to all that the proper place of the London School of Economics is in London.'

Alas! the decision to move back to London could not be implemented. 'No sooner had it been taken than raids began.' The lorries, loaded but not yet dispatched, were unloaded again. The School asked Peterhouse to be allowed to reverse its decision, which led to a crisis in relations, though fortunately one that quickly passed. The Ministry of Works allocated the School buildings to the Air Ministry, which stayed there for the duration of the war.

For the School, the consequences were again threefold. The 'Library enclave', that is the highly confined space to which the books of the BLPES had been moved, remained unused by the Air Ministry, though soon also unusable by readers from the School or anywhere else. Mr Wilson, the head porter, looked after the 'Library enclave' throughout the war. Evening teaching, at least in London, had to be suspended; a limited number of part-time students attended classes and lectures in Cambridge. In Cambridge itself, the School had to settle down to a longer-than-anticipated stay. Apart from The Hostel and Grove Lodge, rooms were rented from Peterhouse in St Peter's Terrace and from Corpus Christi in King's Parade. They provided space for the Social Science Department, for a Senior Common Room, and for accommodation for staff and tutorial teaching. After a year of upheavals the School settled down for another five academic years at Cambridge, under the benevolent wings of Peterhouse.

A Different School of Economics

The Director insisted that LSE would remain the same despite the disruption and dislocation of the move to Cambridge, and he was right to do so. Preserving the ethos of LSE under unusual and threatening circumstances was as important as keeping the School going in a practical sense. In fact, however, the Cambridge LSE was a very different place in almost every respect. Above all, it had become an undergraduate college. Evening and part-time students had disappeared; there were few occasional students left; the number of postgraduates had declined from about 300 before the war to 66 in 1940/1. Of 560 regular students in that year, 359 were reading for a first degree, and 135 for diplomas in social science, sociology, and similar subjects. The Director himself said that 'for the first time in the history of the School, it took on the aspect of an undergraduate College'. At the end of LSE's first century we know that it was the first and also the last time.

LSE at Cambridge was not just an undergraduate college, but also increasingly a women's college. At the beginning of the war, men were still in the majority among students. But as the last pre-war generation completed its degrees, and new male students were only allowed one year (unless they had, for one reason or another, dispensation from war service), women came to predominate. By 1944, the pre-war ratio of 70 per cent men and 30 per cent women had almost been reversed; during the session 1944/5, 224 men and 494 women were registered as regular day students.

The Director never ceased to emphasize that this was less than the whole story. For one thing, LSE lectures and classes and Cambridge teaching became increasingly intertwined. Many Cambridge undergraduates could not resist the pleasure of listening to Harold Laski. Also, LSE had brought with it the as yet forbidden subject of sociology, and Cambridge students went to hear Ginsberg. Nor was this a one-way traffic. Law students went to Cambridge lectures, such as those by Lauterpacht, and used the libraries. The Economics Departments were virtually integrated; I. G. Patel, later to be Director of the School, was but one of many Cambridge students who could hardly tell the difference between their chosen university and LSE.

Patel's story is of interest, not only because he was to become the School's ninth Director in 1984. He arrived in Cambridge after a traumatic seven-week voyage in the autumn of 1944, in order to study economics at King's College. Within days he encountered LSE, the students without gowns, the many women, the international set including K. N. Raj, who was to become a lifelong friend of I. G. Patel's; Veerasamy Ringadoo of Mauritius; Ralph Turvey, who later went to the International Labour Office; the German Jew Rudy Goldstein, who initiated him into Western music. Even forty years later Patel remembered the lectures by Hayek, who never let *The Road to Serfdom* interfere with indifference curves, by Kaldor ('scintillating'), Laski ('mostly anecdotal'), and by 'the greatest delight', Tawney. He also remembered that 'LSE was more research-minded, whereas at Cambridge, a Ph.D. degree was still regarded as a weakness of foreigners, particularly Americans and the Indians'.[15]

Many changes happened in the course of the war. They are remarkable for the fact that the School in exile not only kept going but actually stabilized at a new level of activity, and more, grew a little. In his 1943 Report the Director himself expressed surprise. 'If anyone had been asked what the state of the School would be likely to be in the fourth year of the war, if it lasted so long, the answer might well have been that it would probably be closed.' This was all the more likely since students of the social sciences, unlike those in the technical disciplines, were not entitled to state bursaries. The Director had good reasons, therefore, to state in 1942 with a certain sense of triumph: 'Numbers at the School are not, in other words, the result of a diet of artificial feeding but of the deliberate choice by young people who want to study social subjects at the School.'

The number of regular first-degree students stabilized in Cambridge at around 350, though it suddenly rose to 446 in 1944/5 as if in anticipation of the boom which was to follow the end of the war. Other regular students, mostly aiming at certificates and diplomas, increased more gradually from 135 in 1940/1 to 238 in 1944/5. Throughout there were some 65 graduates. Thus the total number of regular students rose from the low point of 560 in the years 1940–2 to 750 in the last year of the war. As importantly, new categories of occasional students were added as old ones faded away. The railway students went, as did those in colonial studies; but the Ministry of Labour sent between 40 and 70 students to be trained in Factory Supervision and Personnel Management, and other Government departments supplied between 28 and 85 candidates for a special course in statistics. Thus the overall total of Cambridge LSE students rose from a low of 683 in 1940/1 to 1,066 in 1944/5.

Even overseas students continued to play a significant part. The session 1942/3 marked the low point with 130, but in most years there were 170 or more, or between 20 and 25 per cent of all students. Many came from occupied Europe, German exiles, Polish officers, refugees from South-Eastern Europe. However, there were significant numbers of Indians, a continuing stream of Chinese, and surprisingly many Africans for whom, alas! it was not always easy to find 'billets' in Cambridge.

Who looked after these students? In the first instance, Vera Anstey, the Commerce Reader who had become the School's accommodation officer and by all accounts did an extraordinary job. She herself has described the experience. 'The need to pander to the proclivities of hostesses or landladies indeed the great inequality of billeting conditions posed very delicate problems to the harassed, but withal amused, billeting officer.'[16] One of many virtues of Vera Anstey was that she was not only amused in adversity but also coped with adversity by amusing others. 'She had a bubbling sense of humour,' Betty (Bond-)Evans remembers, 'and frequently her laughter turned a thunderous clash between a landlady and a lodger into a passing teacup storm.'[17] Thus she was able to achieve her twin goals, 'that no student should have nowhere to sleep; and that no court case should be instituted'.[18] Not even the Proctors had much to complain about when it came to those strangers without gowns and with a lively interest in politics as well as the other sex.

By far the most difficult task was to retain enough teachers for a growing number of students. The number of teachers available declined rapidly from 90 before the war to 37 in 1941 and remained at that level (1942: 35, 1943: 36) until it increased slightly to 42 in the last year of the war. Remarkably, the Director was able to report year after year that these teachers kept the whole range of LSE subjects going. When Lance Beales bought a large house in Cambridge, he offered hospitality to the many teachers who during the first year commuted between London and Cambridge. Lionel Robbins was one of several who remembered the 'long darkened evenings' at Beales's house, where 'the old atmosphere of good talk and stimulating contacts persisted'.[19] By 1942, however, most remaining teachers had settled in Cambridge, though a certain amount of mobility in and out of government service made for variety.

The names of those who taught at Cambridge almost throughout the war include some of the well-known professors: Ginsberg, Hayek, Hughes Parry (before he became Vice-Chancellor of London University), Laski, Plucknett, Tawney (after his stint at the Washington Embassy), and Wolf. Other names, and not just Vera Anstey and Lance Beales, reverberate in the memories of LSE students: S. H. Beaver and F. C. C. Benham, Miss Clement Brown, and D. Seaborne Davies (later briefly a Liberal MP), Herman Finer and Raymond Firth, Jack Fisher and Otto Kahn-Freund, L. Rodwell Jones, Nicky Kaldor and Arthur Lewis, Hermann and Karl Mannheim, Frank Paish and William Pickles, L. G. Robinson and G. L. Schwartz, H. B. Lees-Smith, Dudley Stamp, and May Wallas, and Eugene Grebenik, W. A. Robson and Ivor Jennings, and of course until her death Eileen Power. The Social Science Department headed by C. M. Lloyd was there in strength, including Miss Eckhard, Miss Haskins, Miss Kydd, Mrs Burns, and later Mrs Judd. They all did more than their call of duty and are remembered with affection.

Even research continued, if on a much reduced scale. The Economic Research Division set up in 1936 had to be severely curtailed. On the other hand a new Social Research Division was set up for which the Director foresaw a bright future. *Economica* continued to appear. *Politica* had to be discontinued, but was, in a way, replaced by the 'Journal of Reconstruction', *Agenda*, which the Director called '*Politica* in war-time guise'.[20] The green-covered journal, with important articles if few readers, appeared for three years. This was not a time for journals.

Administering the School during these years required an unusual combination of contingency planning and supportive presence. All sorts of things might happen, and some did happen either in Cambridge or in London; at the same time, the Cambridge 'college' required a benevolent head. Luckily, Carr-Saunders in these years did not have to worry about finance. One of the standard phrases in his annual reports ran somewhat like: 'The financial situation does not at present give cause for anxiety' (1942). In 1943, the Director went even further:

> The financial position of the School has been fortunate since the outbreak of war; the expenses on our London buildings have been assumed by the Government and the School has been accommodated at little expense in Cambridge, while losses in students' fees have been about made good by savings due to absence of members of the staff on war leave. The accounts have thus shown surpluses.

Most of the funding came from the London University Court (that is, government) grant, though if one reads the accounts closely it is evident that the Rockefeller Emergency Fund as well as other benefactions were extremely welcome. Charlotte Shaw's gift, in 1941, of £1,000 to purchase general literature for the Shaw Library in Grove Lodge gave much pleasure; the money could not possibly have been found from general funds.

Still, the main problem of administration was not money. It was people, and the management of the unpredictable. Moreover, the Director was largely alone in this. At the beginning of the war, the Standing Committee of the Court of Governors had taken the unavoidable decision: 'That the Director be authorised to exercise the powers of the Standing Committee subject to report to the Standing Committee either at a meeting of that Committee or by correspondence to all members of the Committee.' Needless to say, Carr-Saunders never abused this enabling legislation. Whenever possible, Governors' meetings were convened. From time to time, the Director worried about the fact that so few members of the School were taking decisions potentially affecting so many. 'Even if the absent members of the staff cannot participate fully in the discussions, it is hoped that means may be found of keeping them informed and of ascertaining their views.' Usually, means were found. The Director consulted quietly but widely. He even set up, ahead of his time, a staff–student committee.

The Director's job was not made easier by two departures. In 1940 the Chairman, Lord Stamp, retired (alas! only a few months before he was killed in an air raid.) The new Chairman, Sir Otto Niemeyer, then a Director of the Bank of England and known as an international financial wizard, was to be, along with Sidney Webb and Arthur Steel-Maitland, one of the long-serving lay leaders of the School, but while he occasionally came and visited Cambridge it took a few years before he fully assumed his role. In March 1941 the Secretary, Walter Adams, was granted leave to go into government service, and Eve Evans took his place. The Director remained unruffled. Norman MacKenzie, a stern judge of people, wrote later:

> Carr-Saunders was a remote but likeable Director who ran the School in what would now be called 'low-profile' style. Even when he disagreed with you he was reassuring rather than patronizing. . . . The School's survival through the strains and shortages of war owed much more than was then apparent to his patient resourcefulness.[21]

Grove Lodge

Norman MacKenzie, who was a student at the Cambridge LSE, has described the meetings, the artistic occasions, the dances, the walks, and the punts as well as 'the visual reward of Cambridge itself'. 'This description makes it sound like a student Utopia. So it was.'[22] In 1941 the young German exile Dr Anne Bohm replied to an advertisement for a research assistant to L. G. Robinson. Despite the fact that she was stuck in snow in Wales and missed the interview, she got the post, and thus began a lifetime at the heart of LSE. L. G. Robinson was not just a reader in international history but also Dean of Postgraduate Studies. Within a year Anne Bohm became his assistant in that capacity and thus embarked on the journey which led her, after the return to London and with a little help from Harold Laski ('I took fortune by the forelock and spoke to the Director and it is all arranged and you don't need to worry') to her distinguished career as Secretary of the Graduate School. Anne Bohm confessed much later, 'I always say we had a very happy war'.[23] To outsiders this may sound strange, but those who were in Cambridge at the time agreed. John Griffith, then a law student in his last year, and not given to undue praise of the circumstances around him, said, 'Cambridge was fun, like Cambridge is,' and at least before Dunkirk, 'it was a very happy period'.[24]

Even apart from the beauties of Cambridge, the reasons are not difficult to see. Here was a kind of island community of like minds surrounded by the warm and pleasant sea of an ancient university; the distant thunder of world events came only slowly closer and actually never reached the idyll. The LSE group 'was a very small community,' Anne Bohm tells us, 'everybody knew everybody else and everybody was very friendly'.[25] The Director made this point repeatedly in his Reports. 'The tendency towards uniformity in age, aim and experience among the students, while involving loss in one direction, has meant gain in another; the student body has become more of a unity' (1940/1). And again: 'The much more homogeneous nature of the student body at Cambridge is not without beneficial consequences' (1942/3). There has been 'a clear emergence of a sentiment of unity which was not noticeable before . . . It is to be hoped that something of this spirit may be carried back to London.'

The focal point of all this happiness was not The Hostel, let alone the Cambridge Union, but Grove Lodge, the all-purpose mansion which LSE had acquired. 'I conceived of Grove Lodge, our Students' HQ, as having a lightning conductor that worked in reverse, sending our intellectual electricity out to the world at large,' George Brand, by then an international civil servant, wrote in 1983.[26] Norman MacKenzie remembers it as vividly, if in different metaphors, as 'a liner at the start of a long voyage' where, 'uprooted and crowded together, we found it easy to make friends', though it was also a 'little market-place of ideas'.[27] Grove Lodge had something for everybody, in due course even 100 lunches prepared by students. The Shaw Library was in Grove Lodge. When Betty Evans, who looked after it, wrote to Charlotte Shaw to thank her for her generous donation, G.B.S. himself replied—his wife being 'invalided and unable to write at present'—and asked on her behalf whether Upton Sinclair's novels had been acquired.[28] They had not; the Director disapproved of them. The Lodge was full of 'signs and slogans like a float in a May Day parade', all pointing to clubs, societies, and meetings, meetings, meetings. 'Grove Lodge had indeed been a cosy retreat from the world which had begun to bang away beyond its gate.'[29]

Not everyone was as sensitive, or as political, as Norman MacKenzie, and some of those who were did not share his politics. Barbara Sternberg, for example, described herself as 'a minority member of that brilliant, serio-comic, Communist-dominated student union board' which found itself confronted with 'Carr-Saunders wearily asking our deputation whether we didn't think we could focus on some matters over which we might conceivably have some influence'.[30] But among all the descriptions of LSE life in Cambridge, none is more vivid than Norman MacKenzie's. He describes the move of the extreme left from hostility to the war to strong support after the German attack on the Soviet Union, the 'dramatic personality' of the Communist Party as against the 'unexciting' Labour Party, and also the tolerant attitude of the teachers. 'The School staff were comfortably relaxed about both politics and sex.'[31]

In the event, this was as well because more marriages per capita emerged from LSE at Cambridge than even from LSE at Clare Market. Norman MacKenzie (B.Sc. 1943) met Jeanne Sampson (B.Sc. 1943) and they got married. Mary Oxlin encountered Claus Moser (B.Sc. 1943); they married after the war, in 1949. John Griffith, then a young law student (LL B 1940), married Eileen Power's disciple, Barbara Williams (B.Sc. 1940) in 1941. Stephen Wheatcroft (B.Sc. 1943), President of the Students' Union in 1941/2, married his successor, Joyce Reed (B.Sc. 1943), the President of 1942/3. This is not even mentioning those earlier LSE graduates, like David Glass, Nell McGregor, and Huw Wheldon, who found their spouses at the Cambridge LSE.

Norman MacKenzie was, of course, a Laski boy. 'Harold Laski was the focal point of my LSE.' But he looked beyond to many of the other teachers at the School. 'Despite wartime conditions they gave us nothing second-rate, no specifications pared down to "utility" standards. They created an environment in which it was easy to learn, and fun to learn—and these are the conditions in which students flourish.'[32] MacKenzie remembers his fellow students, those who went on to become professors and those who did not, O. R. (Lord) McGregor and (Sir) Claus Moser, Ralph Miliband and (Sir) Kenneth Berrill, Steve Wheatcroft and (Sir) Gordon Brunton, and Arnold (Lord) Weinstock.

In December 1940 D. Seaborne Davies, Reader in English Law, typed out a long 'alumnus letter' full of informative irrelevances to all members of the School far and wide.[33] It describes the Cambridge scene ('It was beautiful beyond description here. Even a Celt could not fail to appreciate anew the nostalgia of a Saxon poet away from England in the spring'), goes on to the events of the session, and then comments on, first, staff, then, students in alphabetical order. The Director receives good marks. 'Despite some very anxious and difficult times, he has weathered the storms very well.' Others get shorter shrift. 'Of Dr Dalton I need not speak. Allen is in the Treasury. Dr Anstey is here with us and I would like to take this opportunity of paying her the very warmest tribute I possibly can.' Seaborne Davies is as good at leaving things out as he is at saying them. 'Adams is living in Peterhouse. Miss Evans is with us and just the same as ever, full of fun.' Brigadier Young, the Careers Officer who never really made a mark, has retired. And Eileen Power has died, suddenly. The shock reverberated through Cambridge and the School wherever she lived in people's hearts and minds. Her totally unexpected death at the age of 51 moved people even at a time of war.

Indeed, 'Clare-Market-on-the-Cam' was not all roses. Some students did not share what one of them, Jack Hendy, later called 'a roseate view of their sojourn at LSE'.[34] Others, like Desmond Hartley, agreed. 'And like Jack Hendy, I don't remember any "creative ferment". Almost all the "intellectual" activity of the students consisted of sparring from entrenched positions between communists and orthodox Labour Party supporters.'[35] Occasionally a more traditional Cambridge expressed its displeasure with the invasion from Clare Market. An author in the 'C. U. Conservative Review' reminded the strangers of manners, such as not 'airing your own opinions on public affairs, or your host's pictures, or your hostess's cook, on first sitting down to table'.[36] The 'refugee visitors', student unions of LSE, and of Queen Mary and Bedford Colleges, which had also been relocated to Cambridge, sent a measured reply which emphasized distance rather than reducing it.

For LSE, more serious events mattered more. Every year now, the Director had to list those who were killed or maimed in action, or as a result of the war. In the end, seventy-one students of LSE had to be entered in the Roll of Honour. There were other deaths to report. Somehow, the Cambridge years mark the end of the early London School of Economics. Lord Passfield, having had a stroke and resigned from all offices, survived the war, but his wife, Beatrice, died in 1943. Carr-Saunders gave her fulsome praise in his Report—'her wonderful charm of manner made her presence welcome on all occasions'—and rightly suggested that her memory would live on. Charlotte Shaw, formerly Payne-Townshend, died in the same year. So did Miss Mactaggart, the first Secretary of LSE, who had retired twenty-two years earlier because of ill health. Or was Lord Snell the first Secretary? He too died a year later in 1944. The first head porter, E. J. Dodson, had passed away in 1943. The School, not quite 50 years old, had lost virtually all its founding members, with Sidney Webb surviving, alone but alert, at Passfield Corner.

The Other War

The Hostel, Grove Lodge, and the beauties of Cambridge provided the setting for one-half of the wartime story of LSE. The other half took place in London and in many parts of the world of allies, and was written by those members of the School who went into government service. A record compiled after the war lists forty-six members of staff, more than there were LSE teachers at Cambridge (though some of them are included, and some in government service gave occasional lectures). The record, in alphabetical order, is another remarkable roll of honour.

The list begins, appropriately, with Walter Adams, the Secretary of the School, who was on the verge of enlisting as a private soldier when his old friend from student and Bloomsbury days, Hugh Gaitskell, now an official in Hugh Dalton's ministry, persuaded him otherwise. Adams served first in the Special Operations Executive of the Foreign Office, then, in 1942, went to Washington for two years as Deputy Head of the British Political Warfare Mission. On his return to its Political Intelligence Department the Foreign Office would not let him go until 1946, so that he had to resign the secretaryship of LSE. Others too went abroad, notably to Washington, like Professor R. G. D. Allen (British Supply Council and Combined Munitions Assignment Board), Frederick Brown (Head of Information, New York, then British Food Mission), Ronald H. Coase (Central Statistical Office Representative), Professor D. H. Robertson (UK Delegate to Bretton Woods), Professor R. H. Tawney (Councillor at the British Embassy), Professor Sir Charles Webster (British Library of Information and UN Delegate), E. H. Wyndham White (First Secretary, British Embassy). Some were posted elsewhere, to the West Indies (Professor F. C. C. Benham), to Cairo (F. J. Fisher, P. A. Wilson), to Turkey (C. Parry, F. Chalmers Wright).

Perhaps the most striking aspect of this catalogue of service is the extent to which teachers in virtually all disciplines of the social sciences turned out to be useful. Some worked in or near their chosen fields: statisticians responsible for records and statistics, or (like Eugene Grebenik) as Statistical Officer of the Admiralty; psychologists (like J. M. Blackburn) on selection tests for Air Defence; geographers (like Professor L. D. Stamp) as Chief Adviser in Rural Land Utilization; lawyers (like Professor H. A. Smith) as Legal Adviser to HM Procurator General, or (like Professor Chorley and Mr Seaborne Davies) in the Nationality Division of the Home Office. On economists more will have to be said presently. Others did general chores of considerable importance, working in the Foreign Research and Press Service, for example (Dr Lucy Mair, Professor C. A. W. Manning, Professor T. H. Marshall), or broadcasting to other countries (William Pickles). In some cases, the link is not evident, as with the anthropologist Professor R. W. Firth in the Postal Censorship Department, or the political scientist K. B. Smellie as Principal in the Clothes Rationing Department.

For the most part, these members of LSE were either advisers, or involved in intelligence in the widest wartime sense of the term. Some, however, rose to senior positions either within the Civil Service or close to the seat of power. Professor Chorley became a Deputy Commissioner for Defence. Evan Durbin ended up as Personal Assistant to the Deputy Prime Minister, Mr Attlee (and in 1944 as an MP), Ronnie Edwards was for five years Assistant Secretary in the Ministry of Aircraft Production. The economic historian A. V. Judges became Deputy Secretary (and Labour Adviser) in the Ministry of Production. Arthur Lewis served as a Principal first in the Board of Trade, then in the Colonial Office. Professor Arnold Plant in the end became Chairman of the Materials Committee of the War Cabinet. W. A. Robson was Assistant Secretary in the Air Ministry.

This is a long list, too long, perhaps, if it was just a roll of honour. But it is more. It is indicative of one of the fundamental developments of the century affecting the School, the discovery of the uses of social science. Some academics had been drafted into public service during the First World War. William Beveridge and John Maynard Keynes are the obvious examples. In the inter-war years, commissions, Royal or otherwise, played a part in advising governments; the 1931 economists' group has been mentioned. But the systematic exploitation of academic knowledge in law and history, geography and anthropology, international relations and psychology, political science and sociology, and of course statistics and economics was new, and was both cause and effect of the blossoming of the social sciences in general, and LSE in particular. It also formed a new facet of the ever tense and ever fruitful relationship between social science and values, the School and public policy.

All this was nowhere more evident than in the story of the Economic Section of the Offices of the War Cabinet. Lionel Robbins, who from 1940 to 1945 was Director of this Section, has described the experience in his *Autobiography of an Economist*, and Sir Alec Cairncross, in a book-length study written with Nita Watts, *The Economic Section 1939–1961*, has put this innovation in 'economic advising' in a wider context. There is also the edition, by Susan Howson, Lionel Robbins's biographer, and Donald Moggridge of *The Wartime Diaries of Lionel Robbins and James Meade, 1943–45*.

'British governments have long made use of economic advisers but have only recently come to employ them on a full-time basis.' The very first sentence of the book by Cairncross and Watts sums up the issue, at least in terms of social science and public policy-making. The Economic Section was in fact a late child of the Economic Advisory Council of 1930 (on which Susan Howson has written an informative book with Donald Winch). The small co-ordination committee set up at the beginning of the war and headed by Lord Stamp, which included the Manchester professor Henry Clay and the economic journalist Hubert Henderson, created the Central Economic Information Service (CEIS) as a resource of facts and advice. Despite such antecedents, the birth of the Section was painful, largely because Whitehall was simply not used to having academic strangers in its midst.

The Economic Section was initially put together by the civil servant who had assisted the Stamp Committee, Francis Hemming. The group was distinguished enough, including as it did, apart from Robbins, among others Alec Cairncross, Norman Chester, Ely Devons, John Jewkes, James Meade, Austin Robinson, and Harold Wilson; but Hemming's idiosyncratic ways and the lack of a clear addressee in government made it ineffectual. Mistakes were made; on one occasion Churchill complained about 'the confusion created through the use of different measures of shipping tonnage: gross, net, and dead weight'; the morale of the advisers was low.[37] Then, in mid-1940, soon after the end of the 'phoney war', several things happened which changed the scene. Sir John Anderson (later Lord Waverley) became Lord President of the Council and proved to be an effective co-ordinator of policy as well as a man willing and ready to absorb advice. Statistical work was hived off from the Section into the newly created Central Statistical Office. John Jewkes of Manchester and, when Jewkes went to the Ministry of Aircraft Production, Lionel Robbins, were in turn appointed head of the Section. Five years of intensive and effective wartime work began, after which the Section was absorbed by the Cabinet Office.

The substance of the work done by the small Section—'eight economic assistants of various grades'—focused, over the years, on three issues. The first was the organization of the wartime economy. Cairncross later wrote that their tendency to take 'a bird's-eye view of the economy' meant that economists 'were, in a sense, at home in a war economy while others were not'.[38] The more liberal Robbins, on the other hand, has described the extent to which wartime problems differed from those which economists normally dealt with. 'They differed as regards aims and they differed as regards means.' The aim of a 'minimum provision of personal consumption necessary to maintain health and morale; everything else for the service of victory' was certainly unfamiliar, and the means were distasteful, 'a degree of collectivist control which would be highly inappropriate for a would-be liberal community at peace'. The irony of history turned the leading representative of LSE's free-market Economics Department into the pacemaker for the more collectivist arrangements which his colleagues in other departments of the School had long desired.

The second issue with which the Economic Section was concerned was indeed planning for the post-war period. Lionel Robbins never got reconciled to his old *bête noire*, 'the unhappy Beveridge, his own worst enemy';[39] however, he made peace with Keynes, the winner of the economic battles of the 1930s and now the leading light of the Treasury; and Cairncross tells us that in fact the Section contributed much to the Beveridge Report. It also defined, under the leadership of James Meade, who was to succeed Robbins briefly in 1945, post-war approaches to unemployment and other problems.

Then, increasingly, there was the international side, and there were notably the American negotiations following the idea of world rules and institutions for commerce and money first mooted as a part of the Lend-Lease Agreement. As a UK Delegate to the conferences of Hot Springs in 1943 and Bretton Woods in 1944 Robbins was at his best, suave in style, yet firm in purpose, full of verbal and at times substantive imagination, equally at home in the conference hall and at the dinner table. Many of those in and around the Economic Section had some LSE connection, but on the international scene the LSE card could be played to even greater effect. 'The final negotiation was conducted behind the chair between the ex-professor of Economics in the University of London and his former pupil, the representative of Mexico (the conference was just littered with the alumni of the London School of Economics).' When the Indians were particularly difficult, Robbins took their delegate A. D. Shroff, a Director of Tata Sons Ltd., to lunch.

I discovered that Mr Shroff, who has been the chief agitator regarding the balances, was a contemporary of mine at the School of Economics just after the last war. This has made all the difference. Shroff and I converged as one old LSE man to another and settled many outstanding problems of the world before touching on outstanding Indian questions.[40]

Not all LSE acquaintances met with Robbins's approval. When John Kenneth Galbraith—'I knew Galbraith in the old days; he sat for some little time in my seminar'—was dismissed from the US Office of Price Administration, Robbins showed himself 'not altogether surprised'. 'I always thought him a dull fellow, well intentioned enough, but a sort of pedant of New Deal economics.' Lionel Robbins used the occasion for a more general comment. 'On the whole, I doubt whether the arrival of the academic world in public administration over here [i.e. in the United States] has been as successful as it has been at home.' He gave an interesting reason. The lack of a proper civil service tradition in America meant that academics were nothing special, whereas in Britain the issue was to link two distinct professions. 'Should we have fitted in, if there had not been something pretty solid to fit into?'[41]

Since then, the subject of 'economists in government' has been thoroughly examined, not least in an international comparative study under that title edited by A. W. Coats in 1981. Actually, analysis began almost as soon as advice had become institutionalized. In November 1943, a special 'Report of the Official Committee on the Machinery of Government' included a section on 'the role of the economist in the machinery of government'. Members of the Economic Section had given evidence to the Committee. Its report made much of a proper division of labour. Accessible ministers ('there is no point in offering advice if there is nobody to listen to it'), and respect for the Civil Service ('another lesson, rubbed in by Lionel Robbins, was the importance of carrying departments with us'[42]) are preconditions of effective academic input into public policy. 'Willingness to become part of the machine and accept its logic rather than pretend to some special status' (as Robbins put it) is a long way from Harold Laski's *Daily Herald* articles, let alone the Fabian dream of a brave new world run by experts. It is different also from Hugh Dalton's or Clement Attlee's road into electoral politics at one end, and Hayek's pure academic concerns at the other. During the war, a new purpose of social science was tried, and it succeeded sufficiently to stay; perhaps, social science as policy science is the right name.

Lionel Robbins enjoyed this venture so much that he found it hard to return.

> It lifted me out of the rather narrow academic atmosphere . . . It gave me the opportunity of rethinking the fundamentals of my subject in a *milieu* of applications radically different from anything with which I had previously been confronted. It gave me inside acquaintance with the machinery of government and the techniques of practical administration.[43]

In fact, Robbins admits, he did not find it possible to return to the pure joys of academic analysis, and one suspects that not a few others who had gone into government service during the war had the same experience. Policy science has a price, and some members of the School, pointing to its demanding puritanical motto *rerum cognoscere causas*, would argue that this was too high a price to pay for an academic institution.

London is Calling

'L.S.E.'s war-time experience formed a complete episode in the life of the School.'[44] Vera Anstey was more right than she may have known when she wrote this. The Cambridge story was complete even if that of policy advice to governments was not. Cambridge LSE had a beginning and an end and represented a rounded capsule in time. It was also an episode, not a core part of the history of the School. Pleasing though it is to hear the student from Colorado tell us, 'my years at LSE were among the most magnificent and precious of my life', it is also shocking to discover as she goes on: 'I never visited Houghton Street.'[45] How can there be a London School of Economics without London?

The Director certainly did not think that there could be. For one thing, his responsibilities did not end at the river Cam. At first he had to look after the remaining evening students. Their numbers were rapidly declining but even the last forty, then nine, needed some attention. Then he had to negotiate with the Office of Works the rent, the rates, and more importantly the conditions of repossession of the School's buildings. In 1940, a seven-year agreement with a three-months notice clause was concluded. Throughout there was the worry of air raids. The School was lucky.

> On one occasion [the Director reported in 1944] a large unexploded bomb lodged for some days in the building on the other side of Clare Market in a position which made it certain that, had it gone off, it would have destroyed a good part of our Library. When that phase ended, we had only some dozen broken windows. Our luck has held during the second phase; the 'near misses' were even closer and one of them destroyed a large amount of our glass; but loss of glass was an event hardly worthy of mention during the period of flying bombs.

Even the sports ground in Malden did not suffer unduly; Walter Stern has described its use as 'the LSE Farm' during the war; in the end, a few bomb craters provided an opportunity for creating a more level playing field. The Library was the Director's main concern. At first, he tried to keep it together and even available to readers. But this definition of the 'Library enclave' did not last. By 1941, most books had been stored in the basement, and other materials sent to safe destinations, to Oxford, to Aberystwyth, to Carlisle. Thus the treasures of the British Library of Political and Economic Science survived the war.

Above all, the Director never tired of reminding the School of its London home. Every year his Report began with a section on London, events around Houghton Street and Clare Market, and the need to return at the earliest possible moment. Absence from this lifeline is bearable as long as the war lasts, but not a moment longer. 'Continued enforced exile would then be a catastrophe.'[46]

As early as 1941, Harold Laski had produced a memorandum on the School after the war.[47] In it he identified long-range as well as short-range problems. The short-range problems had to do with 'the order of the full resumption of the various activities of the School', including the return of the Library, the resumption of evening teaching, and the leave entitlements of staff. Long-range problems included the scope and method of teaching, the relative balance of undergraduate, evening, and graduate work, the function of the Library, and 'public relations including those with other educational institutions and public bodies at home and abroad'. Both the General Purposes Committee and the Standing Committee agreed that policies with regard to the short-range problems had to be agreed before the end of the war. A Post-War Developments Sub-Committee was set up for the purpose. To deal with long-range problems, and notably the first one of the scope and method of teaching, a research assistant was attached to the Director.

In the following years, many options were considered. The most radical suggestion came from Dr E. C. Rhodes, who was Reader in Statistics and, characteristically perhaps, a Cambridge graduate. In a paper written in 1943 he suggested the relocation of the School to 'semi-rural surroundings', for example to the sports ground in Malden or somewhere in Hampstead, the provision of residential accommodation for staff as well as students, and the renaming of LSE as 'Webb College'. Dr Rhodes found few supporters; in fact only one vote was cast for his proposal at the crucial and well-attended Professorial Council of 7 January 1944.

Instead, the committees of the School took a more traditional line, though one suggesting considerable expansion after the war. The School, it was agreed, is and should remain a School of the Social Sciences. New subjects were not required, though better equipment for research and an improved staff–student ratio were essential. Moreover, 'the School must anticipate that it will be called upon to meet the needs of a very much larger and more varied number of students than before the war'. Detailed proposals for staff development were made. All this required funding. Since the 'long-range programme' was presented to the University Grants Committee in 1944, it included a request for new capital grants and a massive increase in recurrent support.

One question remained, that of the cohesion of the School. The Professorial Council re-emphasized the desirability of a weak departmental structure in which academic disciplines do not seek self-sufficiency but co-operation. This, however, was only one side of the picture of cohesion. The other had to do with student accommodation. Two reasons converged in this concern. One was that traditional student lodgings had suffered disproportionately from bombing; one estimate suggested that in the

Bloomsbury area, 74 per cent of the pre-war lodging houses were no longer available. The other reason had to do with Cambridge. People had discovered the advantages of a residential college and wanted to preserve some of its community spirit. What the School needed therefore was hostels. Since the University had far too few of them, 'college hostels' were required. In 1943, the Director reported that the University Grants Committee had shown interest in the idea. When the School returned to London, it was its declared policy 'to offer hostel accommodation to all students who wish for it', but it took years to begin the implementation of this policy, and even half a century later this is far from complete.

Perhaps this is not a cause for great concern. The experience of community, *Gemeinschaft* even, in Cambridge was itself an episode; the special intimacy of LSE was always more of a *Gesellschaft* character, an association of grown-ups, voluntary, cerebral as much as emotional, not necessarily lasting, a thoroughly modern form of human relations. Still, Cambridge had left its imprint. In his 1945 Report, the Director thanked Peterhouse, Cambridge University, and all its members and institutions for their 'unrivalled hospitality'. 'The School will always recall their unexampled generosity.' The School presented the Combination Room of Peterhouse with a handsome engraved 'standish'. Peterhouse had already made Carr-Saunders a Fellow with full governing body rights. When in September 1989 a plaque to commemorate LSE's years in exile was unveiled over the entrance to The Hostel, the Director, I. G. Patel, and the Master, The Revd Sir Henry Chadwick, were able to renew the friendly relationship.

In 1945, however, Carr-Saunders went a step further. Much as he had pressed for an early return to London, he allowed himself to muse about the relevance of the Cambridge experience for the future of the School.

> To attempt to draw out the lessons to be learned from these six years would be to go beyond the scope of this report. That there must be lessons is clear. To take a modern, co-educational and non-residential college from the metropolis and place it under what may be called residential conditions in a town of no great size is an experiment which provides material for the sociologist. When that town is the site of an ancient university the experiment must be especially rich in interest. It may be hoped that before the memory of these experiences has grown dim, they will be recorded and made the subject of some analysis; did the students, it may be asked, get more out of their studies than in London; did they benefit more than before from social intercourse; did they improve in health? When the changes were for the good, is it possible to take steps to better conditions in London and so retain the particular advantages of Cambridge life? Such matters should in any case be made the subject of careful thought; it may well be that we are too apt to accept the conditions of pre-war life at the School as unchangeable, while in fact the disadvantages of some aspects of that life could be mitigated or removed.[48]

Disadvantages or not, it was certainly wrong to take pre-war life at the School for granted. Much had changed, and the post-war School displayed many new features in a new environment, though in its old location, cramped, untidy, noisy, and metropolitan as it was and still is. There was little if anything to be brought back to Clare Market from Cambridge. In the summer of 1945, the episode was closed. It is easier to follow the Director in his peroration, in which he mentioned that the 'return of the School to London will coincide with the fiftieth anniversary of its foundation' and expressed his hope that this 'will prove to be the opening of an even more successful period of work'. But perhaps Friedrich von Hayek caught the role and the problems of the School after its return to London best when he concluded his essay on the first fifty years of LSE (published in February 1946) in a suitably sceptical vein. The services by the staff and the students of the School during the war form a highly creditable story, Hayek said. But following up this story is not so easy:

> That during the coming period of reconstruction the minds [LSE] has trained in the understanding of social and economic problems will be needed as urgently as during the war there can be little doubt. The main concern of the School must now be to see that preoccupation with immediate problems does not result in neglect of the training of a new generation and that the drain on academic talent caused by the great success of the 'professors' as civil servants is not so heavy as to impair the future stream of similarly qualified men and women.

RETURN TO A WORLD OF CHANGE

Houghton Street Again

Returning to London was in all practical regards harder for LSE than leaving it, but the obstacles were also easier to overcome, since the School was coming home. Home meant Clare Market, of course, though more and more people now referred to Houghton Street, the little thoroughfare linking Clare Market and the Aldwych which was dominated by the Old Building of the School on the west side and the new East Wing across the road. Actually, neither was available when LSE had left Cambridge at the end of August, and when they were handed back six weeks later they were in a sorry state. The first building which the School was able to occupy was The Anchorage in Clement's Inn Passage, the rambling and charming old vicarage of St Mary le Strand since its construction around 1800. The School leased it for two years in the first instance, and the administration moved in. In later years it became the residence for Directors before it was turned to administrative use again. Other buildings in Clement's Inn

Passage and Houghton Street were made available one by one, though the School had to wait until 15 October to regain the Old Building. 'When we obtained possession of our front door key and could get into our own premises, we realised what an immense task it would be to make them usable' in view of 'the deplorable state of dirt and dilapidation' in which they were found.

The Director, not given to hyperbole, spoke of the 'difficulties, perplexities and anxieties' attending the return to Houghton Street, and Vera Anstey of the 'tremendous physical task to render the buildings habitable and to restore them to their original uses, whilst at the same time cases and cases of books had to be dealt with in the Library'. But everyone rolled up their sleeves and joined in. 'Valiant work was done by our trusty band of porters, and also by student volunteers, who swept and scrubbed with the best.'[49] By 29 October, only four weeks late, the School opened its door to students for the new academic year. And they came, in large numbers, 2,151 of them, twice as many as in the last year at Cambridge.

Even the characteristically low-key, largely statistical character of the Director's Report on the year 1945/6 cannot conceal the triumph of the return to London. The triumph was somewhat muted, perhaps, not just by the state of the School's buildings but by the general drabness and austerity of post-war London, yet none of this could detract from the obvious attractions of the School. The number of regular students had doubled compared with the last year in Cambridge; there were now 1,584; higher-degree students had nearly trebled, from 66 to 172. The evening school was revived and attracted 259 first-degree students. One price of such change was that the School immediately returned to the pre-war ratio of men to women 'of about seven to three'. Many of the students were ex-servicemen, including a significant number of American GIs, who bolstered the overseas student numbers to 490 in the first year after the war.

Why did students come at the time? John Watkins had trouble disentangling himself from the Navy and also persuading his parents to support him, but he had read Hayek's *Road to Serfdom* as a naval officer stationed in Shanghai, and also the reference in Evelyn Waugh's *Put Out More Flags* to the 'red-headed girl in spectacles from the London School of Economics'; now he wanted to see for himself.[50] He came in 1946 and stayed, with interruptions, until his retirement as a distinguished Professor of Philosophy in 1989. Alan Stuart also came in 1946 and stayed to become Professor of Statistics as well as Pro-Director from 1976 to 1979. As a sixth-former before the war he had been chosen by his headmaster to work on the standardization of some mathematics tests at LSE. He remembered with affection the half-crown which he got every time he came so that the friend who suggested, as he left the army, that he go to LSE had no difficulty persuading him. Robert McKenzie joined the School as a postgraduate in 1947. He was 29 then and had worked as an officer with Canadian Military Headquarters in London 1944–6, but had 'fallen under the spell of LSE' much earlier, in 1939, listening to a lecture by Harold Laski in Vancouver.[51] He too was to stay until his premature death in 1981. When Ron Moody came in 1948, things were almost back to normal, though his Squadron Leader had never heard of LSE and his Commanding Officer initially only let him go in the evenings. LSE for him was 'the only place of Higher Learning [at which] I had ever wanted to be. I was a guest in the house of Bernard Shaw and the Webbs, the Citadel where Laski held court.'[52]

From the other side, as it were, the teachers and administrators, the experience was equally varied and exhilarating. Vera Anstey praised the Director for having brought the School back intact from exile. 'To him, more than to anyone else, is due the praise for the war-time survival and post-war recovery of the School's precious personality, traditions, and potentialities.'[53] Lionel Robbins too had many good words for the (to him) new Director and credited him with 'the creation of a sense of co-operative purpose and responsibility which had seldom existed before'. He still hankered after the practical life of government, but 'if I was to be academic, I would rather be there [i.e. at LSE] than in any other university institution in the world'.[54] Anne Bohm remembered the poverty of post-war London but also the sudden 'explosion' of the School into life and rapid growth, and then the ex-servicemen.

> They practically all had grants from England, America, Australia, Canada, New Zealand. They were all extremely glad to be alive, to be allowed to study and they took their studies seriously—unlike fifteen years later when everybody came in and said 'it's mine by right to study, somebody else has to pay for it' when they hadn't done anything to deserve it. . . . For everybody it was a serious but a very joyous time, these first five years after the war.[55]

In 1945 the School was 50 years old but had little time to celebrate. Only the Students' Union managed to publish a gilded 'Jubilee Number' of *Clare Market Review* in Lent 1945. The special issue included some rather touching features. The editorial ended: 'Finally, we wish LSE and its students as successful a half-century as the one that has passed. We could not wish them better.' The Director, in his careful and reserved way, suggested that there may be 'a case for more readiness to adapt the constitution [of the Union] to changing conditions and for more interest in possible new activities, as, for instance, in the development of debates'. The demand for extending democracy beyond political institutions was developed by several authors, with regard to the Students' Union, to the Beveridge Report, and to the state of Europe, on which Peter G. Richards expressed his own radical optimism. 'In 1945, the nationalities of Europe, with one chief exception, are more mature. The supreme issue is now the future of property relationships.'

The Director, meanwhile, had made his own arrangements. In July 1943 he had written to a long list of former members of the School asking them to put down their recollections. 'If the history of the School is to be more than a bald record of its growth and activities we must rely upon personal accounts of some of the more memorable events which have happened.'[56] The resulting 'Reminiscences' were first used extensively by Janet Beveridge in her *Epic of Clare Market*; they have enriched the present account at many points. The Director had also asked Professor von Hayek to write what the author called, a 'brief sketch' of the history of the School. It turned out to be a highly informative, if perhaps slightly stiff, account of academic developments between 1895 and 1945. 'The full history must wait for a more propitious moment.'[57]

Hayek himself had waited for a propitious moment to publish his sketch in *Economica* in February 1946; he was able to include a brief reference to Labour's election victory. Its effect was evident at the Fiftieth Anniversary dinner which eventually took place in June 1946. The Prime Minister and the Chancellor of the Exchequer, Mr Attlee and Dr Dalton, were guests of honour. (Hayek mentions two other members of the Attlee Government who had been on the LSE staff, the Secretary of State for India, Lord Pethick-Lawrence, and the Minister of State, Philip Noel-Baker, as well as three new Peers, Lord Chorley, the long-standing Professor of Law, Lord Piercy, an alumnus and lecturer in commerce in the early 1920s, and Lord Pakenham, briefly an assistant in economics in 1931/2.) The dinner was a 'great occasion', enhanced rather than marred by a letter from the ageing Sidney Webb read by the Director, according to one source 'almost certainly the last letter Mr Webb wrote to and about the School before he died',[58] in 1947:

Dear Director
I was glad to see that you are celebrating the fiftieth anniversary of the foundation of the School, and thank you for your invitation to attend the dinner. Alas! I am an invalid and confined to this neighbourhood, which confines me to dining alone here on Thursday the 6th of June. I should have been glad to have met so many representative people. I assure you I shall bear the invitation in mind for my lonely dinner.

The School has been a wonderful success.
Yours very truly

PASSFIELD[59]

The Sad End of Harold Laski

One other person present at the fiftieth anniversary dinner had many reasons for rejoicing. Harold Laski, the School's most widely known and most loved professor, had been Chairman of the Labour Party during the election year, 1945. Thus he had presided over a famous victory, and done so with gusto. Yet in his case the victory was marred by an incident which seemed small at first but then grew to unfortunate and ultimately fatal dimensions. Indeed, his entire Party chairmanship had not gone very well. When the election campaign began, he had advised the leader, Clement Attlee, to resign because he was the wrong man for the job. Attlee replied with one of his famous laconic notes: 'Dear Laski, thank you for your letter, contents of which have been noted.'[60] While the count was going on in July (it took several weeks because of the forces' vote all over the world), Churchill had asked Attlee to accompany him to the Potsdam Conference; but Laski asked Attlee publicly to desist, thus causing questions about who ran the party. Once the Labour Government had been formed, Laski was disappointed not to have been offered the ambassadorship in Washington. Instead, he went on lecture tours first to America, then to Russia and gave to many objectionable speeches attacking Americans for leading the world on the 'road to serfdom' (Hayek's attack on him rankled), and praising Stalin for his wisdom. One can thus understand Attlee's letter to the 'wandering minstrel': 'I can assure you there is widespread resentment in the Party at your activities and a period of silence on your part would be welcome.'[61]

None of this, however, reached the dimensions of the incident which happened during the campaign, at a public meeting in Newark in Nottinghamshire on 16 June 1945. Challenged by Tory hecklers about his war record and his attitude to violence, Laski defended his record and 'as for violence, he continued, if Labour could not obtain what it needed by general consent, we shall have to use violence even if it means revolution'. This at any rate is how at least two newspapers, the *Newark Advertiser* and the *Daily Express*, both belonging to Beaverbrook, reported Laski's reply under the banner headline 'New Laski Sensation: Socialism Even if it Means Violence'. Laski denied that he had used the phrase 'we shall have to use violence'; his words had been that 'great changes were so urgent in this country that if they were not made by consent they would be made by violence'.[62] He had writs for libel issued against the Beaverbrook papers.

It is hard to tell what motivated Laski to pursue the libel suit after the election victory, when everybody advised him to drop it. Probably Sir John Simon, the outgoing Lord Chancellor, had upset him with the suggestion that the action would be quietly forgotten after the election. Possibly, he saw it as a question of honour to go on. In any case, he fully expected to win and even went so far (according to Kramnick and Sheerman) as 'telling everyone at the LSE that he would endow "the Beaverbrook Chair of Political Science" with all the damages he would get from the libel action'.[63]

The jury trial on 26 November 1946 became a humiliation for the great teacher and an object lesson in the clash between academic argument and political campaigning. As one reads accounts of the day in court, one has to remind oneself that this was not 'Laski's trial' in which he was tried for any misdemeanour or crime, but on the contrary Laski's lawsuit against the Beaverbrook papers. However, onlookers might be forgiven for forgetting this fact. Laski's counsel, G. O. Slade, could not match the ruthless cross-examination by the defence counsel, Sir Patrick Hastings, himself a man who had moved from left activism to right-wing conservatism. It was an awful occasion. Isaac Kramnick and Barry Sheerman have done full justice to it in their biography of Laski,[64] even if their understandable partisanship sometimes gets the better of their report of events.

Hastings's strategy was simple and effective, especially in front of a 'special jury' selected from a panel of property-owning jurors. Counsel cited gleefully and at length Laski's books, in which the author had used the buzz-words 'violence' and 'revolution' frequently and stated that, if the democratic process did not bring about social change, there was a risk of violent revolution. Laski got himself into a hopeless tangle in the attempt to explain the difference between 'advocacy' and 'analysis'; he had analysed a condition of potential violence, not advocated the use of violent means. A vicious counsel had no problem impressing the jury by dismissing such sophistication as sophistry and throwing in a whiff of philistinism as well as subliminal xenophobia if not anti-Semitism for good measure: 'Forgive me, but for the moment could you forget you are a professional historian, forget everything, and just remember you are an ordinary Englishman and I am asking you a question in ordinary English?' When Laski tried to argue that the whole point of his political activity had been to bring about the consent of the voters for change in order to avoid the violence of revolution, and that he had never said that 'we', the Labour Party, were going to use violence, he was already hopelessly on the defensive. The jury did not like the words, 'violence' or 'revolution', nor the man who used them so readily. All that was left for Hastings to do was to produce a witness who claimed to have heard Laski say the fateful words. He was a journalist, Mr James Wentworth Day, and the public relations adviser of the Conservative candidate for Newark.

The trial judge, Mr Justice (later Lord Chief Justice) Goddard, did not help. His summary was certainly open to charges of bias, as the judge himself seems to have admitted to Laski's brother Neville later. Yet whether the statement that 'the expression of abstract academic opinion in this country is free' added what Kramnick and Sheerman call a 'bizarre twist' must be open for debate. For one thing, Laski was not tried for his views; in fact he was not tried at all, and no one would or could have put him in the dock if he had in fact said what the *Newark Advertiser* reported. For another thing, and on a more subtle level, there is a difference between the lecture room and the market-place, and this is precisely the difference between analysis and advocacy. A strict Marxian analysis suggesting the inevitability of violent revolution may be historicist and mistaken, but it must be allowed in an academic context. It does not even identify the speaker with the prospect which he paints. A similar set of statements before an audience of voters, let alone potential revolutionaries, can be incitement and certainly reflects on the democratic credentials of the speaker. All this is highly relevant for Laski, LSE, and the great bogeyman of the history of the School, its politics.

But such reflections would themselves have been academic in the demeaning sense of the word on that 26 November 1946. It took the special jury forty minutes to dismiss Laski's libel suit and to award costs against him. 'Laski was devastated. According to Frida, "he bore up well till he got home and then wept as I have never seen a man weep". Diana thought "it broke his heart", and most of his closest friends insisted that Laski never recovered his buoyancy after the verdict and was never really well again.' He offered to resign from LSE, and from the National Executive Committee of the Labour Party. Both offers were of course rejected. Indeed his friends at home and abroad began to collect contributions to cover the cost of the case. More than 5,000 people contributed, humble members of the Labour Party as well as American and international celebrities like Albert Einstein, Edward Murrow, and Henry Wallace, and of course many of his colleagues at LSE.

Harold Laski died on 24 March 1950. His students in the last years of his life were as enthusiastic about their teacher as earlier generations had been. Also there never really was to be that famous 'period of silence';

Laski on India or on Palestine was his old self. Yet his life energy, always more cerebral than visceral, began to diminish. 'Exhausted he was in the last years,'[65] his favourite pupil, Ralph Miliband, found, and in the end Laski himself admitted, to the dismay of his friends, that he felt very tired.

The School in 1950

In retrospect, the first five years after the war were a time for the School to sort itself out and prepare for another age, the age of the social sciences in the ascendancy, of consequent academic expansion, and of spreading corporatism with its opportunities for graduates of a school of economics and political science. The sorting-out time was if anything helped by the presence of so many ex-servicemen. They straddled the early School for mature students who for the most part came in the evenings, and the Cambridge college for undergraduates. They also faded away. By 1950 the School was beginning to find a new normality, if one of rapid and, to many, painful quantitative developments. People never like rapid growth, especially not when they look back; life was so sweet in the good old days. Actually, LSE found a viable balance in the remaining decade of the Carr-Saunders—since 1946, *Sir* Alexander Carr-Saunders—directorship.

But changes there were. Laski's death was a watershed. He left a gaping hole for students who came afterwards. Even forty years later one can encounter the question 'Did you hear Laski?' which neatly divides those who came before and after. It was a watershed also because his successor, Michael Oakeshott, while outstanding in his own right, was so utterly different from his predecessor in almost every respect. Oakeshott was conversational where Laski was forever the orator; he was concerned with detail where Laski preferred the great sweep; he dug deeply into the past, preferably that before the seventeenth-century revolutions, when Laski could never get enough of the present, the day, almost the minute; and of course Oakeshott was a true and profound conservative thinker. His tribute to his predecessor, in the famous inaugural lecture, was handsome enough, and also betrayed Oakeshott's irony in recognizing his own position. 'And it seems perhaps a little ungrateful that [Laski] should be followed by a sceptic; one who would do better if only he knew how.' What he did know about politics sent a *frisson* through his audience in the Old Theatre on that March day in 1951:

> In political activity, then, men sail a boundless and bottomless sea; there is neither harbour for shelter nor floor for anchorage, neither starting-place nor appointed destination. The enterprise is to keep afloat on an even keel; the sea is both friend

and enemy; and the seamanship consists in using the resources of a traditional manner of behaviour in order to make a friend of every inimical occasion.[66]

One must doubt whether LSE had ever heard such words before. But then, Oakeshott never dominated the mood and the mind of the School as Laski had done, though his presence was a daily reminder of the changes since the early 1950s.

There was much coming and going among senior staff in the decade after the war, on balance fortunately more coming than going. Examples can be given from almost every department. In 1950 Friedrich von Hayek decided to leave for the University of Chicago after eighteen years at the School. However, not only had Professors Plant and Robbins returned from war service, but others, including Professor R. S. Edwards, Professor F. W. Paish, Professor E. H. Phelps Brown, Professor R. S. Sayers, as well as the future Nobel laureates Professors James Meade, Dr Arthur Lewis, and Dr George Stigler, had joined them in 1948. What was later to be hived off from Economics, as the Department of Accounting, grew rapidly after the appointment of David Solomons (later professor successively at Bristol and the Wharton School) in 1946, Professor William T. Baxter in 1947, and Harold Edey in 1949. Basil Yamey may fairly be said to have straddled economics and accounting and contributed greatly to both.

Important appointments were made in other departments as well. Karl Popper was enticed to the School from his New Zealand exile in 1945 to the post which Professor Wolf had held for nineteen years until his retirement in 1941. Popper soon became one of the focal figures of the School as well as an author of ever growing fame. When, on returning from a trip to Australia, he wanted to stop over in Hong Kong, he cabled to Carr-Saunders for help: 'Please tell Vice-Chancellor who I am.' Carr-Saunders was up to the task: 'Karl Popper Reader Logic and Scientific Method stop One of the greatest living philosophers.'[67] W. A. Robson became Professor of Public Administration in 1947. In geography, R. O. Buchanan got one chair in 1949 and Dudley Stamp the other in 1945. While the lawyers H. A. Smith and Lord Chorley left to retire or to go into government service, Sir David Hughes Parry returned from the Vice-Chancellorship of the University of London, and in 1948 L. C. B. ('Jim') Gower was appointed to the Cassel Chair in Commercial Law. Together with younger lecturers like John Griffith, John Mitchell, Stanley de Smith, and Bill Wedderburn and under the lasting influence of Professor Kahn-Freund, Gower began a remarkable new tradition of technical rigour combined with reformist zeal. In 1953, on the retirement of Sir Charles Webster, W. N. Medlicott was appointed to the Stevenson Chair in International History. Other departments, notably sociology and social administration, will be dealt with separately.

Before the war, the School had had an academic staff of ninety, twenty of whom were professors. By 1950/1, the total had grown to 154, mostly by an increase in the number of lecturers and assistant lecturers. There were now twenty-nine professors, twenty-four readers, fifty-nine lecturers, forty-one assistant lecturers, and one in the newly created category of senior lecturer. Numbers remained roughly stable for the following decade. This meant that the School now had a staff–student ratio of 1 : 14, or if part-time students are taken into account (as the Director insisted on doing) 1 : 16. The ratio was better than before the war but still indicative of the characteristic if at times unsatisfactory LSE method of teaching by large lectures and classes.

So far as students are concerned, everyone talked about growth, and there was certainly a sense of overcrowding, but in fact the main changes during the remaining years of Carr-Saunders's directorship were not so much in overall numbers as in the composition of the student body. To be sure, the overall total of 2,151 in 1945/6 quickly rose to 3,225 in 1946/7 and 3,742 in 1947/8, thus nearly quadrupling within four years in the experience of those who had come back to Houghton Street from Cambridge; but there it roughly remained, fluctuating in fact between just under 3,400 and just over 3,800 in the next ten years. Even the composition of the total did not change dramatically, though there was a steady and deliberate increase of regular first-degree students from about 37 per cent of the total in 1947 to 41 per cent in 1957, and also of regular higher-degree students from 9 to 12 per cent. The proportion of overseas students rose more significantly during the same period, from 15 to 25 per cent or 901 students in 1956 (to take 1955/6 as a not uncharacteristic year), with significant numbers coming from India (123) and from Asia in general (320), from Africa (144), and from North America (190). The number of Europeans seems modest by comparison (142).

Among the occasional students important changes did take place. The Railway Course was briefly re-established after the war but finally discontinued in 1949. A special course for officials of the Exchequer and Audit Office ended in 1950. The courses for Colonial Cadets and Officers were discontinued in 1950 and 1954 respectively. Evening students raised new questions. They were still an important feature of the life of the School but their numbers had declined by about 100 since the pre-war years, fluctuating now around 450. Moreover, about one-third of these were now working for postgraduate degrees, and about 15 per cent for law degrees. It is not surprising therefore that as early as 1948/9 the Director reported that 'the question of evening teaching at the School has given rise to discussion'. In view of new opportunities offered by the 1944 Act and notably by Local Education Authority grants, 'it might become a question whether the effort expended in offering first degrees in the evening is justified'.

One notable, perhaps unsurprising, yet unfortunate change was the relative decline of women students not just by comparison to the war years, but even to the 70 : 30 proportion of the pre-war period. The proportion now was more like 80 : 20, and lower still among those who read for the B.Sc.(Econ.). The Social Science Department, on the other hand, had more than three-quarters women students. The Director suggested in 1952 that the reason for the overall decline was that growth had happened above all in traditionally 'male' departments; but this is merely a statistical explanation. Probably the post-war professional world was one dominated by men, especially by those who returned from the war, whereas women, who had contributed so much to the war effort at home, returned to the hearth and to menial jobs.

Two new diploma courses were introduced at the end of the war, one in Personnel Management and one in Trade Union Studies. The Business Administration course was started again but languished, with fewer than twenty students a year. In 1948/9 the complex process of fusing the B.Sc.(Econ.) and the B.Com. was undertaken; in the end it worked but its completion demanded much academic time. Moreover, the desired effect of a unified first degree did not last; by 1951 a new B.Sc. in Sociology had been introduced which regularly attracted more than 100 students a year until it was reintegrated into a new B.Sc. of numerous options in 1962.

Why did students after the war come to LSE? What did they do after completing their degree courses? From 1948 the School had a Careers Adviser again who put together figures at least for that third of all students who took a B.Sc.(Econ.) degree or at any rate those who answered his questionnaires. Of these, the largest proportion, more than 20 per cent, went into 'Industrial and Commercial Management (Private)', followed by Accounting (15 per cent), Teaching (10 per cent), and the Civil Service (10 per cent). Actually, these figures fluctuate too strongly from year to year to be wholly reliable, but they probably give an indication. Academic research and university lecturing account for another 10 per cent. Smaller though still significant numbers chose market research, banking, public management, and social work.

The Director meanwhile was preoccupied with two issues: one very much his own, student hostels; the other shared with almost all his predecessors and successors, accommodation generally and overcrowding in particular. Neither preoccupation led to great improvements in Carr-Saunders's time though he prepared the ground for his successor. In fact, the hostel story started with a disaster. The School acquired, in 1947, a number of houses in Cartwright Gardens (of Canterbury Hall memories), only to discover certain 'defects' a year later: 'ceilings fell in, the roof leaked, much plumbing was required, and more troublesome still the electrical installation had to be rewired'. When the District Surveyor 'called for the demolition and rebuilding of certain outside walls', the School gave up. Fortunately, new and better houses became available a year later in Endsleigh Place in Bloomsbury, where gradually a School enclave emerged in which over a hundred students were housed ('Passfield Hall') and offices were found for research activities.

Accommodation at the central site was much harder to find and led the Director to ever new dramatic statements as well as a question which accompanied the School throughout the rest of its first century:

> The buildings are greatly overcrowded; there is overcrowding in the Refectory, the Library and in every part of the School. It follows that, unless we increase our accommodation, numbers must be reduced. Even if the accommodation problem could be solved, there would be the problem of the best size of the School, having in mind that permanent expansion beyond a certain limit may bring inescapable loss to certain aspects of the life of the place.

Thus in 1947. And again in 1948, when the Director noted that 'the Library reading rooms, the lecture rooms, the study rooms, the Students' Union premises are grossly overcrowded; the members of the teaching staff have no longer each a separate room', while a refectory with 220 seats has to cater for 1,200 lunches. And so it went on. 'The situation at the end of the session was therefore most grave' (1949). 'The working conditions of members of the staff were deplorable' (1950).

But Carr-Saunders had little progress to report. Most of the time, palliatives were all he could offer. A few rooms were built on top of the existing buildings. Some space was found in Portugal Street and in Clement's Inn Passage. The Royal Statistical Society moved out of the Smith Memorial Hall, which helped. In 1955 the Director was at last able to report success. The Governors had acquired the lease of St Clement's Press on the other side of Clare Market, and the Government Chemist next to it had at last decided to move out. This meant an addition of two-thirds of the 55,000 square feet which, by its own calculation, the School needed. But alas! all this was still in the future and would take three years at least, in any case until after Carr-Saunders's departure.

Finance and Other Certainties

During this period, the Director did not report any financial problems. In 1947 the School had benefited from the first post-war Quinquennial Grant, a fixed allocation in annual instalments by the University Grants Committee (UGC) through the Court of the University of London, which allowed detailed planning. The grant, the Director commented in its last year, 'was on a lavish scale and made possible a great expansion of the universities'.

However, unlike Beveridge, Carr-Saunders had a tendency to be a little too sanguine in financial matters, and he was fortunate that in the climate of public expenditure in which he operated this pardonable sin did not rebound on him. Actually, in his first ten years, including the war, the School's accounts regularly showed a slight surplus. Children of a later age, rather less appreciative of universities, note with surprise the generosity of government in wartime, when the income of the School declined by no more than one-third whereas its staff shrank by two-thirds and its students by even more.

The School's accounts between 1940 and 1945 regularly showed income and expenditure (other than unappropriated or undisposed income) at about £100,000. Nearly half the income came from 'parliamentary and local authorities' grants', plus another 20 per cent from tuition fees. Of the remainder, another 10 per cent came indirectly from government, for example the Ministry of Works' rent payment for the London site. This left 20 per cent of the income from a variety of trusts and with differing grants from year to year, but including the Cassel, Leverhulme, Stevenson, and other foundations, donations by the Commonwealth Fund, and the still important Rockefeller Endowment income which amounted to 6 per cent of the total.

On the expenditure side, somewhat surprisingly, salaries for teaching staff accounted for rather less than half of the total, though if one adds the cost of departmental and research staff, of administrative and library salaries, as well as the payments to members of the School on war leave, a slightly more likely proportion of more than two-thirds emerges. However, considerable expenditure was incurred not just for scholarships, but for printing and stationery, travelling and billeting, books and book-binding, and of course the maintenance of premises. Still, the School was easily able to cope with its tasks.

After the war, in 1947, the first post-war Quinquennial Grant came into operation. By that time, the School's income and expenditure were more than double the war-time level. The session 1947/8 was also the last year of surpluses for some time. A typical year during the quinquennium shows a very different set of accounts from the wartime or even the pre-war experience. In 1950/1, of a total income of £340,000, 70 per cent came from the UGC, another 2 per cent from other government sources, and 19 per cent from tuition fees, many of which were paid for by government as well. The accounts showed a deficit of £16,000, but no one was worried. The School, like other universities, had become publicly financed, although the nearly 10 per cent of its income from trusts, foundations, and one-off donations was still important. On the expenditure side, 51 per cent was now spent on academic salaries, plus another 15 per cent on administrative and library staff.

The new pattern was to last for a long time, though in the 1960s capital grants added an important element. When the School made its Quinquennial Statement for 1952–7, it put the accommodation problem first. Beyond that it emphasized its desire 'to cultivate more intensively those fields of study that are already within its territory, especially history and philosophy, rather than to enlarge the area of study'. The Director made a special point of insisting on an improvement of the staff–student ratio. This time, his statistical artwork not only denied the validity of official UGC calculations which gave the School fourteen students for every teacher while he substituted 16 : 1 as the, to him, correct figure, but he argued that if all part-time, intercollegiate, and other irregular students were taken into account, the figure should be 20 : 1. When the Quinquennial Grant was announced Carr-Saunders concluded that 'the action of the Government was most generous'. The School received £300,000 in 1952/3, rising by annual increments to £355,000 in 1956/7.

Carr-Saunders's pleasure was premature, if only because of the new phenomenon of inflation. Between 1913 and 1945 the retail price index had no more than doubled; for significant periods since the early 1920s prices had actually fallen. Now a new age had begun, with prices rising by 50 per cent between 1945 and 1955. In 1956/7 alone the School was in receipt of supplementary UGC grants amounting to 10 per cent of the total nominal subsidy. But the principle that universities including LSE should be publicly financed had if anything been strengthened; in Carr-Saunders's last year the School received 75 per cent of its recurrent income from the UGC and 15 per cent from tuition fees. It was nevertheless fortunate to have a number of major additional sources of income from trusts and foundations, some contributing more than 1 per cent of the School's total income each: the Rockefeller Endowment, the Cassel Trust, the Ford Foundation, the

Leverhulme Trust, and the Nuffield Foundation. Still, this was a different and much easier game from the one which Beveridge had to play when a mere 40 per cent of the School's income came from government grants.

The picture emerging from such facts is above all one of normality, overcrowding and all. The School had found its post-war balance, and had done so quite early, by 1948, perhaps, when the Director felt it 'correct to say that the School is firmly re-established'. Teaching was back to normal. Research could now at least partly be financed out of general funds, though foundations like Nuffield and Carnegie gave generous help. The Students' Bookshop had been left behind in Cambridge; but in 1946 the *Economist* newspaper and LSE had set up the Economists' Bookshop, for which the School provided premises. For decades, and perhaps to the present day, the Economists' Bookshop provided the first port of call for overseas alumni as well as a welcome resource for members of the School. (In 1991 the company was taken over by Dillons.) Student life was back to normal too. The Students' Union was active; it now published the news-sheet *Beaver* as well as the *Clare Market Review*. Numerous clubs and societies organized their events. Given the preferences of the Director, there was more music, and mountaineering, of course. Important visitors came from all over the world, including the new leaders of former colonies. A Commemoration Ball in the newly built Royal Festival Hall commemorated the sixtieth anniversary of the School in 1955. If all this was pleasingly familiar to the School's long-standing members and friends, the academic centre of activity had shifted somewhat, though even that almost imperceptibly and undoubtedly depending on the vantage point of the observer.

Before we return to this shift, one event needs to be mentioned. In March 1947 the second Director of the School, Sir Halford Mackinder, died, and the Director found gracious words to remember the 'lecturer and teacher of great distinction and influence'. A few weeks later, on 13 October 1947, the founder, Sidney Webb, passed away. R. H. Tawney gave an address, and the Professorial Council passed a resolution remembering the man whom they simply called 'the founder of the School'. Carr-Saunders added a characteristically thoughtful and warm comment of his own:

In the nature of things a founder follows with great interest the fate of the institution which he has brought into being. It is a familiar fact, that a founder is often led in his life-time, to attempt to direct the steps of the institution which he has founded upon the path which he thinks it should follow. Our founder was wiser than most and less self-regarding. Having set the School on its feet, he never attempted to impose his own views. He retained a very real and deep interest in all the doings of the School to the very end, and was ready to help and advise if called upon to do so, as I know from my own experience and to my own benefit, but he was free from that vanity and self-importance, which has so often led other men in a similar position to take advantage of their special relation to an institution. The School was indeed fortunate in its founder.[68]

APOGEE OF THE SOCIAL SCIENCES

Sociologists at LSE

Cambridge was not the only university in which sociology as a subject of study was absent if not anathema in the 1940s and beyond. The hybrid word had been invented a century earlier, and the heroes of the discipline had taught, or more often written, at Bordeaux and Turin and Heidelberg half a century ago, but by the end of the Second World War few European universities went beyond allowing one philosopher or historian or just departmental misfit to call himself by that newfangled name. English booksellers listed books like Hirshfield's *History of Sex* under the heading sociology. The exceptions to such uncertainties were mostly to be found in North America, at Harvard, Columbia, Chicago for outstanding examples, and they were most impressive where protagonists of social theory and social research co-operated, even converged. Talcott Parsons and Samuel Stouffer at Harvard, Robert Merton and Paul Lazarsfeld at Columbia provided the models.

And then there was LSE. More by accident than by design it was about to move into the same small premier league. The tradition, had of course, been started by one of the lesser heroes of sociology, Hobhouse, and supported by generations of anthropologists from Westermarck onwards. Morris Ginsberg had followed Hobhouse in more than the technical sense of succession. 'Morals in Evolution' was still a central theme. But then T. H. Marshall had joined, with his combination of social history and interest in social policy, and an almost uncanny knack of intuitively hitting the sociological level of discourse. The volume on *Class Conflict and Social Stratification* which he edited in 1938 is an example, and his 1950 lectures on *Citizenship and Social Class* a classic in empirically informed sociological analysis. Karl Mannheim may have been an uncomfortable bedfellow but he too left traces of the combination of theory and research. Jean Floud, who worked with him, tells the story of students being sent to Fleet Street 'to observe social reality', and then trying to make sense of it in Mannheim's seminar.[69] But the key figure to make the department jell (if not always to enhance its human cohesion) was David Glass, who had come from demography and was now, at the end of the war, especially active in the various research divisions which the 'practical sociologist' (as Lionel Robbins called him[70]) Alexander Carr-Saunders had set up.

The availability of funds helped. A large donation by the mother of a former student, Charles Skepper, who had lost his life during the war, was used to acquire a building for social research in 1949, first in John Adam Street, then in Endsleigh Street. For many years, Skepper House was to be the seat of social and demographic research. In 1950 the old friends of the Rockefeller Foundation returned with a grant for the co-ordination of sociological research. Immediately after the war, the Nuffield Foundation had given a major grant, extended for five years at the rate of £4,000 a year in 1947, which was instrumental in producing the path-breaking empirical study of *Social Mobility in Britain*, which was not actually published in book form until 1954, though articles had been written and drafts were circulating long before that date.

This product of teamwork was path-breaking in more ways than one. It established the centrality of the study of social stratification for almost two decades. It added substance to the newly founded International Sociological Association, of which Morris Ginsberg was a Vice-President and David Glass the leading light on the Research Committee. Even a British Sociological Association was founded nearly fifty years after the first meeting of the old Sociological Society at LSE. All this activity persuaded the rest of the School that a *British Journal of Sociology* was the appropriate successor to the now defunct *Agenda* as well as the earlier *Politica*; the first issue of the *BJS* appeared in 1949. It helped further to bring the names of LSE sociologists to the attention of the academic world. The School soon had three professors, Ginsberg in the distinguished Martin White Chair, Marshall as Professor of Social Institutions, and Glass as Professor of Sociology.

When Glass was appointed, the widely read, thoughtful theorist as well as conversationalist, Edward Shils, returned to America, though his presence continued to be felt and often seen. Jean Floud, a former student of the School, had joined the department as a lecturer in 1946, and Donald G. MacRae in 1945. They were soon followed by T. B. ('Tom') Bottomore, Rosalind Chambers, Ernest Gellner, Julius Gould, Robert McKenzie, and the persistent and persuasive one-woman department of social psychology, Hilde Himmelweit. Others offered classes in sociology, the criminologist Hermann Mannheim, the statistician Claus Moser. There were also relevant neighbours, in anthropology, in political science, in social science and administration. Then there was Karl Popper.

Thus the post-war generation of students found a formidable array of talent in a discipline almost unknown elsewhere. A. H. ('Chelly') Halsey has described the experience in an article entitled 'Provincials and Professionals: The British Post-War Sociologists', and again in an oral history conversation with Colin Crouch. The latter gives, almost incidentally, a moving account of the post-war School from the perspective of an ex-serviceman who had come in 1947 ('I'm leaving out the ten per cent of girls who had come straight from school') to study sociology:

> I had been six years in the Air Force—and I believe that I wasn't the only one who was a kind of chaotic autodidact as a result of that experience. I had been around a good deal, read a lot of *New Statesman*, Left Book Club, Rationalist Press Association material, but had not had any kind of systematic education during that period. So that what the School offered, particularly the structure of the B.Sc.(Econ.) degree as it then was, was apparently a very orderly map of existing knowledge and existing debate about social affairs, economic affairs, political affairs. It was a kind of systematising experience, or an invitation to that. It did offer a kind of intellectual code or key to a world which was already rather well known to the people who were there.

Actually, if it had not been for the eye-opener of life in the forces, the desire to learn, to find out the causes of things, and of course the more prosaic reason of scholarship money, Halsey would never have gone to university. 'Now that describes quite a few people, and therefore gave the School a kind of emotional, sentimental, romantic, home-like feel from the beginning, as well as being as it were a great feeder of an intellectual thirst that was carried rather chaotically into it by those who entered it at that time.'

Sociology was important in this context; indeed 'it became an article of faith that sociology was at the centre of the social sciences',[71] though as long as it was a part of the B.Sc. degree it remained firmly embedded in the wider context of economics, politics, history, philosophy, and statistics. Actually, Halsey himself, though sometimes called a 'sociologist's sociologist', kept the breadth of the LSE social sciences alive in what Colin Crouch and Anthony Heath in a book in his honour called 'his three central preoccupations':[72] social policy including social work, the recognition of sociology as an academic discipline, and sophisticated empirical research. All these were for Halsey anchored in an ethical socialism which he derived from Tawney and about which he has written a book with one of his fellow students at the early post-war LSE, Norman Dennis.

What then was peculiar about LSE sociology in the narrower sense? The combination of what Crouch called 'evolutionary sociology, citizenship sociology as it were', and 'political arithmetic',[73] of Ginsberg, Marshall, and Glass. In his earlier article, Halsey discusses the intellectual influences on LSE sociology students in the late 1940s and early 1950s. They are three,

Marx, Parsons, and Popper; but all three (so Halsey claims) were in a certain sense not really needed, not at any rate as authors of 'theories of society as a totality'.[74] He is certainly right that there was an air of ruthless eclecticism about LSE sociology at that time, though this was helped rather than hindered by Karl Popper's critique of all dogma, propounded with gusto and a strong admixture of polemic in the basement lecture room of the Old Building on Houghton Street.

Halsey also offers a sociology of this generation. They had come from the provinces and learnt in the metropolis the tricks of professionalization which gave them lasting recognition. The list of graduates on which he bases his theory includes J. A. and Olive Banks, Michael Banton, Basil Bernstein, Norman Dennis, David Lockwood, Cyril Smith, John H. Smith, Asher Tropp, and 'three others who were foreigners, Cohen from South Africa, Westergaard from Denmark, and Dahrendorf from Germany'. Halsey has generous and thoughtful things to say about my role at the time and the later tendency to 're-enact' the dilemmas of Max Weber; but my experience has little in common with that particular British generation. When I came to the School in 1952, most of them, including Chelly Halsey, had in fact left, though David Lockwood became a close and influential fellow graduate.

'Native or migrant, they were all initially sleep-walkers,'[75] Halsey says, and that certainly applies to the 23-year-old Dr.phil. from Hamburg who arrived in T. H. Marshall's (his supervisor's) room on the first floor of the East Wing in October 1952, and was almost immediately asked to identify a thesis subject in a discipline of which he knew nothing. I had studied philosophy and classics, written a dissertation on 'the concept of justice in Karl Marx's thought', and decided that I wanted to go to England, and to find out about sociology, whatever that was. Marshall made me read Hobhouse, which I found hard and unrewarding; Jean Floud held an exciting seminar on social stratification; Popper first puzzled, then thrilled me. Before long, I had revised my early choice of a thesis subject, intellectuals, and moved to the other end of the social scale, 'Unskilled Labour in British Industry'. Above all, we, that is the graduates of the years 1952–4, did what Halsey and the post-war generation had done five years earlier: 'We ran our own seminar—very quick institution building for us in those days!'[76]

The key to the 'Thursday Evening Seminar' was a certain dissatisfaction with the prevailing approach to social stratification, coupled with serious doubts about Talcott Parsons's 'structural-functionalism', which was then in the ascendancy. (Parsons came to speak to us on more than one occasion.) We wanted to resurrect a more dynamic, change-orientated sociology without falling into any Marxist trap. For a term, if not longer, we discussed what we called the 'interest group theory of social conflict'. It led, later, to David Lockwood's brilliant analysis of the 'market position' of *The Blackcoated Worker*, and to his empirical research, with John Goldthorpe and others, on *The Affluent Worker*; it also led, by way of *Unskilled Labour*, to my *Class and Class Conflict in Industrial Society*.

Without doubt, sociology had its historical moment in the post-war LSE. For a decade or so, it was 'at the centre of the social sciences', even if not everyone there at the time would admit it or did even notice the fact. Then things began to change. By 1952, almost all the sociologists on Halsey's list, including himself, had already left for the 'diaspora', Birmingham and Liverpool, Edinburgh and Oxford. Soon after, if we follow Halsey one last time,

> the L.S.E. itself was passing through one of its phases of institutional self-doubt, and the sociology department was somewhat fragmented. Shils had gone back to Chicago, and was in Manchester in 1952–53. Ginsberg retired, Jean Floud moved to the London Institute of Education, and the unifying and civilizing influence of T. H. Marshall was absent from 1956 to 1960.[77]

Marshall was then head of the Social Science Division of UNESCO.

Perhaps we should not follow Halsey, but the great polymath and also oracle Donald MacRae, the last Martin White Professor of Sociology until his retirement in 1986, who said that 'the history of the Sociology Department at LSE was for a very long time the history of sociology in the United Kingdom.'[78] The department certainly remained attractive and lively throughout the 1950s and 1960s, and while there was no one dominant figure, many—Robert McKenzie and Tom Bottomore, Norman Birnbaum and Percy Cohen, Ernest Gellner and John Westergaard, Ronald Dore and Alan Little, Asher Tropp and Emanuel de Kadt—would have been, and often became, the pride of other departments elsewhere. Moreover there is that other point which Donald MacRae made in a conversation with his colleague, David Martin, and which applies to sociology in particular as well as to the School in general: 'Did we gain many stars? I don't know. But we gained an enormous number of admirable first-rate scholars and on the whole good teachers who contributed enormously to the life of LSE and to its international reputation.'[79]

Titmuss and Social Administration

The Department of Social Science and Administration, which had a natural though not always visible affinity to sociology, underwent its own process of reconstruction after the war. When Richard Titmuss was appointed to the Chair in Social Administration in 1950, a new era began. Titmuss's immediate predecessor as head of the Social Science Department had actually been the sociologist T. H. Marshall, but despite his interest in social policy he had been a kind of viceroy for a not-quite-developed colony of the School. T. H. Marshall had taken over in 1944, when C. Mostyn Lloyd retired (he died two years later). Despite his long tenure of fifteen years, Lloyd had for the most part been chairman in name only, dividing his time at the School with the *New Statesman*, where he was Assistant Editor, and leaving the running of the department to his deputy, Edith Eckhard. She too was now ready to retire and left in 1952. The popular 'Minnie' Haskins had retired in 1944. Sibyl Clement Brown, another influential figure, the specialist in mental health care, resigned in 1946 to go on to other positions; she lived to the age of 93.

By 1950 the Department of Social Science and Administration had a remarkable, almost forty-year history at the margin of the real School. In one sense, this statement is unfair. When it looked outward, the department certainly did not regard itself as marginal. The authors of the 'Highlights in the History of the Department, 1912–1973' in the 1981 volume *Changing Course* make much of the way in which the Department both followed and influenced social reforms. 'The Social Administration Department played a leading part in the developments of this teaching, research and contribution to public policy.'

However, from inside the School the department looked different. One wonders whether it would have been so marginal had it then been called Department of Social Policy (as it was renamed in 1993), or even Department of Applied Social Studies: but Social Science was chosen when the School absorbed the old School of Sociology and the Charity Organisation Society in 1912, and Administration added to the name. Later, it came to be known as the 'Social Admin' Department within the School, though Titmuss was the first to hold a chair with that title, indeed the first to hold any chair in the department. The thirteen members of its staff in 1950 had been given proper lecturer status only a few years earlier. They were for the most part teaching specialist courses in mental health care, child care, other forms of social work, as well as personnel management. Their teaching involved fieldwork as well as lectures and classes; it led to diplomas and certificates.

In the year in which Titmuss came, no fewer than 240 students were enrolled in these courses, but the rest of the School took little notice. Was it because 70 per cent of the students and all but one of the teachers (John

Spencer) were women? Was it still, as T. H. Marshall described it for the 1920s, 'popularly regarded as a convenient place for wealthy mothers to send their daughters to when disturbed by the dawning of a social conscience'?[80]

Even Titmuss had to endure questions in the Senior Common Room from his colleagues, 'how my "good-looking midwives were getting on"'.[81] But after his arrival change was not slow in coming. When the department celebrated its fiftieth anniversary in 1962, Titmuss could point to a proud record of teaching at all levels, including postgraduate degrees, of research and publications, of international influence. There were now thirty teaching members of the department; they included his younger colleagues and pupils Brian Abel-Smith, David Donnison, John H. Smith, and Peter Townsend, as well as the familiar names of Christine Cockburn, Helen Judd, Jessie Kydd, Kay McDougall, Katherine ('Kit') Russell, and Nancy Seear. Student numbers had risen to 300, including a growing proportion of men and many from all over the world.

The department has been fortunate in finding fine and sympathetic historians, including the protagonists themselves: Titmuss, Abel-Smith, Donnison, later Robert Pinker, and *Howard Glennerster* and also John H. Smith, as well as the indefatigable Kathleen Jones, the authors of *Changing Course*, Kit Russell, Sheila Benson, Christine Farrell, Howard Glennerster, David Piachaud, Garth Plowman, and professional historians among younger staff, Martin Bulmer, Jane Lewis, Jose Harris. Margaret Gowing wrote a moving and highly informative British Academy memoir of Richard Titmuss after his death in 1973. From these accounts it is clear that much of the post-war change in the 'Social Admin' Department was the work of one man, but he was responding to two major trends of the time. One of these was the welfare state. Social policy had long ceased to be a question of charity, but it was only during and after the war—not least in the wake of the Beveridge Report—that institutions were set up to correct the social iniquities of the market and to provide every citizen with the opportunity of a decent life. Richard Titmuss had actually written the official history of social policy during the war. His brilliant and much-acclaimed book, *Problems of Social Policy*, was published in the year of his appointment at LSE, and laid the foundation of his own belief in universal entitlements, moderated by discrimination in favour of the most needy, and combined with a high degree of voluntarism.

The other major trend can also be described by one word, professionalization. Pressures for professionalization are always complex, mixing the real need for replacing benevolent amateurs by competent experts with the inexorable force of organization along guild or trade-union lines. The mere existence of academic centres of training plays a part, because within them recognition is based on the professional standing of their products. All this, plus a generous admixture of strong and not always compatible personalities, played a role in a conflict which Robert Pinker has described as 'part of the folklore of social work'.[82]

It began with Eileen Younghusband, the unusual, flamboyant, yet deeply committed social work teacher as well as policy planner. She had been a student at LSE. In the late 1920s, Kathleen Jones tells us, 'Eileen and her fellow students knew (in rough outline, anyway) the sort of society they wanted: it was just a question of getting there. But Eileen had no taste for macro-solutions; politics had no appeal. For her, the answers lay in personal contacts, small groups, steady day-to-day work on a practical level.'[83] It is one of the paradoxes of this story therefore that twenty years later, Eileen Younghusband should have been the one, in her Carnegie Reports of 1947 and 1950, to espouse 'generic' training. By this is meant a 'common core of knowledge' for all social workers. Robert Pinker is kind when he says that 'genericism is not so much a theory as a voyage of discovery in search of the essential core and boundaries of a professional identity.'[84] It could be argued that it is simply the replacement of casework in the field by classroom teaching.

This process may well have been necessary but it led to an unholy row in the department and to what Eileen Younghusband called a 'mucky situation' for Richard Titmuss.[85] The story makes good reading, especially in the semi-dramatic version of David Donnison's chapter 'Taking Decisions in a University', in which he mysteriously gives the two main protagonists, Kay McDougall and Eileen Younghusband the (acronymic?) labels, MH and ASS respectively. It was all very confusing, and probably fully comprehensible only to those who were involved. It appears that MH was the traditionalist with modern sentiments whereas ASS advocated modern methods out of traditionalist predilections. The 'Professor of Social Administration' (as Donnison calls him) may therefore not have been wrong to adopt the proposals for a more professional social work course, and to put Kay McDougall in charge of it. Before this finally happened, however, much blood was spilt. 'Richard Titmuss was clearly one of Eileen's imperfect sympathies,'[86] Kathleen Jones tells us, but so was Kay McDougall. And Eileen Younghusband was not the easiest of people either. As she became more famous, she let others feel that she knew about her standing. An element of rivalry may well have crept into her view of Titmuss and also into Titmuss's attitude to her. Eileen Younghusband was, moreover, like Lilian Knowles in another field and an earlier generation, a *grande dame* among upstarts, which is bound to make for a difficult chemistry. In any case, in 1957 she resigned, though her ideas became the new orthodoxy of social work education.

In the mean time, Richard Titmuss was on his way to becoming much more famous than his colleague. His appointment had been unusual, for he had had no formal schooling, no A or O Levels or their equivalent, let alone academic training. The son of a small farmer and haulier who went bankrupt and died early, he became an insurance clerk and looked after his family, while taking a growing interest in social problems. He published books on population and on poverty, including one on *Poverty and Population*, and finally resigned from the County Fire Insurance Office when he was asked by Keith Hancock to write the official war history of social policy. His appointment at the School was suggested by T. H. Marshall and turned out to be a great success. In the twenty-three years to his premature death from cancer at the age of 66, he not only built up his department, but got involved in the development of the Health Service and other aspects of the welfare state, as well as assisting other countries, notably Mauritius and Tanzania and also Israel, in developing their social policy. He served on many committees, including the One Parent Family Committee chaired by the LSE alumnus and lawyer Morris Finer. All the while, he remained a scholar. His books, especially *The Gift Relationship* on ways of blood giving for transfusions, were models of strategic social analysis, taking a specific instance, studying it in great detail, in order to make a general point, in this case about markets and altruism.

The chair which Titmuss held was designated as one in social administration. On the Continent, the word social policy would have been deemed appropriate, *politique sociale*, *Sozialpolitik*. Titmuss liked the term; his course of lectures published after his death is therefore entitled *Social Policy*. Social work was never his central interest; he wished he could have integrated it into a wider notion of the institutions of distributive justice. Unfortunately, such integration did not happen, nor did 'social administration' become a part of public administration as taught in the Department of Government. In the end, the greatest weakness of Titmuss's strength may have been that the Department of Social Science and Administration became a microcosm of the School at large, but no more than a *micro*-cosm: it taught everything from its own perspective, economics and political science, sociology and social history, but its parts remained the little siblings of their bigger brothers and sisters in the other departments.

Richard Titmuss has been called a saint. But A. H. Halsey was right, 'in fact, he was no saint, but a secular agnostic', though also, 'in Sir Edmund Leach's phrase, "the high priest of the welfare state"'.[87] It is thus more appropriate to describe him as a social democrat in the best sense of that word. Politically, Titmuss started, before the war, as a Liberal, but soon turned to the Labour Party, to which he remained committed. But party is not the point. 'He had faith in man himself and in the possibilities of social democracy,' Margaret Gowing tells us. This is clearly important: 'he was sceptical, even cynical sometimes, but was never a sceptic much less a cynic.'[88] Titmuss the social democrat had compassion and translated it into a practical commitment to reform. He was not guided by a utopian dream but by sense of what constituted a decent life in the real world of men, indeed of Englishmen. (With all his international exploits, his country, England as much as Britain, mattered a great deal to him; he strongly opposed British membership in the Common Market.) In his own words, Titmuss wanted to show 'what a compassionate society can achieve when a philosophy of social justice and public accountability is translated into a hundred and one detailed acts of imagination and tolerance'.[89] He accepted and when necessary defended institutions. Those who were surprised by his stance in the 1968 'troubles' at LSE (when he was a member of the Standing Committee of the Governors) had not understood him; he saw the very existence of the School threatened and rallied to its defence. In the process he even made common cause with the much-disdained 'economists', though the social democrat Titmuss continued to believe in the benevolent state and human altruism rather than the market and self-interest.

The department which Titmuss left behind still offered Diplomas in Social Administration, Social Work Studies, and Personnel Management, as well as Certificates in Mental Health Studies. It had, however, become a serious research department as well. Ernest Gellner was therefore indeed inaccurate and discourteous (as Martin Bulmer pointed out to him) when he revived the old canard of the 'Distressed Affluent Folk's Aid Society' in an article in 1980. Gellner makes a more important point when he says of the 'important, cohesive and quietly influential department' that it 'probably comes closest to the Founders' intention . . . of approaching Socialism through patient research rather than through abstractions and propaganda'.[90] Even this, however, is—or was, because much change has happened since Titmuss both in the world at large and in the Department of Social Science and Administration—at the most true in fact, and not in intention. It could be argued that the very ideas of social work and social administration are a political statement, and one that is close to the foundations of LSE. The riddle remains why this great department of Applied Social Studies has never captured the heart let alone occupied the core of the School. The answer is probably that Sidney Webb was but one of the founders, and even he was of two minds, of which the other, non-Fabian segment, the one of *rerum cognoscere causas*, proved the more durable.

Models and Values

All social sciences did well after the war, but none more so than those called by an unkinder age, twenty years later, the 'soft' disciplines, sociology, social policy, social psychology, social anthropology. Anything including the word 'social' came to be regarded as 'soft', which left economics, law, history, even politics (especially 'government') on the 'hard' side.

Yet it was a sociologist of sorts, A. W. H. ('Bill') Phillips, who produced the most memorable piece of hardware at the School in 1950. The New Zealander and trained engineer had come out of the war at the age of 32 very much in the Halsey mood, so that in 1946 he registered for the B.Sc.(Econ.) with sociology as a special subject. 'He embarked on it', his wife Valda explained later, 'partly as a result of being unsettled about his future at the end of the war and intended it merely as an adjunct to future engineering work.' He did not do well, ending with a pass degree, though Lionel Robbins excused this lapse by 'his somewhat lamentable experiences during the war when confined to a Japanese [POW] camp, he contracted so strong a habit of chain smoking that, without cigarettes in an examination room, he was completely at a loss after an hour'.[91] It is as likely that he lost interest in the theoretical imprecisions of sociology as he discovered the temptations of equilibrium economics. In any case, in 1949, and with encouragement by the imaginative and always unorthodox James Meade, he built the first Phillips Machine of national income flows.

The Machine, later improved and produced in about a dozen exemplars at the price of £1,536 in 1952 money, was a masterpiece in engineering and economics. Red liquid running through its tanks, pipes, and valves demonstrated the flow of income as well as the factors which diverted and stopped it.

In terms of simple national income accounting, total income, Y, enters at the top; taxes are siphoned off leaving disposable (i.e. after-tax) income; saving flows out of the central column, leaving consumption spending. To consumption, C, is added investment, I, (flowing in from the right) and government spending, G, (flowing in from the left) to give total domestic spending, from which imports, Q, are then deducted and exports, X, added. The machine thus shows visually the equilibrium condition

$$Y = C + I + G + (X - Q)$$

which should raise at least familiar echoes for those who once-upon-a-time did first year economics.[92]

Nicholas Barr not only gave this description but also helped, many years later, reconstruct the machine which, with computers coming in, had fallen into oblivion. At the time, that is in the early 1950s, it clearly was both a source of merriment and of instruction. It even helped reveal the dispute between Keynes and Robertson about the determination of interest rates as spurious. It also got Phillips first a lectureship, later the Tooke Chair in Economic Science and Statistics; though the inventor of the Machine, as well as the Phillips Curve linking inflation and unemployment, returned to the antipodes in 1967, where he died, at the age of 60, in 1975. 'The Phillips machine, like some strange creature striding through the life of the School from 1949 onwards, has left footprints involving some of the LSE's greatest names: James Meade, Lionel Robbins, Abba Lerner, Bill Phillips himself, and in later years Harry Johnson.'[93] A caricature of it even made *Punch* in 1953, after a budget which no one understood, 'while all the time, tucked away in Houghton Street, W.C.2, is a creature capable of clarifying the whole situation before the man in the street could say John Maynard Keynes'.[94]

While the economists were calculating and building models, the lawyers were doing nothing of the kind. On the contrary, they were beginning to form a Law Department unlike all others in the country. The process had started before the war. 'No department which in the 1930s contained people like Otto Kahn-Freund, Ivor Jennings, William Robson, Theo Chorley, and Theo Plucknett was likely to be conventional.' John Griffith, who wrote this, was himself one of the main advocates of the assumption 'that law is one of the social sciences and that law teaching should embrace the whole range of legal regulation and provision'.[95] When Sir Alexander Carr-Saunders passed the directorship to his successor in 1957, L. C. B. Gower had been a professor for nine years, Griffith himself was a reader, as were D. H. N. Johnson and Stanley de Smith; Cyril Grunfeld, Eryl Hall Williams, Michael Mann, and Olive Stone were lecturers; and of course Professors Kahn-Freund, Hughes Parry, and Plucknett were still active.

Even five years earlier, in 1952/3, the Calendar of the School makes exciting reading for the historian, as it did for the student at the time looking for opportunities of instruction or even edification. Twenty-five of its 166 pages announcing 'lectures, classes and seminars' were devoted to economics. In geography, Mr Dan Sinclair and Dr Michael Wise had arrived to support Professors Buchanan and Stamp. History, Constitutional, Economic, and International, was now taught by Professor Ashton and Miss Carus-Wilson as well as Messrs Lance Beales, Jack Fisher, George Grün, Arthur John, Walter Stern, and others. Professor Popper had been joined by Dr John Wisdom to teach 'The Methods of the Natural and of the Social Sciences'; John Watkins was soon to join them. Professor Manning still dominated the subject of international relations, though Geoffrey Goodwin and others helped him. In what was now called 'Politics and Public Administration', the names of Oakeshott, Greaves, Pickles, and Smellie were familiar enough, though new ones had appeared, like Laski's pupil Ralph Miliband, the Australian scholar Peter Self, the student of trade unionism Ben Roberts, the Commonwealth Studies champion Morris-Jones. Anthropology (listed under 'Sociological Studies') was now represented by Professors Firth and Schapera as well as Mr Maurice Freedman, Dr Edmund Leach, and Lucy Mair. Statistics and mathematics continued to be a strong subject, with Professors Allen and Kendall as well as Jim Durbin, Eugene Grebenik, Claus Moser, and Alan Stuart, all of whom were to become professors before long.

Thus the School was well prepared for the apogee of the social sciences in the 1950s and 1960s. There still was no other place in Europe which could effectively compete with LSE in the breadth and depth of subjects available to students and to teachers. In fact, there was no better place to go for anyone interested in social science. But in one important respect, Sir Alexander Carr-Saunders differed from his predecessor. William Beveridge's School had been outstanding above all by its teaching. Beveridge himself kept on hankering after research opportunities but the great School which he built was at its best in the lectures of Laski, the classes of Robbins, the fieldwork excursions of Malinowski. Carr-Saunders from the beginning showed a systematic interest in research. He developed the Research Divisions, gave much thought to the need for sabbaticals for members of the staff, and showed special pleasure at the success of research projects.

Under his directorship, the annual Calendar included a section on research. 'Until 1947 the School was not in a position to finance research out of its own funds, and was thus dependent on the generosity of benefactors.' The statement, in the 1952/3 Calendar, sounds like a slightly grudging recognition of the Rockefeller Foundation, though foundations continued to be important sources of research finance. The 1952/3 Calendar lists some important projects: Professor R. S. Edwards's studies of the links between industry and research (financed by the Manchester Oil Refinery); the follow-up of the Social Mobility study into educational selection

(Nuffield); the Population Investigation Committee and its journal, *Population Studies* (Rockefeller); Professor Kendall's investigations into research techniques (Nuffield); the electoral survey in Greenwich conducted by R. T. McKenzie and others (Elmgrant Trust); Labour Party history (Passfield Trust).

The School's publications reflected the research done by many individuals. Both *Economica* and the *British Journal of Sociology* were doing well. The *London Bibliography of the Social Sciences* listed new acquisitions by the Library and became a standard source for others. The series of books sponsored by the School was now published by Longmans, Green & Co. Departments had their own series of Studies, including the important 'Monographs on Social Anthropology'. The annual Hobhouse Memorial Lectures were also published. The list of publications by members of staff attached to the Director's Report lists some 200 titles in a single year, which would have pleased even a more quantitative age of academic assessment. Publications ranged (from August 1950 to July 1951) from 'Kinship and Marriage among the Tswana' to 'The Equalization of Factor Prices', from 'How Many Can Climb the Social Ladder?' to 'The Classification and Use and Misuse of Land', from 'The Papacy and World Peace' to 'The Revocation of Testamentary Appointments on Marriage', not to forget 'Zwei Erzählungen Franz Kafkas: Eine Betrachtung' and 'Le Plan Schuman devant l'opinion britannique'.

A GOLDEN AGE FOR STUDENTS

Fun, Games, and Hard Work

Whatever the 'soft' social sciences did not do, they did provide the magnetically attractive setting for a decade and a half in which it was fun to be a student at LSE. When Lionel Grouse contrasted, in the Lent 1949 *Clare Market Review*, 'the drab, dour seriousness of contemporary student life' with 'the extravagances of mind and body of the 'twenties', he was perhaps a little romantic. What he had in mind was that the 1920s generation had memories of better times, it 'bathed in the autumn sunshine of the Golden Age', whereas after the Second World War students only remembered 'the twin shadows of war and social unrest' and sensed that 'our known world was crumbling about us'. The hegemony of the old middle classes was drawing to a close. As one reads the accounts of LSE students at the time, other and different sentiments are equally striking. Many shared Robert McKenzie's 'sense of excitement' at finding himself 'in London *and* at LSE immediately after the war',[96] when a government of social reform had come to power. Later in the 1950s, people benefited from a muted but none the less pleasing economic upturn. In any case, Ron Moody spoke for many when he sighed that 'LSE was a buffer against the grim reality outside, my deep sentimental attachment to the buildings and rooms and students gave me a self-contained and complete way of life'.[97]

Almost forty years later, Anne Bohm looked back on her early years as Secretary of the Graduate School and wrote: 'Altogether, looked at from my standpoint and my age, it seems to me that those years and the whole period from 1945 to the late fifties was a golden age for graduate studies at LSE.'[98] She might well have added undergraduate studies as well. Students flocked from the four corners of the world to read for the B.Sc.(Econ.) (New Regulations), with a heavy dose of Laski, Smellie and Oakeshott, Ginsberg, Popper, and of course Robbins. Quite a few stayed on to begin, and some even to complete, graduate degrees. They worked, and they had fun. These were the years in which the myth of LSE was born, and like any good myth it had a solid base in reality.

Joan Abse's book *My LSE* assembles a notable variety of accounts of the period. Robert McKenzie came as a mature student in 1947 and stayed as a teacher until his premature death in 1981. From this detached view he sees the intellectual excitement of LSE concentrated on the 1930s and 'the great moment of change just after the Second World War'. He notes the essentially conservative nature of the School, which defies appearance and reputation. Not unnaturally, given his own predilections he also remembers as 'one of the great joys of LSE' the ease with which one could try out journalism, broadcasting, even practical politics.[99]

Ron Moody tried out other skills. He came as an undergraduate in 1948 to read sociology, and loved it despite the ultimate disappointment of a Lower Second (which did not prevent him from staying on for a while as a postgraduate in psychology). However, by then his artistic temperament had got the better of his scholarship. As 'Juma Krash' he had drawn cartoons for *Beaver* before he got hooked on the idea of Continental-style cabarets or revues. On 15 November 1949 *Place Pigalle* was first performed before an enthusiastic crowd in the Old Theatre. 'The cheers and laughter seemed to go on forever and probably did.'[100] Bernard Levin's impersonation of Laski was especially remembered. Later Levin did one better and performed a marvellously amusing debate of Laski with his several other selves. In fact, Laski was omnipresent during these years. Moody captured the sense of many in his moving cartoon of the ghost of Laski in a room full of students at work. In the decade after his death, his presence dominated not just LSE but the image of the School outside.

Jacqueline Wheldon—Jackie Clarke before her marriage to the LSE alumnus and Chairman-to-be Huw Wheldon—was lured to the School by Harold Laski, who told the young Labour Party activist that she should take on a secretarial post at the School and work for admission as a student in the evenings. This she did, and the School captured her mind and soul. She wrote the wittiest account of how 'LSE pulsed with the pulse of the capital': the House of Commons (Richard Crossman, Ian Mikardo); the *New Statesman* (Kingsley Martin, Krishna Menon); the presses of Fleet Street; the Law Courts and Lincoln's Inn; 'Dickens's London was all round us'; walks into the City; Bush House and the foreign service of the BBC; second-hand bookshops and pubs and theatres, concerts and art galleries; 'Johnson's House, Sir John Soane's extraordinary and endearing museum, the British Museum'. At night, when only a few evening students were left, the School felt like the heart of all these riches. 'Then the bones of its history and purpose seemed to shine out in its dusty, littery, unexalted corridors and almost deserted library. It was a time to be grateful and pleased not in a spoken way.'[101]

Not all were equally pleased with the 'dusty, littery, unexalted corridors'. When Bernard Crick was faced in 1949 with the choice between LSE and University College, he chose the latter. LSE 'looked to me just like my father's insurance office. The portico at UC looked what I expected a university to look like.'[102] Carr-Saunders, to whom this admission was made, did not dissent. Chaim Bermant, coming from Glasgow in 1955, went even further than Crick—'I hated the place'—and used the same imagery. 'The first shock was the building itself, a charmless pile, like the head-office of a minor insurance company, hidden away in a back street.' But then LSE never was about its buildings. Even Bermant eventually 'came upon what is, of course, the redeeming feature of the LSE, the quality of its teaching'.[103] Bernard Crick suddenly checks the flow of anecdotes about Robbins and Beales, Smellie and Robson and the fellow students of the Chess Club with Bernard Levin as secretary, John Stonehouse as chairman, and Ken Watkins as treasurer. 'This essay has become all persons. But that is how LSE seemed to me.'[104] As one ponders the names of these persons one wonders about the young New Zealander who, arriving at the School in 1952, found that 'LSE educated me, but it did not excite me vastly'. In fact, 'LSE then cooled me down rather than heated me up.'[105] Perhaps he too felt that the 1930s and the immediate post-war years were intellectually strongest. Over the years, however, Ken Minogue warmed to the School, where he became a Professor of Government in 1985.

Those golden years also saw the flourishing of student societies. By no means all of them were political, but the political ones made a splash. The post-war period was that of 'Popular Fronts' in many countries; the ubiquitous Socialist Society represented the fad at LSE. But as early as 1948 John Burgh, later Sir John Burgh and Chairman of the Governors, helped to organize a Labour slate for Union elections. They did well, except that one of those elected actually turned out to be no more than a bogus name and photograph smuggled into the ballot by Bernard Levin. David Kingsley, who never runs out of ideas to improve the quality of people's lives, was in the end 'grateful for my 2.2 degree considering I spent most of my time at LSE involved in politics—Chairman of the Labour Society, the Students' Union (Carr-Saunders gave me a sabbatical when I was President, as he said "to sort it out a bit") and the National Union of Students, countless other societies and their activities'.[106] The Conservative Association may have been so small that its membership could be accommodated on the stools of the Seven Stars pub in Carey Street, but one of its effective and able leaders was later a Governor and Chairman of the Finance Committee, Ian Hay Davison, who left *Beaver* in no doubt about his views in 1952. 'On the economic position Mr. Davison remained essentially conservative. (No Butler's boy he.)'[107] It was all great fun. When the 'Clare Market Parliament' elected a 'Government' 'Rt. Hon. David J. Kingsley, M.P.' became Lord President of the Council, 'Rt. Hon. John Burgh, M.P.' Minister without Portfolio, and 'Rt. Hon. Bernard Levin, M.P.' Minister with Half-Portfolio.

Were there only men in these golden years? Certainly not. We have heard from Jacqueline Clarke-Wheldon. John Burgh was introduced to Ann Sturge 'on the third-floor landing outside the lift. Ann and I married in 1957.' Rosemary Ellerbeck, also known as the novelist Nicola Thorne, remembers that 'we cared about study, though there was no unemployment, and we also had a lot of fun . . . Each age has its *mores* and whenever I think of those days [1951–7] I think of fun: skipping lectures, debates in the union, trysts in the library or the corridor outside and endless talks late into the night.' Deborah Manley (1951–3) finds as she returns to the School thirty years later that 'it wasn't like that in our day'; everything is now so shiny and modern. She was, like her mother a generation earlier, a student of social administration, 'a close-knit caring department as befitted its calling' yet 'drawn into at least the social life of the School'. Liz Wheeler-Ainley recalls no sit-ins or marches. 'We did have a conscience, though, and went in droves when, organised by David Kingsley, we tried to give practical help to the flood victims on Canvey Island.' Beyond that, 'LSE in the early 50s was a great place for parties.' Liz Wheeler is slightly self-conscious. 'To mention mainly the lighter side of LSE life may seem flippant, but in my case it did have a lasting effect. It was at a party given by Alan Tyrrell that Bob Potter introduced me to another LSE graduate, Barry Ainley, to whom I have been happily married for nigh on 30 years.'[108]

Clearly, a lot of bonding took place at the School in the post-war Carr-Saunders years. A benevolent Director presided over a School which found a rare balance between work and fun. Some of the bonding was personal; it would not be surprising to find that the number of LSE marriages reached an all-time high during this period. Some of it was social, or more often political. All of it led many to remember those years in a warm glow of friendship coupled with intellectual exertions. Few of those quoted in these pages were predominantly 'hard' social scientists. Sociology and social administration, political science and psychology dominated not only the curriculum but extracurricular activities as well. These were not the parents but the older brothers and sisters of the generation of 1968. Perhaps Bernard Crick is right to suggest that if one adds what he calls *les folies de grandeur*, that is sudden growth, to the combination of old and new social sciences, the mixture becomes explosive. As long as it lasted, however, the age was golden, and it has contributed much to the memories and views of LSE which were to dominate the next decades.

A Quiet Transition

Compared to the long farewell of his predecessor, the last years of Carr-Saunders's directorship were uneventful and tranquil. In 1954 Eve Evans decided to retire; her work as Secretary was praised by everybody. Her successor, Harry Kidd, had been Assistant Registrar of Cambridge University; he was to become one of the key figures in the 'troubles' more than a decade later. Every now and again the Director made moves to replace the School's Articles of Association by a proper Charter with Privy Council approval, but when the Privy Council wanted to limit the powers of investment of LSE Ltd., the plan was quietly put in abeyance. The enormous needs and modest successes on the accommodation front have already been mentioned. In his last year, Carr-Saunders conducted a survey about student demand for hostel places. Of 614 men who replied, 305 did not want to live in a hostel; 'dislike of institutional life, dislike of restrictions and of fixed meal times, difficulty of working in hostels' were more effective deterrents than the attractions of college-like community living.[109]

On 31 December 1956, two weeks before his seventieth birthday, Sir Alexander Carr-Saunders retired from the directorship of the School. In his nineteen years as Director, LSE had survived the enormous upheaval of the war undamaged if not altogether unscathed. It had returned to Houghton Street and rediscovered its personality under changed conditions. For changes occurred but they were neither loudly proclaimed nor widely advertised. One such change concerned the financial basis of the School. For the next quarter-century it was to be guaranteed by predictable public subsidies to the institution and its students. Even inflation was usually compensated by additional grants. The School, like other universities, had become a part of the growing public sector. Another change was the gradual fading-away of some of the special features of the early School, courses like those in railway studies, evening students, even mature students, in favour of a normal university structure of regular undergraduates and regular graduates. Something else was new. Throughout the post-war years, social sciences were popular, with applicants to the School, with their parents, with governments, with public and private employers. Occasional shortages in graduate job opportunities turned out to be brief hiccups rather than conjunctural downturns.

All these changes were in fact so gradual, incremental even, that those who lived through them almost failed to notice them. Again, the personality of the Director helped. We must beware of painting it in grey. Carr-Saunders was not a boring man, 'his character was full of surprises'. Professor Phelps Brown added: 'The mien of the saddhu masked keen powers of observation and an exceptional capacity for rapid work.'[110] Carr-Saunders used these to live at least three lives. The first was the long period of varied preoccupations which his son Edmund called, 'indecisive'. 'He couldn't decide what to do.'[111] He travelled, discovered his love for the visual arts as well as for mountaineering, was put off course by the First World War; yet somehow he found a path of scholarship which led him by detours and zigzags to the real social biology, the study of population in all its biological, statistical, and social aspects. The appointment to the Charles Booth Chair of Social Science at Liverpool in 1923 may have surprised him, but had its logic and led him deeper into social research and the analysis of basic facts of social structure.

The second life was that at the School. But a third one began in the middle of the war, in 1943, when Carr-Saunders was appointed to the Commission chaired by Sir Cyril Asquith on Higher Education in the Colonies. The Commission reported in June 1945 with a blueprint for university development in British Africa and Asia. Carr-Saunders himself remained deeply involved in the process, through two institutions which he had created and helped direct until long after his retirement, the University of London Senate Committee, which gave many of the new university colleges London accreditation and more, and the Inter-University Council, which did the same for the whole country. It is worth mentioning that Carr-Saunders was instrumental in setting up, among others, the universities at which his two successors at LSE were Vice-Chancellors when they were appointed Directors, the University of Malaya (Sir Sydney Caine), and the University of Rhodesia and Nyasaland, later of Southern Rhodesia (Sir Walter Adams).

Carr-Saunders had strong views of what universities are about, and they shine through the Asquith Commission Report. 'The education provided must be neither rigidly directed to the training of recruits to the professions nor so disdainful of practical needs that its products are unequipped for useful service to the community.'[112] At home, he insisted even more emphatically on academic autonomy from practical demands. 'If I were asked what the true purpose of a university is, I could only echo Newman, and say that, if a practical end must be assigned to the university course, it is the making of good members of society.' Even in 1959, Carr-Saunders disapproved of 'the position into which our universities have drifted', which 'is hardly such as to promote this end, so busy are they and so anxious to justify themselves in ways which the public can understand and appreciate'.[113] As Director of LSE, his sole concern was the integrity of the School as a university institution. He had no axe to grind, nor did he have to suppress any hidden or obvious political motive. If it is true, as Lionel Robbins suggests, that the High Anglicanism which he acquired in middle life made him 'persistently undervalue the contribution of the Philosophy Department', this is really one of his 'few blind spots'.[114]

There is no better description of Carr-Saunders's approach to the directorship than that by his own hand, in response to a grateful letter by one of Esther Simpson's brilliant younger-generation exiles, the 1949 President of the LSE Students' Union (and later Director-General of the British Council), John Burgh. Carr-Saunders's letter of 28 March 1957, three months after his retirement, is worth quoting in full:

Dear Burgh,
It was most kind of you to write and what you said was very interesting and welcome. In one aspect a college is a great machine which must be made to run smoothly. There should be no jolts, no situations; happy in one sense is a college which has no history so long as absence of memorable events is not evidence of low vitality. In another aspect a college is a community, mostly of young people, who are liable to suffer from strains to a greater extent than any other section of society. In spite of outward appearances, healthy, carefree exuberance and so on, there are all the pains of the discovery of self by the intellectually gifted and all the

anxieties about the future arising from the peculiar situation of those whose entry on a career is postponed far beyond the normal age. Probably students in general enjoy more intensely and suffer more acutely than others of their age. All this is very much in the minds of those having responsibility in colleges. For my part I think it important that the atmosphere of a college should be equable so that students should put down roots; it should be such that they come to realise that the college authorities recognize responsibility towards them, but are opposed to paternalism, and have some compassionate understanding which must be made evident otherwise than directly. It is one thing to set this out, and another to bring it into being. I certainly hope that in some measure some students of the School have experienced something of this. Apart from the smooth running of a college which benefits staff and students alike, more can be done, I believe, by the head of a college for students than for staff. Staff can be helped and encouraged, but their most formative period is over. It is the students who are in the mind of a head of a college, and he knows that all he can do is very little and that it is best confined to creating, so far as he can, the sense of a society whose members recognize their special responsibilities which derive from their several situations.

Nothing gives me more pleasure than to hear of students who have benefited from their time at the School, and that is why I was so delighted to get your letter.

Yours sincerely

A. M. CARR-SAUNDERS[115]

It is hard to say what to admire more, the sentiments expressed in the letter, or the words which its author found to express them. As anyone who met the Director can testify, these were not empty words.

Carr-Saunders's equanimity sometimes made him seem aloof. He liked to keep a certain distance from people; 'dear Burgh' was his normal form of address; Anne Bohm remembers that the Director 'did not like people being called by Christian name at all'.[116] Some would say that too little emotion shone through his calm. His children were certainly a little puzzled; but then Carr-Saunders seems to have liked institutions as much as his family. He certainly served the School with distinction. 'His authority was based on a confidence unalloyed by superiority, and on objectivity and patience in all personal dealings.'[117] Professor Phelps Brown had many nice things to say about his former Director, as did others, almost without exception. Carr-Saunders died nearly ten years after his retirement, on 6 October 1966, 'after pushing his car', as the obituarist cautiously put it. It appears that the octogenarian had actually tried to push his car uphill after a breakdown in his favourite Lake District.

Sir Sydney Caine, Sir Alexander Carr-Saunders's successor, had in fact been appointed in March 1956, nine months before Carr-Saunders retired. This was the first time that the directorship was handed over without drama and hurried last-minute arrangements, indeed without Sidney Webb. There was some drama behind the scene, to be sure. Other names were considered by the selection committee, which is proper. Some wanted Geoffrey Crowther of *The Economist* to come. But when Sydney Caine agreed to leave the Vice-Chancellorship of the University of Malaya and come to the School, there was also pleasure at having for the first time a former student and current Governor, a true LSE man as Director.

7

THE SCHOOL IN FULL SWING

INFLUENCE AROUND THE WORLD

Normality Reigns

The decade of Sir Sydney Caine's directorship of LSE was marked by a striking and, in the end, explosive contrast. At the School itself the years from 1957 to 1967 were, at least until the mid-1960s, uneventful. The Director found nothing notable to highlight in his annual reports. '1956–57 saw no major events.' 'The year 1957–58 has seen no major new developments.' 'In 1958–59 the School has continued its growth and development on lines already established.' 'The session 1959–60 has seen no major developments or important changes within the School itself.' '1960–61 has not been a year of great changes.' In the world around, on the other hand, these were years of profound if initially almost subterranean change. The 'wind of change' which blew over the crumbling colonial world transformed the international scene. The national scene was also transformed when in 1964 a Labour Government got into power after thirteen years of Conservative rule. As importantly, the 'swinging sixties' pulled and tore at the moral texture of advanced societies. Dismantling institutions, dissolving rules, and abandoning long-held assumptions became the order of the day. When the two met, the uncanny normality of LSE and the sea changes in its environment, there was likely to be trouble. Sydney Caine arrived in the morning mist of a nice English summer day and departed to the first rumbles of the all-encompassing thunderstorms of '1968'. This is the story which needs to be told even if a quarter-century later it still evokes pain for some.

The new Director was pleased with what he found when he came in 1957. His two long-serving predecessors had bequeathed to him a School which he described in his first Report in 1957: the world-wide reputation of a unique centre of the social sciences; effective teaching for the B.Sc.(Econ.) as well as the sociology, law, and history degrees; extensive research activities supported by the Nuffield and Ford Foundations among others; maintenance of 'the essential unity of social studies' in a structure with open departmental boundaries and much unforced interdisciplinary co-operation; an unusual atmosphere of vitality and debate within an essentially friendly community. 'Perhaps nothing is more typical of the School than the bustle and ordered confusion of the entrance hall throughout the day.'

Sydney Caine saw no reason to change this heritage. In part, such reluctance reflected his personality. He had some of the best virtues of the traditional public servant. He was straight, clear-headed, eager to assemble all relevant facts before reaching a judgement, suspicious of flights of fancy, and conscious of the task to preserve a tradition which transcends all temporary hiccups and deflections. Caine was also deeply committed to LSE. Born in 1902, he had been a student at the School from 1919 to 1922. He then embarked on a civil service career which he spent partly in the Colonial Office, and largely at the very heart of Whitehall, in the Treasury. In 1952 he was made the first Vice-Chancellor of the University of Malaya. Throughout, he stayed in touch with the School; in 1945 he had become a Governor. His account of the *The History of the Foundation of the LSE* as well as the sections on the School in his book on *British Universities* show both his commitment and his style.

Caine's reluctance to allow 'major events' to disrupt the normal flow of daily life also resulted from his view of the directorship. When he was asked, on leaving the post in 1967, what his plans had been and whether he had succeeded in realizing them, he seemed surprised and a little bemused. 'In retrospect I recall in 1957 no dramatic plans of reorganisation or schemes of development. . . . So far as I have had personal objectives, therefore, they grew out of the existing situation.' This may not be a viable attitude when one is engaged in setting up a new university, like Malaya, but LSE after all had been there for over sixty years and was strong. 'Very dully, no doubt, in a world in which novelty is so often taken as the supreme value, I accepted—and accept—the basic character of the School.' It would be quite wrong and in any case ineffective for the Director to try and order people around; all he can do is to go out and find funding to increase the 'margin available for discretionary use'. 'Results are then to be achieved by a series of pulls and pushes, giving this project tactful support in the appropriate committees and pouring cold water on the other, rather than by dramatic and large scale initiatives.'[1]

Discounting Sydney Caine's personality and his view of the directorship, the question remains exactly how 'normal' life at LSE was during his early years. After all, more of the same often involves as much change as loudly hailed innovations. For one thing, Sydney Caine began in a climate of public finance which was favourable both to expansion and improvement. The quinquennial allocation to universities from 1957 to 1962 foresaw an increase of 20 per cent in the UGC grant to the School. Much of it went into academic salaries and as a result the Director was able to write of the 'contentment' of academic staff, an unfamiliar condition in later years. Student numbers topped 4,000 for the first time in 1959/60. Chairs were filled when they fell vacant and new ones were created. In the early years of Sydney Caine's directorship familiar names appeared among the School's professors, including David Donnison, Maurice Cranston, Imre Lakatos, A. W. H. Phillips, Michael Wise, Ely Devons, and John Griffith. Of the research units created in these years, the Greater London Group set up by Professor W. A. Robson was to last for the rest of the century. It was, and is, one of the many links of the School with the capital city to which it belongs.

Other changes concerned individuals. In Sydney Caine's first year, the last major figure associated with the foundation of LSE, Sir Arthur Bowley, died. In the same year Sir Otto Niemeyer relinquished the chairmanship of the Court and Lord Bridges, the former Cabinet Secretary, took over from him. Thus two civil servants were in charge for the next decade. The man who probably brought Lord Bridges to the School and was to succeed him ten years later, Lionel Robbins, was also much in the news at that time. In 1959 he was appointed one of the early Life Peers and adopted the style Lord Robbins of Clare Market. The choice demonstrated his deep attachment to LSE. For this, if for no other reason, the decision to retire him prematurely when he had become Chairman of the *Financial Times* became, in Lord Robbins's own words, 'one of the more painful episodes of my career'.[2] The Standing Committee decided that a professorship and the chairmanship of a public company were incompatible. Professor John Watkins organized a plea signed by forty-three (of sixty-three invited) teachers asking the Standing Committee to reconsider. The Director explained that academic jobs required a full-time commitment but that the School sought to convert the chair into a part-time professorship. This satisfied the protesters but the University Senate rejected the proposal. Soon after these unhappy events Lord Robbins was appointed Chairman of the Commission on Higher Education which was so closely bound up with the expansion of the 1960s. A dent in the relationship between Robbins and LSE remained, though both sides tried to contain it.

The Director cared about the tradition of the School. To give it tangible reality, he established the institution of the Honorary Fellowship. The first twenty-one Fellows were elected in 1958, including two former Directors, Lord Beveridge and Sir Alexander Carr-Saunders; national leaders like

Clement Attlee and Hugh Dalton; well-known scholars like Professors Gregory and Tawney; international figures like Professor Sir Arthur Lewis, G. L. Mehta, and Chief Justice Olshan.

Such stories confirm the impression of normality characteristic of the School around 1960. The fact that the concrete was not allowed to set in the early, or indeed the later, Caine years holds no surprise either. The St Clement's Building on the north side of Clare Market was at last acquired in 1959 and converted to School use in the following three years. The block of property north of Portugal Street called the Island Site was added, as were some other developments in Clement's Inn Passage and elsewhere around the School. The bridge connecting the Old Building and the East Wing was built as the first of the confusing links between the discrete though uniformly undistinguished edifices which make up the warren of the modern School. Caine was probably the first and perhaps the only Director to leave 'reasonably happy' about the facilities of the School.

More importantly, and against the declared wishes of the Director, new trends asserted themselves in the academic life of the School. The most important among these concerned the oldest characteristic of LSE, evening teaching. In the late 1950s it came under pressure and a decade later it had disappeared. Not only the Director, but the Academic Board had resisted suggestions by the University Grants Committee in 1960/1 to reduce evening teaching in order to improve the staff–student ratio. The School defended its tradition, at least 'until it was satisfied that adequate alternative arrangements were available in London'. This was a reference to Birkbeck College, which had taught mature students ever since the days of the old London Mechanics' Institute and had become a school of the University in 1920. The 1960s also saw the establishment of the Open University for those who had either missed out on higher education or wanted to pursue further courses alongside their employment. Both grew and may well have absorbed some of those who would otherwise have come to LSE. Yet it is hard to avoid the conclusion that other forces contributed to the decline in evening teaching at the School. When Sydney Caine arrived in 1957, there were already only 247 evening students for the B.Sc.(Econ.) as against 901 day students. Five years later, the number of evening B.Sc.(Econ.) students had shrunk to 191, and ten years later there were a mere 117 (and 1,067 day students). The end of evening teaching for undergraduates (and the partial replacement of evening students by part-time students) was not so much a conscious decision as the effect of profound changes in educational opportunity following the 1944 Education Act and the general encouragement of education for all by post-war governments. Caine himself, with a Treasury rather than a sociological mind, later attributed the change to pressure for orthodox degrees exercised by government grants, so that the School 'is becoming more like every other university, essentially a degree factory, and a factory producing overwhelmingly the distinctive British kind of "honours" or specialist degrees'.[3]

Winds of Change

The appearance of normality and stability was a little deceptive even at LSE, but totally absent in the world around it. This was notably true for the world in the literal sense of the international scene: 1956 was the year of the Hungarian Revolution, and of Suez. The School was affected by both, welcoming refugees from Hungary, and divided over the politics of Suez. Many years later, in 1969, Ernest Gellner would argue that the fateful year had more lasting effects as well: '1956 is of course a crucial year in the history of contemporary protest and rebellion.' Some went so far as to claim that 'the twin shocks of Hungary and Suez awakened a generation, hitherto somnolent under the influence of the beginnings of prosperity, and brought home to it the need for political action.'[4]

In 1957 Harold Macmillan became Prime Minister. This was also the year in which Malaya, then Ghana, gained independence. Three years later, in February 1960, Macmillan made his famous 'wind of change' speech in Cape Town. 'Will the great experiments in self-government that are now being made in Asia and Africa, especially within the Commonwealth, prove so successful, and by their example so compelling, that the balance will come down in favour of freedom and order and justice?'[5]

For LSE, such words had special poignancy. Sydney Caine was right to point, in his initial stocktaking, to the world-wide reputation and role of the School, and to add in 1958: 'The School has long had a great, perhaps unique, reputation as an international centre of social studies; we have many students from other lands, and many contacts with teachers in other countries.' In the same year, a $250,000 grant by the Ford Foundation made the extension of teaching and research on developing countries and international relations more generally possible. This was a case of building on strength, for the School had long been international from top to bottom, and its influence was if anything greater in the countries now caught by the wind of change than at home.

LSE's internationalism told in the curricula vitae of the Directors themselves. Sir Sydney Caine turned out to be the first of a sequence of four Directors with extensive experience of the world outside Britain. (All four had also been, in their younger years, students, or in the case of Walter

Adams, staff members of LSE.) Caine came to his post from the Vice-Chancellorship of the University of Malaya; Adams from that of Rhodesia. His successor was German-born, and had been a European Commissioner at the time of his appointment. He in turn was succeeded by an Indian with wide experience in the United Nations system. When the British Prime Minister on a visit to Delhi said to a startled I. G. Patel, just before the Director-elect set off for London in 1984, 'Why does the LSE always have to appoint foreigners?' she was more than a little disingenuous. The first two were actually very English Englishmen, and the last two hardly characteristic 'foreigners'. Moreover, from the benevolent imperialism of the early days through the welcome for academic exiles in the 1930s to the support for the winds of change in the 1960s the School had always looked outward as well as to the LCC, Westminster, and the welfare of the British Isles.

This was strikingly true for the students and their origins. Four groups among them stand out: Americans from the United States, some of whom came initially for one year only, as 'General Course' students during their 'junior year abroad' or as research fee and other one-year graduates, though quite a few stayed or returned to take higher degrees; students (and staff) from what came to be called the 'temperate' Commonwealth countries, Canada, Australia, New Zealand, and also white South Africa; a steady trickle of Continental Europeans whose LSE degrees did not seem to harm their own progress to positions of prominence; and of course the future élites of those huge pink areas on the world map, Britain's colonies, which were now, between 1947 and the 1960s, gaining independence. Krishna Menon of India, Jomo Kenyatta of Kenya, Kwame Nkrumah and Hilla Limann of Ghana, Veerasamy Ringadoo of Mauritius, Goh Keng Swee of Singapore, Kamisese Mara of Fiji, Erroll Barrow of Barbados, Eugenia Charles of Dominica, Michael Manley of Jamaica, Shridath ('Sonny') Ramphal of Guyana and the Secretary-General of the Commonwealth—the list of leaders of new nations with an LSE past is long, all the way to the unfortunate Maurice Bishop who, as Prime Minister of Grenada, was killed during the 1983 coup on his island.

Many have written about this role of the School, in their memoirs, in appreciative accounts of particular teachers, in scathing attacks by mostly right-wing publications; but no one has analysed the role of LSE, and of British socialism, in the world more intriguingly than Daniel Patrick Moynihan in his 1975 article in *Commentary*, 'The United States in Opposition'. Moynihan, later for many years US Senator for New York, had just come back from a stint as Ambassador to India, and now watched with 'discomfiture and distress' the extent to which the United States found itself in a minority position in the United Nations. What had happened, he asked, and replied that the cause was nothing less than a revolution, the 'British revolution'. The French and the American revolutions had become history; they were not the model for the new nations of the post-war world. Instead, another model had come to the fore. Ever since India in 1947, new nations had not only been set free by British parliamentary decisions but also adopted a uniform ideology, British socialism, 'as it developed in the period roughly 1890–1950'.

What are the ingredients of this British socialism? Independence, of course, and democracy, of sorts; but beyond that (according to Moynihan) a strong anti-capitalist bias. This meant on the one hand the insistence on redistribution rather than production. Since there was little to distribute in the new countries themselves, a transposition of Marxian class theory to the world scale was involved; the rich owed the poor retribution for a long period of exploitation. Thus British socialism meant, on the other hand, anti-Americanism. (Britain itself had been ruined sufficiently by British socialism not to be able to pay many 'reparations'.) This is why America found itself in such a hopeless minority in the UN.

The *New Statesman* receives a special mention as a source of British socialism, before Moynihan makes his crucial point: it is all the fault of LSE.

> Has there ever been a conversion as complete as that of the Malay, the Ibo, the Gujarati, the Jamaican, the Australian [*sic*!], the Cypriot, the Guyanan, the Yemenite, the Yoruban, the sabra, the fellaheen to this distant creed? The London School of Economics, [Professor Edward] Shils notes, was often said to be the most important institution of higher education in Asia and Africa.

Moynihan goes on to quote Beatrice Webb's claims of Fabian hegemony and states that when Oakeshott was appointed to Laski's chair in 1950, it was too late. 'By then not Communists but Fabians could claim that the largest portion of the world's population lived in regimes of their fashioning.'

Moynihan's answer to all this was sweeping in theory, if modest in practice. He made a strong case for the 'liberty party' in the place of the 'equality party', quoting in the process several LSE godfathers of the idea. He also argued that the United States had to change its stance in the United Nations. From the 'extraordinarily passive, even compliant' attitude to developments like the World Social Report of 1970, the Stockholm Conference on the Human Environment of 1972, the World Population Conference of 1974, with all their hypocrisy and hidden, or not-so-hidden, support for the Romanias and Chinas of this world, the US should move to explicit and argumentative opposition. 'It is time we asserted that inequalities in the world may be not so much a matter of condition as of performance.'[6] Within months of publishing this article D. P. Moynihan, the Democrat, had been appointed US Ambassador to the United Nations by a Republican President.

This does not invalidate the point made by the imaginative and often liberal author-politician. It has of course been made by others. In essence it is that from the 1920s to the 1950s at least, the London School of Economics has had a powerful influence at home and, even more so, abroad. This influence is variously described as socialism, Fabianism, the Welfare State; perhaps social democracy would be a more appropriate notion if it was not so alien to British political discourse. What is meant is the combination of Westminster-style democratic institutions with a benevolent interventionist government guided by a view of the good or just society. A little Laski, so to speak, a little Beveridge, some Tawney, and a lot of Pierre Trudeau, the long-serving Canadian Prime Minister with an LSE past. At home, the influence of this view was always tempered by the strong presence of alternative political philosophies and parties; but abroad British socialism, and the LSE as its prophet, had in many parts a near-monopoly. As one author interpreted Moynihan in *The Times*, 'it was an inspired cartographic convention, Mr Moynihan wryly comments, which decreed that the British Empire . . . should be coloured pink'.[7]

Sophistication in the Establishment: The First World

What then is the truth about the influence of the LSE on the world which had emerged from the Second World War? For one thing, it was never simple and one-dimensional. Pat Moynihan failed to inform his American readers in the article on 'The United States in Opposition' about his own past. It was only after the *Times* author had picked up his argument that he wrote a letter to the Editor, protesting that his *Commentary* article had actually not been an attack on British socialism, let alone the London School of Economics. He was after all an Honorary Fellow of LSE. In fact, he had been a postgraduate at the School from 1950 to 1953. He shared the LSE experience with other distinguished American leaders of the post-war world, like the banker David Rockefeller, or the Federal Reserve chairman Paul Volcker, or the Republican Senator John Tower.

It is not easy to make general statements about even this small band of alumni. Were they all perhaps not quite classifiable as political party people? Did they all combine an occasional if erratic tendency to move quite far to the right with, to many, surprising bouts of understanding for unorthodox views held by difficult young people? They had certainly all kept a soft spot for Britain—and for LSE—in their hearts. They were evidently not 'British socialists', nor does this label describe distinguished North American academics with LSE backgrounds like Will Baumol, Eveline Burns, Arthur Earle, Clark Kerr, H. W. Lyman, L. Rasminsky, P. A. Samuelson, E. A. Shils (to mention but a few of those who in due course became Honorary Fellows). If there is one feature which these men and (one!) woman have in common, it is a sophisticated view of the world. Most of them may have been centre-right rather than centre-left in their politics, defenders of institutions who were not greatly concerned about inequalities in society, but they were not naïve. They had encountered the other view and learnt to respect it. They were able to argue their case. One is reminded of old Ambassador Joe Kennedy, who wanted his sons to be exposed to 'the enemy' in the form of LSE so that they would be better able to fight him.

On the European Continent as in North America one is more likely to find among LSE people bankers and classical economists like Otmar Emminger, Wilfried Guth, Herbert Giersch in Germany, Romano Prodi in Italy, right-of-centre politicians like Joseph Luns of Holland, or middle-of-the-road public intellectuals like the great Raymond Aron of France. Perhaps European socialists, potential or real, could not afford to come to the School. At the time of the German elections of 1953, a group of LSE and other London Germans offered me a seat on their private plane chartered to vote (as was then necessary) within the borders of the Federal Republic because they thought that what they assumed to be my social-democratic vote stood in appropriate proportion to their twenty-odd Christian Democrat ones. In fact, North Americans and Europeans were neither attracted nor deterred by the School's 'socialist' reputation. They came to a renowned school of the social sciences and one which was known to be different, out of the ordinary, and of course located in the centre of the great metropolis, London.

This to be sure is not the whole story. There is Frank Meyer to think of and later David Adelstein and Paul Hoch, as well as their allies from Berkeley, Nanterre, and Frankfurt who found at the School such a fertile ground for their revolutionary fantasies. (As one mentions their names, one also remembers that few of them became a threat to the establishment in later years, and most joined it.) More importantly, there are those whom Moynihan had above all in mind, the leaders, and perhaps even more the administrators and intellectuals of the newly independent countries. They brought their resentments of the colonial power to LSE and returned home with a theory for the new order of their countries after independence.

However, before we turn to the dreams of these new élites, a word of caution about them is in place, for they too were by no means homogeneous. The people with whom Lionel Robbins as the British negotiator for the Bretton Woods agreements found it easy to strike deals in the glowing light of common LSE memories were hardly revolutionaries. A number of alumni went on to become either ministers of finance or central bank governors in their countries: Don Pedro Beltran of Peru, John Compton of St Lucia, Michael Colocassides and Andreas Patsalides of Cyprus, Felipe Herrera of Chile, Bisudhi Nimmanhaemin and Puey Ungphakorn of Thailand, N. M. Perera of Sri Lanka, Sir Veerasamy Ringadoo of Mauritius, to give but some outstanding examples. Those responsible for finance are rarely among the more avid reformers of their countries; they have to say no too often to keep the support of the left. Whether they like it or not, they represent the reality principle. As a result, they too may be called an establishment with sophistication. As students they would probably have been horrified if anyone had predicted such a future for them; yet they owed LSE both their technical competence and the cultural experience which made them serve their countries so well. Certainly one of the lasting influences of the School has been the education of sophisticated members of the establishment in many countries of the world.

New Élites and the Good Society: The Third World

Many of those who have given the United States a sense of being in opposition to the United Nations will not recognize themselves as a sophisticated establishment. This is notably the case with the politicians of Africa and the Caribbean. While the LSE was and is a name to conjure with in the anglophone new nations of these regions, two countries stand out as examples of the School's influence around the world: India and Mauritius.

The story of LSE and India runs deep in both places; it is, in a sense, a story of soul mates. From the earliest days, India—then undivided—was unmistakably present at LSE. The first hockey team included two Indians. An Indian was the first non-European President of the Students' Union in 1912. The gift by the Ratan Tata Foundation made the development of applied social studies possible. By the time Sir Sydney Caine took up the directorship, many hundreds of students from what by then was India, Pakistan, and Sri Lanka had been at the School. The relationship was no longer quite as easy and natural in the 1960s. When the School accepted a special Ford Foundation grant to promote exchanges and joint research in

1962, the need for formalization demonstrated the end of unforced informality as well as effect of exchange controls and the increasingly prohibitive cost of LSE for Indians. Still, the Indian influence remained. Whole dynasties showed its reality, notably that of A. K. Dasgupta, the economist father and LSE graduate of 1936, his economist son Partha, who became a professor at the School in 1978, and his artist daughter, the wife of the ninth Director of the School, I. G. Patel.

The presence of LSE in India is equally tangible. Granville Eastwood does not tell us the name of the 'well-known Indian political leader' who said it, but the statement rings true: 'There is a vacant chair at every cabinet meeting in India. It is reserved for the ghost of Professor Harold Laski.'[8] Lord Pethick-Lawrence, an LSE man, had been Colonial Secretary at the time of independence, when Prime Minister Attlee, another LSE man, saw his long-standing advocacy of Indian self-government come to fruition. (The constitution of independent India had been drafted by the economist–lawyer B. R. Ambedkar, the 'father of the untouchables', who had learnt his economics at LSE in the early 1920s.) Harold Laski was commemorated in many ways, not least by the Laski Institute of Political Science in Ahmadabad. When Vera Anstey died at the age of 87 in 1976, many remembered her devotion, as an economist and a teacher, to India. Her *Economic Development of India* had gone through several editions. Ever since she returned from India after the death of her husband in 1920, she had taught at the School, including the time of her important dual role as teacher and accommodation officer in Cambridge. 'For generations of students at the L.S.E.,' Tarlok Singh wrote after her death, 'from India and the sub-continent, she had been a friend and a guide who had cared for each of them.'[9] Later, other members of the School continued the intimate relationship, notably Professor David Glass (and his wife Ruth Glass) and Professor Tom Nossiter. The Delhi School of Economics is only the most obvious attempt to replicate LSE in India. The long list of Indian alumni includes administrators, businessmen, and academics from all parts of the country.

This last statement is important. Some LSE graduates did go into politics, notably at the state level, but Nossiter is right to point out 'that LSE produced mandarins (and academics) rather than politicians'. One thinks of the economist Tarlok Singh, who was Nehru's secretary, of the chairman of the Indian Investment Centre, R. S. Bhatt, of the founder of the Indian Institute of Public Administration, V. A. Nanda Menon. This means that the School became more deeply entrenched in Indian life than a few highly visible names could achieve. In fact, political élites without the backing of 'mandarins and academics' can be blown away in no time, by distinctly non-academic indigenous colonels, lieutenants, or sergeants, for example. Nossiter remembers hearing often the old quip 'I'm Oxford, you're Cambridge, he is LSE' on his travels to India;[10] but Oxford and Cambridge have long ceased to be central to Indian life whereas LSE remained that for a long time.

The reasons are many, or perhaps this must be put in the past tense—they were many because the perspective of the 1990s is very different from that of the 1950s and 1960s which informs the LSE story at this point. For many decades, a deeper affinity between India and Britain in general as well as LSE in particular played a part. For one thing, both India and Britain had a caste system of sorts. The practical question was how the distinct social estates could live together in peace, and the ideological question was how the divisions could be overcome. Lord Desai of St Clement Danes, the economics professor with an LSE-related title, and Dr I. G. Patel, KBE, are two answers. Britain and India also shared the penchant for a rigid division of labour, with over-defined jobs serving to protect people's identity though not always bringing about Adam Smith's productivity effect. Furthermore, a certain general appreciation of public service united the two countries, unless it was the Raj which brought the British idea of the Civil Service to the subcontinent. Combined with the division of labour it produced that great bane of Indian life, 'red tape', vainly fought by the entrepreneurs of Bombay (who had of course also been at LSE).

It is hard to refute D. P. Moynihan's claim that for these and other reasons a similar attitude to economic success characterized the two countries. This has something to do with the good society and LSE. Two of the best-known economists who taught for years at the School before they emigrated to Cambridge, Massachusetts, and Cambridge, England, respectively—Amartya Sen and Partha Dasgupta—tell the story in their works. Both are distinguished as theorists in the strictest technical sense; but their deeper motives have to do with the values behind Sen's interest in equality and justice, and Dasgupta's espousal of 'social well-being'. The combination may be unfamiliar in American business schools and unusual even in post-1980s Britain, but it is a reminder of an older search for both economic success and social values. One may wonder whether this was invented by LSE, or merely found a natural home at a School which was set up after if not against Gladstone and in an uneasy truce with Alfred Marshall, one which had never really come to terms with Hayek's politics though it lived happily with his economics, which gained him the Nobel prize in 1974. The answer will not satisfy conspiracy theorists, but probably the natural affinity between Britons and Indians was merely reinforced by a School which may have seemed counter-cyclical in 1900 but was certainly very much in the swing of things half a century later. LSE taught administrators and intellectuals who wanted respect for citizens within democratic political and just social institutions first, and economic growth and personal advancement second.

All this is now history. In the first place, partition meant the beginning of the reduction of LSE influence on the subcontinent. Pakistan, with all its emulation of upper-class English ways, slipped away from British and notably LSE influence. In Sri Lanka and also in Bangladesh the School still has a name though it has never been dominant. More importantly, India itself has begun to change, as has Britain, and in both cases it was not the old LSE which defined the direction but either strictly local forces or American values..

Remaining for the moment in the 1950s and 1960s, there are other countries in which LSE had a direct and even deliberate influence. Among them, the overcrowded, multi-ethnic, two-cultural island state of Mauritius stands out. Remarkably, given the Franco-British heritage, it was LSE lawyers who played a major part in preparing the island for independence, guided by the Professor of Constitutional Law, Stanley de Smith. Before he became Chief Justice, Ramparsad Neerunjun was one of the collaborators of the first Prime Minister, Sir Seewosagur Ramgoolam (a medical doctor who had been a student at University College London but attended lectures, notably by Laski, at LSE). Other LSE lawyers, including Renganaden Seeneevassen and Veerasamy Ringadoo, later Minister of Finance and Governor-General, helped him set up both the rule of law and a Labour Party which for many years dominated Mauritius politics.

The Commonwealth Welfare and Development Act in the wake of the Beveridge Report enabled students like K. Hazareesingh to come to LSE. Reciprocal visits by James Meade, Richard Titmuss, and, later, Brian Abel-Smith became equally important. Meade's report on the economy of Mauritius and Titmuss's on social policy were both submitted in 1960. The Titmuss report in particular turned out to make a major difference. It persuaded a Roman Catholic country to promote family planning, and helped introduce a health service as well as a system of technical training. The report 'combined Titmuss's demographic insight with historical understanding, social accountancy, an unerring grasp of detail, keen observation and, as always, the ability to ask new questions'.[11] Brian Abel-Smith supervised its implementation, which meant that Mauritius became a model welfare state. 'There is hardly any other Asian or African country which has introduced welfare services of such long-term benefit to the entire population.'[12]

Hazareesingh, who reached this conclusion in his memoir of LSE and Mauritius, made a point of the School's influence on the private sector in the island as well. The former Social Welfare Commissioner left little doubt that it was 'the ideals of socialism' and a 'social justice that would be within the reach of the humblest members of society' which constituted the heritage of LSE. This is true for the neighbouring African countries also. Independence for many meant self-government plus the welfare state. Justice had both an international and a social aspect, as in Joseph Chamberlain's days, when imperialism and social reform concluded their tenuous marriage. LSE had something to do with both.

The marriage was tenuous, but did it work at all? It was not until the Asian dragons awoke from their slumber in the 1970s and 1980s that independence came to be associated with economic growth rather than social justice. LSE alumni were a part of this development too. Lee Kuan Yew of Singapore is sometimes claimed by the School but had really been at Cambridge, though his long-time deputy Goh Keng Swee was a loyal LSE alumnus. So were the Hong Kong businessman Sir Y. K. Khan, the Korean newspaper owner Kim Sang Man, to say nothing of many younger B.Sc.s, M.Sc.s, and Ph.Ds in the Far East and also in Latin America. No welfare state by or for them, though they all looked back to their LSE days with gratitude and pleasure!

The story of the influence of the School is complicated further by the fact that economists as a rule did not join governments. They wanted to earn better salaries or be less visible, or merely pursue their professional lives. Many of them went into international organizations which were anything but agents of socialism. There was never a shortage of LSE alumni in the World Bank, nor in the regional development banks. By the 1970s the central banks of new countries also had their planning departments in which young Ph.Ds were poring over long-term development models without paying much attention to the human cost of growth. Some, very few, managed to keep both concerns on their minds and in public debate. One of them was the Nobel laureate of 1979, the long-time LSE lecturer from St Lucia, Sir Arthur Lewis. But along with James Meade he was one of very few exceptions.

What, then, was the influence of the London School of Economics around the world? It was above all, as befits a great academic institution, not all of one piece. LSE trained economists and lawyers who helped set up the rule of law in their countries and advised policy makers in their capitals as well as in international organizations. Increasingly, the School educated educators; it was, especially in the decades of expanding social sciences, a school for professors. Some of the alumni of LSE brought their new-found sophistication to the establishment of their countries, others remained radical reformers in opposition. For many, politics was an occasional pastime rather then their main preoccupation. Everywhere, the School was seen as a model of research and, perhaps even more so, of teaching in the social sciences.

Yet there was more to the evocative letters LSE, especially in the newly independent countries after the war. The School was a congenial place for those who wanted their emerging nations to be different. Britain after the war, as much as any particular LSE professor even including Laski, had a strange attraction for those who did not regard wealth and wealth creation as their first priority. With all its drabness and austerity it was a kind country, a country in which people queued rather than using their elbows to get ahead, one in which health and other provisions left much to be desired but were part of a national service for all, a country which mixed old and new in ways which even those who objected to its anachronisms secretly enjoyed. The School was always too diverse to be identified with any one side in the disputes of the time, but those who liked the old Britain and wanted reform without losing social cohesion felt at home here. For them it was actually neither Laski nor Robbins who mattered but the tradition exemplified by Tawney which lived on in the Law Department, in social science, in social administration, in applied economics, and among many younger members of staff.

It was not a tradition destined to last. From one point of view—which came to be dominant in the 1980s—it represented a weary Britain which had found it hard to come to terms with modernity and the industrial revolution. Somehow LSE had managed to turn *English Culture and the Decline of the Industrial Spirit* into the hope for *The Future that Works* (to play on the titles of two influential books of the 1970s on the subject). For a while, this conjuring trick in fact seemed to work but then its economic fragility aggravated the *Decline of Nations* (in the words of another 1970s title), and Britain, along with India and the rest of the world, tried to break out of traditional rigidities. The price was high, not least for LSE. Resourceful as ever, the School found its way into the new world, but the transition hurt. This too is an aspect of the troublesome years which began in the later 1960s.

WIDENING AND DEEPENING THE SOCIAL SCIENCES

Familiar Subjects, New Twists

Only with hindsight and from a distance can LSE in the 1960s be called a unitary school of the social sciences. The proliferation of degrees, diplomas, and certificates was multiplied by the number of options available for most of them. One-third of all students sought a graduate degree or did graduate work as research fee students; even they could choose between eleven alternatives ranging from a Ph.D. to the Postgraduate Diploma of Social Studies in Tropical Territories. Among undergraduates, the B.Sc.(Econ.) was still the preferred choice for three out of every five, but many roads led to this particular Rome. Very little economics was required to attain this economics degree, and in the 1960s economics was not the most popular subject. One out of every five undergraduates read for the Bachelor of Laws or for a BA Honours degree in Sociology, Anthropology, History, Geography, or even—a recent innovation—the combined Philosophy and Economics BA. The remaining fifth (and these are rough figures designed to indicate orders of magnitude) aimed at a Diploma or Certificate in Social Administration, Applied Social Studies, Personnel Management, Mental Health, Business Administration, Trade Union Studies, or half a dozen other subjects.

Undoubtedly there were teachers and, even more so, students, who lived in one small or not so small niche of the maze called London School of Economics without taking any notice of what went on in other such niches, in the recesses of the Old Building, the less accessible corridors of the East Wing, the refurbished St Clement's Building, elsewhere in Clare Market, Houghton Street, or Clement's Inn Passage. And yet the School was still run as if it were a single faculty, with weak departmental structures and strong central institutions including the Academic Board and above all the Appointments Committee. More importantly, many of the names and lectures listed in, say, the 1962/3 Calendar were familiar then and are still recognized thirty years later. Lord Robbins was teaching 'The History of Economic Thought' and Sir Arnold Plant 'The Structure of Modern Industry'; Professor Phelps Brown 'Applied Economics: The Labour Market', and Professor Yamey 'Economics of Industry and Trade'; Professor Wise 'Introduction to Geography'; Dr Barker 'Industrialisation and the International Economy 1850–1939'; Professor Medlicott 'Diplomatic Relations of the Great Powers'; Professor Griffith 'Administrative Law' and Professor Kahn-Freund 'Industrial Law'; Professor Oakeshott 'Political Thought' and Mr Schapiro 'The Government of Soviet Russia';

Professor Roberts 'The Political History of Trade Unionism'; Professor Firth the 'Outline of Economic Anthropology' and Professor Glass 'Population Trends and Policies'; Dr Himmelweit 'Theories of Personality' and Professor Titmuss 'Introduction to Social Policy'; Professor Ginsberg 'Social Philosophy' and Professor Gellner 'Comparative Morals and Religion'; Professor Allen 'Basic Methods' and Professor Moser 'The Nature and Sources of Social Statistics'.

Was there then nothing new under the sun? Along with a number of American universities, LSE had established the social sciences in the inter-war period. Elsewhere in Europe, developments were both slower and more tentative. Even economics did not easily find a place in the older universities; and new foundations concentrated more on trade and business studies than on general economics. Other social sciences had to squeeze a precarious existence out of faculties of law or philosophy. The feeling was, to be sure, not unknown in England, where Oxford and Cambridge (though not some of the civic universities) resisted recognition of the social sciences—other than economics at Cambridge—until the 1960s and beyond. It took Nuffield College and later St Antony's to get the newfangled disciplines at least half-established. LSE did not share this problem. It was there, it was good; its students went all over the country and the world to teach the social sciences. Developments at LSE were therefore remarkable even when they were undramatic.

Some departments developed their traditional academic strengths under new management, as it were, or with a combination of old and new. This was notably the case in areas which branch out from the social sciences into neighbouring fields in the sciences or humanities. Geography (Professor Emrys Jones tells us) underwent a double revolution in the 1960s which pulled it hither and yon. On the one hand, quantitative methods gained ground; 'some university departments began to demand "A" level mathematics as a prerequisite to a degree course'. On the other hand, 'Geography has never been happier than when leaving the ivory towers of learning and making a foray into the real world.'[13] The 1960s with their penchant for town and country planning offered many an opportunity.

The two parts of the history department continued to flourish, though judging from Jack Fisher's account they also had their problems. 'Recent intellectual fashions, of course, have not been highly favourable to historical study of any sort.'[14] Still, Fisher himself, with Theo Barker and Donald Coleman, Charlotte Erickson and Arthur John, Jim Potter and Walter Stern, formed a formidable group of economic historians. At the same time, Professor Medlicott's team of international historians added to the weight and influence of history in a school of the social sciences: Matthew Anderson and Kenneth Bourne, George Grün, Ragnhild Hatton, James Joll, and Donald Cameron Watt.

The Quantitative Department (as one is tempted to call it) benefited from new trends of formalization and computerization, but it also split into separate, tenuously related subgroups. Even in the 1960s it was hard to find a unifying name for the department. 'Statistics, Mathematics, Computational Methods and Operational Research' was chosen for a while; this did not even include demography. Mathematics and statistics in the strict sense were coupled with econometrics, survey methodology, actuarial science, computer programming, management analysis, linear programming, and other new specialities. This may have been a far cry from Bowley's days, but it still confirmed the empirical bent of many departments of the School.

Many, but not all. In Sydney Caine's time, the Department of Politics—now Government Department—turned from Laski- and later Oakeshott-dominated political theory tempered by Robson and the study of public administration into that set of outstanding individual scholars which was to characterize it for the next twenty-five years. The names defy any common denominator other than the absence of empirical political science. Maurice Cranston pursued his own subtle history of political thought. With William Letwin and Kenneth Minogue he gave further support to a shift in intellectual focus in a crucial department of the School which would have surprised the founders. Those who remained reform-minded, like Bernard Donoughue and George Jones, certainly sought another Labour Party from that of Laski. Professor Elie Kedourie, while a student of Laski's, showed (in the words of Maurice Cowling in his obituary in the *Daily Telegraph*) 'how critical historical writing could sustain a conservative mentality'. Kedourie, who taught at LSE from 1953 to 1990, was a Middle East scholar and political theorist, but above all an unorthodox and effective supervisor of research students. Then there was Leonard Schapiro, whose two earlier lives as a barrister and a Major in the Intelligence Division during the war made him a late academic, though one who soon acquired distinction and influence as one of the foremost analysts of Communist Russia, and of totalitarianism generally. Undergraduates and graduates alike loved the teacher, who from 1950 to his retirement in 1975 and beyond made his mark on the School. Other senior and junior academics adorned what had become one of the attractive parts of the School if perhaps not a homogeneous department.

At least two departments reached high points of distinction in the 1960s: international relations and law. International relations had been taught (initially as international studies) even before Charles Manning was appointed to the chair in 1930. When he retired in 1962, he was replaced by Geoffrey Goodwin, though others, notably Martin Wight and Hedley Bull, taught alongside him. Hedley Bull later described how he arrived at the School from Oxford in 1956 to take up his position as assistant lecturer in international relations: 'I had not done a course of any kind in International Relations, nor made any serious study of it, and as I arrived in Houghton Street I wondered how I was to go about teaching the subject and even whether it existed at all.'[15] Charles Manning sent young Hedley Bull to Martin Wight's lectures, where he learnt two characteristic LSE lessons, at least so far as his subject was concerned. One was that international relations at the School was more a philosophical and historical than a behavioural and empirical subject. The other was that it had a preference for what Bull, with Wight, called a 'Grotian' approach, following the early seventeenth-century Dutch jurist and father of international law, Grotius. He was neither a realist *à la* Machiavelli nor a 'revolutionary' concerned with social and economic forces but sought binding rules for a community of states. In some ways this tradition removed LSE International Relations from the (American) mainstream, but it also made it special, and made the changes of the late 1970s difficult.

The Law Department already had a distinguished and unusual tradition to which reference has been made earlier. Its members, ranging from Otto Kahn-Freund and L. C. B. Gower to the younger lecturers in almost all specialist fields of the law, enjoyed teaching their discipline as one of the social sciences. In the 1960s, the Saunders Committee reforms within the University of London enabled LSE lawyers to devise their own syllabus, and they made full use of the new-found freedom. The relevance of this shift for the intellectual history of the social sciences has yet to be explored. While other disciplines turned more and more to what one might call the sub-institutional realities of social life, the LSE Law Department preserved an older interest in the interplay of institutions and patterns of social behaviour. As a result, few departments if any have been as influential in the wider debates within LSE in the 1960s and 1970s as the Law Department.

Not all innovations came to fruition immediately. Hilde Himmelweit, since 1964 the first Professor of Social Psychology in Britain, had to struggle to secede from the Sociology Department, though her rare combination of charm and tenacity in the end brought success. Ben Roberts got a chair in industrial relations in 1962, and by 1964 he had succeeded in setting up a department to accommodate his subject as well as trade union studies, though not yet personnel management. The thrust of all these developments was later well put by Jack Fisher in his characteristic style:

The social sciences have become more professional and more specialised; the School has become more departmentalised; Directors have confined themselves to directing (although the Pro-Director still teaches economic history); Librarians confine themselves more strictly to library matters; economists have abandoned history to whore after mathematics; the international historians have declared their independence; the number of economic historians has multiplied.[16]

Monks and Technocrats

In the midst of such changes, some familiar problems reappeared. From its earliest days, LSE had found it hard to decide between its scholarly vocation and its practical inclinations. Professor David Martin, in writing about 1968, spoke of the difficult relations between the 'monks' and the 'technocrats'. The technocrats ran the Great City, but the monks 'were careful scholars and in accordance with an ancient vow of intellectual chastity bent all their energies to make their studies pure. . . . Anything which was "applied" smacked of the Great City and contravened the ancient vow.'[17] John Alcock, Registrar from 1958 and from 1967 to 1983 Academic Secretary, became one of the guardians of the pure school of thought, so that his memories may be somewhat tinted. 'Webb had talked to the LCC with one voice and to the Ministry of Education with another. He believed that we should have research into the social sciences to better the human race and he believed that we should have courses for city clerks to better the turning of the wheels of business and so on.' In Alcock's view, 'it was the academic work which by 1958 had really won out'. 'I'm quite sure that the School for most of my time, that is most of the period between 1958 and 1983, was steadily as it were purifying itself and getting rid of the very worthy but perhaps less interesting chores.' Then the defender of monasticism has an afterthought about his technocratic Director. 'Sydney Caine was one who tried to preserve "applied" courses, but in the end it seems to have been the need to raise money that brought them back.'[18]

The Registrar and Academic Secretary has seen such developments at close quarters and usually drafted the relevant committee minutes. His central observation is important. Expansion financed from public funds under conditions of university autonomy encouraged academics to pursue their subjects purely and simply. Sidney Webb had to address not so much the LCC and the Treasury (the Ministry of Education emerged rather later) as the LCC and the Chamber of Commerce, from which the combination of economic theory and railway studies ensued. The Directors following

Sydney Caine and Walter Adams had to find high-fee-paying students and other private sources of funding; again, applied subjects gained in importance.

Yet throughout its history, including the 1960s, the tug of war between pure scholarship and application went on. (David Martin's belief that 'for many years . . . the monks and the technocrats lived in peace' may describe Cambridge, and possibly Oxford, but never LSE.) We have seen Emrys Jones's account of the Geography Department. In political science, Michael Oakeshott was hardly concerned with application (and probably surprised when he was dragged into the Great City during the Thatcher years), but public administration went its own way, led by William Robson, Peter Self, and later George Jones. The School's central department, economics, actually offered little theory in the early 1960s and after Lionel Robbins's departure it had almost a gap in pure economics. Ely Devons in particular made many efforts to fill this gap. Terence Gorman and Frank Hahn were attracted to the School. Then the econometricians arrived in the late 1960s, who were followed, in the 1970s, by a new generation of theorists. In the mean time, applied economics as it was called ('including money and banking, international economics, business administration and accounting, and transport') dominated the scene. Professors Arnold Plant and Henry Phelps Brown, Frank Paish and Basil Yamey, Ely Devons and Richard Sayers taught a whole range of applied subjects and were often involved in the doings of the Great City, not least the City of London.

Applied concerns entered in other ways. From 1946 Professor (later Sir) Ronald Edwards had held his Tuesday evening seminars for businessmen, civil servants, and academics. The 'Ronnie Edwards Seminars' became famous in the 1950s and 1960s and were quoted even later as precursors of a possible think-tank quite apart from providing grist for the mill of the Business History Unit. In fact they were a training ground for a particular kind of businessman who was more inclined to follow the precepts of Andrew Shonfield's social capitalism than those of Milton Friedman's neo-liberalism. One thinks of important corporate figures of the 1960s and 1970s with an LSE association, Sir Arthur Knight, Sir Anthony Burney, Sir Gordon Brunton, and Sir David Orr, who all combined business interest with a sense of public service. Later they would be denigrated as representatives of 'corporatism', though the attempt to run companies as well as the national economy by a consensual rather than adversarial approach was perhaps dismissed too lightly.

The 'Ronnie Edwards Seminars' epitomized this approach even by their organization. They were based on papers prepared by industrialists and civil servants. These papers were circulated and then discussed. Most of them had to do with business administration. Ronald Edwards published them (with Harry Townsend) in three volumes entitled *Business Enterprise* (1958), *Studies in Business Organization* (1961), and *Business Growth* (1966). His LSE obituarist Basil Yamey tells us that 'Ronnie regarded the seminar not only as a necessary link between different worlds but also as part of his continuing inquiry into business organisation and policies.'[19]

In Professor Edwards's case this was not just enquiry. In 1961 he became Chairman of the Electricity Council and later Chairman of the Beecham Group. He believed that universities in general and a school of the social sciences in particular had a special role to play in educating managers. His 1951 inaugural lecture on 'Industrial Technologists and Social Sciences' showed a particular concern with training business leaders who have technical as well as financial skills. This led Edwards, with the support of the then Director, Carr-Saunders, to an initiative to create, with Imperial College, a Joint School of Administration, Economics, and Technology. The project did not advance very far though it left behind a small economics unit at Imperial College and a course in Economics for Engineers at LSE.

Others pulled in the same direction. The new Department of Industrial Relations has been mentioned already. Transport Studies became the updated version of the railway studies of old. Above all, the sub-department of business administration and accounting—soon to be a separate Department of Accounting—had grown in significance since Professor Baxter's appointment in 1947. By the early 1960s, Professor Baxter had been joined by Professor Edey and several lecturers. Teachers in neighbouring fields, like Professors Yamey and Wheatcroft, contributed to teaching and research. Thus the School broke the path for academic accountancy until, when others followed suit, most chairs in the country were filled by LSE graduates.

Behind many of these developments in applied, especially business studies stood one man, Professor Sir Arnold Plant. For twenty-five years he had taught the course in business administration. Ronnie Edwards and Basil Yamey were among his early students, along with Ronald Coase, Arthur Knight, Arthur Seldon, and others. His view was unequivocal: 'Universities are the home of vocational education.' The view was not widely shared at the School nor did the majority of members of the Academic Board like it. Thus Plant was not universally popular, especially not in the post-war years in which 'his brilliance subsided' and he 'was becoming very difficult'. Actually, Ronnie Edwards had his detractors too ('he had some difficulty in hiding his contempt for the rest of us'),[20] though the accountancy professors were liked and appreciated, and played an important part in both the scholarly and the administrative life of the School. But then they were technocrats seeking admission to the monastic order whereas Plant, and Edwards, remained unreconstructed defenders of the Great City.

The consequences were important. By the mid-1960s the course in business administration ceased to have much support; most wanted to integrate it into the M.Sc.(Econ.). When Plant retired, the course was discontinued; an attempt to substitute a Diploma in Business Studies in 1966 did not survive the first year. At the same time, wider discussions of business education and a possible graduate degree in business studies had taken on a new urgency from 1961. Once again, it was felt that, in order to sustain economic growth, business leaders and managers needed better training. The School was well placed for such discussions. Professor Edey took the initiative, and he was supported by Professors Baxter, Durbin, and Yamey, by Ben Roberts and Nancy (later Baroness) Seear, and of course by the Director.

The University Grants Committee showed an interest in these developments and responded favourably despite its declared view that earlier efforts by the School 'lacked unity and needed to be focused on a few major topics'. Once again, however, it was suggested that LSE and Imperial College should get together to sponsor a two-year graduate course, and possibly a joint Institute of Management Studies. The two colleges did not find it easy to agree, especially not on a location somewhere between the Aldwych and South Kensington, with ready access to the BLPES and yet in easy reach for students from Imperial College. While the dithering went on, two business groups—both advised, naturally, by members of the School in their personal capacity—moved ahead with their own plans and in the end invited Lord Franks, the great national ombudsman of the period, to adjudicate.

The Franks Report on *British Business Schools* was published in 1963. It recommended the creation of two graduate business schools, one associated with Manchester University, and the other jointly with LSE and Imperial College within the framework of the University of London. The Franks Report was followed in 1964 by the Normanbrook Report on *British Business Schools: The Cost.* Already, the Federation of British Industries and the British Institute of Management had created a Foundation for Management Education. With its help, sufficient funds for setting up two free-standing business schools were found. The London Graduate School of Business Studies opened its door in 1965. It was and still is funded partly from private and partly from public sources; LSE and Imperial College are represented on the Board; but the creation of the Business School meant that (apart from a few Fellowships attracted from the Foundation for Management Education by Professor Edey) LSE had lost out on this area of study.

This at least is what Lionel Robbins thought. He had not been able to attend a critical meeting in 1963 to save the earlier plan of a joint school or an institute. Robbins sent his sceptical comments in writing. He had been in favour of the plan, would indeed have liked to see the School 'do this thing itself. But we have muffed too many opportunities in the past to be able realistically to hope for that.'[21] Robbins may have been thinking of Arnold Plant's course, of Harold Edey's project in the 1950s, even of the saga of the National Institute of Economic Research in the 1930s, and, of course, the story of the B.Com. which never quite made it. But not all agreed with his assessment. Ronald Coase refers to an earlier period but his comment is none the less relevant: 'The main thrust of the work in economics at LSE was the development of pure theory and did not reflect Plant's interests.'[22] Even Sydney Caine, when he looked back over his ten years as Director in 1967, wrote to the Clerk of the University Court that 'there is a tendency outside the School to take a narrow view of education for management and to forget the part that the general study of the social sciences in their different aspects can play.'[23] In other words, an M.Sc.(Econ.) may be better than an MBA even for business leaders.

David Martin, with some benefit of hindsight, was gloomy about the fate of the old monastic university in the 1960s. When the technocrats wanted to know things, they turned to the monks. For that they needed more of them, so that they filled the monasteries to overflowing. But the roots of frustration were built into this process. The monks disappointed the technocrats while allowing their pure world to be taken over by the Great City. Nor did the new order of 'friars'—Martin adds a laconic footnote: 'social scientists'—help. In the end the monasteries were dissolved, the Great City was no wiser, vagrants and protesters took over. Hence 1968.

This is a good story but at the 'London School of Friars' things were not quite so clear-cut. There is John Alcock's point to consider; at a time of ample public finance the School if anything purified itself. Also, the pure school of thought tended to win by default. When committees dithered, outside opportunities faded away, or rather went elsewhere, and the School remained if not monastic then certainly bent on *rerum cognoscere causas*. Moreover purity was a strangely ambivalent concept for LSE. When Sidney Webb had pleaded for a school of the social sciences which remained untainted by vested interest he certainly knew that scholarship needs autonomy. But he also hoped that in this way a new élite would be educated which was itself above vested interest. The monks were, in his view, the ideal teachers of technocrats. Mandarins were after all no less pure than monks. The enemies were not so much in the Great City as in the muddles of markets and the dirty tricks of politics.

It is not easy to tell what all this added up to, yet it is important to try. For a variety of reasons the School has had, throughout its history, a tendency to tilt towards the pure side of social science. Yet application was omnipresent within its walls; it was in a sense inevitable. The dilemma epitomizes the great conflict between (social) science and values, and the School's solution. In theory there is no solution but in practice there is. Like a mature person, a strong institution can live with the dilemma, especially if it refuses to give way to temptations. From time to time, LSE drew a line: it embraced many aspects of business studies but did not want a business school. Later it accepted the relevance of its work for public policy but did not want a British Brookings. The 1960s and 1970s were in this sense a time of strength.

Falsification and its Discontents

If application and the new twists of familiar subjects widened the social sciences at LSE in the 1950s and 1960s, philosophy deepened their understanding. At least, it was intended to do so, since it was for the most part philosophy of science. In 1941 Abraham Wolf, the Spinoza scholar and historian of science, had retired from his long and half-time if not half-hearted tenure of the philosophy post at the School, held in conjunction with University College. After a long and difficult search, a telegram was sent to Christchurch in New Zealand, where the Austrian émigré philosopher Karl Popper had found, with the help of Walter Adams and Esther Simpson of the Academic Assistance Council, a post at Canterbury University College early in 1937. He had used the war years to good purpose; indeed he himself later described his books on *The Poverty of Historicism* and *The Open Society and its Enemies* as 'my war effort'.[24] Popper's friend Ernst Gombrich brought the books to the attention of Friedrich von Hayek, in whose seminar Popper had given a paper in 1935. Lionel Robbins, for Popper 'the uncrowned king of LSE', had met the philosopher and taken a liking to him. Laski also claimed to know Popper, though perhaps he was once again 'lying out of nicety'.[25] Anyway, Popper was offered the Readership in Logic and Scientific Method, accepted with alacrity, and set sail for London in the summer of 1945. He was to remain a loyal member of the School until his retirement in 1969 and beyond.

Popper is one of the great minds of the century. Some would say that no one in the history of LSE surpasses him in intellectual stature. Like other great men he stood for one simple but infinitely fertile idea. In a world of uncertainty, we cannot know the truth, we can only guess. No amount of evidence will prove our guesses right, but often one fact suffices to prove them wrong. We have to try again and make better guesses. Some later mocked the simplicity of this idea, but when first formulated it flew in the face of current assumptions, notably those about induction as the method of science. We can never (Popper argued persuasively) find the truth by amassing facts and trying to squeeze theory out of them. The approach was first developed in technical detail in the 1935 German edition of *The Logic of Scientific Discovery*. Scientific progress through falsification ('fallibilism') and the 'hypothetico-deductive method' became the hallmark of Popper. In *The Open Society* Popper applied the principle to social and political theory. What matters above all is to set up institutions which check dogma and its tyranny, open institutions which allow of change by regular means. In history too we try and we err and we try again.

Why would Karl Popper be at LSE? The question has a subtle and a simple dimension. Those who think that Popper's philosophy went nicely with the prevailing 'positivist' tradition of the School overlook critical differences. Popper was never in any sense a positivist, though Hegelians and other enemies have at times called him that because for them all who are not 'one of us' look alike. His view of science, and of history, has unique features which defy labels. What is more, it is in fundamental conflict with two of the traditions of the School. One has to do with induction. Popper had nice words to say about Beveridge and his plan ('I was very interested in that and I got his book as soon as possible') but then he had nice words about most things at LSE at least in so far as they happened outside his own department. In terms of scientific method Beveridge represented everything that Popper rejected. He epitomized the endless and often thoughtless collection of facts and the rejection of theory. Moreover, in this Beveridge was not far away from the prevailing mood of the School, a handful of economists excepted. Sidney Webb, and even more so Beatrice Webb, thought that finding out the causes of things meant above all gathering facts. Popper poured scorn on such attitudes, though he probably knew too little about LSE to associate them with his new environment.

The other School tradition which was deeply alien to Popper was the belief in the good society run by mandarins, by a well-educated élite of experts. Much of his *Open Society* is a diatribe against philosopher-kings who arrogate to themselves not only power but knowledge, and thus try to make their power unassailable. With one of his less felicitous phrases, Popper spoke of the desirability of 'piecemeal engineering'; but technocracy, expertocracy were for him a form of tyranny. Like Immanuel Kant, he preferred less appealing but more real forces. 'There can be no human society without conflict: such a society would be a society not of friends but of ants.'[26] The School never recognized quite how subversive its Professor of Philosophy was to some of its traditions.

Why, then, would Karl Popper be at LSE? The answer is simple really and Popper was quite open about it. 'Oh yes, I really loved it, I had a real love relation with the LSE, but at the same time I also felt that, as far as I and my ideas were concerned, I would have been better in a science area in a university.'[27] Popper never ceased to be grateful to the School for having rescued him, but intellectually the London School of Economics and Political Science was an accident of his life. He saw himself as a scientist, a physicist for the most part and perhaps a biologist later (though such disciplinary descriptions were themselves alien to Popper's thinking). 'Yet the social sciences never had for me the same attraction as the theoretical natural sciences.'[28] In 1961 the attempt was made, at a Tübingen meeting which has since become famous as the origin of the *Positivismusstreit*, the 'positivism dispute', to confront Popper in debate with the Hegelians of the Frankfurt School. Compared to Theodor W. Adorno, however, Popper remained somewhat pale and reluctant to engage. The dispute took off only when Popper's German disciple Hans Albert, an economist by training, involved the leader of the second Frankfurt generation, Jürgen Habermas.

Yet Popper was undeniably a part of LSE. In the basement of the Old Building he lectured to eager if sometimes shocked students; his personal manner did not always bear out the openness which his theories advocated. Bernard Levin recalls one example:

> It was not, I must say, an entirely smooth passage into enlightenment; our guru had a sharp tongue. One day, at a seminar, a fellow student offered an opinion couched in terms greatly lacking in coherence. The sage frowned and said bluntly: 'I don't understand what you are talking about.' My hapless colleague flushed, and rephrased his comment. 'Ah,' said the teacher, 'now I understand what you are saying, and I think it's nonsense.'

The style notwithstanding, Levin, like many others, adored the philosopher. 'His *Who's Who* entry should be set to music by Beethoven, no less; all over the world his former students remember him with gratitude, awe and affection. How glad I am to be one of them!'[29] One early Popper student at LSE, George Soros, did not so much set his teacher's achievements to the music of Beethoven as to the tune of the many millions of dollars which he made on financial markets and then devoted in significant part to Open Society Funds promoting liberty in emerging democracies.

The community of Popper fans is large, but was there ever a Popper School? There should not have been, John Watkins argued plausibly.

> So there ought to have been a Popperian critical tradition and not a Popperian School. But I think that, at least from the middle 1950s to the late 1960s, there was something like a Popperian School. Among its members were Joseph Agassi, Bill Bartley, Imre Lakatos, and myself. We regarded Paul Feyerabend as a brilliant, but erratic and fitful ally.[30]

Others spoke of a 'circle'. Either way, it was a highly fissile group, at least in later years. John O. Wisdom, who held a readership after Popper had been promoted to a chair in 1949, became a stranger when he devoted himself ever more to psychoanalysis, and in 1965 left for the United States. Another lecturer, Bill Bartley, followed him after he had tried in vain to prove the rational foundations of Popper's rationalism. John Watkins was the only one who, by his combination of straight thinking and courteous manners—the latter rare in the Popper circle—managed to remain on good terms with most members of the circle. Then there was Imre Lakatos. The explosive, humorous, brilliant, larger-than-life philosopher had left Hungary after 1956, and after dramatic years first as a communist reared by György Lukács and then as an outcast hiding in the mathematics section of the Academy. He was 34 when he came to Britain. Not only was Popper a revelation to him, but he also brought his own ideas on the philosophy of mathematics, so that he got another readership in the growing department. There he came to be much loved. After his early death in 1974, Ernest Gellner described his lectures as 'intelligible, fascinating, dramatic and above all conspicuously amusing',[31] and John Watkins remembered that 'the room would be crowded, the atmosphere electric, and from time to time [there would be] a gale of laughter'.[32] Not surprisingly, Paul Feyerabend had the most generous words for Lakatos, 'a fascinating person, an outstanding thinker and the best philosopher of science of our strange and uncomfortable century'.[33]

Inevitably, perhaps, Popper and Lakatos fell out. This was ultimately as much a matter of style and temperament as of substance. Lakatos himself wrote, in the two massive tomes edited by Paul Schilpp and devoted to *The Philosophy of Karl Popper*: 'The reason is not that our disagreement is too big; but that it is so very small.'[34] Popper did not share this view. In his reply to Lakatos he comments characteristically:

> After having been my student, he became my colleague in 1960, and he is now one of my successors at the London School of Economics. It is for this very reason that

I feel, unfortunately, obliged to warn the reader that Professor Lakatos has, nevertheless, misunderstood my theory of science; and that the series of long papers in which, in recent years, he has tried to act as a guide to my writings and the history of my ideas is, I am sorry to say, unreliable and misleading.[35]

One can see why John Watkins later spoke of the 'personal cleavage between Popper and Lakatos' from about 1970. 'Relations between them became very bitter.'[36]

Perhaps it is unwise for the outsider to meddle in these ancient and often arcane disputes. However, two wider issues of importance to the history of the School arose. One has to do with what Lakatos called 'scientific research programmes' and their methodology. Popper's methodology is simple, striking, and also very austere. It does not give much comfort to those who toil away in day-to-day research. For Popper, research without theory seems pointless, and research with theory has a point only if it is geared to falsification. In the social sciences in particular this makes most of the work people do appear superfluous. Some have therefore tried to rescue the meaning of empirical research, notably Thomas Kuhn and Imre Lakatos. The disputes about 'demarcation' and 'induction' to which this attempt led them seem themselves rather spurious in retrospect. The real question is much more practical: what are scientists and scholars doing when they are not inventing a theory of relativity or devising an experiment to test it?

The other issue has to do with truth. If we can only guess and never really know—what is it all about? Lakatos argued that the later Popper inserted in his methodology the notion that we are trying to get ever closer to the truth, though Popper held against him that he had said this ever since he encountered Tarski in 1935. Here is indeed a small difference. The more important point is that for Popper, truth is the ultimate objective of science, even if we can never tell whether we found it. At the other end stands Paul Feyerabend, who tries to make fun of such 'dogmatism' and plainly speaks of science as a 'game'. Western science, is moreover, only one of many games of the mind, and no one will ever know which is 'better' or 'worse'. Indeed, 'anything goes'. Lakatos had little time for Feyerabend's 'anarchism', but his irony also cast doubt on Popper's underlying belief in something that can never be demonstrated or proved.

All this has a great deal to do with the mood of the 1960s even if few at the School realized the full subversiveness of the Department of Philosophy, Logic and Scientific Method. Popper towers above the others. His position, however, tinged with Lakatos-style irony and combined with a profound knowledge of social processes, was later developed by Ernest

Gellner. In the 1960s, when he wrote *Words and Things* and *Thought and Change*, Gellner still had the curious title 'Professor of Philosophy with special reference to Sociology'; in the 1970s he was to join the Philosophy Department, which was then led by Watkins. Gellner's belief in the 'uniqueness of truth' coupled with his recognition of the varieties of its configurations in human life in society gave him firm ground to stand on in the troubled late 1960s. The other side, as it were, was taken by Paul Feyerabend. His relativism driven to excess—he liked to describe himself as 'Dadaist', which simply means that he loved chaos—describes more than Herbert Marcuse or anyone else the confusions, the hopes, and the limitations of the makers of trouble.

MORE BUT NOT ALWAYS MERRIER

The Robbins Report

When he reported on his seventh year as Director, the session 1963/4, Sydney Caine found himself for the first time unable to say that nothing particularly noteworthy had happened. 'The most absorbing activity of the School during the last session has been the re-examination of its general position and prospects, consequent on the publication in October 1963 of the Report of the Committee on Higher Education, generally known as the Robbins Report, after its Chairman, Lord Robbins.' The subjects of this re-examination were themselves absorbing: needs, or otherwise, of expansion; relations with the University of London; the longer-term future of the School. The first of these required a rapid response, which was given by special meetings of the Academic Board and the Court of Governors. For the longer-term needs, the Director set up a 'special Research Group'. Apparently it was called by some 'Sydney and his dreamers',[37] though in fact its fifteen members included some of the hard-headed senior members of the School as well as others renowned for their imagination: Professors Devons, Durbin, Edey, Goodwin, Moser, Roberts, and Wise, Dr Bridbury, Dr Corry, Mr Diamond, Dr Little, Mr McKenzie, Mr Minogue, Mr J. H. Smith, and the Secretary, Harry Kidd. The Group played an important part in the next few years, beginning with the School's response to the Robbins Report.

So far as reactions to a major committee report to government by the institutions involved and affected are concerned, responses to Robbins must be a record. But then the Robbins Committee was unusual. Lionel Robbins has described how reluctant he had been to accept the offer of the chairmanship made to him by the Home Secretary, R. A. Butler, in the summer of 1960. He wanted to write a book on economic theory; and he changed his mind only when an old Treasury friend pointed out to him, a little unkindly perhaps, that the Committee on Higher Education might have a longer-term effect than the book he was planning to write. After that, he tackled the task with characteristic vigour and thoroughness. The report produced three years later was an immediate sensation. What is more—and contrary to the Beveridge Report—within twenty-four hours Government published a White Paper on it. The report included six volumes of statistical analysis prepared by the research group which Professor Claus Moser had led and of which Richard (now Professor) Layard was a member. This material, Lionel Robbins wrote later, 'seems to me to be one of the most notable achievements of social studies in our time'.[38] The group, led in any case by LSE people, was transferred to the School in 1964 as the Higher Education Research Unit (HERU).

The reasons for setting up the Robbins Committee were many. New kinds of academic or para-academic institutions in advanced technology and teachers' training had sprung up whose status was unclear; the relative responsibilities of government departments, notably the Treasury, needed to be sorted out; the general penchant of the 1960s for planning had reached education; but above all there was the problem of the 'bulge'. A generation of secondary school-leavers was about to arrive at the doors of universities, but their keepers were unprepared. At this point, the Robbins remit tied in with wider international concerns. In many countries the question of university expansion had been raised. As early as 1961 the Organization for Economic Co-operation and Development (OECD) had held an influential symposium on 'Ability and Educational Opportunity'. An economic organization concerned with questions of ability? The reason was the growing belief that more higher education would stimulate further economic growth. Robbins displayed a healthy scepticism with regard to 'crude correlations between higher education statistics and gross national product'. But in another way he arrived at the same conclusion that expansion was necessary. 'It is reasonable to suppose that, in general, society is likely to be more efficient, more progressive, more humane, the larger the proportion of people who have had the opportunity of developing their intelligence in this way.'[39]

And so it happened. 'Apart from electronics and natural gas, higher education has probably grown faster than any other major industry in the 1960s.' Richard Layard and John King also note that the readiness of universities to expand, and to do so quickly, 'considerably surprised those who hold a general belief that universities are impervious to social need'.[40]

Clare Market on the Cam

28. A cosy retreat from the world. Grove Lodge, 1940.

29. Fifty years on. Alumni gather, September 1989. Dr I. G. Patel and the Revd Professor Sir Henry Chadwick, Master of Peterhouse, with the commemorative plaque

A different School in wartime Cambridge

30. Morris Ginsberg and students, 1940

31. F. A. von Hayek, and students including *seated extreme l* Stephen Wheatcroft, *2nd from r* W. Arthur Lewis, 1942

32. Open-air class, and more women than men in the war years, 1940

33. Department of Social Science and Administration, 1949, *front row l–r* Pleasance Partridge, Ruth Griffiths, Kitty Slack, Eileen Younghusband, Richard Titmuss, Christine Cockburn, Helen Judd, Kit Russell, Lesley Bell, Muriel Patten.

Back to London. Late forties and fifties. A rare balance between work and fun.

34. A Research Students Association Dinner, late 1950s. *Back row centre* Anne Bohm, *2nd r* Charles Manning, *seated l* F. Jack Fisher, *3rd l* Leslie Wolf-Phillips, Hugh Dalton, *r* Sydney Caine.

35. Sports Day at Malden, late 1950s. The staff tug-o-war team. *From the front* David Donnison, Harold Edey, J. A. G. 'Dixie' Deans, Raymond Chapman, (*part hidden*) Keith Panter-Brick.

A golden age for students

36. Fun, games and hard work: mock parliament 1948

37. The teacher lives on: H. J. Laski as drawn by student Ron Moody in a memorial issue of Clare Market Review, Michaelmas 1950

38. A masterpiece of engineering and economics. Professor James Meade introducing the restored Phillips Machine, STICERD, 1989. The audience includes, *r*, Mrs Valda Phillips, the widow of Professor A. W. Phillips.

39. The Troubles

31 January 1967. Prior to the 'banned meeting' in the Old Theatre. Sydney Caine (*centre*) with Harry Kidd (Secretary) and in front student Marshall Bloom

24 October 1968. Dr Meghnad (later Lord) Desai, Honorary President of the Students' Union, chairs a meeting

5 December 1968. Oration Day. At the lectern Lord Robbins; *seated l* Dr Walter Adams, *r* Professor Hugh Trevor-Roper

25 January 1969

Towards new horizons

40. Lord Robbins introduces benefactors to HM Queen Elizabeth the Queen Mother, Chancellor of the University of London, 1979, *from l* Mr Arthur Earle, Mr Peter (later Lord) Palumbo, Lord Rayne, Mr Eric Sosnow

41. At the Honorary Fellows Dinner, 1982. The Director introduces Professor Leonard Schapiro to Her Majesty. *Behind*, Professor Sir Karl Popper; *at the back*, Canadian Prime Minister the Rt Hon Pierre Trudeau

42. Under the Nicholson portrait of the Webbs, the Suntory-Toyota Foundation is set up. Sir Huw Wheldon signing the trust deed, 2 June 1978, with *to his right* Dr Shoichiro Toyoda (Toyota) and H. E. Mr Tadeo Kato, Ambassador of Japan; *to his left* Mr Keizo Saji (Suntory), and seeing that all is well Mr John Pike, LSE. *At extreme right* Professor Michio Morishima, originator of the project.

43. Two Chairmen looking up. Lord Robbins (1968–74) and Sir Huw Wheldon (1975–85) with Miss Jennifer Pinney, at the naming of the new Library building, 27 July 1978.

44. Five Directors, 1990. In front of the portrait of W. A. S. Hewins; John Ashworth, Sydney Caine, Ralf Dahrendorf, I. G. Patel.

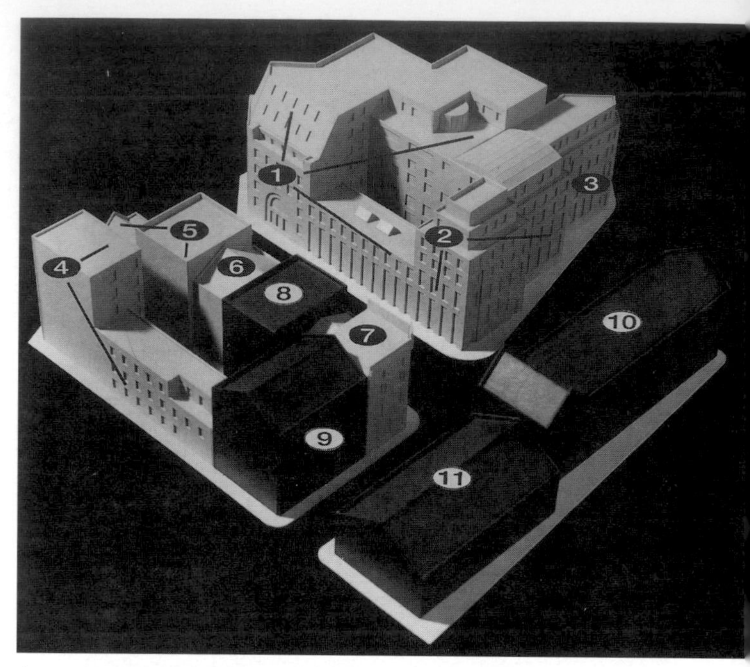

LSE The Buildings 1935

1 The 1922 building, extended and heightened 1928
2 The Houghton Street/Clare Market corner added as part of the Library development completed 1933, incorporating…
3 The Passmore Edwards Building and the 1925 Cobden Wing extension
4 Rear section of the East Wing, erected 1931
5 The former St Clement Danes Grammar School, and…
6 Nos 17 & 18 Houghton Street (shops/dwellings)—acquired 1929; adapted for LSE use until demolition in 1936 to make way for front section of East Wing
7 The original Three Tuns public house and No 15 Houghton Street, acquired 1932 though not in LSE use until 1946/7

and indicating

8 Holborn Estate Charity Office (acquired 1954)
9 8–12 Clements Inn Passage, incorporating The Anchorage (in LSE use from 1945)
10 St Clement's Press Building (acquired 1959)
11 Government Laboratory (acquired 1964)

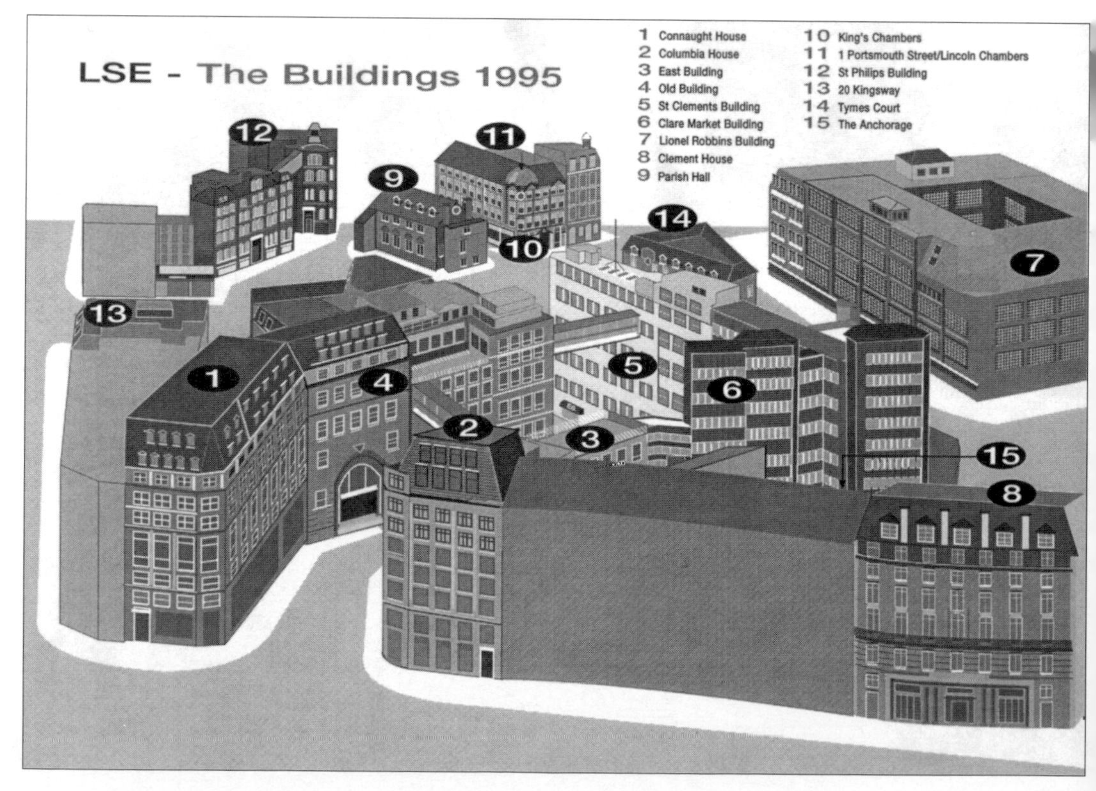

Nationally, student numbers had already risen by about 40 per cent in the six years preceding the Robbins Report; now plans were made to raise them by another 50 per cent within a mere four years. LSE, one might have thought, was overcrowded even without Robbins. But the School too rose to the occasion. It decided to increase its undergraduate numbers by 20 per cent by 1967 while continuing to open its gates to postgraduates. Almost half the total increase was to be effected in the next academic year, 1964/5. Overall, the Director reported in 1964, 'we proposed that our total number of regular full-time students should expand from a figure of about 2,450 in October 1963 to a total of about 3,000 at the beginning of the session 1967/68'. In fact, this figure was overshot by several hundred.

Not surprisingly, such changes provoked lively debates. The obvious question was: does more mean worse? This was the local version of the wider question whether there is a sufficient 'pool of ability' for further expansion. Both questions turned out to be less dramatic in practice than in theory, though doubts remained. For the most part, debates concentrated on practical matters. Obviously, more teachers were needed and also more staff. This depended crucially on additional finance. In the mid-1960s public funding presented no insuperable problem. In fact, between 1963 and 1967, the staff–student ratio at LSE improved from 11.1 : 1 to 9.7 : 1, and the number of full-time students per seat in the library went down from 3.1 to 2.8. Over time, however, money became scarce. The significant fall in student numbers, at least at LSE, which began in 1969, reflected demographic changes and perhaps even uncertainties about the School's public image after the troubles, but above all pressure on public funding.

The less obvious question was not asked at the time: will more mean merrier? Yet in the mid-1960s, and despite the relatively small increase in student numbers at LSE, almost everyone began to feel the pressure of growth. It is as if a threshold had been crossed beyond which the familiar cramped and somewhat chaotic life of the School became unbearable. Harry Kidd, then Secretary of LSE, later cited expansion as the first and possibly crucial factor in the background to the troubles of 1967. 'The life of staff, then, was anxious and full, not easy'. The library could not keep up with requirements. Above all, 'the School acquired the habit of taking in more students than it could comfortably provide for, of stretching its resources to the limit and perhaps beyond'. As a result, it 'felt more than usually overcrowded' and people began to complain.[41]

Once again, it is important to beware of retrospective determinism and describe 1963/4 as if the events of 1967/8 were an inevitable consequence. Still, when expansion began to be experienced as a problem, several things happened. The unity of the School began to crumble. People had to take refuge in their departments because the School as a whole no longer provided a home. At the same time, procedures became more formal, even bureaucratic. Questions were raised about the governance of the School because the old informal ways no longer commanded support. The whole syndrome of frustrations described so vividly by Fred Hirsch in his *Social Limits to Growth* came into being: more people wanting the same means that 'the same' is no longer and everybody feels cheated as well as frustrated.

The Graduate School

Not all proposals by the Robbins Committee met with a similarly quick and effective response. Lionel Robbins himself regretted that the organization of government continued to be what he regarded as inappropriate by lumping the whole of education together in one ministry. More importantly, later Governments—now of Labour persuasion—did not accept the proposal to expand one unified university sector, which entails a 'spectrum' or 'continuum' of institutions. Instead, Anthony Crosland insisted on creating polytechnics and with them the 'binary system' which existed until the early 1990s. Crosland had his reasons. 'Universities are unwilling to develop applied studies at a high level on a large enough scale, and are correspondingly remote from industry and commerce.'[42] Perhaps Lord Robbins let his aspirations for LSE (for example) get the better of his knowledge of reality when he described this as 'very unfortunate'.[43]

Another concern of Lord Robbins had to do with specialization. He felt strongly that undergraduate education should be geared to a corpus of general knowledge and he liked to quote the Scottish system as an example. In fact, as he well knew, his own School had been moving further and further in the direction of specialization at the B.Sc.(Econ.) level. For Robbins, the Graduate School was the appropriate locus for specialization. Here at least his view coincided with those of LSE. As the School reflected about its longer-term future, it firmly decided that expansion should be 'concentrated on the postgraduate side'. 'Sydney and his dreamers' suggested that 'in about fifteen to twenty years' time' the School should have 4,500 students 'of whom rather more than half would be postgraduate'.[44]

Committees of the School agreed. This was clearly a major policy decision. It is therefore sobering to look at the facts. Fifteen years after 1963/4, in 1978/9, the School had a grand total (including part-time and occasional students) of 3,696 students. Of these 1,739 or 45 per cent were higher-degree, higher-diploma or research fee students. Among full-time students the proportion of postgraduates amounted to 40 per cent. Twenty years after 1963/4, in 1983/4, total numbers had risen to 4,333 (of whom only 3,613 were regular students), and 1,935 or 45 per cent (39 per cent of the regular students) were postgraduates. Another ten years later, total numbers had risen significantly to 5,500 students, but postgraduate proportions remained roughly the same.

A policy failed? A policy which led the School in the right direction even if it did not achieve its objective fully? No policy at all but a series of market-led accidents? Such questions point up all the difficulties of theory in the social sciences. They also remind us of the limitations of the 1960s belief in the do-ability of social change. Above all they tell a story about LSE. The desire to increase graduate numbers in the wake of the Robbins Report was serious enough. The Director raised a number of questions about the process. What would it mean for staff? Were there better methods of selection? Was teaching and supervision of graduates adequate? How about the balance of the new 'taught' Masters' courses and traditional research? Would additional resources be needed? In the best LSE tradition, a study was commissioned. It was carried out by Howard Glennerster with the assistance of Anthea Bennett and Christine Farrell, and with material gathered by HERU, and it was published in 1966 as a book entitled *Graduate School: A Study of Graduate Work at the London School of Economics.*

The best LSE tradition means that data were gathered on every aspect of graduate work—admission and performance, wastage and course length, future demand and resource requirements—but that the authors abstained from taking a stance. There are one or two points which make one wonder whether a particular view was propounded after all. Glennerster showed that the intake of overseas students could easily double within a decade while that of LSE graduates and those from other British universities might treble. Thus why not turn LSE into a Graduate School? There was also an astonishing table on resource implications showing that a Ph.D gained in three years is significantly cheaper than an undergraduate degree which takes the same time. But to go further it took John King and Richard Layard of HERU, who had no inhibitions about making a definite case in their article 'The LSE as a Graduate School?'

King and Layard begin their article with the partisan question: 'If LSE became a predominantly graduate institution, would this imply that it took an impracticably large share of the national pool of postgraduates?' They wonder why so far (the article was written in 1970) the graduate intake had risen more slowly at LSE than elsewhere in the country, but give no clear answer. Could the 'undulating plateau' of graduate numbers be turned into an upward curve? King and Layard think so. It is Government policy to increase undergraduate numbers; many undergraduates will read social studies. If LSE's share of postgraduates from this rising pool remains constant at 9 per cent, and overseas graduates increase slightly, the School could have between 64 per cent and 83 per cent of its 1970s numbers in the Graduate School by 1981. 'What is clear is that, even if graduate numbers grew no faster at LSE than in the rest of the system, LSE would become a wholly graduate institution some time during the 1980s.'

Is this desirable? King and Layard think so. Postgraduate numbers in the social sciences (they are convinced) will grow. This is costly. 'To minimize the costs of achieving the desired result', concentration is needed. Economies of scale and comparative advantage tell in favour of an institution which already has a great deal of experience of graduate work as well as a Library and a favourable location.

> But for LSE people perhaps the most persuasive argument is that the LSE was founded to plug a gap in British education. It was extraordinary and unique in those days, but it has since become more and more like everywhere else, as everywhere else has started teaching undergraduates the social sciences. But there is still a gap. We lack first class graduate schools, and many of our best graduates still go to America to do their graduate work. Here is a social need which a more graduate LSE could satisfy.[45]

These are powerful arguments. Why, then, did the transformation of LSE into a Graduate School not happen? Another factual study in the best LSE tradition was conducted by an administrator, Dr Diana Sanders, in 1981. Among her many telling tables there is one of graduate admissions between 1966 and 1980. It shows that during these fifteen years the applications received had doubled. With certain fluctuations, about half of those who applied were offered places. Of these, again about half took up the offer. The resulting numbers fall well short of the extrapolations by King and Layard. When numbers rose in subsequent years, the increase was driven by the need to collect fee income and made possible by degrees which do little to justify the name Graduate School.

John Alcock, under whose supervision the Sanders study was undertaken, adds other factors. Contrary to earlier advice, the UGC began to discourage the School from moving further in the graduate direction after 1967. Also, 'the scale of the School's overall activities may have carried it beyond the point where it can pursue a single aim'.[46] The Academic Secretary might have added that while School committees decided that 'rather more than half', according to some, 'well over half' the students should be postgraduates, there never was even a significant minority on these same committees for abandoning undergraduate teaching altogether. Free-standing graduate schools are in fact few and far between anywhere. Where they existed, like the New School for Social Research in New York, or even the London Graduate School of Business Studies, they soon tended to go out and recruit unorthodox students: mature, short-term, part-time. Short of the unreal project of turning the School into an Institute of Advanced Study with but a few doctoral candidates, the only alternative was to separate undergraduate and graduate studies sharply. Some of the great American universities have done that and created a graduate school next to the undergraduate college. In the process of discussions about a new location for the School, the proposal was advanced by a section of the Academic Board. But to vary the Academic Secretary's point, the School's overall activities have never been carried to the point at which such division could make sense.

To be sure, had the School turned graduate and done so quickly, it might have saved itself a lot of trouble. Ernest Rudd has studied 'the troubles of graduate students' and found that they are many. There are, of course, all kinds of graduates, 'dedicated scholars', 'would-be academics', career-oriented Masters' students, 'drifters', even 'Peter Pans' who are 'unwilling to grow up and leave the university'. All of them are liable to suffer from 'social loneliness and intellectual isolation'; they often find the staff remote and facilities unsatisfactory. Yet 'they are not in a constant state of revolt'. Peter Pans may become the 'shop stewards' of student protest but they are the exception. 'Graduate students do rebel against their working conditions, but they do it in a different way—they drop out.'[47] Certainly Howard Glennerster found among all LSE higher-degree students a 'wastage rate' of over 40 per cent. It would seem that on top of everything else a Graduate School without undergraduate infrastructure would be a recipe for institutional consumption in the sense of the old wasting disease.

Croydon School of Economics

The Director's initial reaction to the Robbins Report contained an ominous statement. It would be all but impossible, he said in his 1964 Report, to accommodate an institution twice the size on the present site of LSE. 'Accordingly, we have started an examination of possibilities of removal to other sites, whether in central London or in the periphery of the greater London area.' This is how LSE almost became the Croydon School of Economics.

The School has been itchy about its site throughout its history. No single issue seems to have occupied successive Directors more than how the

School can make do with its buildings and whether it would not be better to move somewhere else. In fact, LSE moved only once, in 1901, when it crossed the Strand to take possession of the newly erected Passmore Edwards Building in Clare Market. After that, dreams of total relocation were usually accompanied by realities of piecemeal extension at or around Clare Market and Houghton Street. Beveridge wanted a Left Bank in Bloomsbury with LSE as one of its central attractions. Judging from the one edifice put up through his good offices, Senate House, it would have been more like Brasilia than St Germain. In any case, Rockefeller money helped expansion within, notably on Houghton Street. Carr-Saunders had to move the School all the way to Cambridge at the beginning of the war. Some were happy in Arcadia but the Director himself had a very strong sense of the essential importance of the bustle and even the inconveniences of Clare Market for the vitality of the School, which is why he put much effort into the speedy return of LSE to its familiar environment. Caine played his cards closely to his chest, so that it is not altogether clear where his preferences lay, but in reading the committee papers of 1964 one gets the sense that he would not have minded the green field in the south of London.

In any case, the Director asked his Research Group to investigate 'the possibility of moving to a site which could provide room for expansion and buildings designed for its own use on an unrestricted site, as against the present situation in which there is a chronic shortage of space and use has to be made of such buildings as become available'. The Committee started its exploration near home, at 'Covent Garden, the South Bank, the Euston Station development, and the Dockland area near the Tower of London'. But it soon became clear that the chances of finding an appropriate site in central London were small. London's government, the Greater London Council (GLC) which had recently emerged from the ashes of the LCC, was of course firmly established in its majestic monument to bureaucracy, County Hall.

At that point in autumn 1964, the Chief Education Officer of what was then the County Borough of Croydon, who must have heard of the search begun by LSE, approached the Director to offer a large greenfield site, Heathfield. The old farm had passed through various hands before Croydon took it over from the estate of an 'insurance magnate'. Faced with this prospect, the Research Group did its name proud. It not only inspected the site, but investigated several of its aspects with scholarly thoroughness and precision. Geographers drew maps of the distance of the site from various central London as well as country locations. Accountants found out the precise cost not just of the site but of relocation and of the new buildings needed. The Secretary of the School, Harry Kidd, went into considerable detail in his letter of 16 October 1964 to the Borough: 'We have been accustomed in Central London to finding plot ratios and angles of light the principal limiting factors in planning new buildings.'[48] A geographer-cum-historian described the history and brave new present of Croydon. 'An immediate impact was made [on the Research Group] by the new office building, the re-development of the shopping centre and by the new civic centre.' But were the 45 acres of Heathfield big enough? Michael Wise thought that in the light of the fact that new universities generally sought an area of about 200 acres 'it might well be desirable to seek additional land for development either on the Heathfield estate itself or on some closely adjacent site'.[49]

By May 1965 the Group was ready to report through the General Purposes Committee to the Academic Board. The Group was divided and put both 'the case for moving' and 'the case for remaining in Houghton Street'. Croydon was an opportunity to be seized. 'If this opportunity is allowed to pass, it is felt that the School will be taking a decision to remain in Houghton Street indefinitely.' Improvement and expansion were possible on an attractive site. Why should 'visiting speakers and lecturers' not come to Croydon, since they were prepared to go to Oxford, Cambridge, and the new University of Sussex in Brighton? It will be easier to find non-academic staff in residential Croydon than in central London. The move itself may be disruptive but the upheaval might be cushioned by another decision, slipped into the report almost as an afterthought: 'It is suggested that it would be quite acceptable to move the undergraduate work to Croydon and to leave the postgraduate side in Houghton Street.'

The case against hinted at possibilities of expanding on the central site, including the (still very tenuous) hope to acquire Strand House, the warehouse of W. H. Smith's, for library purposes. It saw the upheaval of the move as a major negative factor. Then the advantages of the central site were listed, including its attractiveness to visitors and students from overseas. Several greenfield social science universities (presumably Sussex and Essex) were already being created; one at least had to be in the centre of a large city. 'It is recognized both by those who favour a move and by those who want to stay that a School of Economics in Croydon would be a rather different kind of place from a School of Economics in Houghton Street.'

When the Academic Board assembled on 26 May with a record turnout of 125 members, the Director summarized the situation with exemplary neutrality. Yet his list of the hurdles of finance and planning which would have to be overcome if the School were to move cannot have helped the Croydon party. In any case, the minute states, 'it emerged from the debate which followed that the Academic Board was almost unanimously opposed to a complete move of the School to the site at Croydon'. Possibilities within central London were invoked, including the new library building, as well as planning perspectives which might emerge from the new GLC. Above all, 'members of the Board expressed the opinion that the School's present site was as good as any which was likely to be offered and that there was little point in seeking for an ideal which did not exist'.

A separate motion put by Professor Self and Dr Tropp to move undergraduate activities to Croydon and keep the Graduate School at Houghton Street was opposed by Donald Cameron Watt and lost by a large majority. Thus LSE remained as it was as well as where it was. Heathfield became the home first of a teacher's training college and later of the staff training centre of the Borough of Croydon. But the whole affair had given LSE a chance to have a good look at itself. The result was that the School on the whole liked what it saw. The door to the future was left slightly ajar even in the Academic Board resolution by 'not ruling out the future possibility of [the School] considering an alternative site in central London, should such a site ever become available'. Basically, however, the advantages of proximity to many relevant institutions and of accessibility won the day.

Few made the perhaps more romantic points that any change in location can damage the soul of an institution, and indeed that the cramped and crowded conditions of Houghton Street have their own charm. The Secretary with his Oxford past (and future) had a clearer notion of what the move would have meant than teachers like Robert McKenzie and Alan Little. A move would affect the intellectual life of the School. 'We might be physically much more beautiful, but we should, before very long, be a much dimmer lot.'[50] It may be better for students of social science to be exposed to the nervous pleasures and pains of city life than to green fields and rectangular office blocks. When porters were interviewed by David Kingsley about the Croydon affair, they saw the point. '"LSE is *here*," said one porter, "and anywhere else it would become a different place."'[51]

The Machinery of Government

One of Sydney Caine's main preoccupations in the later years of his directorship was the structure of academic administration. His enthusiasm for this to most somewhat arid subject was shared by the Secretary of the School, Harry Kidd. Kidd had succeeded Eve Evans as Secretary in 1954 at the age of 37. He had by then been First Assistant Registrar of Cambridge University for nine years, and before that an official in the wartime Ministry of Labour. Caine later described him as a paragon of the new profession of university administrators. The two worked closely together. 'Above all I at least, as Director, have looked to the Secretary for advice on any and every subject that has come up.'[52] Thus Director and Secretary tried to reform the 'machinery of government' together. They began with fairly modest ideas about a more effective academic committee structure in 1961/2. In 1963/4 a new impetus was given to administrative reform by the Robbins Report, notably so far as relations with the University were concerned. Since 1966 the 'Machinery of Government of the School' was the subject of a special committee. When it reported in February 1968, much had changed. Harry Kidd had left for the untroubled world of St John's College, Oxford. Walter Adams had become Director. The report ran to 25 pages, but to these were added 10 pages of 'reservations' and 'dissent' by various members of the Committee, and a 58-page Minority Report by two student members, Dick Atkinson and David Adelstein: 'We wish to challenge the assumptions and limitations of the report, to raise deeper problems, ones relating to educational aims. It follows that our proposed solutions will differ in fundamental ways from those of the Majority Report.'

The early proposals by Sydney Caine were straightforward enough though nevertheless of considerable significance. The Appointments Committee, consisting of all professors, had become too unwieldy to deal effectively with appointments, promotions, and other staff-related matters of academic policy. It was therefore proposed that a Standing Sub-Committee should be established, consisting of the Director and eight members of the Appointments Committee. These eight should be made up of two representatives each of four departmental groups. When the Appointments Committee and Academic Board adopted this proposal in June and October 1962 respectively, they took two decisions simultaneously. One was to recognize at last the existence of departments, including department heads, though these were called Conveners in order to avoid labels suggesting entrenched power, like 'head' or 'chairman'. The departure from a long-standing tradition nevertheless merely recognized reality in most areas of the School. The other decision pointed in a sense in the opposite direction. By keeping a single committee for the whole range of subjects represented, the School reaffirmed its intention to remain to all intents and purposes a single-faculty school and not to allow the complete compartmentalization of the disciplines taught at LSE. The combination of devolution and co-ordination worked and has served the School well in subsequent years.

The Robbins Report raised less tractable issues. Since 1901 LSE had been a college of the University of London. The relation between the School and the University had its ups and downs; one might be forgiven for comparing it with that of member states to the European Community, like London University a federal arrangement *sui generis*. In the early days, LSE needed the University—rather like Germany needed the early EEC—in order to gain recognition and be a part of the University scene. The Beveridge years were more turbulent, with William Beveridge temporarily taking control of the University but making himself thoroughly unpopular in the process. In the 1950s and 1960s, a sense of friendly detachment characterized LSE attitudes to the University. Few members of the School took an interest in Senate House; but the benevolent reign of the Principal, Sir Douglas Logan, whose wide experience included a stint as an LSE law lecturer in the 1930s, meant that the University was accepted. John Alcock 'never felt that my presence at Senate House was particularly useful' and thought that the Director shared his view. 'My impression is that Sydney felt that the School was the School and should look after itself and what went on at Senate House was really of no consequence so far as we were concerned.'[53] On the academic side this was evidently true. While degrees, apart from certain diplomas, were University degrees, teaching and examining took place at the School. Appointments would have involved external assessors wherever they were made. On the financial side, however, the University offered additional protection from government interference. The Court of the University was like a second-tier University Grants Committee, receiving and dispensing block grants within London.

Still, the Robbins Report suggested that it was time for the University of London to review its rather confusing structure, and notably its relationship with constituent schools. The suggestion led to a decade, if not more, of if anything aggravated confusion. 'An almost continual salvo of reports'[54]—'Saunders', 'Murray', 'Phillips', later 'Flowers', 'Swinnerton-Dyer'—set in motion tidal movements of decentralization and centralization in quick succession, until in the end, in the 1990s, the University began to fall apart, never to see its centenary as the institution which Haldane and Webb had invented. The School's initial position was well put by Sydney Caine. 'The School did not wish to propose any separation from the University but felt it to be essential that it should have greater freedom within a revised federal structure.'[55] This did in fact come about, and it remained LSE policy until changes in funding patterns led a later Director to take a much tougher stance. In 1992 John Ashworth wrote:

In the LSE, only one-third of our income now comes from Senate House. So clearly the links which bind us to Senate House and thus to the Federal University are going to weaken . . . we no longer see the point of having the kind of bureaucratic machinery that Senate House represents, it no longer seems to us to be functionally useful.[56]

Sydney Caine saw a link between the devolution of functions within the University and the need to review the machinery of government within the School. In his view the new responsibilities of the School in the development of degrees and in other academic matters had led to a growing

overhauling so that we may be able to ensure that all the issues arising in connection with projects of new academic development, including the syllabus of any new degree, its relationship to other degrees of interest to the School, and the staff, accommodation and financial implications of the development can be considered together.[57]

This is a rather low-key statement of the far-reaching proposals which the Director incorporated in a memorandum to all members of staff and to Governors in August 1966. Here, Caine tried to tackle three major issues, the relations between the Court of (essentially lay) Governors and academic committees, the efficiency of the committee structure, and the involvement of students.

To deal with the first two issues the Director proposed a structure not unlike that of other universities outside Oxford and Cambridge. Most of the functions of the Court should be taken over by a 'Council' with ultimate financial authority. The Council should also have greater academic representation than the present Standing Committee of the Court. On the academic side, a 'Senate' should combine the functions of the Appointments Committee and the Academic Board in one widely representative body. When the Machinery of Government Committee finally reported in February 1968, it did in fact propose a Council and an Academic Senate in somewhat modified form. Two members of the Committee, Dr Miliband and Professor Wedderburn, wrote an important note of dissent. They saw no point in having a Court of lay outsiders at all, and they wanted to preserve the all-inclusive 'democratic' Academic Board. In the event, disagreements went much wider. When Walter Adams reported on his first year in 1968 he began with a statement of management philosophy quite unlike that of his predecessors: 'The School has always enjoyed the benefit of an unwritten constitution and the quite exceptional flexibility that this has given it in its internal government and administration.' He then reported that the Machinery of Government Report 'was widely discussed throughout the School during the Lent Term. In the event, neither the Academic Board nor the Students' Union approved the Committee's report, and further discussions are proceeding to find agreed proposals.' There are, to be sure, needs for change, but 'it may be better to proceed by a series of practical experiments and trial and error, than by the theoretical formulation of new constitutional schemes and the unending debates to which these give rise in an academic community'.

And so the Report sank almost without trace. Two minor though not insignificant changes resulting from the exercise were the creation of the post of Pro-Director and of a Dean of Undergraduate Studies. In 1967, Professor Harold Edey became the first Pro-Director and Dr (later Professor) Percy Cohen the first Dean. Beyond that the School kept much the same administrative structure for the next twenty-five years, until Dr John Ashworth returned in 1993/4 to some of Sir Sydney Caine's ideas, if without the terminology of Council and Senate.

With all its apparent complexity the structure was fundamentally simple. The Court of Governors remained the ultimate authority and was strengthened by eight elected members of the academic staff and six students among its nearly 100 members. Its executive committee, the Standing Committee, is in fact the Council of the School. It has equal numbers of academic and lay members and a lay chairman. On the academic side, the Academic Board is the assembly of all members of staff. Its work is prepared by committees, notably by the General Purposes Committee, to which an Academic Policy Committee was added later. This Committee, advisory to the Director in general policy matters, first grew and proliferated, then became cumbersome, and finally was discontinued in 1982. Appointments, promotions, and related matters are dealt with by the Appointments Committee and its Standing Sub-Committee as devised by Sydney Caine. The Director is Secretary of the Court, chairman of the Appointments Committee, as well as the Academic Board (and he is an *ex-officio* member of all committees).

Vexations of Power

Where, then, does the power lie? Power is an elusive concept, especially in academia. A later Director's Report makes the point.

One former Pro-Director used to tell the story of how, as a lecturer, he suspected that all power rested with the professors. When he became a member of the professoriate he thought that perhaps in the Standing Sub-Committee of the Appointments Committee, or in the Standing Committee of the Court of Governors lay the real seat of power. In due course, he became a member of both and decided that power must reside somewhere around the Director 'on the sixth floor', until as Pro-Director he saw yet another suspicion evaporate before his eyes.[58]

Perhaps Alan Stuart (the Pro-Director in question) underestimated the extent to which power in academic institutions is the ability to block change rather than that to initiate change; even so, it is certainly diffuse.

Power also shifts. The Director continues to have a central position if only by virtue of his full-time commitment and unrestricted access to information (though not always rumour). In its own area and at times beyond it, since it is composed of senior professors, the Standing Sub-Committee of the Appointments Committee is very important. More generally Harry Kidd noted a steady shift 'from the Governors and their Standing Committee to the academic staff'.[59] Still, the Standing Committee remained the ultimate authority in matters of finance and general policy direction for many years, with the Academic Board at the other end the seismograph of legitimacy whose indications of mood all others would ignore at their peril.

Wondering why the Machinery of Government Report disappeared almost without trace, one might be tempted to dig deeply into the traditions of the School which were closer to Walter Adams's praise of the unwritten constitution and of trial and error than to Sydney Caine's search for efficiency and formalization. At a less fundamental level there is the fact that Sydney Caine made his proposals for major change shortly before his own departure. 'I feel a certain embarrassment in making this point,' Lord Bridges, the Chairman of the Governors, wrote to him, 'but it seems to me there would be a certain awkwardness in a new scheme being introduced by you at the very tail end of your Directorship.'[60] What if the successor has different views? Above all, however, a sea change in subjects of dominant concern occurred during the very years 1966 to 1968. In Caine's initial proposals, student participation in the governance of the School was but a minor feature on a much wider canvas. Two years later, it had become the central issue. Hence the Minority Report and the 'notes of reservation' by Colin Crouch and Peter Watherston, by David Kingsley and by Professor Ben Roberts.

The Director's annual comments on the Students' Union and students more generally were almost as predictable as his statements about the absence of noteworthy events in the early years. 'Relations between the administration and student representatives remained as they have been for some years past, cordial.' Thus in 1963/4, and in 1964/5: 'Relations between the School and the Students' Union officials have continued to be cordial.' However, in 1965/6, the tone changed. A long section of the Director's Report on Student Affairs introduced new themes as well as a new style. 'During the last year or two a number of issues have arisen in discussion with the officers of the Students' Union and other student representatives in which there has been evidence of wide differences in the approach to staff–student relationships.' Such discussions are 'by no means confined to the School', they are 'widely spread among universities in the Anglo-Saxon world'. The only difference is that the School has 'more quickly become aware' of a need to review relations and consider change than others.

The issues have to do with the status of the Students' Union, its financing, the range of its activities, the demand for a 'sabbatical' year for its President, but also with 'representation of students on the governing body or important committees'. Caine refers to the 'staff–student liaison committee' and other methods of consultation, but notes that there is 'no provision of representation of students' and that the Academic Board 'decided that no such right could be given'. Moreover, the School had not accepted the sabbatical for the Union President. 'It has been clear in the discussion of these various matters that, while the student representatives have failed to convince the School authorities of the justice of all their proposals, we on our part have failed to convince the student representatives that their proposals are not reasonable'. There is thus 'a basic divergence of view' at least about the role of the Students' Union.

The Director then gives his version of the points at issue. It is debatable whether Students' Union activists represent student views generally. At the same time, many members of the teaching staff feel that questions of the position of the Students' Union 'are really the outcome of a much more deep-seated malaise affecting the whole relationship between students and teachers'. The Academic Board had therefore set up a committee to examine staff–student relationships. However, special LSE features apart, 'it is clear that any malaise which exists in the School is part of a malaise affecting most, if not all, universities in the country and we venture to hope that if we are able to discern ways of improving the position in the School as a whole we may also be able to help in the solution of the wider problem.'

Rarely has a pious hope been more rudely shattered. The following year, 1966/7, turned out to be one of the more traumatic years in the history of LSE. The Director made it the subject of a long appendix to his last report entitled 'The Student Disturbances'; in the course of the year, his hands-on yet even-tempered involvement in the troubles had been accompanied by numerous open letters, reports, and memoranda from his pen describing the events. In July 1967, the School's Publications Officer P. D. C. Davis published an account, 'The Troubles: A Chronology with Observations' in the *LSE Magazine*. Other, more partisan accounts appeared at the same time. The Director's formal report on the year mentions that 'Mr. H. Kidd, who had occupied the post of Secretary with great distinction since 1954, was offered the office of Bursar of St. John's College, Oxford and resigned with effect from 1 August 1967'. Kidd had been at the heart of the disturbances, and within two years his detailed account, *The Trouble at LSE 1966–67*, appeared in print. A group of teachers of sociology and social administration conducted an extensive survey of student views in the summer of 1967; it was reported in 1970 in an LSE Monograph by Tessa Blackstone, Kathleen Gales, Roger Hadley, and Wyn Lewis, *Students in Conflict: LSE in 1967* (with an important preface on university governance by Professor Donnison). The view from the other side, as it were, was put by Colin Crouch, President of the Students' Union in 1967, in his adventurous mixture of description and analysis, *The Student Revolt* (1970). Crouch may have been a radical by some standards but students regarded him as a moderate compared to the militant left and also the 'direct action' eccentrics. Their perspective on the second troubles of 1968/9 was put by Paul Hoch and Vic Schoenbach in *LSE: The Natives are Restless* (1969). If one adds the dozens of pamphlets and articles and the hundreds of newspaper reports of the time it is clear that there is a story which needs to be told.

THE TROUBLES

A Director is Appointed

Normality is often deceptive; at LSE in the 1960s it certainly was. The contrast between tranquillity at home and winds of change all around sharpened every year. With the Robbins Report, change had reached the School as well. The mood of expansion created a sense of pressure and of anomie at the same time. Staff and students found the School a less homely place; they felt lost but also exposed to masters whose bureaucratic power they could neither permeate nor control. Teachers became more competitive; students wondered about their opportunities. The triumph of the social sciences was also a moment of doubt in their future direction, and their uses in the real world. When assumptions began to be questioned, the process turned to issues of governance, especially the participation of hitherto excluded groups.

At the same time the social and political environment impinged on internal debates. President de Gaulle's antiquated authoritarianism in France and the opposition-less Grand Coalition in Germany had created their own extra-parliamentary protests. In Britain, the great expectations of many which greeted the Labour Government of 1964 soon turned into disenchantment and a feeling of betrayal which encouraged a 'new left'. Everywhere the question was raised: what can we do to change things? How can democracy be made more real? Increasingly the answer was: by helping ourselves, by direct action. The American Civil Rights campaigns but also the 'free speech' student movement at Berkeley set the tone. In Germany, the Emergency Powers legislation brought hundreds of thousands out into the streets. The demand for nuclear disarmament had much the same effect in Britain.

International issues were soon mixed in with more parochial concerns. For one thing, people moved across borders: CND activists all over the place, Berkeley (and other American) students to Britain, not least to LSE. More importantly, events in seemingly distant places were brought home, literally to homes, by television and other means. The Vietnam War was an American war which took on world-wide dimensions. Tens of thousands of people were killed in South-East Asia at the very time at which civil rights legislation was introduced by President Johnson. Nearer home, South Africa had been made to leave the Commonwealth in 1961 because of its racial policies; apartheid remained a source of moral outrage for the next thirty years. Rhodesia, then still a colony, increasingly grabbed the headlines. In 1964 Northern Rhodesia became an independent republic called Zambia. Southern Rhodesia, however, remained under white minority rule, and in November 1965 Prime Minister Ian Smith unilaterally declared independence. Years of battle with this 'rebel against the crown' (Harold Wilson) began.

It is thus more than retrospective determinism to say that all that was needed was a focus to set the diffuse discontents alight. This was not a matter of a revolutionary powder keg lit by a spark, for there was no revolutionary powder keg. The various discontents were not only diffuse but also unconnected. 'There is no unity in their angers.'[61] However, they were temporarily connectable by virtue of a widespread latent readiness for direct action against established authorities in the name of democracy. What was needed was some incident or fact which made the overcrowded library, the betrayal of the people by its government, and racial discrimination in a flash look like one big conspiracy of the powers that be. The appointment of a new Director at LSE achieved this to perfection. Abstract questions of governance without participation suddenly got a name, and more, it was the name of someone who had spent the last decade holding office in Southern Rhodesia of all places. The opportunity was too good to miss, though one wonders whether those in charge could have foreseen and prevented it.

What followed was far from a revolution. It was not even a revolt. It was trouble, to be sure, *the* trouble, as it soon came to be called. In fact, as we shall see, it was two troubles with a year in between and the two were quite different in character. The plural, 'the troubles at LSE' is therefore strictly correct. But the explosions which occurred, while infinite grist to the mills of interpretation, had little direction or purpose. They left charred remnants and scars on many. They heralded a less happy age for the social sciences and their practitioners, not least for LSE. But they did not change the world. They did not even achieve a single one of their declared purposes. The troubles at LSE were ultimately an episode, if one that went close enough to the raw bones of the School to tell us a good deal about the vibrating, exasperating, simultaneously wide open and strangely closed, confident yet endearingly uncertain, unique London School of Economics and Political Science.

The story begins before the end of the session 1964/5, when Sir Sydney Caine confirmed his intention to retire in July 1967. In December 1963 his appointment had been extended beyond the age of 62 which Caine was to reach in 1964. The Director insisted, however, that he did not want to continue for more than another three years. 'Administrators, the Director had said, can lose their efficiency rapidly at this sort of age and 65 was old enough, in his view for a definite commitment.'[62] Accordingly, a Selection Committee was set up in June 1965. It was chaired by Lord Bridges, the Chairman of the Governors, and included the Vice-Chairman, F. E. Harmer, the Vice-Chairman of the Academic Board, Professor Michael Wise, six lay Governors ranging from Sir Geoffrey Crowther of *Economist* fame to the trade union leader Vic Feather, and five professors nominated by the Academic Board, Professors Allen, MacRae, Oakeshott, Schapiro, and Titmuss. In October 1965 the Committee began its extensive trawl of possible candidates.

The procedure adopted at the end of the trawl was probably unique. From a long list of seventy suggestions the committee produced, on 9 December, two shorter lists of fourteen external and ten internal candidates. It then began a complicated process of weighted voting which yielded four clear front-runners, one of whom was later replaced by the quasi-internal candidate Sir Ronald Edwards, who of course was no longer a professor but Chairman of the Electricity Council. The other three were Alan (later Lord) Bullock, then Master of St Catherine's College, Oxford, (Sir) William Deakin, Warden of St Antony's College, Oxford, and Sir Fraser Noble, Principal and Vice-Chancellor of the University of Aberdeen. In February the *Observer* carried a story which was at least partly correct. It surmised that 'most of the staff of the School as a whole are strongly, and even passionately, opposed' to Sir Ronald Edwards.[63] In any case, Alan Bullock was asked and declined; he had set up his College in 1960 and wanted to stay, quite apart from his involvement in the University of Oxford, whose Vice-Chancellor he became in 1969. William Deakin also had other plans; while he left St Antony's two years later (and the College did not 'collapse without him' as the *Observer* had suggested), he did not want another major administrative job. Sir Fraser Noble was happy in Aberdeen.

The Committee met again on 27 April and 17 May 1966. It did not revert to its various lists but considered two entirely new candidates, one of whom was quickly agreed, Dr Walter Adams. Adams was, of course, a familiar name at least for older members of the School. After the short stint as the Secretary of LSE he had served in the Foreign Office during the war and then become Secretary of the Inter-University Council. In 1955 his chairman, Sir Alexander Carr-Saunders, as well as the Foreign Office, regarded him as eminently suitable to become Principal of the new University College of Rhodesia and Nyasaland. Carr-Saunders wrote in his recommendation:

> Adams does feel very strongly that Africans should have the same educational opportunities as Europeans, but this you must either know or suspect already . . . He is not the sort of man who gives expression to large views on racial policy; his attitude would find expression in and around practical questions such as admission to educational facilities on equal terms.[64]

The shrewd description of Walter Adams's character encapsulates both his strengths and his weaknesses. In Rhodesia he was obviously much in the news. During the LSE disturbances, students re-published the 26 May 1956 issue of the *New Statesman* which had reported attacks by Hugh Gaitskell and the Labour Party on 'the liberal-minded principal'. While Adams had made sure that by statute no test of race would be imposed for admittance, he had also tolerated separate residential and dining facilities for black and white students. The *New Statesman* asked, 'is it justifiable to insist on the principle of no segregation at all costs, even when it means rejecting the advice of a man of the calibre of Dr. Adams?', and it gave the principled answer, yes, whereas Adams had found a pragmatic way to accommodate black students.

In March 1966, at about the time at which the LSE Selection Committee had gone back to the drawing board, more complex and serious issues arose at the University College of Rhodesia which gained publicity in the London press. The College had been in turmoil for several months over various matters involving race. Adams was on the one hand his courageous pragmatic self; he publicly refused to give the police any information about a black 'restrictee' who had been seen on campus. On the other hand, a report commissioned by the University Council from the former headmaster of Eton and steadfast campaigner against apartheid, Robert Birley, described Adams as somewhat remote in crises. In fact he had been absent in London for financial negotiations during one critical period. Adams offered his resignation but withdrew the threat when he was fully exonerated by the Council. He clearly had both critics and friends. 'As principal,' the *Guardian* of 18 April 1966 reported friends as saying, 'he has always been on a tightrope, and they think of him as never rattled. "I've never seen him worried," said one. "He is so used to living in crises."'

Within a few weeks of these incidents, on 17 May, the LSE Selection Committee decided to invite Walter Adams to be Director from 1 October 1967 until he reached the age of 65 in 1972, that is, for five years. The invitation, Adams told the press, 'came out of the blue'. By 7 June Adams had agreed to let his name go forward. It remained for the Court of Governors to take the formal decision. Meanwhile, Harry Kidd had engaged in correspondence with the Chairman about the long-standing practice of the School to delegate most powers of the Court of Governors to the Director on appointment. Did the Articles of Association really allow this? Lord Bridges refused to take the Secretary's legalistic point seriously. More significantly, no one took any notice of a proposal made by the Students' Union to the Staff–Student Committee meeting on 15 June: 'that the Students' Union should participate in the final decision of the appointment of the Director of the School'. On 16 June the Court of Governors met and accepted the recommendation of the Committee. It would be more correct to say that it ratified a decision which to all intents and purposes had been taken. Members of the staff were informed on 17 June, and the press acclaimed the appointment the next day, with even the Communist *Morning Star* referring to Adams as a man who had been 'defying police attempts to recapture an African student'.

Death of a Porter

On 19 August 1966 *Private Eye* carried comment on Walter Adams (as well as Sir Douglas Logan) and Rhodesia. 'No one would call him racist. But he has exhibited a constant willingness to compromise, and accept the status quo, even in an unconstitutional de facto regime.'[65] A detailed account of attacks on academic freedom at the University College of Rhodesia followed. Members of the LSE Socialist Society ('SocSoc') which had split off from the Labour Society in 1965, probably read this. In any case, a number of them began to put together material on Walter Adams. They claimed that their twenty-page *Agitator* publication, *LSE's New Director: A Report on Walter Adams*, had been prepared by 'a small group of about twenty students', though Colin Crouch tells us that in fact only two were responsible. The sixpenny paper which was to spark the trouble at LSE was in some ways a shot in the dark. The authors could hardly predict its effect (as the underestimate of its print run of 750 showed) nor is its political thrust clear save for the conclusion

> that Adams—a Principal unprepared to defend the freedom of his staff and students—is not a suitable person to be placed in charge of any centre of higher education. Nor, especially, is he suitable as the Director of a multi-racial college like L.S.E., since his belief in multi-racialism does not seem to extend to actions in its defence.[66]

Compared to later 'revolutionary' literature, the pamphlet is an almost scholarly attempt to describe the alleged weaknesses of Walter Adams as the head of a university. Some of these sound strange from the pens of the far left: 'avoidance of important decision making', 'extreme isolation from staff and students', 'administrative inefficiency'. The latter was later to be the main reason for Lord Robbins and other academic Governors to turn against Adams. But the thrust of the pamphlet had to do with Adams and his unreadiness to oppose Rhodesia's Unilateral Declaration of Independence (UDI) outright as well as his general attitude—contrary to specific cases—to racial discrimination. Here, the paper was based on the July 1966 Birley Report to the Council of the University of Rhodesia and a report to Amnesty International made by the human rights lawyer Louis Blom-Cooper in August 1966. Both were quoted out of context, and both protested later. Blom-Cooper said that his 'comments about Dr. Adams formed only a small section of my report': 'I was not, and am not, concerned with the propriety or otherwise of Dr. Adams's appointment as director-elect of the London School of Economics.' Robert Birley went even further: 'Dr. Adams's work in founding the College has been one of the few constructive pieces of work on behalf of multi-racialism during recent years on the continent of Africa.'[67]

But when these letters to *The Times* were written, it was too late. The *Agitator* pamphlet, sent to the media on 14 October and formally published three days later, had done its work.

On the day of the pamphlet's publication Adelstein [the President of the Students' Union, a sociology student of South African extraction] wrote to the Chairman of the Governors, enclosing a copy of it, and asking how much of the information contained in it was true, and had been known at the time of Adams's appointment; what had been the reasons for overlooking the criticisms, and whether there was now a case for reconsidering the decision.[68]

Lord Bridges replied in constitutional terms; appointments procedures were confidential and there could be no public debate of the merits of the case. His letter, by mishap, reached Adelstein late, too late to be known to a Students' Union meeting on 21 October which instructed the President to obtain from Dr Adams a reply to criticisms within eighteen days, failing which Union would oppose the appointment.

The authorities did not like this demand at all, but in view of widespread public interest they had to react visibly. As a result, Lord Bridges wrote to *The Times* on 25 October. He expressed the 'indignation which many of us feel' at the campaign against Adams but insisted that it would be 'neither necessary nor indeed proper that any reply should be made by L.S.E.'. In any case, enquiring into Dr Adams's role in Rhodesia would mean meddling in the internal affairs of another college. The Students' Union regarded this as a direct attack and asked its President, David Adelstein, to reply. Adelstein was a moderate man, active in student politics, inspired by a strong sense of justice, but respectful of institutions and far from 'SocSoc' militancy. When he was told that the letter he wanted to send to *The Times* required, under School regulations, the permission of the Director and that this would not be granted, he was inclined to write as a private citizen. But the Union, supported by two junior staff members of the Law Department, Lee Albert and Alexander Irvine, who doubted the School's interpretation of the regulations and dismissed them as inappropriate anyway, instructed its President to write in his official capacity. The letter was published on 29 October. It stated in measured terms that no 'personal denigration' had been intended by the students' request for information about Adams in Rhodesia. Indeed the Union had rejected a motion condemning his appointment to LSE in order to find the opportunity to consider the evidence. Surely one cannot 'avoid discussing a man's record as an administrator in one college when he is being considered for the post as Director in another'.

The School reacted swiftly. David Adelstein was told that he had offended against regulations and disobeyed the Director's instructions. Disciplinary proceedings would have to be instituted against him. And so, for the first but not the last time—though not unlike the Beveridge incidents of 1934—a substantive difference was turned into a dispute about authority and its use, coupled, to make matters worse, with a question of the right to express a view freely. From this moment onward, the snowball threatened to turn into an avalanche.

Adelstein's 'trial' turned out to be a bit of a farce. The constitution of the Board of Discipline (which had not met for fifteen years) proved difficult; in the end, the Hon. C. M. Woodhouse as chairman, Dame Mary Green, Professor Donnison, Professor Wheatcroft, and the Director were to serve. The students raised questions of natural justice, concerning legal representation, minute-taking, and above all the involvement of the Director, whose order was after all the issue. In fact Adelstein threatened a Court injunction until the student conditions were met. On 21 November the Board assembled before the background of a noisy boycott of lectures decided by 516 to 118 votes at a Union meeting. The Board found that Adelstein had been in breach of regulations but had acted in good faith. His mistake was no more than an error of judgement. 'We have decided to impose no penalty.'

The students claimed victory; the School did not. Indeed, the Director tried to meet wider student demands by setting up the 'ten–ten committee' with equal numbers of staff and students in order to review disciplinary procedures and staff–student relations generally. But the truce at the end of Michaelmas Term was uneasy. Machiavelli might have noticed that both students and staff were divided but then, perhaps to the School's credit, there was no Machiavelli about, there were only honestly fumbling mortals who got things wrong without ill intentions.

Divisions among staff became evident at a dramatic Academic Board meeting on 2 November. The meeting ended in a published resolution of support for the appointment of Walter Adams and the School's procedures. But in the process, harsh things were said. Dr Morris started the onslaught. The School 'had lived through many crises; this was perhaps the worst one'. Staff–student relations were bad; there were real questions about Adams; Bridges had been wrong to prevent discussion. Professor Greaves was one of many who 'did not welcome the appointment of Dr. Adams' but did not want to rock the boat by reopening the question of Adams's appointment. However, 'there had been a lack of leadership in the course of recent events', and the Chairman of the Governors should retire. Professors Titmuss and Wise defended the School's procedures and decisions, but Professor Gellner supported Dr Morris. Miss Seear underlined the good faith in which students had acted and hoped that they would be told so. Professors Phelps Brown and Roberts spoke against Dr Morris. Several speakers sought compromises, but Professor Watkins 'said that he was against the appointment of Dr. Adams' though 'the student affair' now made it difficult to pursue the question. Dr Miliband demanded a statement on how the appointment had been made; 'we should not reject Dr. Adams but candidly he hoped that he would withdraw'. Professor MacRae spoke of the integrity of Dr Adams and compared 'the present attacks with a swarm of stinging insects. They were reminiscent of the techniques of McCarthy and Vishinsky.' Mr Newfield reported that an Association of University Teachers' questionnaire had had a 46 per cent response rate; of those responding, 61 per cent (105 local AUT members) had said they 'would welcome Dr. Adams's voluntary withdrawal'. Professor McKenzie 'agreed with Professor Greaves that a mistake had been made' but he also agreed that it was too late to try and unravel things. Professor Wedderburn shared this view but 'seriously questioned the behaviour of Lord Bridges'.[69] Mr Westergaard stressed the need for Lord Bridges to tell the Board whether the Selection Committee had known of the Birley Report. Several junior members complained that they had not been consulted. After four hours of discussion the Board rose, no doubt in a state of confusion and much unhappiness.

In a confidential 'Note on the Adams Affair' Sydney Caine later summed up his impressions of the meeting: Had he 'asked the simple question, "Do you like the Adams appointment?" a majority would have said "No"'. Had he asked whether staff were 'entirely happy' with the selection procedure, 'an overwhelming majority, including some of the academic members of the Committee, would have said "No"'. Indeed 'a small group in the staff' were still prepared to grasp 'any opportunity to reverse the understanding reached at the Academic Board'.[70] 'In January 1967 militancy was very much in the air.'[71] Government had decided to introduce differential fees for overseas students, a measure much resented at LSE. The formation of the Radical Student Alliance at the national level formalized calls for direct action; David Adelstein was among the twelve signatories of the initial manifesto. Also, Marshall Bloom arrived on the scene, an American graduate student who 'belonged to the libertarian wing of the new left'[72] and who almost immediately got himself elected chairman of the (usually dormant) Graduate Students' Association. With the help of the student newspaper, *Beaver*, and using his experience of the American civil rights campaign, he reopened the Adams question. It was his planned 'teach-in on sit-ins' which led to the meeting called to assemble in the Old Theatre on 31 January at 4 p.m. under the banner heading 'STOP ADAMS'.

Again, substance slipped away and issues of authority came to the fore which changed the mood, united the disunited, and led to trouble. The Director had reluctantly decided to let the meeting go ahead until the Secretary showed him, at midday on 31 January, the 'yellow leaflet' which contained the paragraph:

> We must make it clear that we still don't want Adams and are prepared to take direct action to prevent his becoming our Director. Come to a meeting on Tuesday, at 4.0 p.m., in the Old Theatre, to discuss what can be done to stop him.

The Director decided to ban the meeting, put up notices to that effect, and at 3 p.m. informed Bloom, who apparently said, 'I understand your position'. In fact, neither he nor anyone else was quite clear why exactly the meeting had been banned and whether the Director would tolerate a meeting at an alternative location. Anyway, the Union Council was incensed. New leaflets were produced, though apparently not by the Union: 'This is an attack on free speech—it must be opposed. This meeting shall take place.'[73]

It did, in a manner of speaking, for the Director had not only posted porters at the doors of the Old Theatre and instructed them to refuse students entry but also removed the light fuses for the Old Theatre so that the windowless room was pitch-dark. Students in growing numbers assembled in the entrance hall, on the stairs, and in every niche of the ground-floor area of the Old Building. At first, 'the atmosphere seemed light-hearted, and there was badinage between students and porters'. But soon this changed. Small things can make a difference in such situations. Caine came to explain his decision and was asked whether students do not have a right to be in the Old Theatre. He replied: 'Students have no rights'—and before he could finish, 'to the Old Theatre', his voice was drowned in the roar of the crowd. The subsequent discussion was confused, until a vote was taken whether to go to the students' bar or 'storm the Old Theatre'. Suddenly the crowd surged for the doors of the Old Theatre.[74] Porters were pushed and at least one was hit. 'The scene inside the darkened theatre was extraordinary. Some students had lit candles, and small points of light illuminated agitated and gesticulating human forms as the theatre gradually filled.'[75]

One porter not on duty was Mr Edward Poole, who had his tea in the Porters' Room. He was a frail man, known to have a weak heart; but 'he hated to be out of things, and came to the help of his colleagues'. After a minute or two in the crowd he was seen, 'grey in the face, slipping down to the floor'. He was carried to the Porters' Room; an ambulance was called; but by then he was dead. He died from heart failure. 'No one had molested him in any way.' Still, it is hard to contradict Kidd when he remarked that 'it remains my opinion that if it had been a normal day Poole would have had his tea and gone home'.[76]

The Director went into the candle-lit theatre and told the students. The shock was immediate. 'The mood changed dramatically. It fell quiet, and students began slowly to leave.' Caine closed the School for the rest of the day and went himself to the students' bar. 'In the bar Caine, to his lasting credit, was comforting weeping students, telling them they should not feel responsible.'[77] Indeed a joint press statement was issued: 'The School and the Students' Union share the deepest regret that the chain of events should have had this tragic end.'[78]

'The weeks following were extremely unpleasant.' Colin Crouch, who wrote this, was then a student activist, although it was the events of 31 January which 'removed me from the company of the mass-participatory left'.[79] The students were as split as the staff. Meanwhile the Director had instituted an inquiry by three professors, Wise, de Smith, and Edey. The committee, after numerous interviews, found that there was a prima-facie case for disciplinary proceedings but was divided on whether such action should be taken. The Union Council issued their own account. Both Adelstein and Bloom came under pressure from their respective constituencies. When the 'ten–ten committee' met a few days later it agreed on a statement:

> Student representatives were of the opinion that the Adams issue would die of inanition provided that disciplinary actions arising out of last Tuesday's events did not exacerbate feeling. The staff members of the committee agreed that there was a real danger that strong disciplinary action would worsen the situation.

However, no one listened. The Secretary of the School, Harry Kidd, decided that disciplinary proceedings should be instituted against Adelstein, Bloom, and four other members of the Union Council who had been at the meeting which allowed the 31 January meeting to go ahead. The Director had been disobeyed; authority had been defied; support by students and staff was no reason to waive the rules. 'If that concession were made, the result could well be anarchy.' Moreover, 'I did not think it right to be deterred from disciplinary proceedings by the fact that disorder might be the consequence.'[80]

It was the consequence. The trial before the Board of Discipline, this time chaired by Lord Bridges and including Leslie Farrer-Brown (both of whom had been ruled out of the earlier Board for possible bias) as well as Professors Donnison and Wheatcroft, dragged on for weeks, from 16 February to 13 March. During these weeks, moderates and militants among the students were battling for supremacy. The study by Tessa Blackstone and others makes it abundantly clear that those who saw only an active minority at work were wrong. Activists played an important role, but more than 40 per cent of all students took at least some part in sit-ins and boycotts during these weeks. There were differences between departments, some expected—more than 60 per cent of all sociologists were active—some less so. Law students, for example, had a high proportion of non-participants (54 per cent) but also a comparatively high proportion of very active participants (28 per cent). Generally, LSE students clearly tended to the left of the political spectrum; 53 per cent of the British students were Labour supporters and a further 9 per cent adherents of left-wing groups (though among overseas students 'liberals' were the largest group, with 39 per cent). But it would be wrong to infer from such data a highly organized cadre type structure. Student politics was fluid, situational, and often personalized. During that spring of discontent, Peter Watherston, chairman of the Conservative Society, was elected to succeed David Adelstein as President of the Students' Union. Student politics was also a direct mirror of the attitudes of the School authorities.

At first the meetings of the Board of Discipline seemed to lean towards leniency. Four students who had taken part in the Union Council in the early afternoon of 31 January were found not guilty. There remained Adelstein and Bloom. The complaints were that by taking votes at the confused meeting in the entrance hall they had disobeyed the Director's orders not to hold a meeting; moreover, Adelstein had also encouraged the storming of the Old Theatre. Their tutors spoke in mitigation. Percy Cohen described Adelstein as 'a good student, doing his work regularly and some of it well. He was not a man who would advocate violence.' Ernest Gellner said 'that these were not dishonourable young men'. Their punishment would harm staff–student relations. Bloom used the occasion for a major statement in which among other points he was minuted as saying: 'They had, in fact, close and devoted allies among the staff who were as concerned as they were with the unpleasant results of the cultivation of an expanding twentieth-century bureaucracy and a nineteenth-century view of authority.'[81]

One of these allies was Professor John Griffith, their counsel for the defence. He made a powerful case in terms of the moral character of the accused, the confusion of the situation, and the role of others who had, for example, produced the incriminating leaflets. Griffith played a crucial role both during the first troubles and subsequently. 'I was less on the students' side than against the Administration', he said later, though he also remarked that he was 'trusted by them because they knew I was on their side really'.[82] He minded the 'incompetence' with which the School had handled the affair and deplored the 'fairly mutual incomprehension' that prevailed. He also worried about the underlying causes, the 'paternalism' of the authorities coupled with the 'laxity' of academics who did not care much about students and teaching. At times, John Griffith could appear to be that paradox, a lawyer who dislikes institutions; at times he may well have dreamt of a world in which sanctions would not have to be used because mutual comprehension prevailed. His devotion to LSE was great, which made his criticism of School policy all the more weighty. Those who were exasperated by him may not have realized his indispensable contribution to the sanity of LSE.

However, counsel for the defence did not prevail on 13 March 1967. Adelstein and Bloom were found guilty and suspended until the end of the summer term—a conclusion which gave Harry Kidd 'an uneasy feeling that the sentence was an unsatisfactory and ineffective compromise'.[83] It did not take the Union long to call a mass meeting of over 800 students and decide on a sit-in which was later described as 'the first major student strike this country has known':

> The boycott and sit-in continued for the next eight days, until the end of term. During those eight days the School presented an extraordinary spectacle to people used to its normal ordered appearance. The lobby and corridors of the main building were filled with students sitting on the floors, holding seemingly endless discussions, listening to distinguished outsiders who thought they had the solution to the problem of L.S.E., reading, singing, eating or just sleeping. The blackboards normally displaying neatly written notices of school functions bore boldly drawn slogans, and banners were draped from the walls. In the Old Theatre the Union seemed to be in permanent session. Outside the main entrance TV teams rubbed shoulders with the pickets. No attempt was made to block the passage of staff or students who wanted to go about their business, but many classes and lectures were cancelled as a result of low attendance.[84]

The last two points are important. There was no violence during the long sit-in. The only major incident occurred when on 15 March a small advance party gained access to the administration building, and the police were called to carry them out, one by one—and to the delight of the media—to Houghton Street. Beyond that, the sit-in was actually a good-humoured affair. Colin Crouch speaks of the 'early revivalist' atmosphere at meetings, and makes much of the sense of community generated by the common experience. Several people have commented on the 'fun' which students had. 'It was great fun for them,' said John Griffith, who knew what 'they' felt.[85] Perhaps one may also suggest, from a more remote perspective, that such fun was always particularly welcome at the time of year at which

Catholic countries symbolically turn the social order upside down in their carnival season, when the fancy-dress jesters rule and the rulers stay at home. The winter has been long; spring has not quite arrived—how can one break out of the grey routine of life for a short while?

The point about TV teams is more serious. March 1967 is the time when LSE got into the news in a big way, and not with favourable connotations. 'Rebellion at the School for Rebels.' Fleet Street, then still the pulsating centre of the capital's press, was too close for comfort and the Independent Television News headquarters at the corner of Kingsway and the Aldwych even closer. A part of the conflict was in fact fought out in the letter columns of daily papers. Pictures of students sitting in, or being carried out, of a beleaguered Director, of banner slogans went literally around the world. The School would not easily recover from its 1967 image as a hotbed of revolution, the less so since it found it difficult to come to terms with the events itself.

In the spring of 1967 these events took a curious turn. On 17 March the Standing Committee of the Court listened to an appeal on behalf of the two suspended students, dropped the charge of 'encouragement', and allowed Adelstein and Bloom earlier access to the School under certain conditions. The students were still unhappy, though on 20 March they decided to suspend the sit-in until the next term. In subsequent weeks, attempts were made to reach what was called a 'settlement', with John Griffith 'actually taking forms of words around' from Union to Governors and back. On 13 April, Adelstein and Bloom signed a dignified statement. 'We are intent on working with the School authorities through constitutional processes.' The Governors responded: 'In the light of the undertakings given, the Court of Governors have decided on an act of clemency, and have accordingly, during the good behaviour of Mr. Adelstein and Mr. Bloom, suspended the penalty imposed on them.' At the beginning of the summer term, the Students' Union decided formally to end the boycott and sit-in.

The 'Quiet Year'

There followed what came to be called the 'quiet year' at LSE. This was notable because it was also the year in which students got restive elsewhere in Britain and all over Europe. 'Red Dany' Cohn-Bendit in fact came to the School several times, but even his inflammatory speeches did not rekindle the flame. Sydney Caine left during this lull, a much chastened man. Whether he should ever have allowed the two incidents of 'disobedience' to arise may be debatable, but once the troubles had begun, he behaved with a helpful sense of proportion and undiminished sympathy for students. He also never lost his commitment to teaching as the School's main obligation. Still, as he conveyed to Simon Jenkins of *The Times* in March 1967, 'his own troubles have clearly surprised and slightly bewildered him'. He did not believe in a '"consensual" university, without authority' but 'he sees little prospect of the situation getting any better'.[86] Sydney Caine had presided over the changes which many saw as the root cause of the trouble, though he did not do so either willingly or knowingly. In the end they swamped him and the School with him.

Yet he never ceased to try and find answers. This was above all true for staff–student relations. The 'ten–ten committee' turned out to be short-lived. 'Once the Adelstein–Bloom issue had been dispatched within the committee, the revolutionaries' only objective was to ensure that the committee would soon disintegrate.'[87] Before Caine left, the students had walked out on this committee as well as on another one set up to consider the reform of disciplinary regulations. However, there was still the Machinery of Government Committee. In the summer of 1967, five students were elected to it, with Union backing though against a boycott by militants. They were the successive Students' Union presidents David Adelstein, Peter Watherston, and Colin Crouch, as well as the Secretary of the Union, Chris Middleton, and Richard Atkinson. Committee meetings became at times acrimonious. In January 1968, Middleton and Atkinson walked out of what had been (in Atkinson's words) 'an essentially ignorant, insensitive and superfluous charade in which power counts more than ideas'.[88]

The Majority Report agreed in February 1968 noted that 'systematic channels for a two-way flow of information between students and other members of the School are inadequate' but proposed only minor changes in student representation. There should be, it suggested, from four to eight student members on the Court (with its membership of 60–100). There should further be three student members of Council and five of the fifty-strong Senate, elected by secret ballot and single transferable vote, and present for 'non-reserved subjects' only. The Committee wanted these changes reviewed after three years because 'there are many risks in embarking on such changes'. The moderate student members of the Committee, Colin Crouch and Peter Watherston, accepted the proposals; their reservation concerned numbers, which in their view should be higher in order to make student representation effective. Professor Roberts, on the other hand, entered a reservation on student members of Council. 'I do not believe that students are equipped by experience or knowledge acquired to be useful participants in most of the work of the Council.'

In the event neither Council nor Senate came about. The Standing Committee of the Court, instead of accepting student members, later adopted the practice of regular meetings with Student Governors. The General Purposes Committee of the Academic Board (though not the Board itself) co-opted four student members of the Union. This, plus membership in a few student-related committees, was the whole outcome of the 'revolution'. True, the Minority Report by Atkinson and Adelstein had introduced, with reference to 'Berlin, Regent Street, Berkeley, and many other Colleges', the notion of 'parity':

> We feel very strongly that a situation of parity or near parity would create a situation through which over time, a mutual trust could develop, where the recognition of common interests, by staff as well as students would allow members of each 'side' to cross regularly to the other 'side' without fear of 'persecution' from their own ranks.

The School as a whole, however, was not inclined to take such proposals seriously.

Other changes in the administration happened during the 'quiet year'. The most important concerned the secretaryship. In the summer of 1967, Harry Kidd was about to leave for Oxford. Some mystery remains about his precise role in the first troubles. Was he, as some surmised, 'bitterly anti-student', even 'seen as the evil spirit'? Or was he, on the contrary, a 'victim' who had acted 'with the utmost uprightness and vigour', 'at times tough' but always 'understanding'? He certainly took the trouble of 1967 very seriously. His book is a warning of things to come which ends characteristically with a chapter on discipline. Shortly before he left he sent a 'private' paper to the Director in order 'to suggest to those who are to come after us that they ought perhaps to be a bit more systematic about political intelligence than we have been'. The demand for 'parity', for example, was bound to lead to further trouble and required a better 'armoury':

> In this connexion, I would wish to draw attention to the problem of identification of offenders. Anyone who has to deal with an offender who refuses to give his name must have at his ready disposal a camera and flash-bulb equipment comparable in quality and performance to those that are used by press photographers.

Moreover, in order to secure its survival 'in certain difficult circumstances' the School 'should take care to be able, if circumstances warrant it, to get rid of students without due process'.[89] The militants, who were in any case liable to suspect conspiracies behind every School decision, would have had a field-day if they had known of this document. Needless to say, no cameras were handed out to members of staff. But it is just possible that Mr

Kidd was not the only one to be pleased when the offer from St John's College came.

Minor matters of style apart, many members of the School had come to the conclusion that the Secretary carried too heavy a burden. Some would put this differently and say that he, or she, was too powerful. Intriguingly, Walter Adams had been the first to suggest a split secretaryship when he considered returning to the job in 1945. He had recommended a 'division of duties . . . aimed at entrusting to Miss Evans responsibility for most of the internal academic machinery of the School, leaving to Mr. Adams responsibility for the non-academic machinery, for external relations and the "non-recurring" tasks that cannot now be fully foreseen'.[90] Twenty-two years on, and with Adams about to arrive as Director, this is almost exactly what happened. John Alcock was promoted from Registrar to Academic Secretary, with responsibility for the Academic Board and the Appointments Committee, and John Pike was appointed as Financial Secretary, whose province included all matters concerning the Court of Governors. Pike's colonial experience was to stand him, and the School, in good stead when 'the natives got restless'; his steadying influence was much needed throughout the second troubles. However, the subdivision of the secretaryship did not last; after Alcock's and Pike's retirement in 1983 the unitary post was re-established; yet while it lasted the uneasy balance may have been difficult for the two Secretaries but turned out to be beneficial for the School.

One other important change occurred at the end of the 'quiet year' in 1968. Lord Bridges, who had been Chairman of the Governors since 1957, retired. Some have called him remote, but one might argue that this is precisely what the Chairman ought to be except in times of crisis. However, on this point views may differ. Certainly, Lord Bridges's successor was anything but remote, for the Court chose Lord Robbins, who thus assumed his penultimate major role for the institution which was his life.

When the chairmanship changed hands, the new Director was already in place. 'Walter Adams took his place as director with scarcely a murmur from the student body.'[91] Looking back over his ten years, Sir Sydney Caine expressed 'a real disappointment and real apprehension about the future' so far as the 'variety, flexibility and informality in teaching' was concerned, but otherwise declared himself satisfied.[92] Lord Bridges congratulated the Director for his understanding 'of the L.S.E. with its so strongly entrenched libertarian and egalitarian traditions' (by which he meant too much discussion and too little decision). 'His countless friends deeply deplore that Sir Sydney's last year as Director should have been clouded by the events of the Lent term.'[93] But William Pickles, in a cordial toast at the

Senior Common Room farewell dinner, spoke for many: 'In my eyes and in those of many others, the Director emerges from all our recent troubles with his stature enhanced.' Pickles added a thought which sums up Caine's directorship. 'The picture in my mind is of a long effort at reforming and liberalizing, in the process of adaptation of the School to a background that was changing—and becoming indeed steadily more difficult to live with—at a speed which in itself raised new problems all the time.'[94]

Walter Adams arrived on the scene quietly, as was his way. He always remained a kindly man, firm in his deepest convictions yet reluctant to impose them, or indeed himself, on others. Some called him shy; others, not least Lord Robbins, regarded him as irresolute, even weak; again others described him as obstinate, stubbornly if quietly pursuing his goals. Adams had served the Academic Assistance Council under the chairmanship—and leadership—of Beveridge, and the Inter-University Council under that of Carr-Saunders. Was he now, at LSE, ready to step into the shoes of his authoritative predecessors? He certainly looked forward to the task and was undaunted by the prospect. Problems? 'A School without problems would be dead, and most are superficial to the continuing constructive achievements that compose nine-tenths of the School's life.'[95]

In his first Report the new Director was able to say: 'In a year in which many universities in the United Kingdom experienced student disturbances, there were few incidents or moments of tension within the student body or between students and School authorities and none which merits recording.' Actually there had been some tension during the 'carnival season' of 1968 which was also the time at which the Machinery of Government report came out. 'The union was split three ways', and Colin Crouch emerged victorious as President. A speech by Enoch Powell of 'rivers of blood' fame aroused a certain amount of emotion. On 23 May a solidarity vigil for the students of Paris took place. However such 'half-hearted one-night stands' failed to 'recapture the atmosphere of March 1967'.[96] Thus the new Director had a peaceful start, which was as well considering that his next Report had to begin with the words: 'The session 1968–69 was gravely disturbed by student troubles.' What is more, this second time round the undercurrent of civility, common purpose, and even good humour, which had never been wholly absent during the first trouble, came under serious strain and for a while seemed to collapse.

The Gates

The 'unhappy period, perhaps the unhappiest in the history of the School',[97] started two weeks after the beginning of the Michaelmas Term,

1968. At a Students' Union meeting on 17 October, an emergency motion changed the subject of debate from student participation in School committees to a more immediate issue. On Sunday 27 October there was to be a national demonstration against the war in Vietnam. Members of the Union wanted to make their special contribution and moved that 'sanctuary, medical aid and political discussion' for the many thousands expected from all over the country should be offered in the School buildings. 'The idea seemed so extraordinary that I did not take it seriously,' the then President of the Union, Colin Crouch, wrote later. But it found a majority on the day so that it took a mass petition to reopen the matter at a second meeting on 23 October. Like others of its kind, the meeting ended in confusion. A majority of sixty overturned the earlier decision; when Dr Meghnad Desai, who as Honorary President of the Union was chairman of the meeting, ordered a recount, there was still a majority of six against action; but the pandemonium over a second recount led to a statement by the Union President that, if action was taken, the Students' Union would be neutral, that is it would neither authorize nor condemn the action. 'It was a shabby result.'[98]

The next morning the transformation of a substantive issue into one of authority and its acceptance occurred once again. The School authorities were clearly and understandably worried: 1968 was a restless year; violence had been used in connection with Vietnam demonstrations. Shopkeepers and others along the route of the planned march were preparing for a difficult weekend. Rumours had reached Governors that there would be trouble at the School and many felt that the Students' Union would not be able to control outsiders even if it wanted to do so. Thus the Director put up a statement on behalf of the Court of Governors on his notice board. The School buildings would have to remain closed even to academic staff on Saturday and Sunday; 'without my express consent' no one would be allowed in. 'The Governors have conferred on me at my sole discretion power to close the School.' The Director had every intention of using this power without notice if forced to do so.

There are numerous accounts of the events of subsequent weeks and months by the Director himself, by members of the staff, by some of the apparently omnipresent journalists. The two extensive histories by students involved in the events agree that the Director's statement changed the mood, especially since it was made on behalf of the Governors. Colin Crouch states this with dismay: 'People who previously had opposed the occupation now supported it for no other reason than that the Governors had intervened.'[99] Paul Hoch and Victor ('Vic') Schoenbach were young

American natural scientists. Schoenbach was actually registered for a while for the M.Sc.(Econ.); Hoch had a Ph.D. in theoretical physics and had not been admitted to LSE, but as a Bedford College student he had the right to use School facilities, which he certainly did extensively. Hoch and Schoenbach describe in clinical detail how the Adams statement helped them get the occupation of the School going.

For the activists the weekend of 25–7 October 1968 turned out to be a trial run, 'Committee of Public Safety' and all. The story of how the SocSoc factions united to try and organize lectures and film shows because 'Mao's followers'—presumably from among the 'Piccadilly beats, Maoists, anarchists and other "outside agitators" '—'would wreak havoc if they were not "kept busy" ',[100] has its sinister humour. Even Tariq Ali was invited to serve this tranquillizing purpose. For Colin Crouch the loss of initiative to the left meant that he felt he had to resign. Another non-militant, Francis Keohane, was soon elected in his place, though before long he too gave up and made way for the SocSoc candidate, Chris Pryce. The job of School committees was made easier by the fact that the occupation forces decided to clean up before they left on Sunday night. It was decided that no disciplinary action would be taken. Nevertheless the Court of Governors' statement of 31 October did not go down well. Students minded the 'having regard to the immaturity of those concerned' in the decision not to take action, and staff were concerned that in future even those attempting 'to encourage' action would be liable to have their contracts terminated.

At this point the institutional line-up characteristic of the following six months began to emerge. Academic opinion was represented by the Academic Board under its clear-headed and active vice-chairman, Professor Alan Day. The Board's General Purposes Committee assumed a central role and at times met almost daily; it had student members. The Board and its GPC tended to view the Governors as outsiders whose interference was often unwelcome and sometimes unhelpful. However, the Court and its Standing Committee were chaired, not to say run by a man who could hardly be described as an outsider, Lord Robbins. During the 1968/9 troubles Robbins turned out to be a more hands-on Chairman than even Sidney Webb had been. He was supported in this by the Standing Committee, and not least by some of its academic members, notably Dr Bernard Donoughue and Professor Ben Roberts. The Director cut a lonely figure in this constellation. Lord Robbins thought little of him; he even told one of his academic friends 'that his support for Adams's appointment as Director had been a major error of judgement'.[101] The Academic Board, on the other hand, tended to set him aside; the minutes of its most dramatic meetings show virtually no contribution by the Director other than reports of events. The main result of all this was that in its critical months LSE lacked a focus for what in similar circumstances at Columbia University in New York were called 'the loyalists', leading academics who were respected and trusted even by those who regarded them as either too 'liberal' or too 'rigidly institutional'. The 'ad hoc joint committee' of non-student members of the General Purposes Committee (GPC) and academic Governors convened by Professor Alan Day was the nearest to such a loyalist focus, but it could not change either the style of the Director or the preferences of the Chairman.

The uneasy quiet was next disturbed on Oration Day, 5 December 1968. It was to be the last such occasion which actually ran its course for almost twenty-five years. The students wanted the guest speaker, Hugh Trevor-Roper, to discuss with them his apparent support for the Colonels' regime in Greece, and so they occupied the reserved seats in the Old Theatre. Lord Robbins handled this occasion with consummate skill and good humour. He accommodated guests in another room, persuaded Trevor-Roper to answer questions about Greece later that day at a separate meeting, and promised to come and talk about LSE to the students next term. 'Robbins received a personal ovation at the end; and groups of revolutionaries held an urgent inquest on what had gone wrong.'[102] Thus Crouch; and Hoch and Schoenbach: 'Many of us felt cheated. We had sat through an incredibly boring speech—and still no opportunity for blood.'[103]

Lent Term began, for the student activists at any rate, with another go at Rhodesia and South Africa. The occasion was the imminent Commonwealth Prime Ministers' Conference, the subject the Director's past, and then, the investment of School funds in companies breaking the sanctions against Rhodesia, or trading in South Africa. A 'teach-in' was arranged to which the Director was invited. Even when they wrote their book, Paul Hoch and Vic Schoenbach made no attempt to conceal their glee at the discomfiture of others, especially of sensitive people in authority: 'As the teach-in opened the Director was up on the platform looking mildly terrified.' The ensuing 'discussion' was more a series of threats. 'On the stage, Adams was quaking. He stuttered as he began his reply, but in some confusion promised to return to the teach-in at the end of the week to report on progress toward meeting the demands.'[104]

He did not return, notably not when he was asked to do so that same evening. If one has experienced similar situations one can sympathize with Walter Adams. On the other hand, anyone who has ever confronted authorities also knows the surprise and even the *frisson* of satisfaction at seeing the supposedly powerful at a loss for words. Yet there is an all-important boundary between this *frisson* and what a Göttingen student–author called the 'furtive pleasure' at seeing powerful men like Aldo Moro or Hanns-Martin Schleyer captured, humiliated and finally killed. Stressing this boundary is important: there were no potential, let alone real, Brigate Rosse or Red Army Faction killers at LSE. Hoch and Schoenbach are at times distasteful but neither they nor those who were regarded as beyond the pale even by them, ever really threatened the life and limb of others. This is not to uphold the distinction between 'violence against persons' and 'violence against objects'. 'Objects' in conditions of conflict are never just objects; they are institutions even if they are iron gates. And gates became the issue at LSE immediately after Rhodesia and South Africa had ceased to excite the crowd.

During the summer preceding the fateful academic year, the School authorities had, without consulting any committee, begun to install exceedingly ugly and, as it turned out, fairly flimsy iron grille gates at a number of strategic points. They were obviously designed to cut off access if necessary. At first, comment by staff and students was muffled, 'little more than ironic grumbles' (to quote Ken Minogue),[105] though even the Secretary of the Graduate School, Dr Anne Bohm, who belonged to those who later took a hard line on discipline, wondered:

> I remember going up to the Senior Common Room, and on the stairs between the third and fourth floors I suddenly saw gates. They weren't closed, but there were gates at the side and I remember to this day [in 1990] I said 'Good Lord, are we now going to live in a prison?' And that is of course what a lot of people felt.[106]

During the Michaelmas Term, more gates appeared, including some in the basement which looked as if they were designed to block access to the ladies toilet among other facilities. When notices were put up to explain that the purpose of the gates was to protect property, especially on occasions when School rooms were used by outside bodies, some, like Colin Crouch, accepted such arrangements for physically separating the parts of LSE as normal. But as collective paranoia grew, notably after a series of GPC proposals (later rejected by the Academic Board) for the protection of the School in cases of emergency, the gates acquired a symbolic and soon a real significance in the minds of students.

On 17 January 1969 Lord Robbins made good his December promise and addressed the Students' Union. The atmosphere was no longer so friendly. Robbins had to field many hostile questions, notably on South Africa and the involvement of companies with which he and other

Governors were associated. In the end he was asked about the gates and, when pressed, admitted that they were also meant to prevent '"unauthorised" access to parts of the buildings, for example, during occupations. The sharks smelled blood and drew nearer,' is how two of them, Hoch and Schoenbach, described the student reaction.[107] After Lord Robbins had left, the Union meeting passed, by a large majority, an emergency motion that the gates would have to be removed within seven days, otherwise the students would do it themselves.

> The gates were an act of authority; the possibilities of interpretation were infinite and advantage was taken of them. Nicholas Bateson, a lecturer in social psychology later dismissed from the School for his part in the events, said that tearing down the gates was required of us if we were to show proper solidarity with the Africans in Rhodesia, the guerillas in Thailand and the Arabs in Palestine [*sic*]. Robin Blackburn, a lecturer in sociology also later dismissed, said that the gates were the material expression of class oppression.[108]

When Colin Crouch wrote this, his disaffection with the militants was already abundantly evident; later, the precise time and place of the statements made by the two lecturers would become a matter of legal inquiry. As on earlier occasions, the Students' Union reconsidered its initial militant impulses. The Director was asked to explain the gates. On 20 January he did so in a long letter in which he also repeated that they might be used if 'there were unauthorized occupation by any persons of part of the buildings'. However, the School would remove or move elsewhere some of the unnecessary gates (including the one which the Secretary of the Graduate School had found so objectionable).

The Union meeting on Friday 24 January was not as well attended as the militants had hoped. Moreover, the majority for direct action was initially embarrassingly small: 242 for, 236 against, 76 abstentions. Again, a recount had to serve to produce a more militant result: 282 for, 231 against, 68 abstentions. While even SocSoc members still hesitated, some others had stormed out to start dismantling the gates. When they found senior academics in position ready to defend the gates, the mood got angry. Suddenly, weapons appeared. One philosophy lecturer, Alan Musgrave, claimed much later that 'building workers [from 'a very large building site nearby'] came along with their sledgehammers and bashed down the gates'.[109] However, the Barbican site (to which Musgrave referred) was hardly 'nearby', and more importantly, if this had been true, it would have been too good a story of student–worker solidarity to miss for the revolutionaries. They on the contrary observed: 'Two wild-looking, bearded anarchists, who nobody had ever seen before, had somehow appeared for the occasion, and were flailing away with sledgehammers with all their might.' Thus Hoch and Schoenbach, who otherwise simply say: 'Someone produced a crowbar', and 'someone arrived with a sledgehammer he had brought especially for this occasion'.

One of the defenders of the gates was Professor John Watkins, who clung to the third-floor gates. He testified later about a female student: 'She had on her person the head of a pick-axe and the shaft was under her coat. She had a large bag. She was holding the pick-axe head.'[110] Did someone really shout: 'Don't hit him! He's a Professor'?[111] Anyway he was eventually prised away by five women. Others, including Professor Alan Day, were similarly 'neatly prised away by several lithesome ladies, then the gate was wrecked by a man with a pick-axe'. The work of destruction began. It did not take very long, little more than an hour, and the frightening and the absurd were never far apart. This is how Hoch and Schoenbach experienced the end of the gate-smashing scene:

> We then went off to the St Clement's Building. As we ran through the refectory, brandishing pick-axe and sledgehammer, people barely looked up from their dinners. On the way over the St Clements bridge, we could see the members of the Chess Club, bent over and intent, locked in struggle. But then came reports that the cops had arrived.
>
> There was an almost immediate wave of panic: tools and implements were wiped clean of fingerprints, and dropped in all directions. Nobody wanted to get caught with a sledgehammer.[112]

The self-serving story sounds more romantic than the real experience of most of those present, yet it was too weird to be total fabrication. Clearly, the students did not know what to do. Most of them eventually dispersed, though a fair number repaired to the students' bar. By then, two things had happened. The Director had, at 9.30 that evening, declared the School closed indefinitely. By ten o'clock, the police had arrived in sufficient numbers to close off the buildings.

There followed the episode which even the supporters of its objective, including the Director, later described as 'distasteful'. In order to identify perpetrators among the one hundred or so students in the bar, these were ordered to come out one by one and run the gauntlet of staff lined up to discover familiar faces. 'It was absolutely ghastly,' the Academic Secretary remembers. One senior professor 'was hopping up and down on his toes with sheer embarrassment and horror'. On the other hand, 'one rather tough member of the academic staff put a half-nelson on a female student. That was very very nasty.'[113] In the end, three were identified. The scene moved to Bow Street, with incidents along the way which meant that some thirty students were arrested and had to spend the night in uncomfortable cells before they were either discharged or released on bail. In the end, the School took out injunctions against thirteen students not to enter the School without special permission. Four of them were actually not current students at the School; of the thirteen, six were British, three American, two South African, and one each Australian and Italian. The School also sent the two lecturers, Bateson and Blackburn (as well as one other lecturer whose case was later dropped), notice of impending disciplinary action.

The School remained closed for twenty-five days. During those weeks, a desperate attempt was made to uphold some degree of normality at what came to be called 'LSE in Exile'. Lectures were held all over Bloomsbury, and beyond, at Bedford College. The Academic Board meetings in the Beveridge Hall of Senate House, and the Students' Union meetings in the University of London Union (ULU) were, to be sure, anything but normal. Closing the School is the kind of definitive act which is easier done than undone. What would have to happen for the School to open again? One way was, of course, to persuade the students to support a more moderate approach. The Director tried, in a sensible and conciliatory 'Dear Student' letter, the first of several. But the attempt did not work.

One of the consequential—no doubt by some, deliberate—errors in the students' analysis was that they saw all authority as a monolith. If Walter Adams was Director of a school which had Governors serving on the boards of companies suspected of breaking Rhodesian sanctions, then a capitalist conspiracy was responsible for the gates. But many of those involved in the governance of the School made the corresponding mistake about the students. They had no sense of the deep internal divisions among even the active portion of the student body, and were incapable of seeing the motive forces behind the wording of resolutions. Crouch and Watherston lost their motion at the Union meeting on 3 February which was attended by 1,200 to 1,500 students, because the text was regarded as too conciliatory in its condemnation of violence and acceptance of fair and just disciplinary procedures. But the Socialist Society also lost its entirely unaccommodating motion. The middle road taken by the majority did not look good to literal analysis; it failed to condemn violence and put all blame on the School's authorities. Yet in the circumstances this was a rejection of militancy. Even among the militant there were many divisions; International Socialists (IS) were more militant than the bulk of SocSoc; 'two score Maoists and anarchists', joined by 'a new group of lunatics', would have liked to 'wreck something today' every day but they did not cut much ice with the rest. In the event, no bridges were built between School authorities and students.

Admittedly, the atmosphere was tense, and all eyes were on LSE. The media went on their own rampage of sensationalism. More importantly, public authorities got increasingly involved. In 1967, Shirley Williams, then a junior minister in the Department of Education and Science, had refused to be drawn into the conflict. This time, on 29 January 1969, the Secretary of State, Mr Edward Short, made a speech of unusual toughness in the House of Commons:

> The real perpetrators are a tiny handful of people—fewer than one-half of 1 per cent of the 3,000 at L.S.E. Of these, at least four are from the United States. They are subsidised to the extent of between £1,000 and £2,000 for their one-year master's degree course by the British taxpayer. These gentlemen are clearly not here to study, but to disrupt and undermine British institutions.
>
> This small group of less than half of 1 per cent are the thugs of the academic world. Already they have succeeded in closing L.S.E., whose former free and easy and delightful relationships are remembered by so many hon. Members. . . .
>
> They are out to destroy and disrupt. I hope that no one in this House or outside it will underestimate the long-term effect of this kind of activity. It can only result in the slow rotting of institutions like the London School of Economics.[114]

Ted Short represented a sense of outrage and betrayal which was present among Labour supporters at the School as well. The advances of a policy of social opportunities and the sacrifices of taxpayers had allowed more students from all classes to come to the cherished institutions, universities—and now there were wreckers in their midst. It was unthinkable that these were anything other than a tiny minority, and probably foreigners as well. One appreciates the feeling, especially that of 'scholarship boys' who had worked their way out of working-class disadvantage, but in fact the world had changed, and with it the class base of the Labour Party, the prevailing mores, and the commitment of students to universities.

A debate in the House of Lords on 23 April 1969 was more balanced, though harsh things were said there as well. On 30 April a Select Committee of the House of Commons which was investigating relations with students in universities came to LSE. On the first day it was heckled, contrary to a Union decision, until in the end the Chairman suspended the session, and police had to escort the Members of Parliament out. What happened at LSE had ceased to be a little local difficulty.

The Academic Board was exposed to all these pressures. Its numerous motions and amendments tell the story. One group, including John Griffith, Ralph Miliband, John Westergaard, George Morton (a statistics teacher who left the GPC in anger and soon afterwards the School as well), and

Meghnad Desai saw above all the School authorities at fault and favoured conciliation. They had the support of between thirty and forty members of the Board. Numerous attempts were made to mediate between what may be called the Standing Committee line and the students. Adam Roberts and Robert Cassen, K. W. Wedderburn and Frank Hahn, Donald MacRae and Dan Sinclair, Robert McKenzie and Brian Abel-Smith, Ronald Dore and Emanuel de Kadt all made proposals to end the quandary. There was a clear majority for reopening the School sooner rather than later. In practice, however, the line taken by the Standing Committee, including its academic Governors, prevailed.

This was notably true with respect to the vexing issue of 'victimization' which from a certain point onwards obscured all others: who should pay what price for participating in the acts of trespass and violence? Most universities find it impossible to deal with disciplinary issues; straightforward criminal matters belong in the courts, and all others generate confusions rather than conclusions. The sense of academic community allows of many divisions but not of exclusion. Had the Academic Board been sovereign, those in favour of strong disciplinary action against students and members of staff would probably have been in the minority. In the event, the measures taken were strong. The contracts of two lecturers, Bateson and Blackburn, were terminated for misconduct, and their dismissal was upheld on appeal. A number of students had to appear before the magistrates court and later before the High Court; four were found guilty. Three were suspended from the School for varying periods. At least one other student, Dick Atkinson, co-author of the Minority Report on the Machinery of Government, suffered indirectly when his appointment to a lectureship at the University of Birmingham in 1970 was overturned because of his LSE activities. Two of the foreign students, including Paul Hoch, were deported.

The dismissal of the two lecturers obviously drew much attention. There is no doubt that they supported the action taken by the students. Equally, they did not actually take part in the gate-smashing; the issue was what they said. But when and with what effect? Robin Blackburn was lecturing at a conference at the School while the gates were demolished, though he did support the action in a speech which he made later that evening at Central Hall, Westminster, and again in an interview with BBC Television News on the next day. Also, he and Nicholas Bateson were not exactly friendly members of the School. 'Dear Adams,' they wrote to the Director on 6 May, 'we are in receipt of your insolent letter of 1st May.' They then spoke of the 'insincerity and dishonesty' of the tribunal and the 'arbitrary and authoritarian' ways of LSE.

It is obvious to everyone why you have made this new move [of setting up an appeal tribunal]. Your own job and that of Lord Robbins, indeed the position of the entire clique of self-appointed capitalist manipulators on the L.S.E. court of governors, is in grave danger. The students have been enraged at the attempt to victimise some of the individuals who supported the Union decision to remove the gates you so arbitrarily errected [*sic*!]. You hoped to cow them into submission. In fact they have shown great courage and fortitude in standing by their principles, perhaps at the expense of future comfortable careers.[115]

Not surprisingly, the Appellate Tribunal set up by the School found that Blackburn 'is a committed believer in violence as a means of reforming the School, at any rate where the School authorities do not meet student demands', though he 'appears to support violence against inanimate objects rather than against persons'.[116] Mr (now Lord) Justice Ackner, Professor George Keeton, and Professor H. W. R. Wade upheld the decision to terminate the contracts of Bateson and Blackburn.

The decision to reopen the School was announced by the Standing Committee on 12 February with a statement which not only members of the Academic Board but also Governors, among them leading Conservatives like Lord (Rab) Butler and Reginald Maudling, found 'unconstructive'. At least one lay member of the Standing Committee, the social reformer and judge Morris Finer, had tried to uphold a more 'liberal' stance throughout. The statement mentioned impending legal proceedings and disciplinary action, notably the action against the two lecturers, and threatened that if there was any further direct action 'the School will be declared closed and grant-awarding bodies will be notified'. The idea of re-installing gates, though this time 'collapsible' ones, was dismissed, but a number of less unsightly solid-oak doors took their place. The subsequent Summer Term became a period of frequent disruption. Lectures were boycotted, lecturers were verbally and at times physically attacked; the reinstated gates were damaged; metal glue was used to block doors all over the School; stink and smoke bombs were dropped into meetings; slogans were painted on the doors of those who had taken part in the identification ('academic spies'); the Senior Common Room was occupied on several occasions; Lord Robbins's portrait was taken down; false fire alarms were set off; all deliveries to the Senior Common Room and the refectory were 'blacked'.

The issue throughout was 'victimization' and above all the sacking of the two lecturers; but the actions taken made the students involved no new friends among the staff. This time academic freedom—the freedom to teach and to learn—really was at risk. Also, the 'state of virtual chaos' in which the School found itself was clearly brought about by 'a small if ubiquitous group', no more. Eventually, everyone got tired of it all. 'In the midst of inter-revolutionary vituperation, tense and strained staff–student relationships, and a general weariness and boredom, the student protest at LSE, and the summer term, choked to a halt.'[117]

Postscript with Hindsight

The troubles at LSE were over. What if anything did they accomplish? The protagonists certainly did not go on to spread revolution; not one of them became a terrorist or otherwise fell foul of the law. Marshall Bloom went to a 'communal farm' in America and in November 1969 took his own life by inhaling the fumes of a car. David Adelstein disappeared from the public eye; he worked as a researcher on educational matters for the media and for various institutes. Paul Hoch died in 1993 at the age of 50; he had been involved in the study of science and technology at several universities in North America and in Britain. Victor Schoenbach became a Professor of Epidemiology in the United States. Nicholas Bateson, one of the two lecturers dismissed in 1969, became a civil servant in the Office of Population Censuses and Surveys. Robin Blackburn remained active on the intellectual left as editor of the *New Left Review* and chairman of its publishing company, Verso. These are not exactly spectacular careers, but normal professional lives like those of many others of the same generation.

The more difficult question is: what did the troubles do to the School? Walter Adams ended his sad account of events in the 1968/9 Director's Report on a note of defiant optimism: 'I am, however, confident that a later and impartial judgment will discover that in spite of all the tensions of the year the School emerged more united and clearer about its purposes and character.' Others would agree. The general weariness in which the troubles ended described a mood rather than a cause of their conclusion. The School had survived their trials and tribulations more successfully than academic institutions elsewhere. Still, even twenty-five years later it is not at all easy to be impartial about what happened in the School's unhappiest year.

Adams—from January 1970 Sir Walter Adams—may have had certain weaknesses. A newspaper profile quotes an academic as saying: 'He is basically very liberal minded, but he seems easily moved by strong people near him.'[118] Robbins above all dominated him visibly. But Sir Walter Adams also had a subtle mind and a stubborn sense of purpose. His numerous notices, memoranda, and letters during the troubles tell their own story of civility and commitment. Adams's main preoccupation was what he liked to call the 'academic community'. The key sentence in his 'Dear Student' letter during the closure read: 'I want to affirm to you my belief that the LSE is an academic community.' He worried that 'confused issues and situations' had 'strained the mutual loyalties within an academic community'.[119] Behind the events at LSE, Adams saw deeper trends. The 'abdication of religion' has left a moral gap. 'Unless the academic community can define more clearly for itself and for others the different loyalties which it serves and will not surrender to competitors', other forces like 'the State' or 'contemporary society' will move in. Both institutional autonomy and academic freedom have to be redefined. This will give new strength but it also involves obligations. 'The academic community, to protect its own freedoms, has a duty to prevent the disruption of teaching and research as a form of protest whether political or domestic, and that duty is shared by all its members, staff and students alike.'[120]

The more vocal students did not see things this way. For them, the School was the community if not the conspiracy of the others, of the Establishment, the authorities. Few went as far as the anarchists in their slogan: 'LSE academics are thought police. Why not make them real police?' But not a few may well have followed Hoch's and Schoenbach's 'campaign to demystify the nature of the university and its authorities'. By their own testimony they did not succeed, but the project tells a story:

> Our failure to demystify the staff is intimately related to our failure to identify the bourgeois content of much of their social science disciplines. In particular, we have yet to convince students of just how much their supposedly objective, value-free teaching and research actually contains a built-in apologetic for the present system, or is deliberately centred on trivial, irrelevant techniques and problems.

In the mean time, the struggle was turned into one about authority rather than community. 'With the first blows of our sledgehammers, some of our "psychological gates" of submissiveness to law and authority began to weaken.'[121] At the same time, it looked as if some members of the School's authorities were playing the same game from the other end, as it were. 'Robbins and all that lot thought that they were standing on the battlements withstanding, as I've said many times, the red hordes from Berkeley on one side and Nanterre on the other.'[122] Ronald Higgins, a former student and a Governor of the School, put it with succinct irony: 'One has sometimes almost fancied the Governors' Standing Committee to be a front organization of Soc.Soc.'[123]

The majority of students were less interested in such battles, but they were also looking for community, their own community. The wearying nights of occupation were so much more intimate than the boring days of lectures. Colin Crouch has described the paradoxes of this 'activist community' in its combination with an emphasis 'on spontaneity, an opposition to structure, on the psychedelic'. Crouch has also given reasons. 'The university has become a community of professionals, and the professional is the last man to be found in a thorough-going community.'[124] The post-Robbins university is no longer the *universitas magistrorum et scholarium*, the academic community, but another exemplar of modern bureaucratic organization, professional management, organized industrial relations, strikes, lock-outs, and all.

This is not the whole story. There is that other feature which, among the hundreds of explanations of '1968' Ernest Gellner has brought out most clearly, the almost existentialist quality of student and more generally of youth protest. One of the rhythmic chants of the times tells all: 'What do we want? Everything! When do we want it? Now!' If the protest is about anything, it is about authority in general. Raymond Firth calls it, perhaps with the generosity of age, 'essentially moral', 'a kind of defence of ultimate moral principles of society',[125] but the language of protest was more one of attitudes than of values, of sincerity rather than justice or civility. 'Its basic idea is that *sincerity* is the key to truth, and above all that any kind of order or structure is a betrayal.' One has to cut through all institutional constraints in order to discover a reality which has got lost in the hypocrisy of organized society. In order to do so one must either 'smash' institutions or opt out; in an odd way, the gates, the lock-out, the exile, then the sit-ins provided both. The hippie and the activist met in the much-publicized night of nude bathing in the ULU pool during the 'exile' (though it is said that most militants kept on their underwear). The public was duly shocked. Ernest Gellner offers another sobering thought: 'Perhaps these rebels simply offer a foretaste . . . of the problems of social control in affluent, liberal, welfare and doctrine-less society.'[126]

All this was, however, less true of LSE than of Berkeley and Nanterre. The School could still be divided, which means that it had not dissolved altogether as a social entity, even a community. In those troubled years the School was divided. While incident followed incident the divisions were many but in the end one difference remained which had to do with the simple question: how did the authorities of the School discharge their responsibility for the precious institution given in their trust? Some would say: they did well, better at any rate than the authorities of Berkeley and Nanterre; they were firm without intransigence, and thereby preserved the governability of the institution which might otherwise well have gone down the path of 'parities', with the result of immobilizing all decision-making, or even of long-term chaos and anomie. At the other end there are those who argue: the authorities were inept and made every mistake in the book; they used their power without sense or sensitivity, alienated many students and quite a few members of staff, and only managed to preserve a deeply non-democratic institution.

Between the extremes there are as usual many, possibly—though who is to count them?—a majority whose feelings are mixed and troubled by questions more than by answers. Ugly things happened, and even uglier things were said—but did the authorities gauge the situation at critical moments correctly? Or did they too, like the 'revolutionaries', fight a rather abstract political battle? Had we not all learnt, as students at the School, to appreciate the complexity of situations of conflict? Had we not also come to appreciate the suspicious quality of putting the blame on 'small minorities'? One may even wonder whether the line which Walter Adams believed in but was unable to enforce might not have been appropriate. But then one remembers Adams's travails in Rhodesia, and the ease with which the 'soft centre' can be squeezed almost out of existence when polarization occurs and extremes take over. One also remembers the benefits of hindsight.

Using hindsight to the full and taking a long view, it is most striking how devoted most of the protagonists of the troubles were to the School. To be sure there were those among the staff who despaired and left or withdrew forever into their shells, and some of the student activists never really got to know the institution which they chose as a platform for their extraneous struggles. Even this needs to be said with care. Colin Crouch was not the only one who turned away from militancy when he saw its destructive effect for the School; and even Vic Schoenbach, Paul Hoch's co-author and companion among the militants of 1969, later made a donation to the LSE 1980s Fund for students in hardship. Certainly, Lionel Robbins acted out of concern for his beloved School just as John Griffith did, or Alan Day as Vice-Chairman of the Academic Board, or Percy Cohen as Dean of Undergraduate Studies. The Academic Governors and the dissentient members of the Law Department did not agree on many resolutions and decisions, but they agreed that the School was worth fighting for, which is not a matter of course. Some, like Ted Brown, the Head Porter, did not show their feelings but shared the concern for LSE during the events of 1967 and 1969. 'I was disappointed really; the fact that it was an upheaval over nothing. I resented a lot of it, although I showed no emotion either one way or the other.'[127] One former student, Theo Richmond, who had been at the School many years before the troubles, reacted to the events with a melancholy and slightly apprehensive comment which many who were there at the time or otherwise feel a part of the School's academic community will share: 'It seems to me that LSE, like the freedom we tend to take for granted, has become more vulnerable than most of us realize. I'd hate to see either of them go.'[128]

NORMALITY WILL NOT RETURN

From Improvement to Economy

One effect of the troubles at LSE is beyond doubt; it was grist to the mills of those who grew increasingly weary of universities in general and the social sciences in particular. The change was partly one of the public mood, but as public expenditure came under pressure for all kinds of reasons the changing climate of public opinion meant that there was little outcry against making life more difficult for the coddled children of the 1960s, the universities, and their students. The squeeze had begun before there was any sign of trouble. Sydney Caine had in fact spent the better part of his last Director's Report describing the effect of the settlement for the new quinquennium which had begun in 1967. 'This quinquennium will be a period of standstill rather than of development, because the recurrent grants from central and local government sources are barely sufficient to meet existing commitments.'

More ominously, in December 1966 the Government minister responsible for higher education, Shirley Williams, announced that from 1967 onwards differential fees would have to be charged for overseas students. The initial level was set at £250 per annum. This was ominous in several respects. It signalled a new approach to those who have no vote in Britain; apart from ignoring the international quality of Britain's best universities, it indicated that in future those who could be made to pay for services would be required to do so. Beyond that, it marked the beginning of a consequential change in university finance. Universities were allowed to retain overseas student fees. This introduced a 'private' interest into what had been public-sector institutions since the war if not before.

Initially the effect on relative importance of the sources of the School's income remained modest. Throughout the 1967–72 quinquennium the School got about three-quarters of its total income from the University Grants Committee through the Court of the University of London, that is from the state. In the early 1970s the state share in School income even rose to 80 per cent. Moreover, these were still block grants with only a few strings attached. The real shocks happened after 1975, and most dramatically in 1977/8, 1980/1, and 1981/2, by which time LSE received almost as much from student fees as from government grants. Since then, the state share of LSE funding has shrunk to less than 30 per cent.

The bane of the early 1970s (as well as the later period well into the 1980s) was not 'cuts' but inflation. As its rate accelerated, supplementation became tardy and tenuous. Salary increases were usually covered, but for the rest the School, like other universities, was at risk for 50 per cent of whatever increases occurred in a given year, and reality fully lived up to the worst fears. Small wonder, then, that the School had a deficit in every year since 1968 with the single exception of 1972/3!

In this climate, LSE worked out its quinquennial plan for the years 1972–7. The School decided immediately to aim at stability rather than expansion though it insisted on what in those days was still described as 'the need to make good some of the deficiencies caused by underfinancing in the present quinquennium, in such spheres as staff–student ratios, staff promotions and Library acquisitions and services'. The only change projected was an increase in the proportion of graduates to 50 per cent of the 1977 target of 3,000 students, and the doubling of the 400 part-time graduates. On this as on many other occasions the Director made a strong case for the expansion of graduate studies in an institution uniquely placed to tend 'the seed-corn on which future harvests in national and international scholarship and real wealth depend'.[129]

The University Grants Committee (UGC) did not agree. In the first place it took its decision late, so that the new quinquennium began in 1972/3 with a 'provisional year, a holding operation in which all universities had to make short-term plans, without knowledge of their recurrent grants and of their student number targets for the quinquennium'. When the allocation was finally announced, it was 'abundantly clear that the epoch of expansion to which the universities had become accustomed in the post-war period has ended'. The UGC explicitly stated that the grants represented 'a change to an "economy factor" from the "improvement factor" which had characterised previous settlements'. A 2 per cent reduction in 'grant per student' was built in, and inflation adjustments would be provided by ad hoc supplementary grants given as a rule a year in arrears.

For the School the real shock concerned student numbers. The letter to the University of London said that 'the [University Grants] Committee have been unable to accept the proposal from the London School of Economics that their postgraduate proportion should increase'. In fact it declined during the years in question. Part-time graduate numbers, instead of going up from 425 in 1972/3 to 800 in 1976/7, went down to 363, and the number of Higher Degree and Higher Diploma students remained exactly the same, while first-degree numbers rose by nearly 10 per cent. The reasons are not just 'instructions' by the UGC; fine-tuned planning was no easier in the 1970s than at earlier times. Graduate scholarships for home students became increasingly scarce. Moreover, some potential students probably were put off by rising fees for overseas students and even the effect of LSE's reputation on their parents, on whom they depended for their living.

In his last year as Director, Sir Walter Adams observed an increasing 'chaos' in financial matters. He singled out 'two blows' for special mention: the 'abrupt withdrawal of all supplementation for rising costs in 1973', and 'the exceptionally rapid rise in prices and wages'. 'We, like others, may survive for a year by eating up our reserves but will then be defenceless against rising deficits.' As we know, this is not what happened. It would be cynical to speak of 'creative chaos', but the confusions of the 1970s became the medium of a fundamental change in British university finance. Not all of it was intended; no one wanted the inflation rates of the 1970s with their peak at 24.2 per cent in 1975. But none of four successive Prime Ministers made any attempt to preserve the publicly funded university system with sensible planning horizons either. First the quinquennial system went, then the whole notion of safeguarding the 'unit of resource' by public funding, and in the end the UGC itself with its built-in guarantee of academic autonomy by block grants.

Strand House

The seventh Director was no less concerned about what one might call the physical geography of LSE than his six predecessors, but when he retired in 1974 he had every reason to be pleased with developments under his aegis. In Sydney Caine's time the School had been able to acquire the office block at the corner of Houghton Street and the Aldwych, Connaught House. While it added to the labyrinthine warren (and also to the need for 'gates') it gave the School's administration a long-term home. During Adams's directorship, and with UGC capital grants, two major building projects, the St Clement's extension and the Clare Market Building, were completed. Public funds were after all still available when they were really needed. Moreover, the anonymous benefactor who during these years did so much to improve the amenities for students of the University of London provided funds for an LSE hall of residence in Rosebery Avenue; after its completion, the School was able to provide accommodation for nearly 1,000 students. There remained, however, the ever more pressing problem of the Library. The BLPES was sometimes described as 'the best and the worst library of its kind in the world' (Sydney Caine adds: 'with some pardonable exaggeration')[130]—an unrivalled collection in the social sciences housed in conditions which were as inadequate for the books, journals, and papers as they were unacceptable for their users. More and more material was sent to the University depository at Egham, which meant further delays in access. Outside users in particular had to be kept on a very short leash in a library that was after all a national and international treasure.

Then a silver lining appeared on the horizon. First Caine, later Adams dropped mysterious hints. These began in 1963 and got broader by the year. 'We believe, however, that if finance were available it would not be impossible to find a solution by the purchase of a site or building quite close to the School' (1965/6). 'Although no definite conclusion has emerged it remains the hope of all concerned that a site not too far from the existing buildings can in due course be acquired' (1966/7). 'Only the radical solution of rehousing the Library on an alternative site will solve its problems satisfactorily. Continuous efforts throughout the session were maintained in pursuit of this radical solution, but a final conclusion cannot yet be reported, and may yet elude us' (1967/8). In the event it did not elude the School, though more mystery and much more effort was required before the Director could begin his final Report in 1974 with the laconic but triumphant statement: 'The School has purchased the freehold site and building, Strand House, in Portugal Street.'

Strand House was the warehouse and central office of the booksellers and newspaper distributors W. H. Smith & Son Ltd. The massive redbrick building had been erected for their purposes and opened in 1916 (when it was immediately requisitioned by Government for the rest of the war). Separated from the rest of the School only by the narrow lane which runs behind St Clement's Building, its over 15,000 square metres of floor area on six levels were ideally suited for a Library which had so far had to make do with less than half that amount of space. When news reached the School in the early 1960s that W. H. Smith's had decided to relocate its head office to New Fetter Lane and move the warehouse facilities as well, those who heard about it, notably Lord Robbins, immediately registered LSE's interest in Strand House. In subsequent years, the company displayed patience and generosity towards an institution which, while technically also a 'company limited by guarantee', took its decisions in more measured ways than businesses normally do. The then chairman of W. H. Smith's, Sir Charles Troughton, became a good friend and, after the completion of the deal, a Governor and member of the Standing Committee.

The problem was, obviously, money. Site and building were to cost £3.8

million; a further £700,000 initially estimated for the conversion of Strand House later turned into £1.5 million. There began that massive and eventually successful effort which came to be known as the Library Appeal. It became a likely proposition when the UGC and the University of London agreed to help; they provided nearly £2 million for acquiring the building as well as, later, £850,000 towards the conversion cost. Still, the remaining task seemed daunting to most: £1.8 million had to be found by 1973 in order to complete the purchase, and another £600,000 or so afterwards to bring the building into use—and all that at a time (as the Director remarked even in 1974) 'when the School's "public reputation" was low in business and industrial circles'.

The man who was undaunted by the task, and set to work with his usual gusto as virtually full-time Chairman of the Library Appeal, was Lionel Robbins. He had retired from the chairmanship of the *Financial Times* in 1971, though he remained Chairman of the LSE Governors until the end of 1973. The success of the Appeal was to be his last great gift to the School. Fortunately, Lord Robbins had much support. Among lay Governors, David Kingsley came once again to the rescue, and John Morgan followed. A Public Relations Sub-Committee was chaired by Professor Ben Roberts and attracted as its secretary Jennifer Pinney. Ambrosine Hurt acted as secretary of the Appeal Committee itself. Professor Harold Edey, having successfully completed his stint as the first Pro-Director during the troubled years, became Appeal Co-ordinator. The experience gathered in this process was of lasting benefit; over twenty years later the diplomat (now Sir) John Morgan became director of the LSE Foundation, and Jennifer Pinney assisted him in the Centenary Campaign.

Many others helped, but perhaps the most helpful factor was the object of the appeal itself, the Library. It made the appeal both plausible and attractive. Strand House would fit the LSE Library like a glove. At the same time, the British Library of Political and Economic Science, BLPES, was known at home and abroad. 'There are in the world few libraries which the students of social sciences can use with so much profit,' wrote Raymond Aron; and Sir Isaiah Berlin confirmed: 'All serious students of recent history and present problems of society must share a common interest in the preservation and continual growth of the facilities of this famous library.'[131]

The BLPES had indeed grown continually. The long-serving Librarian, Geoffrey Woledge, has described the great improvements under his predecessor, W. C. Dickinson, and the 'steady but less spectacular' progress since the war.[132] He had come in 1944 and was to retire in 1966, handing over to D. A. Clarke (who had actually been the first Librarian of the University College of Rhodesia and Nyasaland in the early years of Adams's principalship). At the time of the move to Strand House the Library contained some 3 million items, among them 700,000 bound volumes, relating to 'the social sciences in the widest sense of that term'. Clarke paid special attention to research materials like the Passfield Papers, the archives of Dalton and Beveridge, of Tawney and many others, and then, to the large 'collection of governmental and inter-governmental material'. The combination of size, range, and depth of information sources was unique at least in Europe, which made it all the more imperative to find adequate space for the collection, its users, and the library staff.

The cause of the Library Appeal was attractive, and its proponents were enthusiastic and well organized. However, their greatest success was possibly one which had nothing directly to do with pounds sterling or dollars or even books; it was the mobilization of support for the Appeal and thereby for the School itself. HM Queen Elizabeth the Queen Mother sent a message of encouragement and followed the entire appeal with sympathetic interest, all the way to the formal opening ceremony in 1979 which she attended. The formal launch of the Library Appeal was undertaken at the Mansion House in February 1973 by the Lord Mayor. Important functions included a banquet at Skinners' Hall and a concert at Banqueting House in Whitehall. They not only raised some funds directly and indirectly but brought the School into the news, and this time into the good news. When Sir John Hicks donated his Economics Nobel prize of 1972 to the Library Appeal, the generous act found a resounding echo. A benefactor underwrote anonymously, at the critical moment, the remaining sum so that Strand House could be purchased in time. This was Lord Rayne, a friend of the School, and of its chairman of Governors.

The School had once again captured the imagination of the world as the foremost research university in the social sciences. This return to favour was not confined to London and its establishment. The Appeal provided an opportunity to build, for the first time, effective contacts with alumni all over the world. This is where Anne Bohm's contribution was to become crucial, though other members of the School's staff also travelled to the four corners of the world to remind friends that the old place was very much alive and interested in their contribution. By 1975, a register of some 21,000 names had been drawn up; it has since been extended to 75,000 former students and other members of LSE.

For the purchase of the site of Strand House, £1.83 million was eventually found. Of this sum 22 per cent had been given by LSE groups all over the world; in fact, 40 per cent of the total had come from overseas; first from the United States, then from Germany, Japan, Iran, Canada, Singapore, Hong Kong, Australia, Mauritius, and other countries. A large portion of the total came from foundations, mostly American, British, and German; but no less than 22 per cent was contributed by mostly British companies. The names and even more the long pages of advertisements in the brochure produced when Strand House was finally reopened for the BLPES show that the School had regained its 'public reputation', if indeed it had ever lost it.

Strand House was purchased at the end of 1973, and eventually vacated by W. H. Smith's in 1976. Finding the remaining funds for the conversion turned out to be difficult; the objective was less likely to appeal to donors. However, in the end Lord Robbins achieved this feat as well, and conversion work began. The effects of the effort were many, and they were almost all beneficent, though one which was noted both by Sir Walter Adams and by his successor left niggling doubts in the collective mind of the School, as later events showed. In 1972, the Director said: 'Implicit in the decision to acquire Strand House is a decision that the School with its Library should remain on its present site in the heart of London.' Six years later, when the new Library was opened, another Director assured his audience that the School would still be cramped.

> Indeed, I suspect that so long as there is a London School of Economics, it will be huddled into that collection of buildings around Clare Market, a little too noisy, a little too dingy, much too small, but in circumstances which befit a distinguished academic institution in the social sciences. LSE is, among other things, a training ground for living in the modern world.[133]

Mixed Fortunes

Some senior professors who have seen it all will say that, the brief 'exile' apart, life had never ceased to be normal at LSE. Students were pursuing their studies, teachers were teaching, and a continuing stream of doctoral dissertations and publications by staff testified to the vitality of research. There was the strange scene of one lot rushing through the refectory with pick-axes and sledgehammers while others were quietly having their dinners, or even more quietly playing chess. Not that one group was revolting and the other living normal student lives; many were doing a little of both, and most were working for their degrees. Colin Crouch even wrote that 'in many ways the concerns of the protests mingled fruitfully with those of our studies', and added: 'I remember one of the leading radicals, David Adelstein, infuriating academics by telling them that the protests provided a kind of experimental laboratory for students of the social sciences.'[134] That was no doubt a minority view, more widespread among sociologists than mathematical economists; but serious study continued throughout the troubled years.

On the other hand, when the troubles were over an atmosphere of tension remained in the air. Small incidents would lead to explosions which did not last but which worried the authorities and did not help attempts to improve the School's image. In 1970, Sir William (later Lord) Armstrong, the top civil servant, was prevented from giving the Oration; soon the occasion itself would be suspended for over twenty years. In 1973 a guest lecture by the psychologist Hans Eysenck, whose views on genetics many regarded as 'politically incorrect', was interrupted. So-called rent strikes led to picketing, and, on one occasion, to a nasty incident in Houghton Street, though the Students' Union condemned the violence which ensued.

During these years it became fashionable to interpret all disturbances of the normal routine as failures of communication. Consequently, an Information Officer was appointed in 1969 to improve 'internal and external communication'. Shirley Chapman did an excellent job; she became one of those who were accessible to all and never seen to take sides; but she could hardly make real differences vanish.

One of these real differences continued to concern relations with students. Drastic changes of the machinery of government had now been shelved. The more modest offer of participation in some committees and membership of the Court of Governors was rejected by the Union year after year; when it was finally accepted in 1975, the moment of effective student pressure had passed. In 1972 the Students' Union at last ratified its new constitution, one part of which contained rules which the Union could not amend without the consent of the Court. The Director observed 'a weakening in the representativeness of the Students' Union' and wondered about the need for, and the justification of, compulsory membership. In any case, departmental staff–student committees remained the main channel of communication and co-operation.

Beyond politics, many disciplines were flourishing in the early 1970s. In 1972/3 the Department of Sociology had seven professors (Percy Cohen, Ernest Gellner, David Glass, Robert McKenzie, Donald MacRae, David Martin, and Terence Morris) and sixteen lecturers. However, student demand had dropped dramatically after 1969 and, judging from their examination results, those who came did markedly worse than an earlier generation. Soon the BA in Sociology would be merged with the B.Sc.(Econ.). In the history of sociology, '1968' has a special place; it probably marks the turning-point away from general sociology ('sociological theory') towards a proliferation of specialisms and growing professionalization.

The Department of Economics reached one of its high points in the history of the School during the early 1970s. The political economists were still there in strength: Professors Peter Bauer and Alan Day, 'Hal' Myint and Alan Prest, Peter Wiles and Basil Yamey. They had been joined by the larger-than-life influential theorist who inspired simultaneously LSE and the University of Chicago, Harry Johnson, and the rising star of economic theory and social thought, Amartya Sen, as well as by Professor Alan Walters, who was to become Margaret Thatcher's guru. Then there were the mathematical theorists and econometricians, some of them, like Terence Gorman, Michio Morishima, and Denis Sargan already professors, others, like Partha Dasgupta, Lucien Foldes, David Hendry, and Steve Nickell, soon to be promoted. (They were almost all gripped by wanderlust; four of those mentioned went to Oxford, one to Cambridge, almost all, at least for a while, to the United States.) Moreover, a new generation of combined theorists and political economists was emerging, Nicholas Barr and Richard Layard among them. As if this were not enough, the long-name department, now called 'Department of Statistics, Computing, Demography, Mathematics and Operational Research', notably though by no means only through Ken Binmore, James Durbin, and Colm O'Muircheartaigh, showed a growing interest in economics. Alan Stuart, then a senior professor in the department, saw this cutting both ways: 'Well, in fact the econometrics people are better statisticians than most statisticians are.'[135]

As ever, large generalizations about a complex institution have to be taken with a grain of salt. While these changes happened, the Government Department pursued its own varied, and highly individual interests, with a combination of political theory, area studies, and public administration. The Department of International History had a phase of great strength. Still, two trends were discernible in the early 1970s. One was the resurgence of 'difficult', often technical subjects in the social sciences. If sociology had been the preferred option of the 1950s and 1960s, economic theory and econometrics attracted many of the best students in the 1970s and 1980s. The other trend is professionalization. The days in which social science meant the exploration of uncharted lands of knowledge were over. People sought a solid preparation for high-quality jobs and preferred disciplines which promised just that.

One untypical yet telling example was provided by the Department of Social Administration. When the much-beloved Field Work Tutor in the Department of Social Science and Administration, Kit Russell, retired in

1973, her colleagues—Sheila Benson, Christine Farrell, Howard Glennerster, David Piachaud, and Garth Plowman—decided to offer her a very special present. They conducted (not without Kit Russell's guidance, to be sure) a survey of all students who had taken the Diploma or Certificate in Social Administration between 1949 and 1973, which was later published under the title *Changing Course*. An astonishing 87.8 per cent replied to the written questionnaire, and the findings are of wider interest.

'Our diploma was, and is,' the authors say, 'an introduction to the social sciences with social policy and social administration as a core, and with a vocational slant chiefly towards social work.' Though two-thirds of the students were women even at the end of the period, the proportion of men had increased. The authors were a little surprised by the discovery that a diploma in social administration was actually not an avenue of social mobility; two-thirds of the nearly 2,000 students came from 'professional and managerial' families. Given the fact that during this period women were still at a considerable disadvantage in many professions, 'social administration' clearly offered a way forward for them, notably for married and older women. But now professionalization had set in. Some regretted such developments; at the end of the book a student of the late 1960s is cited who wrote to Kit Russell: 'I was sad to hear at the LSE day that there is some talk of abandoning diploma and certificate courses at the LSE and concentrating on 1st and 2nd degrees.' Change is always sad for some; but the study showed clearly that 'more and more areas of social work have begun demanding the professional ticket'.[136] It was thus only a matter of time before the diplomas were absorbed by Master's degrees.

The year in which Kit Russell retired, 1973, was also the one in which Richard Titmuss died. Other major School figures had passed away in the early 1970s: Morris Ginsberg in 1970, the former Secretary of the School, Eve Evans, in 1971. In the same year, the long-time Chairman of the Governors, Sir Otto Niemeyer, died. In the present Court, Sir Frederick Harmer retired after sixteen years as Vice-Chairman; he was replaced by the lawyer, former student, and ex-President of the Students' Union Sir Morris Finer, who was thus groomed for the chairmanship in succession to Lord Robbins.

In 1971 Lord Robbins had, on behalf of the Court, extended the appointment of the Director by two years, to September 1974. In his penultimate year, Walter Adams had a heart attack which led to an extended absence. Perhaps the 'colleague' was right who was quoted as believing that 'the LSE troubles affected Walter Adams far more than he will ever reveal'. Outwardly he certainly remained 'the incarnation of all those sensitive, decent, civilized and vulnerable virtues of liberal England that have proved so difficult for public men to sustain unsullied in the twentieth century'.[137] This was written by Peter Hennessy, in a long appreciation of Sir Walter Adams on the occasion of his retirement, under the suggestive title 'A Man Uniquely Tossed by the Gale of the World'. Adams himself found characteristically moving words in his last letter to staff and students. The School, he wrote,

> is essentially a group of some thousands of men and women, sharing in certain common interests, enthusiasms, and tensions, seldom acting in formal unity and hostile to articulating definitions of their purposes, pragmatic, and non-evangelising, but with a great vitality which shapes all those who are associated with it. I certainly have enjoyed every moment of my membership of this great company, with all its domestic, national and international associations, and while warmly welcoming Ralf Dahrendorf on his return to the School cannot hide my twinge of jealousy that he will now have that enjoyment. [138]

8

TOWARDS THE CENTENARY

VALUES OF THE ACADEMY

Coming Home

The directorship of LSE came my way as a surprise. When Lord Robbins, then still Chairman of the Governors, first asked me in the course of a lunch in the less-than-intimate grandeur of the Reform Club whether I would accept if invited, I was pleased to be asked but sure that the answer had to be no. Why should a European Commissioner, destined (perhaps) to go back to a political career in Germany, leave both his country and his Brussels position to be Director of LSE? Because, my friends said when I consulted them during the next few days in Brussels, in Hamburg, and again in London, it would enable me to combine my academic and my public interests, my feelings for London and my internationalism, for the benefit of an institution of which I had been fond ever since I first set foot in it in September 1952. Suddenly, to say yes became the obvious answer to the invitation, to me at least, for to some it still looked improbable. 'You must be crazy,' one distinguished Governor, who had been on the Selection Committee, remarked over drinks on the day of my election, 18 September 1973, though *The Times*, then still the newspaper of record, consoled me the next morning with a leading article under the heading 'An Original and Welcome Appointment'.

Coming back to LSE was like coming home, except that now home was no longer a comforting place looked after by a remote and benevolent Director called Sir Alexander Carr-Saunders and a somewhat less remote Ph.D. supervisor called Professor, and after a while, Tom Marshall, but I was in charge, or at any rate in the chair and called Director, so that others, students and staff, could do their thing. Left alone with such responsibilities, one needs help, yet there are few who can give it. My predecessor was generous with his advice but he was also tired. It did not take long to recognize that the School was still divided by fault lines of memory and generation, personality and convictions, and, however unfairly, Walter Adams was a part of these divisions rather than one to overcome them. He was longing to get back to the historical research which he had abandoned forty years earlier in order to assist German exiles. He also wanted to use the tranquillity of his house in Sandwich on the Kent coast to recover from the strains of his septennium at the helm of LSE and the illness of the last two years. Alas! none of this was to be. The only truly happy event that came to pass was the award of an honorary degree by his old University of Rhodesia. In May 1975 he went back to Salisbury (as Harare was then still called). The night after the ceremony he had a terminal heart attack and died. 'In retrospect,' his son Peter wrote on 22 May 1975, 'Wally's death could not have been more perfect, even if he had planned it, as he died while observing and enjoying the tremendous fruition of his labours in Rhodesia.'[1] It was hard to recreate in one's imagination the young idealist of the 1930s, the skilful intelligence adviser of the 1940s, or even the generous colonial vice-chancellor of the 1960s. The memorial service at the University Church of Christ the King on 2 July 1975 gave members of the School pause for thought about rebuilding their community.

Next to the predecessor, the Chairman of the Court of Governors was a source of advice and support. Lord Robbins had retired (though not left the scene; he was still in charge of the continuing Library Appeal now seeking funds for the conversion of Strand House). His successor, Sir Morris Finer, then a judge of the Family Division, Vice-Chairman of the Governors, and, of course, a former student of the School, stepped into his shoes with his own firm, thoughtful, and compassionate style. Looking back, it appears as if dinners to introduce the new Director to Governors and friends of the School took place about twice a week. We were sitting around the splendid table donated by Georg Tugendhat for the Director's Dining Room; Morris Finer would start a serious and yet suitably light-hearted conversation, always the judge and the journalist at the same time; in the end he never failed to ask the guests for a contribution to the Library Appeal. In fact, there cannot have been many occasions of this kind, for in December 1974, eight weeks after the arrival of the new Director, Morris Finer went to hospital for a check-up, never to return. A week later, on 14 December, he died.

A new Chairman had to be found within weeks. The School and its Director were fortunate: they succeeded in persuading another former student, the outgoing Managing Director of BBC Television, (soon Sir) Huw Wheldon, to take on the job. Huw Wheldon remained apprehensive of committee meetings throughout his ten years as Chairman, but he was relaxed and friendly in personal contacts and a memorable speaker on formal occasions. His judgement, or perhaps simply his ability to smell a rat and show discomfiture in his expression when he did, prevented many a mistake by the Director and by the School in general. Others helped. Professor Cyril Grunfeld had, as Pro-Director, in fact directed the School calmly and ably since Sir Walter Adams's first illness. The two Secretaries appointed to share the administrative burden in order to avoid undue concentrations of power, John Alcock and John Pike, formed a relationship of convenience rather than intimacy but were severally and together indispensable tutors for the new Director. John Alcock had an unerring sense of academic needs and values; in his quietly effective way, often by tightly argued written memoranda, he insisted that LSE must not stray from the path of academic virtue. John Pike brought to bear a wide range of experience, and a rare absence of self-interest, on the exigencies and conflicts of the School's administration. He was far more than a Financial Secretary, though he certainly was that too. For the Director, he soon became the indispensable partner in managing the School. His steadying influence in times of trouble came to be appreciated by all groups, academics, staff, and students. Others, especially on the academic side, emerged who for one reason or another commanded respect at the School: Professors Yamey and Wise, Prest and Day, Edey and Griffith. Some in more junior academic positions, like Maurice Perlman or Dan Sinclair or Peter Dawson, had a commitment to the School and a way of getting on with all groups which made them important collaborators.

In due course the circle of supportive colleagues widened, among academics and administrators as well as some students. It included above all the porters. When one of my successors asked me what he should do first on arriving, I remember telling him: talk to the porters! They will not take sides, not even for you, but they are the mainstay of the School, and as you pass the Lodge their demeanour will tell you whether things are all right or not.

A university neither wants nor needs to be run. It may like someone sitting on the woolsack with a smile on his face, but basically it runs itself, by way of its own mysterious 'usual channels'. Interfering with the usual channels should be reserved for extreme situations. Yet it helps if the tone is set in ways which encourage everyone, or at any rate most, and also if someone finds the words both within and without to make people aware of their strengths and their prospects. Of course, there is the occasional crisis in which someone has to be seen to take responsbility who is not playing face-saving games. Such crises may come from inside the institution, like restless student activists at the carnival time of February and March, or explosive feuds which are the bane of academic institutions. In my time, the deepest antagonisms still went back to 1968. Some younger members of staff refused to go to the Senior Common Room because they did not want to encounter certain senior colleagues. Bringing peace to a troubled institution is not easy if the troubles were real. There was no question of facile reconciliation. The School had to learn to live with itself, with all its members whatever their stance had been in the past. Luckily I had not been involved in the troubles. Who knows how I would have acquitted myself in their various twists and turns!

But most crises during the years from 1974 to 1984 had their origin outside. It was fortunate for the new Director that they did not start during his first year. On the contrary, 1974/5 was a hopeful session. Strand House had been bought, and the Library Appeal had come close to reaching its target. Contributions from fifty countries spoke well for the international quality of the School and the loyalty of its members, which was soon to be tested by overseas students' fees designed to deter rather than attract. The conversion of Strand House would provide an opportunity to improve the physical environment of the School. The closure of Houghton Street for through traffic in September 1975 was a great step forward in that regard. 'I sometimes dream of it as an open-air area with trees and chairs and sculptures and plenty of life, linking the ground floor area of all our buildings into one attractive cityscape.'[2] Some dreams come true, this one through the generosity of Peter Palumbo, Governor of the School and friend of the arts, who had the street paved and embellished.

The session 1974/5 was also the last time the School prepared a Quinquennial Development Statement, intended for the period 1977–82. In retrospect it reads as if we already knew that medium-term planning would soon be submerged in annual grants with occasional supplements and increasing government interference. A letter by the Director to *The Times* on 24 May 1975 made the case for the defence of universities in traditional terms. 'Maintaining their ability to teach without dogma or even immediate function is essential for the survival of a free society.' The Quinquennial Statement itself was 'realistic' and thus 'falls far short of the aspirations and indeed the potential of the London School of Economics'. It projected more or less stable student numbers at about 3,500, and demanded the 'unfreezing' of posts which for financial reasons had remained unfilled. It emphasized the need to keep the School's strengths intact, its overseas and postgraduate element, the growing category of 'post-experience students', the Library. Funds were needed to accommodate changing choices by new students. The conversion of Strand House required some support. Research was seriously underfunded. That was it, though 'the School is bound to state that anything short of the modest proposals advanced in this statement would amount to an uneconomical waste of existing resources and ultimately to the deterioration of one of the international centres of learning in Britain'.[3]

A British Brookings?

The new Director had spent the better part of the ten years preceding his appointment in politics rather than in universities. He had been an adviser of governments, an elected parliamentarian, a member of the German government, and a European Commissioner. It is not surprising therefore that he took a special interest in the predicament of his adopted country as well as the world in general, and also that he wanted to make some useful contribution. From numerous conversations inside and outside the School, a twenty-two-page paper emerged in January 1976, entitled 'A Centre for Economic and Political Studies in London'. Thus began the little saga which came to be known under the heading 'A British Brookings'.

The paper was about 'the need and shape of a Centre that helps politicians, businessmen, administrators, professional people and scholars to make sense of the economic, social and political predicament of the world in the 1970s and 1980s, and of Britain in it'. Indeed the Centre was about more than 'making sense'; it was about 'knowledge-into-policy'. 'The London School of Economics is, as it was in the past, a place uniquely suited to provide a basis and forum for such research.' The 'predicament' was then described: pressure on international economic and political institutions; the Helsinki Final Act and *détente*; the oil crises and limits to growth; new social trends and cultural attitudes; and then the specific British problems of low growth, high state involvement, bad industrial relations, a class-ridden society, adversary politics. Too much short-term thinking impeded medium-term solutions, it argued. 'Coming to grips' with such issues required a meeting-place of brains and power.

Three options were then described. An 'institute of institutes' would bring existing institutions together in a federal or confederal structure. A 'centre for the determination of the national interest' would help 'to mobilise and stabilise the restless and increasingly disillusioned mass of middle-ground citizens around a focal point which is independent of the existing political parties'. Then, thirdly, there was the 'socio-politico-economic think tank' for 'the dispassionate, but synthetic study of problems of contemporary politics, economy and society with a view to contributing to the clarification of the horizon of decisions which have to be taken in any sector of the community'. Such a think-tank would conduct research into the predicament, bring together academics and practical people, and disseminate its findings.

The 'operational principles' of such a Centre would begin with a reassertion of impartiality *à la* Sidney Webb; research would have to be conducted without 'extraneous direction'. However, 'one is not talking about a purely academic institution'. Work should be relevant, and should draw on the views and the members of parties and other public organizations. The Centre should be accessible. Yet association with a university would be helpful. In order to do its work properly, such a Centre would need £900,000 annually ('in January 1976 money') to finance a dozen permanent and two dozen temporary fellows as well as the necessary infrastructure. And LSE? The intention was not its 'aggrandisement'; in any case the School had many of the features of a policy research centre already, though perhaps it could do with rather more 'political economy'. However, 'if a Centre for Economic and Political Studies is set up in London, the London School of Economics would be the obvious place to have it associated with'.[4]

The paper was sent to Governors and staff of the School, many of whom responded with long and thoughtful letters. For the most part, these letters agreed with the principle and then voiced more or less serious reservations. Sir Sydney Caine, for example, pointed out (as others did in their own words) that many of the issues listed 'involve basically political decisions or, if one chooses to vary the terminology, value judgements which are not susceptible of determination by economic research'. 'I should prefer to see an organization much less policy-oriented and much more "academic".' Professor Abel-Smith even saw a connection with the incipient discussion of a new centre party. 'The middle ground suggests the type of alliance which *The Times* newspaper has from time to time tried to encourage—the Roy Jenkins/Edward Heath Coalition, which has never been formed—and the attempt of Dick Taverne to found a new party which has fallen flat on its face.' T. H. Marshall shared these doubts, and in addition feared that a Centre of the projected size, if effectively associated with LSE, 'might swamp the School, while spending a lot of time and money deciding what it was supposed to do'. A traditional 'research unit' would be preferable.[5]

Student groups of the left were angry. (Those of the right remained silent.) The LSE International Socialists produced a pamphlet, 'The Noblest Director of Them All'. By nailing his colours to the mast of the centre, the Director had revealed that he was as conservative as his predecessors. '*Et tu, Brute*?' The Director replied with serious arguments against philosopher-kings and for synthetic and practical thinking. The LSE Broad

Left, Labour's own coalition, sent an open letter to the Academic Board urging it to throw out the proposals which were not only 'explicitly conservative' but 'explicitly designed to assist capitalism and the institutional forms of democracy which have grown up along with that mode of production'.[6] A Policy Studies Centre would politicize the School. The Charing Cross Branch of the trade union ASTMS passed a resolution agreeing with this stance. Not being 'political' now served to protect the left from others rather than, as in the days of the founders, from itself.

On 1 March 1976 a well-attended meeting of the Academic Board took place to discuss the Director's paper. Questions were raised and a variety of views expressed in the course of a tense hour. Increasingly the debate turned to the issue of what an academic institution is about. Professor Alan Prest on behalf of the economics professors drew a clear line between politics and scholarship. This ruled out the first two options, the 'institute of institutes' and the 'centre for the determination of the national interest'. The third option contained useful elements, but an outside board to oversee an academic institute was unacceptable. Moreover, it was essential for a university to maintain the link between teaching and research. Probably, LSE could do all that was needed by simply co-ordinating existing resources. After Professor Prest, coming from another department, law, and another persuasion, further left, Professor John Griffith, put the clinching argument in verse:

I rise on the feast of St David
(No day for a Welshman to cringe)
I can hardly support a new centre,
As one of the lunatic fringe.

I admit that the heart of this paper
Is somewhat offset—to the right.
But it keeps coming back to the Centre
And the need for the Best and the Bright.

I distrust 'thoughtful and sensible' people,
I'm appalled by the 'serious Press'
And see twelve professorial Fellows
As apostles of doom and distress.

But above all whatever the politics
Of Institutes, Centres or Tanks
I prefer LSE uncommitted—
And return your paper, with thanks.

The Welsh are a barbarous people
From a land of mountains and fogs
But I hope you'll abjure the Centre—
March 1st is *not* St Rees Mogg's.

Seven years later, in 1983, I wrote in my Director's Report that the Academic Board was the place where 'many years ago the decision was taken that the School as such should not get involved in policy-oriented research'.[7] Not so, wrote one attentive reader. The Director had in fact produced a second paper in April entitled 'Policy Studies' and a Working Party of the School comprising teachers from many departments (including Professor Prest but not Professor Griffith) had met twice. It had agreed that 'the Centre, if established, should be a fully integrated part of the School and should not seek any dominant role', and invited the Director to prepare a draft which might be laid before the Academic Board. Such a draft was never prepared. 'Whoever killed policy studies, it was not the Academic Board.'[8]

Technically, Professor George Jones, the author of this reminder, is of course right. In many ways the meeting of 1 March 1976 was not the end but the beginning of the debate on a British Brookings. (Turning the name of the large, and reputedly Democratic, Brookings Institution in Washington into a generic term was actually one unhelpful twist in this saga.) The idea of a policy research centre, as the Director's Report put it in 1977 with a metaphor from the repertory of strategic arms, turned out to be a MIRV, a Multiple Independently-Targeted Re-Entry Vehicle. Existing policy studies institutes in London felt particularly targeted. The *Times Higher Education Supplement* published an article, 'Why NIESR looks most like a British Brookings'. The National Institute of Economic and Social Research was, of course, the result of an earlier and equally difficult LSE debate in Beveridge's last unhappy years. Another product of the 1930s, Political and Economic Planning (PEP) took a step towards the 'institute of institutes' by merging with the Centre for Studies in Social Policy. A special issue of the *PEP Bulletin* in May 1977 assembled under the by now familiar heading 'A British Brookings?' contributions from many sources. 'St Rees-Mogg', then still editor of *The Times*, was in favour, but the Presidents of NIESR, PEP, and Chatham House feared a dissipation of scarce resources, and the Director of PEP, John Pinder, argued against big institutes which destroy plurality. Later still, and by way of the Ford Foundation, the debate moved to the European plane, though once again without much success. The 1980s, as it turned out, were to be a period of much more partisan policy studies than had been intended by the proposal of a Centre.

At LSE itself, however, the Academic Board debate of 1 March was to all intents and purposes the end of the story. Among many letters addressed to by the Director after the debate, one signed by fourteen mostly younger economics lecturers encouraged me to persevere but above all not to set up a think-tank separate from the School. The point was well taken though I had made it clear from the outset that my first loyalty was to LSE. There remains a question, to be sure, a profound and disturbing question. 'Strangely perhaps, a period of expansion has encouraged academic institutions to become somewhat more remote from "concrete facts" and the "actual working of economic and political relations".' This is how I put it, using words of the original paper, in my Report in 1976. Two decades later, one must ask even more pointedly: Could it be that universities are no longer the places 'where it's at'? Have even the academic social sciences become remote and irrelevant? Does one have to go to 'interstitial' institutions, half academic and half practical, to recapture the excitement of knowledge in its applications?

None of this can detract, however, from the values of the university. In its core, the Brookings debate at LSE was an assertion of these values, and a reminder to the Director that he was no longer in politics, at least in his responsibilities in and for the School. The lesson was clear, and it was taken. 'Scholarship is either free or bad; there is no other alternative.' Free scholarship follows its own lights; in principle it does not recognize the interests or timetables of strangers. Academic institutions are there to protect free scholarship. Certain kinds of involvement with outside interests must be resisted; these perhaps include policy studies, and certainly the study of policies designed to support the 'middle ground'. The Director's Report said it all:

> It was encouraging to see the sureness with which the School accepts the challenge of external problems while at the same time insisting on the need for academic autonomy; LSE—one may conclude—is incorruptible, or at any rate incorrupt. It was an important lesson for the Director to be made aware again of the fact that the strength of a university is in its combination of research and teaching at all levels, so that there is nothing to be gained by separating out one or the other.[9]

Or perhaps this says almost all. The British Brookings saga was not a success story. But it helped clarify the values which govern LSE, the great school of the social sciences, and also those of its Director. In the second paper on 'Policy Studies' written in April 1976 these were spelt out in some detail. They were on the one hand tactical. Acquiescence is a bad response to encroachments on academic autonomy but counter-attack must not ignore the strengths of the invader. There is a case for a contribution by universities to public affairs, though universities should define their response themselves. Autonomy has to be used as well as enjoyed. This is notably true in the social sciences. The values underlying the idea of policy studies at LSE were thus on the other hand substantive. Like all disciplines of scholarship the social sciences are subject to the logic of scientific discovery. This logic is relentless and value-free, with observation refuting the guesswork of theory, or not. But there is also an ethic of scientific discovery. Natural scientists have discovered it late, after Hiroshima; social scientists may have been too obsessed by it from the beginning. What it means is that the responsibility of the scholar does not end with the results of his scholarship. The effects of knowledge cannot be ignored. *Felix qui potuit rerum cognoscere causas* but the happiness of the discoverer may be the source of suffering and dismay among those to whom the discoveries are applied. Social scientists cannot close their eyes to what is done in their name, nor can institutions of social science teaching and research close their doors to policy issues. The question is how to reconcile the logic and the ethic of (social-) scientific discovery. It can only be answered in practice, and the history of LSE is as good an answer as any.

The Environment Turns Hostile

Policy studies soon turned out to be a harmless diversion compared to the new threats emerging from a succession of Governments. These meant that by November 1979 the Director had to state, in an open letter to all members of the School, that 'LSE is faced with the worst crisis in its history. While the crisis looks financial it may affect our international quality, our standards, our unique contribution to higher education.'[10] Three main causes accounted for such developments. One was an unmistakable crisis of public expenditure at a time of high inflation and punitive taxation. All recipients of public funds had to pay a price. Secondly, the darlings of the 1960s, universities, and more particularly the social sciences, had fallen out of favour with the general public, and politicians found it convenient to follow rather than lead the prevailing mood. Thirdly, a new ideology of marketization began to pervade all institutions and replace appreciation of public goods and public service.

Defending a school of the social sciences against such attacks was necessary but not easy. As a (Labour) minister put it at one of the dinners in the Director's Dining Room at LSE: 'You cannot win. The Tories think that all university teachers are red, and Labour thinks that all university students are middle-class. The coalition is unbeatable.' In the event, it had been a Labour Government which had first introduced differentials between fees for home and for overseas students in 1968. The minister responsible then, Shirley (now Baroness) Williams, as Secretary of State for Education and Science in the late 1970s, more than doubled overseas students' fees to £850 by 1979, while at the same time cutting the University Grants Committee payments to universities significantly. In the years from 1975 to 1978, the School not only had to follow suit so far as student fees are concerned, but also lost between 2 and 3 per cent of its real income each year. Leaving aside the implications of what came to be known euphemistically as 'negative growth' for the management of change—it is just not true that reforms take place under severe pressure!—there comes a threshold at which such cuts begin to hurt, and hurt badly. In the hope that the pain would not last, the School rallied and took all necessary steps to keep going.

In 1978 it appeared as if the hope had been justified. In fact, 1977/8 was an excellent year for LSE. In 1978 the new library building came into use (it was officially opened by HM Queen Elizabeth the Queen Mother in July 1979). School committees agreed to name it the Lionel Robbins Building, thereby setting a lasting monument to one of the great School figures of more than five decades. In some ways the event was a harbinger of things to come, for the purchase of Strand House from W. H. Smith & Son and its refurbishment for the BLPES had been made possible by the Library Appeal. Now, the School was able to benefit not just from its funds but also from its staff and the office which they had so successfully run. Before long, it was turned into a full-blown External Relations and Appeals Office, with Jennifer Pinney as its head and Ambrosine Hurt as Alumnus Officer. Also, after her retirement as Secretary of the Graduate School, Dr Anne Bohm was added as a roving ambassador for LSE, travelling around the world, keeping alumni in touch with the School and the School with them, attracting future students, and preparing the ground for further giving. Her charm and determination, coupled with her long experience of the School, were invaluable for LSE.

Other new developments were also the result of private initiative. They benefited from the fact that the UGC and the University Court, in supporting the conversion of Strand House, had told the School that the Library did not require the whole building. LSE made a virtue of necessity and designated one floor, the Fourth Floor (or 'Level 5') for housing research units. The first to move in was the Business History Unit, which owed much to the support and activity of Sir Alistair Pilkington and Sir Arthur Knight as well as Professors Theo Barker and Leslie Hannah. In June 1978 the single most important development of the School's research base since the Rockefeller grants of the 1920s followed. Two Japanese companies, Suntory and Toyota, gave £2.5 million to set up a foundation at the School which finances what is now the Suntory–Toyota International

Centre for Economics and Related Disciplines (STICERD). The Academic Board, as usual, was sceptical at first. Would there be unacceptable strings? At a memorable meeting, the distinguished Economics Professor Michio Morishima, author of the initiative and later the first chairman of STICERD, told the Board that there was no need to worry because he had known the leading donors, Mr Keizo Saji and Dr Shoichiro Toyoda, since their common school-days and could vouch for them. Michio Morishima was as good as his word; the foundation safeguards the autonomy of STICERD totally, and the benefactors never tried to interfere with its work. To its centenary, and no doubt long after, the School has an endowed research base which also attracts project funds from many sources.

Even in terms of public funding, 1978 was a year of hope. It looked as if a new plateau had been reached, lower than was comfortable but still viable. Government and the University Grants Committee even began to speak of a return to medium-term planning, for three years if not for five. Then came the election of May 1979 and Margaret (now Baroness) Thatcher's Government. The budget of June 1979 shook the very foundations of the School's funding structure. Its central feature, so far as LSE was concerned, was the decision that within two years overseas students would be charged so-called full fees, and grants to institutions would be reduced by the corresponding element. Thus the School's grant would be cut by the proportion of its overseas students, which stood at 37 per cent at the time. In addition, other painful reductions were announced. Even apart from the overseas element, university grants would not keep pace with inflation. Academic salaries were set to fall relatively, and in certain years in absolute terms. Graduate scholarships in the fields of what was still called the Social Science Research Council (until January 1984) were reduced massively, from over 2,038 for the whole country in 1979 to under 600 when Margaret Thatcher left office eleven years later.

In later years, the insult of legislation aimed at university autonomy was added to the injury of severe funding cuts; the cuts, however, were enough to leave outstanding institutions like LSE, Imperial College, and one or two others stunned. The School decided quickly on a two-pronged strategy. On the one hand, the case for excellence, for internationalism, and for the social sciences was made wherever and whenever possible. There is no way in which I, for example, could have come to LSE as a student if the fee had been (to use the 1994/5 figures) £7,120, or even the £2,350 for EC students; in 1952, £15 15*s.* 0*d.* seemed quite enough for someone coming from a system in which paying £1 for registration was regarded as an imposition. On the other hand, the School made sure that it would survive, and survive with integrity. Squaring the circle of academic quality and financial viability became the dominant subject throughout the 1980s.

To this end, five sets of measures were taken and explained in detail to the whole School in a series of papers starting with 'LSE: First Steps in a New Situation' in November 1979. Two of these measures were particularly painful because they were so clearly wrong in academic terms. The first and most important was to add 400 full-fee paying overseas students. Numbers tell the story (see the Graph on p. 177). In 1978/9 LSE had 3,696 students, an order of magnitude which had become normal for a decade or so. During the following year the total rose to 3,871, in 1980/1 to 4,404, and then to 4,562, the highest number yet in the history of the School. Home students remained stable throughout this period; the increase was entirely on the overseas side. Did it involve a decline in quality? 'Ambassadors' were sent out, notably to American universities, to try and attract students. The General Course for undergraduates who come for one year without taking a degree was doubled, then trebled. More significantly, the staff–student ratio deteriorated from 1 : 10 in 1974 to 1 : 12 in 1979 and 1 : 14 in 1982. In the newspeak of the 1980s this would be called an 'increase in productivity' or an 'efficiency gain', but in the language of ordinary humans there must have been some decline in standards.

The second wrong but inevitable measure was to reduce the retiring age for academics from 67 to 65, and thereby shift a part of the salary bill to pension funds. Academics should not retire at any fixed age, and certainly not at 65; agreement on this measure by the School's professors was therefore an act of unusual institutional commitment.

Two other sets of measures were less objectionable. To examine current expenditure closely with a view to making savings is at all times a useful discipline, though some short-term savings, like deferring maintenance of plant, can be costly later. The most important new initiative, however, had to do with attracting support from private sources. The Library Appeal had set the tone and created the machinery. The next immediate task was to find funds in order to help students in distress, especially those who for reasons outside their control could not afford their fees. The LSE 1980s Fund was set up, and the projected £2 million were found, and spent, before the end of the decade.

Another measure in response to what might have turned into a catastrophe was more complicated though very important. Living institutions must not stand still. Many other universities responded to the new situation by a freeze on appointments, but the School rightly rejected this idea. Not only did some posts have to be filled, but there had to be a chance for new developments, however modest their scale. The method chosen had much to do with the peculiar strengths of LSE. Whenever a teacher retired or left, his or her salary would go into a 'pot' for appointments. (A similar arrangement was made for non-academic staff.) The School would then decide, on financial grounds, what percentage of the 'pot' could be spent in a given year. This was initially a very small percentage, 30 per cent or so, but it meant that some appointments could be made, even in new fields. The 'pot' principle worked, to be sure, only because LSE still regarded itself as a single-faculty School with one Appointments Committee for all subjects, and no automatic entitlements by departments. In a perverse if not totally undesirable way, the need to take difficult decisions in the interest of the whole School even contributed to its cohesion.

Whenever the Director communicated with the staff in this difficult period, he not only explained what had to be done but added four important principles to which LSE would adhere:

1. There must be no redundancies on account of the savings.
2. Promotion on merit will continue to be possible.
3. No group must bear a disproportionate share of the burden of the cuts.
4. No student should be forced to leave the School as a direct result of a tuition fee increase which takes place after he or she has entered.

These principles may have helped when it came to getting support for obviously unpopular decisions. Certainly academics rallied to see the School through some of its more difficult years. Non-academic staff occasionally got restless, though the no-redundancy promise took the edge off their concern, especially when the pledge was seen to be serious.

Students found it less easy to reconcile themselves to a new climate. Throughout the 1970s and 1980s there were incidents of protest and expressions of dismay. The fashion of the time was to 'occupy' the Director's Office, or even better, the Registry and other parts of the Administration. In line with the prevailing style of industrial relations such actions were legitimized by acclamation at Union meetings and accompanied by self-appointed 'open committees'. They were on the whole good-humoured, even on the occasion on which a court possession order had to persuade the students to leave. Perhaps the most serious incident was a hunger strike in 1979, which went on for too long to be laughed off. It took long nights of discussion with student activists and the hunger strikers to persuade them to desist. All this was in a sense misguided, or rather unguided, protest. The students felt under attack and chose the authority nearest to them, the Administration of the School, to launch their counter-attack.

Perhaps they also felt that their own future as social scientists had become more uncertain. I had much sympathy with their motives and some understanding for their actions, though at times these were testing.

The effects of the shocks of the 1970s and 1980s on LSE were profound and lasting. The changing financial base of the School had academic as well as managerial ramifications. Even in its barest outlines the story is dramatic enough: in its early days, the School experienced an increasing certainty of public support by the LCC and the Treasury, from 42 per cent of total income in 1900 to 55 per cent in 1905 and 70 per cent in 1910. After the First World War, and with the help of the Rockefeller Foundation, a new income pattern emerged. By 1928 20 per cent came from the endowment, 30 per cent from fees, 40 per cent from the Treasury and other public bodies, and the remaining 10 per cent from miscellaneous sources. After the Second World War, indeed throughout the Carr-Saunders years, the public element, that is grants from the University Grants Committee and the Court of the University of London, increased rapidly to 80 per cent of total income. Of the remainder, home student fees were also paid from the public purse. Now this process was reversed with disruptive rapidity. Between 1974 and 1984 the Court Block Grant declined from 81 per cent of total income to 47 per cent. Fee income rose during the same period from 9 per cent to 41 per cent; by 1984, overseas students alone accounted for 29 per cent of the School's income. Income from endowments, research contracts, and other sources grew much more slowly from 10 per cent in 1974 to 12 per cent in 1984.

In abstract, this may look like greater financial autonomy; in fact there was not only the management of rapid transition to think about but also the options foreclosed by financial exigencies. For the foreseeable future, reducing the size of the School was ruled out, though this might well have been academically desirable. Equally, the School was no longer free to go for a healthy mixture of a majority of home and a minority of overseas students. Some overseas students began to complain that they rarely met any British students, and to wonder why LSE should not be on some offshore island. Or was Britain beginning to be just that? Staff clearly had to neglect either teaching or research as long as their numbers declined while more and more students were taken on. When, under I. G. Patel's directorship, the School received the Queen's Award for Export Achievement, the irony was not lost on those who would have preferred academic quality to financial success. The absence of a longer-term perspective, which soon seemed to become a desired principle of 'permanent revolution' rather than an unfortunate necessity, added a hectic quality to the life of the School which could not be conducive to academic progress.

At times, friends within and outside LSE have told me how much they regretted that I had come to the School at such an unpropitious time. They would have liked me to implement my own ideas rather than spend most of my time defending the School against threats from outside. Such friendly comments raise delicate questions of personal strengths and weaknesses. Who knows when one can give one's best since the alternative is rarely tested? Given the chance, I would probably have nudged LSE further in the direction of a graduate school and research university. Even so, no opportunity was missed to affirm the strengths and the values of the School. 'Our future need not be bleak, if an effort is made by everybody to concentrate on what is essential for this great institution which is loved by so many and admired all over the world.' 'LSE's aim for the 1980s is to preserve the character of the School as an international institution, with opportunities for the best scholars in the social sciences both from home and overseas.' These are perorations of financial papers, to which the Director's Report of 1981 added a defiant comment on the social sciences:

> The social sciences have a long way to go; what is more, as they go this way they will annoy many who believe that untutored minds are a safer compass; but unless a large number of people have encountered them at some stage, our society is not likely to remain a civilized and free society.[11]

This is where a feeling of sadness came in, sadness about the School tinged with anger about the country of which I am so fond. How could it be that a Government, and a wider concerned public, were so unaware of some of the greatest strengths of Britain? When has there ever been such a wave of destruction of successful institutions, with the authors of the destruction gleefully viewing their work? Here was the great School of Economics, the envy of the world, and all we could think about was how to squeeze money out of its students and friends and turn a centre for advanced study into an efficient supplier of probably fictitious markets. Under these conditions, leaving behind a recognizable LSE, whose members and friends were proud to declare their allegiance, was nearly the best I could hope for.

FIN DE SIÈCLE

A Caring Internationalist

The appointment of Indraprasad Gordhanbhai (hence I.G.) Patel as ninth Director of the School was appropriate in more ways than one. Dr I. G. Patel had known the School and many of its members as a student in wartime Cambridge. He was the first professional economist at the helm of the School, though Hewins and Beveridge had at times thought of themselves as such. As an Indian he represented one of the important communities of LSE alumni with an unbroken tradition all the way back to the beginnings. Patel was also a man of transition. Almost 60 years old when he took up the directorship in October 1984, he was nearing the end of a distinguished career. Apart from teaching at the College of his home town, Baroda, and elsewhere, I. G. Patel had held high office in his country and in international organizations. From 1977 to 1982 he had been Governor of the Reserve Bank of India. Before that he had been an Executive Director for India of the International Monetary Fund and Deputy Administrator of UNDP, the United Nations Development Programme.

I.G. is a soft-spoken man, so much so that a special microphone had to be installed for him even for small committee meetings. He is also deceptively soft-spoken, a man of determined views and the ability to make them real. In his first Director's Report of 1985 he himself referred to a 'period of transition' but this did not prevent him from taking the reins firmly in his hands and defining new objectives for the School. Above all, he soon concluded that by fending off the threats of a hostile environment in the way in which it had done, the School had come close to jeopardizing its claim to excellence. A staff–student ratio of 1 : 14 was simply not acceptable, to say nothing of a 'pot' policy which made only a portion (albeit by the time Patel arrived 90 per cent) of the funds accruing from departures and retirements available for new appointments and was bound to lead to further deterioration. The new Director put his mind and considerable energies to the task of reversing this trend, and in his first years he was remarkably successful, even if, with characteristic precision and self-deprecation, he had to point out that 'to do relatively well in a deteriorating total environment is not to do well enough'.[12]

Three factors contributed to the 'momentum of consolidation and progress' which the Director noted. The first was a kind of lull in government inroads into university finance. For a short while during Dr Patel's early years a new stability seemed possible. Grants from the Exchequer through the University Grants Committee were still made annually and never totally adjusted for inflation, but universities were invited to engage in medium-term planning and a three-year rolling grant was indicated. Secondly, I. G. Patel reluctantly but firmly decided to charge overseas students more than the minimum fees stipulated by government. As an economist, Patel regarded the 'economic cost' of a student as a not unreasonable yardstick, and now that 45 per cent of the School's income came from fees, he found it necessary to grasp this particular nettle. Thirdly, and perhaps most importantly, Patel used visits by the University Grants Committee and by the Court of the University of London in his first year to impress on them the glaring discrepancy between the 'special mission' of LSE and its financial provision. The 'highest standards of academic excellence', 'interdisciplinary cooperation', an 'academic culture of cohesiveness and cooperation', and the 'international identity' were all used to impress the visitors who controlled the purse strings.[13] The success of this campaign was spectacular; for four years the School received grants significantly above the average for the country and, more remarkably, for other colleges in London.

The funds were used for a small number of objectives close to the Director's heart and supported by the School as a whole. One was new appointments. At the end of his third year, in 1987, the Director was able to speak of the 'very large number of new appointments that were authorised during the year' and had led to a net increase of academic staff by eleven teachers. Another objective was research. New units included the Financial Markets Group (Professors Mervyn King and Charles Goodhart), the Development Research Group (Professor Nick Stern),. the Research Group on the Voluntary Sector (Dr David Billis). Older units like the Centre for Labour Economics (Professor Richard Layard) and of course the Suntory–Toyota Centre, continued to do well. Thirdly, the Director made a special point of trying to bring Library allocations back to a more normal level. Book acquisitions had been cut from 20,000 items in 1974 to 13,000 in 1984; they were now, in 1987, back to 20,000.

One other purpose that particularly concerned the Director was student accommodation. When he arrived he called it 'the most glaring deficiency'[14] and immediately set to work to remedy it. After various unsuccessful searches, a site was finally found at the near end of Docklands, in Butler's Wharf. With a mixture of loan finance, internal funding, and an appeal for £1.75 million, the £7 million needed to build an accommodation block for 281 students was found. In November 1989 HM Queen Elizabeth the Queen Mother opened the Butler's Wharf Residence. The Director had reason to be pleased not just because of the 'elegance' of the building but because it was there at all; for it was his own lasting achievement.

Other accomplishments of these difficult years could be listed, but perhaps the most important was the imprint of the Director's personality. It was evident in specifics. In 1988 I. G. Patel added to his Report a section on 'The Journey Ahead'. 'Today, I have no hesitation in saying that if I have one aspiration for the School it is to enable it to say: We do not turn away any student of high academic merit from anywhere in the world for lack of ability to help him or her financially.'[15] The statement combines two values which were central for the Director: excellence and equality. He never ceased to insist on the need for quality, but equally never forgot the needs of those at a disadvantage. Nor was this confined to students. Like other Directors before him, Patel liked to invoke the unique location of the School between the City and Westminster, Bloomsbury and Theatreland, but he never forgot to add one other 'neighbour', the 'depressed inner city', 'the disadvantaged parts of Greater London'.

For the School itself, the Director found warm words of affection and encouragement. He liked to invoke 'the spirit of coherence, cooperation, and communal loyalty to a cherished institution that has characterised the history of the School now for some nine decades'.[16] 'LSE has always been a more close-knit family than most institutions.'[17] I. G. Patel also found a form of words for the School's governance, which he repeatedly praised for 'the fine balance between participation and decisiveness'.[18] It is of course the ancient dream of all rulers to combine democracy with authority, the reality of which depends as much on people as on structures. Patel the economist might have devoted all his energy to the 'hard' social sciences, and he was certainly pleased about developments in his own discipline as well as in quantitative and information-related subjects, but when he spoke about extending the range of the School he mentioned above all the humanities. He would have liked to see a literature department alongside the afforced departments of history.

I. G. Patel expressed strong views about universities in general and LSE in particular in the 'great debate' which was waging during his years as Director, and which culminated in the 1988 Education Act. He could be scathing, as when he castigated the 'elaborate charade' of a policy from a Government which looked for arguments against universities when it had already made up its mind to cut their funding; he called it 'letting the crime fit the punishment'.[19] He reminded people 'that there is indeed life beyond what the government decrees', and found strong words against 'implicit or explicit interference in the functioning of British universities'[20] as well as 'the mounting burden of imposed bureaucratic procedures of reporting and assessing'. 'The present unease, I believe, arises mainly from attempts to standardise everything, to codify it, to institute procedures for monitoring and assessing and reporting in terms of criteria or indicators laid down from above.'[21]

'In an uncertain and rapidly changing world where we do not hold all the cards, it is all the more imperative to plan ahead if we are not to be blown this way and that by every passing phase of fashion or circumstance.'[22] This is but one of I. G. Patel's civilized beliefs and maxims, but alas! circumstances gave him little chance to practise them. The Director tried to plan, from his first days at the School to the 'Development Perspectives' of his last year. Every one of the plans made good sense, but had to assume some stability of the environment. Such stability, however, was not to be. 'Government policies', Patel wrote despairingly in 1989, 'seem to be changing without any clear sense of direction or a comprehensive and consistent strategy.'[23] As early as 1987, the Director was back to the bad old ways which he hoped to have overcome. 'I had to recommend', he reports, 'that the School should pause in its planned developments towards improved staff–student ratios and other areas of the School's activities while defences were drawn up against the uncertainties of the remainder of the planning period to 1992.'[24] One is reminded of Michael Oakeshott's inaugural lecture, of the 'boundless and bottomless sea' without harbour or shelter, starting-place, or destination. 'The seamanship consists in using the resources of a traditional manner of behaviour in order to make a friend of every inimical occasion.'[25] This is not what I. G. Patel wanted, but he certainly excelled in it when forced to do so.

LSE in 1990

The Director's initial appointment for five years was extended to six; even so he had little time before he welcomed the appointment of his successor during his fifth year. Other transitions happened. Sir Huw Wheldon stepped down as Chairman of the Court of Governors after the new Director's first year. It was a great sadness for his many friends at the School when he died a few months later, on 14 March 1986. In December 1985 a former student as well as Students' Union President, Sir John Burgh, had become Chairman. However, on his election as President of Trinity College, Oxford, the School saw an incompatibility of positions. In 1988 Sir John Burgh was succeeded by Sir Peter Parker. In the mean time the calm and devoted Vice-Chairman, Sir John Sparrow, who arrived in Dr Patel's first year, provided continuity as well as much needed help with the Butler's Wharf appeal.

Many others were new as well. When I. G. Patel came, the new Secretary, Dr Christine Challis, had been there for a year. The new Librarian, Mr Chris Hunt, and a new Academic Registrar, Bursar, and Finance Officer all took up their appointments during Patel's first session. It was as well that one of the most senior and respected professors, Michael Wise, provided some continuity as Pro-Director; he was succeeded by Professor Robert Pinker in 1985.

On the financial position of the School, the gist has been conveyed already. In an inflationary age, absolute numbers tell us little. Thus the fact that the School's administrative budget was not much over £10 million in 1980 but £40 million in 1990 and over £50 million by 1995 does not mean that School activities have quintupled, though it does indicate very nearly a doubling in real terms. Expenditure continued to be concentrated heavily, for about two-thirds of the total, on staff costs, with the maintenance of the growing number of buildings taking up a worrying 15 per cent of the total. Perhaps the most pleasing financial development was the growing inflow of research funds from a variety of sources; in the late 1980s, research grants and contracts amounted to nearly £1.5 million a year, not including income from the STICERD endowment. The School continued to find it difficult to break even, and showed a deficit in 1990/1.

During the Patel years the School was still growing in size. For better or worse, LSE had now left its historic total of between 3,000 and 3,500 students far behind. The deliberate increase between 1977 and 1981 from about 3,500 to about 4,500 students could not be undone. On the contrary, by 1989/90, 4,863 students were registered at LSE. The proportion of postgraduates had declined from 50 per cent in the early 1970s to 44 per cent in 1990. Otherwise, growth was spread fairly evenly over different student groups, though the increase in part-time postgraduates as well as occasional and single-term students may be worth noting. Social Policy and Administration and the subject of Management Science are obviously popular in 1990, though again no massive shift towards or away from subjects is shown by the figures. The proportion of overseas students fluctuated over the years, but if one takes a longer view it had declined from 35 per cent to about 30 per cent in the late 1960s, risen steadily in the 1970s to 43 per cent in 1980/1, and then gone up further to 45 per cent in 1985 and 48 per cent in 1990. There can be little doubt that the reasons for the increase were not only academic.

Dr I. G. Patel enjoyed good—or, as he put it, 'cordial'—relations with the students, though from time to time the activists gave him, like his predecessors, headaches. Students were obviously unhappy about their own economic position as well as the general squeeze on the social sciences, if not the universities. Once again, they expressed their unhappiness by attacking the authorities of the School. Demonstrations, occupations, angry exchanges in committees were bound to happen. More worrying for the School were symbolic acts of defiance, as when the Students' Union elected in 1988/9 as its Honorary President a man who had been convicted of murdering a policeman, though there were then rumours of a miscarriage of justice and the conviction was later quashed. Questions like investment in South Africa-related shares also remained on the agenda, as did the occasional attempt to prevent an outside speaker from making a case which some did not like. The Director coped with such incidents calmly and by putting his case in considered detail. He knew, of course, that in the late 1980s activists were not the tip of an iceberg but a minority without a silent majority to tolerate its antics. This was not 1968 but 1986. Most students had their careers uppermost on their minds, and more often than not, in those days, careers meant money, quick money. The late 1980s were after all the heyday of 'Casino Capitalism' (as Professor Susan Strange, then head of the Department of International Relations, called it).

This is relevant for a question to which it is too early to give an answer as this History is written: if Dr Patel's directorship was a time of transition, what was it leading towards? We know where the School came from and what it stood for. I. G. Patel himself was particularly good at putting it into words. In a farewell message to the School in September 1990, entitled 'LSE into the 1990s', he said: 'As a School of Economics, we have always understood the importance of markets and of the wealth-creating function. But there are things beyond markets and economics—and knowledge for the sake of knowledge as well as for securing social mobility are among those that LSE has always valued.'[26] Patel connoisseurs will note the reference to 'social mobility', which once again aims at the disadvantaged. But 'knowledge for the sake of knowledge' takes up the oldest theme of the School, *rerum cognoscere causas*. Except that the theme sat uneasily on a time in which the Government's view that there are in fact no 'things beyond markets and economics' began to percolate everywhere.

This was true even for Dr Patel's LSE. The Queen's Award for Export Achievement given to the School in April 1990 was in recognition of an increase in overseas students dictated by financial rather than academic needs. The Centre for Economic Performance set up by the then Department of Education and Science took the School in a very practical direction. This thrust was strengthened by the creation of an Institute of Management in order to 'integrate and co-ordinate undergraduate and postgraduate teaching, research and short courses in the area of management'. An LSE Consultancy Service contributed to the School's income but also encouraged members of staff to apply their knowledge directly. In fields like communications, information technology, decision analysis, much emphasis was placed on applicability for the sake of revenue increases as much as the advancement of knowledge.

As important, perhaps, are seemingly small yet important questions of style. The session 1987/8 was the last in which the 'Report by the Director on the Work of the School' in the past academic year appeared. The tradition had been started by W. A. S. Hewins and developed to its lasting form by Sir Halford Mackinder in 1904. It enabled Directors to give a personal perspective to facts and figures and thereby contributed to the cohesion of the LSE family. Now the information was separated from the perspective. The Director added a personal gloss to an essentially bureaucratic account. All this was packaged in a manner intended to appeal to glitzier times. The logo of LSE changed, as did the old black-purple-and-gold colours. Bright red was the order of a bright blue day.

I. G. Patel left in the middle of such changes. He returned to India, and to his international concerns, notably the Uruguay Round of GATT for liberalizing trade as well as financial and other services. He left with the unselfish charm with which he had arrived. 'I am allergic to the phrase—what have I done?' he said to the student newspaper *Beaver* in March 1990. 'Such achievements are the result of efforts by the governors, students and academics. The credit must go to all of them.'[27] Still, he was pleased when he was awarded the KBE just before he returned to his home town, Baroda. I. G. Patel left the School in a sound state financially, aware of its tradition and its academic strengths. It was not his fault that the environment was still riddled with uncertainties, and that LSE was uncertain how exactly to respond.

The Tenth Director

The tenth Director of LSE, Dr John Ashworth, was elected by the Court of Governors on 6 July 1989. He was chosen after an extensive trawl of potential candidates by a Selection Committee chaired by Sir Peter Parker, and consisting of equal numbers of lay and academic Governors, many of whom wanted to set the School on a new course. Ashworth is a biochemist by training and profession. He thus shares the background in biology with Sir Alexander Carr-Saunders as well as with Beveridge's early ambitions. W. A. S. Hewins the founder apart, Ashworth is the first Director not to have any previous connection with LSE, though soon after his appointment he donned shorts and T-shirt and took his yellow pad to an LSE summer school in economics, which he completed with distinction. Ashworth had come to LSE from being Vice-Chancellor of the University of Salford. The ailing University, until 1967 a college of advanced technology, had been close to bankruptcy when Ashworth took over, but he turned it round financially as well as academically.

Before Salford, John Ashworth had spent five years as Chief Scientist on the Central Policy Review Staff (CPRS) in the Cabinet Office under the two LSE men, Sir Kenneth Berrill and Sir John Sparrow. The CPRS provoked a furore in the 1970s by its unorthodox, even radical proposals for reforming established institutions. Ashworth belongs to a small but influential group of public figures who emerged to prominence in the 1980s, and who combine Margaret Thatcher's penchant for leading from the front with a Schumpeterian confidence in the virtues of 'creative destruction' by entrepreneurship, concern about equity though not necessarily equality, a strong sense of Britain's destiny at the heart of Europe, and a very un-Hayekian faith in the do-ability of change.

The new Director did not wait long before he stirred the School with two major projects. Just over four months after his arrival, on 12 February 1991, he addressed the staff of the School in the Old Theatre and circulated his paper 'The LSE: A 2020 Vision' widely amongst the Governors and others. The paper began with an analysis of higher education in general and LSE in particular. 'The elite system based on the values and expectations of Oxbridge is disappearing but the future mass system is not yet in place and its form, financing and even function is not yet clear.' During this transition LSE is threatened by several factors. Its quality 'in terms of the CVCP's performance indicators' is not unchallenged; its financial prospects are gloomy; its location is expensive and unattractive; its facilities at Houghton Street are 'overcrowded, old-fashioned and well below the standards of its UK and foreign (especially US) competitors'. 'What is called for is a new Vision of what the LSE could be—if we, and the wider community of which we, the academics, are part, have the courage to go for it.' This Vision 'must be seen as an opportunity for the growth and development of the School in new and exciting directions and not as the response to a threat'.

The Director then suggested how his Vision might be developed in the debate that he wished to start. First, a Graduate School must be 'planned and managed in ways which are more familiar to those in the natural or the physical, than in the social, sciences'. Institutes and 'corporate' research will increasingly complement and, in some disciplines, even take the place of solitary scholarship. In undergraduate teaching another change of direction is needed: 'With hindsight what the LSE did in its first century was, in academic terms, to take the lead in creating the UK community of social scientists and, in practical or policy terms, to help define, devise, and then train the leading cadres, of the welfare state.' This job is now done. 'What can we do of comparable seriousness, importance and difficulty in our second century?' There are many challenges 'but there is one that deserves special attention. That is to help define, devise and then train, the leading cadres of that integrated Europe which will evolve in the course of our next century.' Such plans have many implications. The School, if it accepts its new 'mission', 'must think of expansion'. This is not possible in Houghton Street. On the old site, the Graduate School has first call, but 'as the Undergraduate School expands we will need to establish it on another site' somewhere 'within a reliable and predictable 45 mins travel time from Houghton Street by public transport'. The removal of the whole School to another central site would, of course, be preferable. Beyond that 'productivity increases' are needed. One simple possibility would be an 'increase in the return on the capital invested' by using the School for two intermeshed student generations simultaneously with alternating populations in the six terms which would replace the present three. 'The future cannot be predicted but it can be created. I hope that you will be as excited as I am at the thought of the future that, together, we might create for the LSE.'[28]

The School was stunned and, as tends to happen when it is, it assembled its Academic Board. On 5 June 1991 the Board met. It recognized 'the need for a review of the School's operations' but removed the proposed phrase 'as contemplated in the Director's paper'. Four Working Parties on Research, Teaching, Organisation, and Finance were set up. They began a year's work before they reported back to the Board on 13 May 1992. By then, however, another subject had come to dominate the attention of the School, its Director, and its Committees to the exclusion of almost everything else. In 1991 the possibility had opened up of the School after all finding another central site by acquiring County Hall, the former seat of the Greater London Council, which would offer three times the space presently occupied by LSE in the Houghton Street area.

County Hall seemed the answer to all prayers, including even the little proviso which the Academic Board had added to its decisive rejection of the move to Croydon in 1964: 'not ruling out the future possibility of [the School] considering an alternative site in central London, should such a site ever become available'.[29] The huge edifice opposite Parliament on the other side of Westminster Bridge had its attractions not only because of its size and location but also because it had been the seat of London's government. The School, after all, was at least in part given its important early blessing by the London County Council of which Sidney Webb was such an active member. In 1986 the Conservative Government had introduced legislation to abolish the successor to the LCC, the Greater London Council. Since its last remnants, notably the education authority ILEA, had departed, County Hall stood empty except for the occasional surveyor on behalf of potential buyers, and the rats which had crossed the river to invade it. Then a Japanese property developer showed an interest. The London Residuary Body which looked after GLC property had until the end of 1992 to make a preliminary agreement with him firm, or accept another buyer.

The months from March to September 1992 saw two parallel processes of activity. They were kept going, sometimes day and night, by a small group which had perhaps a little unfortunately been named the 'war party', and which comprised four Governors, including the Chairman and the Vice-Chairman, four academics, including the Director and the Pro-Director, and, of course, the relevant School officers. One process was to persuade the world that here was a unique opportunity for LSE. This part of the 'war party's' battle was immensely successful and brought lasting benefits. A veritable movement came under way to promote 'County Hall for LSE'. Leaders and letters in newspapers, public and private meetings, statements by the great and the good all supported the bid. In the end even a Government not known for its generosity towards higher education and more particularly the social sciences went so far as to instruct the Residuary Body to wait for the bid from LSE. That, however, is as far as it went. The Chairman of the Finance Committee of the LSE Governors, Ian Hay Davison, was undoubtedly right to point out that the French Government, or any other government in Europe for that matter, would have put at least some of its money where its mouth was. But this was Britain and a Britain intent on squeezing rather than serving public goods. As a result, LSE had to find a large sum on top and above the probable proceeds of the Houghton Street site in order to convert and refurbish County Hall. Some of this would have to be borrowed, and its repayment would depend on the School attracting several hundred additional high-fee paying students. On 25 September 1992, the Secretary of State for Education said in reply to a Written Question in the House of Commons that, having examined the financial plan, 'the Universities Funding Council has written to the LSE today that it did not feel able to endorse the proposal'. County Hall went to the developer from Japan, to be turned into a hotel.

At the School, disappointment was mixed with exhaustion. The Academic Board discussion on the future of the School had led to a number of significant if for the most part procedural decisions. An Academic Policy and Resources Committee (APRC) had been set up to continue the strategic planning process initiated by the Director's 'Vision' paper and to allocate resources for the implementation of plans. The Research Committee was given additional responsibilities, including the possible planning of a Centre of Advanced Studies in the Social Sciences. Decisions were taken promoting the devolution of responsibilities to departments. On 28 September 1992, after the collapse of the County Hall bid, the Director convened all staff and spoke again of the need to create a European centre for the social sciences. He also announced a major new fund-raising drive which was soon to result in the setting up of the LSE Foundation.

In the early 1990s important changes took place, albeit in a piecemeal fashion. The School did in fact add greatly to its space by the purchase of St Philip's Hospital and Clement House on the Aldwych as well as by the generous gifts of the Tymes building and the Royalty Theatre. A European Institute as well as a number of interdisciplinary research centres were set up. Management studies were developed. Research income became an increasingly important element of a continuing difficult financial picture, with contributions from the European Commission rising to nearly £1 million by 1995. The Director's attempt in 1993 to persuade the School and the funding agencies of the need for 'top-up fees' did not meet with much support; even so, fees became by far the largest single source of the income of a School which had now grown to 5,500 students. Though still giving London degrees, LSE, along with other colleges, became increasingly independent from the University of London in making appointments, designing courses, and dealing with public funding agencies which had changed their names at regular intervals, as funding itself remained predictable only in its decline.

The School started to give thought to its centenary. On 4 August 1994 a charabanc of LSE lovers, including founders' kin, the School's top officers, and some of those who helped with this History, journeyed to Borough Farm in Surrey, where the present owners, Tony and Margery Herring, joined them to re-enact the famous breakfast and to toast the second century of LSE.

PAST AND FUTURE

In its first century, the London School of Economics and Political Science has become a precious asset, for London, for Britain, for the social sciences, for a world in need of informed change. The fondness for the School which is felt by many who have shared even a small part of its history rests on more than nostalgic memories of Houghton Street and Clare Market, the Old Theatre and the Library, the Founders' Room, the Three Tuns bar, and the people who brought it all to life. LSE had made a difference to the lives of those who studied, taught, and worked there, and it made a difference to the world at large. What kind of difference? The answer to this question must be the source of the School's ability to cope with the future.

First, LSE has been a pioneer of the social sciences. They were an aspiration based on extravagant dreams and tiny beginnings when the School was founded. In the 1920s and 1930s they came to full bloom in a few places, though in Europe nowhere more strikingly than at LSE. After the war, the scent of their blossoms almost intoxicated a generation of students. The 1960s marked the heyday of the no-longer-quite-so-new disciplines and, as tends to happen at the zenith, also a turning-point. After 1968, the social sciences lost much of their popularity. They also fell apart into a small number of theoretical and a large number of professional subjects. The School lived through all these phases.

The visible, sometimes audible, public story of the social sciences is, however, only a part, and not the most important part of their—and LSE's—history. The social sciences are here to stay, whether Governments like their designation as such or not. Their achievements are moreover considerable. The omnipresence of economists hardly needs emphasis. Modern social statistics has become an indispensable tool for all decision-makers, public or private. Idiosyncrasy is no longer altogether acceptable when people try to make sense of crime, or consider the shape of cities, or devise ways of improving industrial relations. Alongside lawyers and accountants whose ancient trades became professionalized by an input of social science, entire new professions have emerged, social workers, town planners, personnel managers, consultants and advisers on all manner of things. Before projects are embarked upon by public or private agencies, we need research, and such research is more often than not social science.

Application impresses those who are asked to pay but a civilized community abhors philistines and cherishes the life of the mind for its own sake. It also sometimes discovers that such appreciation of the seemingly useless has unexpected pay-offs. Which way should the writing of history go, towards highlighting great events and their heroes, or to a social-science inspired history of everyday life? Has economics lost the sense of political processes and institutions, and become too enamoured with pure theory? Is there such a discipline as political science in any strict sense of the term? Rapid expansion of the social sciences has meant a loss of cohesion. Should this be taken for granted, or should an attempt be made to pull the divergent strands of social science disciplines together again? Do we need one unified theory of the social sciences? What exactly is theory in the social sciences anyway? Is it Popper's, Parsons's, or Foucault's theory, scientific explanation, comprehensive classification, or philosophical interpretation?

There may not be answers, let alone one correct answer to these questions, but whoever wants to pursue them, or find out about the social sciences and their applications, knows that there is one place where this can be done, LSE. The London School of Economics is no longer the only home of the social sciences even in Europe. It may not be the best place in some particular subjects at any given time. In range, in depth, and in the concentration of social-science disciplines it has, however, few if any rivals.

For many, the social sciences are but a vehicle to further destinations. They want to know 'where it's at'. Where what is at? The simplest answer, probably, is the search for the good society. How should we organize our affairs to promote the well-being of most humans? The question before the question is again where one would go to find out, to talk to those who have views, and preferably not just one view. Noel Annan, in defining the 'we' in the phrase which gave his autobiography of a generation its title, *Our Age*, cites the great 'Oxford don and wit' Maurice Bowra, who clearly believed that in order to be 'where it's at' one should 'go to Oxford or Cambridge. He should have added the London School of Economics. These were the three places where ideas fermented.'[30]

The School was, of course, founded around a particular idea of the good society. Its features have often been described and almost equally often, ever since H. G. Wells, been caricatured. Yet the benevolent state and its devoted public service, which are at the heart of this idea, are far from the worst political project on offer during LSE's first century. They were born in reaction to the rampant individualism of Gladstone's age, and intended to pave the way for a kinder, gentler society of justice and solidarity. Social philosophers at LSE developed the underpinnings, political scientists the strategy, and, appropriately, a former lecturer in the School's Department of Social Policy and Administration was in the end called upon to implement both. 'Ginsberg was the major premise, Laski the minor premise, and Attlee the conclusion.'

Even when it came to the good society, the School was never of one piece. Ernest Gellner makes the point about Ginsberg, Laski, and Attlee in a reflection on LSE, the 'contested academy'. 'The distinctive essence with which it was endowed, was in dispute from the start and remains so.'[31] For long periods, one crucial strand of thinking at the School was unambiguously on the political right. On closer inspection, it actually turns out to be two strands tenuously intertwined, indeed more incompatible than its proponents liked to admit. The economists, most notably Friedrich von

Hayek, were classical 'a-social individualists'. Michael Oakeshott and some of his colleagues in the Department of Government, on the other hand, were Burkean, if not Hegelian, communitarians. While both agreed that deliberate interference in human affairs was likely to do more harm than good, they did so for different reasons and to different ends.

The School had few Marxists among its memorable teachers. There actually never was an explicit 'great debate' of the good society, what it was supposed to be, and indeed whether it was a desirable objective at all. This has to do with a pragmatic or, as Norman MacKenzie put it, 'positivist' streak which ran through the history of the first hundred years. Karl Popper conducted the debate in his books, and before a more limited audience in his lectures; but those who engaged with him were not as far away from his views as he thought. Yet the desire to identify the purpose of human endeavour, and the ways to achieve it, seemed almost inherent in the air which people breathed at LSE. Few went away without some notion of the good society.

This has much to do with a third and, for the author of this History, crucial strength of LSE. The School embodies and by its very existence resolves one of the great dilemmas of the life of the mind, the dilemma of the ascetic and the worldly, of detachment and involvement. Sidney Webb and he alone—not Beatrice, not George Bernard Shaw, nor later William Beveridge—was acutely aware of this dilemma. He hoped that its resolution by detachment would help his involvement, that his practical predilections would be confirmed by dispassionate research, but he was prepared to try the value-free course at the risk of failing. This, as much as the practical act of the foundation of the School, was his great achievement.

In terms of Sidney Webb's preferences, the experiment failed. It is perhaps a little unkind to speak of the 'gentle euthanasia' of the Fabian dream at LSE but social democracy did not last. What did last was the ever-present tension between the desire to know the causes of things and the other desire to change them, or even to make a deliberate effort to keep them as they used to be. Containing this tension was difficult and sometimes painful. It had victims. Laski was the most memorable among them, though his profound commitment to students, his kindness, and his sheer intellectual brilliance meant that he was never at risk of being excluded from LSE. John Griffith has similar virtues. Lesser mortals like Frank Meyer and also Robin Blackburn were not so lucky. While they may have deserved better, they did not appreciate that living with the tension between asceticism and worldliness was what LSE was all about.

Were the victims all on the political left? The left–right question has vexed LSE throughout the decades. It was never the right question, and yet if the wrong question is asked with sufficient persistence it begins to sound real. Sometimes I found myself thinking that the secret of LSE was a right-wing faculty confronted with left-wing students; but then I discovered that even this neat formula applies only to a few. If there is such a thing as a 'hegemony of ideas' (in itself a leftist notion) in a university, this was never exercised by the right at LSE. Neither Hayek nor Oakeshott nor, in intellectual-political terms, Robbins dominated thinking at LSE as Laski did. But the real hegemony was that of Webb, almost involuntarily carried on by Beveridge, and preserved, for different reasons though with equal firmness, by subsequent Directors: to be detached in order to be involved, to take the ascetic route to worldly purposes.

This does not work; it cannot work. It is bound to lead to conflict. The history of LSE is the story of such conflict, from George Bernard Shaw's letter to Sidney Webb on the unacceptability of neutrality to the pathetic attempts by anarchists and others during the troubles to turn the School into 'spawning grounds for opponents of the system', indeed a cadre factory of revolution. The more subtle the conflicts were, the more they deserve attention. Almost every one of the ten Directors of the first hundred years represents one answer to the unresolved question.

Perhaps the author of this History will be allowed to make the point which he has tried to put over throughout his intellectual life. Living with conflict is itself a virtue. There is no answer to Max Weber's dilemma of *Politics as a Vocation* and *Science as a Vocation*. At least, there is no answer in theory. In practice, it is possible to live with the dilemma, swaying in this direction and that at different times, accepting its reality but never abandoning the belief in the need for both dispassionate enquiry and committed action. It may be difficult for an individual to be both ascetic and worldly, but LSE throughout its first hundred years has been exactly that, and it has been strong by its successes as well as its failures.

This is heavy language at the end of a story which offers an often gripping tale as well as some food for thought. The strengths of the School are not confined to its ability to contain dilemmas and tensions. One thread which runs right through its first hundred years is that the School has offered opportunities which did not otherwise exist. In the early decades this meant giving people a second chance to study and acquire academic degrees at all. The first Director, W. A. S. Hewins, had come out of the Oxford tradition of university extension teaching. He and several of his successors, not least Beveridge, had worked at Toynbee Hall to bring light to the darker recesses of the East End. For nearly three-quarters of the first century, evening teaching was a cherished and valuable part of the School's life. In its own way the School prepared the ground for the great educational reforms which stand in the name of R. A. Butler in 1944 and Lionel Robbins in 1963.

One group, which was badly in need of greater educational and professional opportunities when the School was founded, was women. Prima facie LSE did well for women. From its early days a significant proportion of the students were women; by the last decade of the first century it stood regularly above 40 per cent. Some of the great and long-remembered teachers of the School were women: Lilian Knowles, Eileen Power, Vera Anstey, Edith Eckhard. Later, influential professors in many departments—with the notable exception of economics and political science—were women: Hilde Himmelweit in psychology, Susan Dev in accounting, Ailsa Land in statistics, Rosalyn Higgins in law, Susan Strange in international relations, Ragnhild Hatton in history, Nancy Cartwright in philosophy, Jean La Fontaine in anthropology, Eileen Barker in sociology, and this is not a complete list. Even more remarkably, for two-thirds of its first hundred years, the School has had women as chief administrative officers: Miss Christian Mactaggart from the foundation to 1919, Jessy/Janet Mair/Beveridge from 1919 to 1938, Eve Evans in fact from 1941 and formally from 1946 to 1954, and Christine Challis from 1983.

In the latter part of these first hundred years, however, women became more vocal and exposed the weaknesses in this story of opportunity. In an obituary of her, Hilary Rose, a lecturer in Social Administration, remembered that Ruth Glass's career suffered from 'the so-called convention against "nepotism"' at the School, which meant exclusion because her husband was a professor there. She adds that 'the Sixties were the peak of sexism in academic and intellectual circles', though one of her male colleagues, Professor Terence Morris, promptly replied that this suggestion was 'outrageous'.[32] Betty Scharf, also a sociology teacher, talked about the difficulties of introducing a course in gender studies. A closer look at the distribution of female members of staff in the early 1990s revealed a familiar pattern. While LSE had a larger proportion of women than the national average in all grades, at the School as elsewhere the share declined dramatically from the lecturer grade (varying around 30 per cent) through that of reader (around 15 per cent) to that of professor (5 per cent). Thus the Governors had their reasons for supporting the participation of the School in the scheme 'Opportunity 2000', which is addressed as much to concealed and subtle discrimination as to open disadvantage.

When I. G. Patel reopened the question of the School's manifest destiny as a place of opportunity in the late 1980s, it was not altogether easy to tell what exactly needed to be done. Providing a second chance for those who had missed out the first time in getting a higher education was no longer the dominant issue. Also, there were others in the field, Birkbeck College within the University of London, or the Open University. Patel was thinking of the disadvantaged in inner cities, perhaps of racial minorities. He did not have the time to tackle such issues; in any case the School, contrary to myths and appearances, does not have a tradition of helping the truly disadvantaged. However, it does have a long history of catering for mature students. The average age of LSE students was probably always higher than that of any other comparable university. Even in the late 1970s, a quarter of all applicants for undergraduate degrees were over 25 years old, and half the postgraduates had some work experience before coming to LSE. As the demand for lifelong learning and post-experience study grows, the School has responded with new courses, which is in keeping with its tradition as a place of opportunity.

Innovation in social science, the tensions of theory and application, detachment and involvement, and the provision of opportunities for those who do not find them elsewhere all leave one big question unresolved: how to create and preserve a community, an academic community, in the middle of London and with students and teachers who come from all over the world. The School's success in answering this question makes it truly special.

There is an École *Nationale* d'Administration in France and there was, in Weimar Germany, a *Deutsche* Hochschule für Politik; but LSE was not and could not have been the British, or the National School of Economics. Its base was London, and its home the world. Frequent repetition of the striking list of LSE's neighbours in inner London—Westminster and Whitehall, Fleet Street and the City, the Law Courts, Theatreland and Bloomsbury, and much else besides, not least the depressed parts of the inner city—makes it no less important. The names stand for institutions, for politics and public administration, for the media, the world of finance, the law, for the cultural centres from the South Bank to the British Museum. Nowhere else in the world is there such a concentration of excellence in all major fields of human ingenuity in a handful of square miles. It is hard to see where else a great school of the social sciences built on the tension between detachment and involvement could find a more appropriate location. The School has to be the *London* School of Economics.

London is a city of villages, but also of transients. LSE has contributed more than its share to the latter class. It has attracted thousands of students, tens of thousands if one considers the whole century, from every corner of the world. Teachers were multinational too, though it is right and proper that a large majority should at all times have been British. The School never aimed at a baseless internationalism; it is easier to be cosmopolitan for those who have passports. Visitors from other countries have found LSE an exhilarating place. LSE has made a major contribution to combating one of the plagues of the twentieth century, the plague of narrow and often aggressive nationalism. Not that every nationality represented at the School loved all others! Some, as we have seen, felt isolated at times; others fought out the battles of their home regions at the School. But they fought them out in seminar rooms, meeting-halls, and bars. They talked. Of all the tensions and troubles at LSE, not one arose from internecine warfare between national or religious or racial groups. The School was united in its opposition to overseas students' fees. They are an assault on the values of a School. As Walter Adams put it so well, imposing 'punitive fees' on overseas students would be 'to deny the essentially supranational nature of higher learning and would prove to be short-sighted even on political grounds by entailing a loss in those immeasurable returns which the School's history of international services so vividly exemplifies'.[33]

Sir Walter Adams was also the most insistent of all Directors on the need to build and maintain an academic community. Building a community at all in the middle of London, and from a great mix of cultures, might well seem an impossible proposition to social scientists. Post-war and then post-Robbins expansion added to the problem. Yet having a community was for LSE a question of survival. The School would have fallen apart without it, perhaps like some of the former polytechnics in inner London which are spread over a dozen sites or more and have no sense of identity and cohesion.

At LSE this did not happen. The fiction of a single-faculty school probably helped. The Academic Board is itself a community in being, ready to spring into action when there is a threat to the integrity of the School. Even the cramped and crowded site helps by generating the need to protect it with invisible borders from the hustle and bustle beyond it. More than that, however, the School gave its members from the beginning a sense of belonging, of being more than passers-by who come and go. One thinks of the friendships struck at LSE, of the hundreds of marriages which began there. The porters have much to do with the community called LSE, not only as the first port of call for those who return to the School as well as those who are there, but as one visible, tangible expression of *the* School.

The School was always more than the sum of its parts. One was not just a student of Robbins or Laski but a student at LSE.

The memorable events of the first hundred years have made their own contribution to the community called LSE. Myth is a strong glue, especially if at least some of it is based on real events. In the run-up to the centenary, dozens of members of the School have contributed to build up a corpus of 'oral history'. Reading through the reminiscences, one finds ever new angles on familiar events: on 1968, of course, on Laski, on the real exile in Cambridge, on William Beveridge and his lady, on the incidents of the early 1930s and the social biology toads later, on the great figures of the inter-war years, Tawney, Hayek, Malinowski, Bowley, on Lionel Robbins in his several LSE incarnations. . . . But we must not begin the history of the first hundred years afresh as we reach its conclusion.

These, then, are the strengths of the unique institution, the London School of Economics and Political Science. At the end of its first century they are all under threat. Size, diffusion, anti-internationalism, and the decline of London as a vibrant capital city put pressure on the sense of community. In response to financial exigencies, the balance of asceticism and worldliness has had to swing far in the worldly direction. Universities in general, and the social sciences in particular, are going through a phase of dissipation and doubt. Thus there is much to do, but for LSE there is also much to build on.

NOTES

Notes to Chapter 1

1. Hutchinson will, 10 Oct. 1893, Hutchinson Trust Minute Book, LSE 1/1; and London School of Economics, *Reminiscences of Former Students, Members of Staff and Governors, with Other Material Relating to the History of the School 1992–1947*, 1947, LSE, R(SR)1101 (hereafter *Rem.*).
2. Beatrice Webb, *The Diary of Beatrice Webb*, ii. *All the Good Things in Life 1892–1905*, ed. Norman and Jeanne MacKenzie (London: Virago/LSE, 1984; hereafter *WD* ii), 56.
3. Notes by Bertram Hutchinson, 'Henry Hunt Hutchinson', LSE History Box, Origins/Foundation, 1.
4. Pease, in *Rem.* 5.
5. Hutchinson will, LSE 1/1.
6. Graham Wallas, 'An Historical Note', *The Students' Union Handbook 1922–23*, LSE CF 116/C.
7. *WD* ii. 56.
8. Janet Beveridge, *An Epic of Clare Market: Birth and Early Days of the London School of Economics* (London: G. Bell & Sons, 1960), 21.
9. Sidney and Beatrice Webb, *The Letters of Sidney and Beatrice Webb*, i. *Apprenticeships 1873–1892*, ed. Norman MacKenzie (Cambridge: Cambridge University Press/ LSE, 1978; hereafter *WL* i), 116.
10. *WD* ii. 57.
11. S. and B. Webb, 'Reminiscences IV: The London School of Economics and Political Science', *St Martin's Review* (Jan. 1929), 25.
12. Michael Holroyd, *Bernard Shaw*, i. *1856–1898: The Search for Love* (London: Chatto & Windus, 1988), 267.
13. Correspondence from Mr A. E. Herring, 5 Feb. 1993.
14. *WL* i. 79 n.
15. S. and B. Webb, *The Letters of Sidney and Beatrice Webb*, ii. *Partnership 1892–1912*, ed. Norman MacKenzie (Cambridge: Cambridge University Press/LSE, 1978; hereafter *WL* ii), 29.
16. *WD* ii. 57.
17. *WL* i. 208 n., 405 n.
18. *WL* ii. 29–30.
19. Sydney Caine, *The History of the Foundation of the London School of Economics and Political Science* (London: G. Bell & Sons, 1963), 39.
20. Beatrice Webb, *Our Partnership*, ed. Barbara Drake and Margaret I. Cole (London: Cambridge University Press/LSE. 1948; hereafter *OP*), 37–8.
21. Terence H. Qualter, *Graham Wallas and the Great Society* (London: Macmillan/LSE, 1980), 168, 99–100.
22. Caine, *History of the Foundation*, 39–40.
23. *WL* ii. 31.
24. L. L. Price, 'Obituary: William Albert Samuel Hewins', *Economic Journal*, 42 (Mar. 1932), 154.

25. F. W. Galton to Carr-Saunders, 19 Nov. 1945, LSE CF 116/C.
26. Mactaggart, in *Rem.* 66–7.
27. Webb to Hewins, 29 Mar. 1895, Hewins Papers, University of Sheffield, 43/134; *WL* ii. 33.
28. Hewins to Webb, 30 Mar. 1895, ibid. 43/138.
29. Albert E. Sloman, *A University in the Making: The 1963 Reith Lectures* (London: British Broadcasting Corporation, 1964), 20.
30. *WL* ii. 178.
31. Marshall to Hewins, 7 June 1895, and Tout to Hewins, 9 June 1895, Hewins Papers, 44/12, 44/19.
32. Trueman Wood to Sir Owen Roberts, ibid. 43/142.
33. Secretary of London Chamber of Commerce to Hewins, ibid. 43/197.
34. Mactaggart in *Rem.* 80; J. Beveridge, *Epic*, 30.
35. M. Holroyd, *Bernard Shaw*, iv. *1950–51: The Last Laugh* (London: Chatto & Windus, 1992), 5.
36. G.B.S. to B. Webb, 1 July 1895, Passfield Papers, BLPES, II.4.a.2.
37. W. A. S. Hewins, *The Apologia of an Imperialist: Forty Years of Empire Policy*, i (London: Constable, 1929), 16.
38. R. B. Haldane, 'The Civic University', in *Selected Addresses and Essays* (London: John Murray, 1928), 115.
39. Caine, *History of the Foundation*, 28.
40. S. Webb to Revd Archibald Robertson, 3 Jan. 1903, in *WL* ii. 177.
41. Three-page prospectus, undated, LSE 1/1.
42. Hermione Hobhouse, *Lost London: A Century of Demolition and Decay* (London: Macmillan, 1971), 96.
43. Hewins to his mother, 27 May 1895, Hewins Papers, 10/40.
44. Hewins, *Apologia*, 27–8.
45. 9 Nov. 1895, Hewins Papers, 44/101.
46. *CMR* (Oct. 1907), 30.
47. R. C. K. Ensor, *England 1870–1914* (Oxford: Clarendon Press, 1936), 304.
48. Bernard Shaw, *Collected Letters 1874–1897*, ed. Dan H. Laurence (London: Max Reinhardt, 1965), 453.
49. Carole Seymour-Jones, *Beatrice Webb: Woman of Conflict* (London: Allison & Busby, 1992), 170.
50. *WL* i. 72 n.
51. Caroline Moorehead, *Bertrand Russell: A Life* (London: Sinclair-Stevenson, 1992), 135.
52. Norman MacKenzie, 'Socialism and Society: A New View of the Webb Partnership', lecture delivered at LSE, 15 May 1978 (London: LSE, 1978), 13.
53. Anne Fremantle, *This Little Band of Prophets: The Story of the Gentle Fabians* (London: George Allen & Unwin, 1960), 41–2; Holroyd, *Shaw*, i. 172.
54. N. MacKenzie, 'Socialism and Society', 2, 3, 9, 25, 5.
55. Holroyd, *Shaw*, i. 412.
56. H. G. Wells, *The New Machiavelli* (London: John Lane/Bodley Head, 1911), 222.
57. Pease, in *Rem.* 9.
58. Alon Kadish, *The Oxford Economists in the Late Nineteenth Century* (Oxford: Clarendon Press, 1982), 26.

59. Ibid., 96, 99.
60. Ibid., 98, 107–09.
61. Haldane, 'Civic University', 124.
62. Gresham University Commission, 'The Report of the Commissions appointed to consider the Draft Charter for the Proposed Gresham University in London', British Parliamentary Papers, 1893/4, 31, 24 Jan. 1894, para. 14, 4.
63. Eric Ashby and Mary Anderson, *Portrait of Haldane at Work on Education* (London: Macmillan, 1974), 33, 40.
64. W. Cunningham *et al.*, 'Methods of Economic Training in This and Other Countries', Report of the Committee, in *Report* of the British Association for the Advancement of Science Meeting at Oxford, 1894 (London: John Murray, 1894), 365.
65. Kadish, *Oxford Economists*, 118–19.
66. A. W. Coats, 'Sociological Aspects of British Economic Thought (ca. 1880–1930)', *Journal of Political Economy*, 75 (Oct. 1967), 719, 723.
67. Gerard M. Koot, 'An Alternative to Marshall: Economic History and Applied Economics at the Early LSE', *Atlantic Economic Journal*, 10 (Mar. 1982), 3.
68. C. F. Bastable, 'Presidential Address to Section F, Economic Science and Statistics', in *Report* of the British Association for the Advancement of Science Meeting at Oxford, 1894 (London: John Murray, 1894), 723.
69. John Maloney, 'Marshall, Cunningham, and the Emerging Economics Profession', *Economic History Review*, 2nd ser. 29 (Aug. 1976), 441.
70. John Maynard Keynes, *Essays in Biography* (London: Macmillan, 1933), 171.
71. Robert L. Heilbroner, *The Worldly Philosophers: The Lives, Times and Ideas of the Great Economic Thinkers* (5th edn.; Harmondsworth: Penguin Books, 1983), 161.
72. Coats, 'Sociological Aspects', 724.
73. Koot, 'Alternative to Marshall', 3.
74. W. J. Ashley, 'A Survey of the Past History and Present Position of Political Economy: Leicester, 1907', in R. L. Smyth (ed.), *Essays in Economic Method: Selected Papers Read to Section F of the British Association for the Advancement of Science, 1860–1913* (London: Duckworth, 1962), 224.
75. Marshall's 80th Birthday Address, Beveridge Papers, BLPES, V14.
76. Kadish, *Oxford Economists*, 277.
77. F. A. Hayek, 'The London School of Economics 1895–1945', *Economica* (Feb. 1946), 6.
78. Lionel Robbins, 'A Student's Recollections of Edwin Cannan', *Economic Journal*, 45 (June 1935), 393, 398.
79. Edwin Cannan, *History of Local Rates* (2nd edn.; London: P. S. King & Son, 1912), 1.
80. G. R. Searle, *The Quest for National Efficiency: A Study in British Politics and Political Thought 1899–1914* (Oxford: Basil Blackwell, 1971), 1.
81. Ibid., 54–5, 59.
82. Bernard Semmel, *Imperialism and Social Reform: English Social-Imperial Thought 1895–1914* (London: George Allen & Unwin, 1960), 63.
83. *WD* ii. 239.
84. Ibid. 237.

85. Searle, *National Efficiency*, 139.
86. *OP* 87.
87. G. M. Trevelyan, *British History in the Nineteenth Century and After 1782–1919* (2nd edn.; London: Longmans, Green, 1937), 403.
88. Fremantle, *Little Band*, 191.
89. Ibid. 15.
90. Ibid. 20.
91. S. Webb, in Bernard Shaw (ed.), *Fabian Essays* (Jubilee edn.; George Allen & Unwin, 1948), 57.
92. Stefan Collini, *Liberalism and Sociology: L. T. Hobhouse and Political Argument in England 1880–1914* (Cambridge: Cambridge University Press, 1979), 41.
93. G. K. Chesterton, *Autobiography* (London: Hutchinson, 1936), 111.
94. S. Webb, in Shaw, *Fabian Essays*, 31.
95. Norman and Jeanne MacKenzie, *The First Fabians* (London: Weidenfeld & Nicolson, 1977), 112.
96. S. Webb, in Shaw, *Fabian Essays*, 47.
97. Fremantle, 37, 98.
98. N. and J. MacKenzie, *First Fabians*, 286.
99. Winston Churchill, *Great Contemporaries* (London: Butterworth, 1937), 72, 61.
100. Ibid. 62–3.
101. Trevelyan, *British History*, 436.
102. Semmel, *Imperialism*, 25.
103. Michael Coren, *The Invisible Man: The Life and Liberties of H. G. Wells* (London: Bloomsbury, 1993), 226.
104. Julian Amery, *Joseph Chamberlain and the Tariff Reform Campaign: The Life of Chamberlain*, vi. *1903–1968* (London: Macmillan, 1969), 981.
105. Fremantle, *Little Band*, 190.
106. London School of Economics, *Brief Account of the Work of the School during Michaelmas and Lent Terms, 1895–6*, 1896, LSE 1/1, 3.
107. Webb to Hewins, 22 Mar. 1896, Hewins Papers, 44/114–115.
108. *WD* ii. 101–2.
109. Mactaggart, in *Rem.* 80.
110. Snell, in *Rem.* 32.
111. *CMR* (1909/10), 64.
112. Mactaggart, in *Rem.* 80, 75, 73.
113. Pease, in *Rem.* 7.
114. London School of Economics, *Proposed Establishment of a Library of Political Science*, Apr. 1896, LSE 1/1, 1–3.
115. Herbert Spencer to Hewins, 24 Mar. 1897, Hewins Papers, 44/146–7.
116. *OP* 90–1.
117. Holroyd, *Shaw*, i. 465.
118. Caine, *History of the Foundation*, 68.
119. H. G. Wells, 'The So-Called Science of Sociology', *Sociological Papers*, 3 (1906), 376.
120. *WL* ii. 98.

121. Passmore Edwards to Webb, 16 Aug. 1899, and Webb to Passmore Edwards, 17 Aug. 1899, Passfield Papers, X.3.a.100–1.
122. Pease, in *Rem.* 6.
123. A. L. Bowley, 'Address to the Students Union, LSE, April 18th 1945', in *Rem.* 45.
124. Mactaggart and Pease, in *Rem.* 84–5, 6.
125. Caine, *History of the Foundation*, 80.
126. Hayek, 'LSE', 14.
127. J. Beveridge, *Epic*, 41.
128. Norman MacKenzie in correspondence with author, 7 May 1993.
129. Bowley, in *Rem.* 43.
130. Caine, *History of the Foundation*, 80–1.
131. Sidney Webb, 'The Provision of Higher Commercial Education in London', in International Congress on Technical Education, *Report* of the Proceedings of the 4th Meeting, held in London, June 1897 (London: Society of Arts, 1897), 208.
132. London School of Economics, *A Brief Description of the Objects and Work of the School*, Mar. 1901, updated to 1910, LSE 1/1, 3.
133. Alon Kadish, 'The City, the Fabians, and the Foundation of the LSE', undated, unpublished typescript, 18, 16; copy in LSE History.
134. Galton to Carr-Saunders, 19 Nov. 1945, LSE CF 116C.
135. *CMR* (Oct. 1907), 30.
136. Mactaggart, in *Rem.* 88.
137. H. G. Richardson to Carr-Saunders, 23 Nov. 1945, LSE CF 116C.
138. Bowley, in *Rem.* 48.
139. Richardson to Carr-Saunders, 23 Nov. 1945, LSE CF 116C.
140. Bowley, in *Rem.* 51.
141. Richardson to Carr-Saunders, 23 Nov. 1945, LSE CF 116C.
142. N. and J. MacKenzie, *First Fabians*, 269.
143. Fremantle, *Little Band*, 138–9.
144. E. R. Pease, *The History of the Fabian Society* (London: A. C. Fifield, 1916), 131.
145. *CMR* (Oct. 1907), 31.
146. *WL* ii. 103.
147. J. Beveridge, *Epic*, 41.
148. Webb to Revd Archibald Robertson, 3 Jan. 1903, Passfield Papers, 2(ii)16–39, 6–15.
149. Hewins, *Apologia*, 30.
150. *WD* ii. 196.
151. Ibid. 258–9, 196.
152. *OP* 247.
153. *WL* ii. 65–7.
154. Passfield Papers, 2(i)56, 60–2, 64–5, 67, 77, 85.
155. Hewins to Chamberlain, Nov. 1903, Hewins Papers, 46/102.
156. Chamberlain to Hewins, 9 Nov. 1903, Hewins Papers, 46/85–6.
157. Hewins, *Apologia*, 77.
158. *WL* ii. 185–6.

159. C. H. Firth to Hewins, 15 June 1903, Hewins Papers, 46/27.
160. Marshall to Hewins, 14 July 1903, Hewins Papers, 46/35–6.
161. *WD* ii. 302.
162. *OP* 329.

Notes to Chapter 2

1. Edmund W. Gilbert, *Sir Halford Mackinder 1861–1947: An Appreciation of his Life and Work* (London: LSE/G. Bell & Sons, 1961), 4.
2. Wilson Morse to Ritchie, 5 Aug. 1952, LSE BF Mackinder.
3. Gilbert, *Mackinder*, 11.
4. Brian W. Blouet, *Halford Mackinder: A Biography* (College Station, Tex.: Texas A & M University Press, 1987), 197.
5. Mackinder to Sir John Scott Keltie, 16. Dec. 1903, Mackinder Papers, School of Geography, Oxford.
6. Edmund W. Gilbert, 'Seven Lamps of Geography: An Appreciation of the Teaching of Sir Halford J. Mackinder', *Geography*, 36 (1951), 27.
7. *WD* ii. 252.
8. Sidney Webb, 'Lord Rosebery's Escape from Houndsditch', *The Nineteenth Century and After*, 50 (Sept. 1901), 386.
9. Blouet, *Mackinder*, 135.
10. *WL* ii. 170.
11. Wells, *New Machiavelli*, 337–8.
12. Coefficients, *Minutes of the Club 1902–06*, LSE, R.(Coll.) Assoc. 17.
13. H. G. Wells, *Experiment in Autobiography: Discoveries and Conclusions of a Very Ordinary Brain (since 1866)*, ii. (London: Victor Gollancz, 1966), 765.
14. Mackinder after-dinner speech, 13 May 1931, Mackinder Papers, c/100.
15. L. S. Amery, *My Political Life*, i. *England before the Storm 1896–1914* (London: Hutchinson, 1953), 224.
16. *WL* ii. 265.
17. Wells, *New Machiavelli*, 351.
18. Caine, 'Brain Power and Co-efficiency', *Daily Telegraph*, 11 May 1973.
19. Wells, *Experiment in Autobiography*, 763.
20. Mackinder Papers, c/100.
21. Semmel, *Imperialism*, 81.
22. Wells, *New Machiavelli*, 351.
23. Ibid. 329.
24. L. S. Amery, *Political Life*, 227.
25. Wells, *Experiment in Autobiography*, 766.
26. L. S. Amery, *Political Life*, 228.
27. Blouet, *Mackinder*, 3.
28. Gilbert, *Mackinder*, 13.
29. Bertrand Russell, *Autobiography*, i. *1872–1914* (London: George Allen & Unwin, 1967), 176.
30. Blouet, *Mackinder*, 202, 197, 200.
31. T. W. Freeman, 'The Royal Geographical Society and the Development of Geography' in E. H. Brown (ed), *Geography Yesterday and Tomorrow* (Oxford: Oxford University Press, 1980), 50.

32. L. S. Amery, *Political Life*, 228.
33. Gilbert, *Mackinder*, 22, 23.
34. Freeman, 'Royal Geographical Society', 16.
35. Bottomley and Wilson, in *Rem.* 153, 149.
36. Halford Mackinder, 'The Geographical Pivot of History', *Geographical Journal*, 23 (Apr. 1904), 422, 433, 436.
37. W. Roger Louis, *In the Name of God, Go! Leo Amery and the British Empire in the Age of Churchill* (New York: W. W. Norton, 1992), 55.
38. Joseph Partsch, *Central Europe* (London: Heinemann, 1903), 8.
39. Halford Mackinder, *Democratic Ideals and Reality: A Study in the Politics of Reconstruction* (London: Constable, 1919), 194.
40. *Geographical Journal*, 103 (Mar. 1944), 132.
41. James Trapier Lowe, *Geopolitics and War: Mackinder's Philosophy of Power* (Washington, DC: University Press of America, 1981), 99, 9, 79.
42. Gerry Kearns, 'Halford John Mackinder 1861–1947', *Geographers Bibliographical Studies*, 9 (1985), 81.
43. Blouet, *Mackinder*, 198.
44. Halford Mackinder, *Money-Power and Man-Power: The Underlying Principles rather than the Statistics of Tariff Reform* (London: Simkin-Marshall, 1906), 21.
45. Halford Mackinder, 'On Thinking Imperially', in *Lectures on Empire 1906/07* (London: published privately, 1907), 42.
46. L. S. Amery, *Political Life*, 238–9.
47. Russell, *Autobiography*, i. 176.
48. Autobiographical notes, undated, Mackinder Papers.
49. Mactaggart, in *Rem.* 67.
50. J. Beveridge, *Epic*, 59.
51. Semmel, *Imperialism*, 186.
52. Blouet, *Mackinder*, 130.
53. Governors, 3 July 1905, LSE 4/3.
54. DR 1905/6.
55. Board of Education, *Reports from those Universities and University Colleges in Great Britain which Participated in the Parliamentary Grant for University Colleges 1908–9*, British Parliamentary Papers, vol. 24, 1910, pp. iii, vii.
56. *WL* ii. 243.
57. J. Beveridge, *Epic*, 76.
58. Governors, 3 July 1905, LSE 4/3.
59. Jose Harris, 'The Webbs, the COS, and Ratan Tata', in Martin Bulmer, Jane Lewis, and David Piachaud (eds.), *The Goals of Social Policy* (London: Unwin Hyman, 1989), 50, 37.
60. Kit Russell, Sheila Benson, Christine Farrell, Howard Glennerster, David Piachaud, and Garth Plowman, *Changing Course: A Follow-up Study of Students Taking the Certificate and Diploma in Social Administration at the London School of Economics, 1949–73* (London: LSE, 1981), 7.
61. Philip Boardman, *The Worlds of Patrick Geddes: Biologist, Town Planner, Re-educator, Peace-Warrior* (London: Routledge & Kegan Paul, 1978), 137.
62. *Nature* obituary, 14 May 1932, 714.

63. Sociological Society, *Sociological Papers 1904–06* (London: Sociological Society, 1904–06), 1904: pp. xiv, 15–16.
64. Ibid. 4, 5.
65. Philip Abrams, *The Origins of British Sociology* (Chicago: University of Chicago Press, 1968), 101, 102, 103.
66. R. J. Halliday, 'The Sociological Movement, Society and Genesis of Academic Sociology in Britain', *Sociological Review*, 16 (Nov. 1968), 381.
67. C. P. Blacker, *Eugenics in Prospect and Retrospect* (London: Hamish Hamilton Medical Books, 1945), 16.
68. Abrams, *Origins*, 105, 106.
69. William Collins, Introductory Address, in *Inauguration of the Martin White Professorships of Sociology, Dec. 17, 1907* (London: John Murray, 1908), 5.
70. Collini, *Liberalism and Sociology*, 51.
71. Ibid. 125, 247.
72. L. T. Hobhouse, *Morals in Evolution: A Study in Comparative Ethics* (7th edn.; London: Chapman & Hall, 1951), 605.
73. Collini, *Liberalism and Sociology*, 209.
74. Ernest Barker, 'Leonard Trelawny Hobhouse 1864–1929', *Proceedings of the British Academy*, 15 (1929), 547, 546.
75. Collini, *Liberalism and Sociology*, 251.
76. Edward Westermarck, *Memories of My Life*, tr. Anna Barwell (London: George Allen & Unwin, 1929), 211, 212.
77. Martin White to Beveridge, 14 May 1926, LSE CF 408.
78. Westermarck, *Memories*, 226, 199.
79. William H. Beveridge, *The London School of Economics and its Problems 1919–1937* (London: George Allen & Unwin, 1960), 65.
80. Mackinder to Carr-Saunders, 25 Nov. 1945, LSE CF 116C.
81. Blouet, *Mackinder*, 133, 132.
82. *CMR* (Nov. 1908), 1, 2.
83. Pease, in *Rem.* 8.
84. Keith Sinclair, *William Pember Reeves: New Zealand Fabian* (Oxford: Clarendon Press, 1965), 311.
85. Webb to Reeves, 23 May 1908, LSE SF Reeves.
86. B. Webb, *The Diary of Beatrice Webb*, iii. *The Power to Alter Things, 1905–1924*, ed. Norman and Jeanne MacKenzie (London: Virago/LSE, 1984; hereafter WD iii), 95.
87. Sinclair, *Reeves*, 278.
88. Ibid. 251.
89. *WD* ii. 324.
90. Harris, 'The Webbs, the COS', 40.
91. Maxine Berg, 'The First Women Economic Historians', *Economic History Review*, 45 (May 1992), 317.
92. Mary Stocks, *My Commonplace Book* (London: Peter Davies, 1970), 81, 82, 84.
93. J. Beveridge, *Epic*, 81.
94. Stocks, *Commonplace Book*, 87, 95.
95. Westermarck, *Memories*, 199, 201, 202.
96. Bowley, in *Rem.* 53.

97. H. G. Wells, *H. G. Wells in Love*, ed. G. P. Wells (London: Faber, 1984), 73, 71.
98. Sinclair, *Reeves*, 317–18.
99. *Railway Review*, 23 Sept. 1910.
100. *Railway News*, 24 Sept. 1910.
101. *Daily Telegraph* and *Manchester Guardian*, 18 Oct. 1910.
102. Webb to Lyall, undated, Passfield Papers II.6.a.1.
103. *Daily News* and *Pall Mall Gazette*, 25 Oct. 1910.
104. *WL* ii. 354.
105. Ibid. 356.
106. *OP* 463.
107. 'The Webbs and their Service', Beveridge Papers, IX.b.46.
108. *OP* 463–4.
109. *WL* ii. 371.
110. Passfield Papers, X.2 (i).
111. Sidney and Beatrice Webb, *Indian Diary*, ed. Niraja Gopal Jayal (Oxford: Oxford University Press, 1990), 207.
112. Harris, 'The Webbs, the COS', 54.
113. Norman Dennis and A. H. Halsey (eds.), *English Ethical Socialism: Thomas More to R. H. Tawney* (Oxford: Clarendon Press, 1989), 159.
114. *Indian*, 10 June 1911.
115. J. Beveridge, *Epic*, 82.
116. Reeves to Webb, 13 Mar. 1917, Passfield Papers, II.6.a.5.
117. *WL* ii. 114.
118. Reeves to Carvill, 2 May 1916, Reeves Papers, Alexander Turnbull Library, Wellington, NZ, MS 129: 6.
119. Mactaggart, in *Rem.* 86, 68.
120. Sinclair, *Reeves*, 333, 328, 331.
121. Richard Clogg, 'Politics and the Academy', *Middle Eastern Studies*, 21 (Oct. 1985), 84, 86, 63.
122. Reeves to Charles Wilson, 9 Apr. 1919, Reeves Papers, 129: 12.
123. Reeves to Steel-Maitland, 10 Apr. 1919, ibid.
124. Reeves to Carr-Saunders, 2 May 1919, ibid.
125. S. and B. Webb, *The Letters of Sidney and Beatrice Webb*, iii. *Pilgrimage 1912–1947*, ed. Norman MacKenzie (Cambridge: Cambridge University Press/LSE, 1978; hereafter *WL* iii), 184.

Notes to Chapter 3

1. W. N. Medlicott, *Contemporary England 1914–1964* (London: Longman, 1967), 81, 79.
2. *WL* ii. 117.
3. John Maynard Keynes, *The Collected Works of J. M. Keynes*, xvi. *Activities 1914–19*, ed. Elizabeth Johnson (London: Macmillan/St Martin's Press for the Royal Economic Society, 1971), 458.
4. *WD* iii. 344.
5. *WL* iii. 118.
6. William H. Beveridge, *Power and Influence* (London: Hodder & Stoughton, 1953; hereafter *PI*), 113, 15.

7. Beveridge to Austen Chamberlain, 25 July 1919, LSE SF Beveridge.
8. Webb to Beveridge, 15 July 1921, ibid.
9. B. Webb to Beveridge, 5 Apr. 1937, Beveridge Papers, II.b.36.
10. Caine to Philip Mair, 18 Mar. 1963, LSE SF Beveridge.
11. Hayek, 'LSE', 19.
12. DR 1921/2, 214.
13. Beveridge, *LSE*, 18, 23.
14. Ibid. 28.
15. DR 1924/5, 18.
16. Kingsley Martin, *Father Figures: A First Volume of Autobiography 1897–1931* (London: Hutchinson, 1966), 154.
17. Jose Harris, *William Beveridge: A Biography* (Oxford: Clarendon Press, 1977), 263.
18. Beveridge, *LSE*, 17.
19. Hayek, 'LSE', 19.
20. *WD* iii. 339–40.
21. Lord Chorley, 'Beveridge and the LSE, Part One', *LSE Magazine* (Nov. 1972), 7.
22. William H. Beveridge, 'Economics as a Liberal Education', *Economica*, 1 (Jan. 1921), 12–13.
23. DR 1922–24, 24.
24. Beveridge, Foreword to the Students' Union Handbook, 1922/3, 3, LSE CF 116/A.
25. Graul, Students' Union Handbook, 1922/3, 6, LSE CF 116/A.
26. DR 1922–24, 25.
27. Webb to Beveridge, undated, in reply to 8 Apr. 1921, LSE CF 751/A.
28. Minutes of the Committee appointed to discuss the School Emblem, 24 May 1921, LSE CF 751/A.
29. Letters Patent from the College of Arms, 26 Sept. 1922, LSE CF 751/A.
30. Beveridge, *LSE*, 61.
31. Patrick Davis, 'LSE's Coat of Arms', *LSE Magazine* (June 1977), 10.
32. DR 1921/2, 215.
33. Beveridge, *LSE*, 63.
34. Ibid. 20.
35. Ibid. 19.
36. Mactaggart, in *Rem.* 86.
37. 'Report of the Director on the Work of the School, Michaelmas Term 1920–21', 11 Nov. 1920, Council of Management, LSE 4/6.
38. Beveridge, *LSE*, 21.
39. 8 Dec. 1919, Steel-Maitland Papers, 88/2/77.
40. Beveridge, *LSE*, 22.
41. Beveridge to Carr-Saunders, 28 Nov. 1945, LSE CF 116/A.
42. Janet Beveridge to Carr-Saunders, 5 Dec. 1945, LSE CF 116/A.
43. Janet Beveridge, 'In Retrospect', *LSE Magazine* (July 1951), 8.
44. Harris, *Beveridge*, 72.
45. Hugh Dalton, *The Political Diary of Hugh Dalton, 1918–40, 1945–60*, ed. Ben Pimlott (London: Jonathan Cape/LSE, 1986), 33.

46. Philip Beveridge Mair, *Shared Enthusiasm: The Story of Lord and Lady Beveridge* (Windlesham: Ascent Books, 1982), 72.
47. Harris, *Beveridge*, 74.
48. Ibid. 19.
49. Ibid. 73.
50. Ibid. 75.
51. Harold Wilson, *Memoirs: The Making of a Prime Minister 1916–64* (London: Weidenfeld & Nicolson), 45.
52. Isaiah Berlin, *The Hedgehog and the Fox: An Essay on Tolstoy's View of History* (London: Weidenfeld & Nicolson, 1953), 1–2.
53. Ben Pimlott, *The Life and Times of Harold Wilson* (London: HarperCollins, 1992), 61.
54. 12 Aug. 1924, Beveridge Papers Supplement, 1/11.
55. 15 Aug. 1924, ibid.
56. Harris, *Beveridge*, 280.
57. 26 July 1927, Beveridge Papers Supplement, 1/12.
58. 30 May 1932, Beveridge Papers Supplement, 1/15.
59. 21 Jan. 1931, ibid.
60. 17 Feb. 1931, ibid.
61. 21 Jan. 1931, ibid.
62. Harris, *Beveridge*, 281; Shehadi Interview, Hayek, LSE Archive, BLPES, 1983.
63. Beveridge, *LSE*, 8.
64. Wilson, *The Beveridge Memorial Lecture 1966*, given at Senate House on 18 Nov. 1966 (London: Institute of Statisticians, 1966), 12.
65. Dr Marjorie Plant, 'LSE's Coat of Arms, Notes', 25 Feb. 1977, LSE CF 751/B.
66. Harris, *Beveridge*, 283.
67. Beveridge, *LSE*, 22.
68. Ibid. 7.
69. 'The Director and the School, June 1919–June 1925', unsigned, *CMR* (Summer Term 1925), 102.
70. Martin and Joan Bulmer, 'Philanthropy and Social Science in the 1920s: Beardsley Ruml and the Laura Spelman Rockefeller Memorial, 1922–29', *Minerva*, 19 (Autumn 1981), 355, 362–3.
71. Ibid. 380.
72. 'Note of interview with H. Laski', 25 Sept. 1923, LSRM, File 592.
73. Bulmer and Bulmer, 'Philanthropy', 394 n.
74. Beveridge to Ruml, 16 Oct. 1923, LSE CF 222/A.
75. Flexner, memo to Ruml, 12 Nov. 1923, LSRM, File 592.
76. S. M. Gunn, 26 May 1936, RF, Folder 943; Mrs Mair, 20 Jan. 1926, LSRM, File 593.
77. Sydnor Walker, note, 20 June 1936, RF, Folder 943.
78. Comment to author by Martin Bulmer, Jan. 1994; J. V. Van Sickle to Rufus (E. E. Day), 19 Mar. 1934, RF, Folder 939.
79. S. M. Gunn, 26 May 1936, RF, Folder 943.
80. 'Memorandum', July 1925, LSE CF 222/B&C.
81. William H. Beveridge, 'The Physical Relation of a University to a City', lecture delivered to the London Society, 16 Nov. 1928, LSE, R(Coll.) Misc. 237 I25, 13.

82. *PI*, 177–8.
83. Note on LSE, drafted by S. M. Gunn, letter, 10 Mar. 1931, final version dated 25 Mar. 1931, RF, Folder 934.
84. Mrs Mair to E. E. Day, 9 May 1931, RF, Folder 935.
85. Fosdick to Ruml, 4 Oct. 1928, LSRM, File 594.
86. J. R. G. Hecht to Rockefeller Foundation, 9 June 1931, RF, Folder 935.
87. Gunn to E. E. Day, 13 Nov. 1931, ibid.
88. Day to Gunn, 24 Dec. 1931, ibid.
89. S. M. Gunn memo, 4 Mar. 1931, RF, Folder 934.
90. Governors, 27 May 1925, LSE 4/6.
91. Beveridge, undated reply to Flora M. Rhind, secretary to E. E. Day, on a letter dated 24 Oct. 1933, RF, Folder 938.
92. D. A. Clarke, '"A New Laboratory of Sociological Research": The BLPES', *Journal of Documentation*, 40 (June 1984), 152–7; Arthur H. John, *The British Library of Political and Economic Science: A Brief History* (London: LSE, 1971); Marjorie Plant, 'The BLPES', *LSE Magazine* (Jan. 1952), 2–5; and Geoffrey Woledge, 'The BLPES', in R. Irwin and R. Staveley (eds.), *The Libraries of London* (2nd rev. edn.; London: Library Association, 1964).
93. John, *BLPES*, 10.
94. London School of Economics, *Review of the Activities and Development of the London School of Economics and Political Science (University of London) during the Period 1923–1937*, prepared for the Rockefeller Foundation (London: LSE, Feb. 1938). Other students' statistics published annually in the *Calendar* in the session following, from 1918/19 to 1993/4, in LSE History Box, Students.
95. Beveridge, *LSE*, 119.
96. LSE, *Review*, 54–61.
97. Beveridge, *LSE*, 119.
98. The graph is based on figures from the School's audited accounts. Some adjustments have been made to the original figures to improve year-by-year comparability. In particular, capital outlays and receipts have been excluded where identified. The first account ran from 1 May 1895 to 30 Sept. 1897: in the graph the figures have been divided equally between 1895/6 and 1896/7. In 1920 the accounting year-end, up to then 30 Sept., was changed at the UGC's behest to 31 July: the 1919/20 figures cover 10 months only. Original summary table of income and expenditure and notes on their preparation in LSE History Box, Accounts.

    The conversion to 1994 pounds has been made following the recommendations of the Central Statistical Office document 'The Purchasing Power of the Pound', Oct. 1990. This links the CSO's Retail Prices Index with Bowley's Earnings and Cost of Living from 1895 to 1914 (R. G. D. Allen, *Statistics for Economists* (London: Hutchinson, [*c.*1948]), table 5).
99. Rowland B. Eustace, 'LSE Historical Notes' attached to letter to the author, 10 Jan. 1974, LSE History, Chapter 3 notes.
100. Beveridge, 'Reflections on the "School of Economics"', circulated at Constitution Sub-Committee, 30 Oct. 1935, LSE 17/5/2.
101. Beveridge, *LSE*, 82.

102. 27 May 1937, LSE 6/18; 6 June 1934, Constitution Sub-Committee of the PC, LSE 17/5/12; 'Internal Government of the School', Governors, 6 Dec. 1962.
103. *WL* iii. 275, 283, 301.
104. Webb to Steel-Maitland, 24 June 1916, Steel-Maitland Papers, Scottish Record Office, Edinburgh.
105. Steel-Maitland to Webb, 25 Oct. 1916, ibid.
106. Steel-Maitland to Gladstone, 3 Feb. 1920, ibid.
107. Lord Robbins, *Autobiography of an Economist* (London: Macmillan, 1971), 71–89, 91, 123.
108. K. Martin, *Father Figures*, 152–64.
109. Judith Listowel, *This I have Seen* (London: Faber, 1943), 48–52.
110. Nehru, in Joan Abse (ed.), *My LSE* (London: Robson Books, 1977), 14–27.
111. K. Martin, *Father Figures*, 154.
112. Robbins, *Autobiography*, 75, 136–7.
113. K. Martin, *Father Figures*, 154.
114. *PI*, 190–3, 205, 210.
115. DR 1928/9, 14.
116. Hayek, 'LSE', 23.
117. Beatrice Webb, *The Diary of Beatrice Webb*, iv. *The Wheel of Life 1924–1943*, ed. Norman and Jeanne MacKenzie (London: Virago/LSE, 1985; hereafter *WD* iv), 157; Beveridge, *LSE*, 98.

Notes to Chapter 4

1. Beveridge, 'Economics as a Liberal Education', 2, 17.
2. William H. Beveridge, 'The Place of the Social Sciences in Human Knowledge', *Politica*, 2 (Sept. 1937), 460, 476, 479.
3. Robbins, *Autobiography*, 141.
4. Governors, 14 May 1936, LSE 4/9.
5. William A. Robson (ed.), *Man and the Social Sciences* (London: LSE/George Allen & Unwin, 1972), pp. xii, xxxvi.
6. Shehadi interview, Fisher, LSE Archive, 1984.
7. Beveridge, 'Place of the Social Sciences', 465.
8. Michael Wise, 'Man in his Environment', in Robson, *Man and the Social Sciences*, 231.
9. Harris, 'The Webbs, the COS', 47.
10. Bryan Carsberg, 'New Directions: Accounting and Finance', *LSE Magazine* (June 1983), 10.
11. W. T. Baxter, 'Early Critics of Costing: LSE in the 1930s', in O. Finley Graves (ed.), *The Costing Heritage: Studies in Honor of S. Paul Garner* (Harrisonburg, Va.: Academy of Accounting Historians, 1991), 138.
12. *PI*, 176.
13. *DNB 1951–1960.*
14. *The Times*, 23 Jan. 1957.
15. Bowley to Beveridge, 16 May 1935, LSE SF Bowley.
16. Maurice Kendall, 'The Statistical Approach', *Economica*, NS 17 (May 1950), 127.
17. Robbins, *Autobiography*, 89.
18. Baxter, 'Early Critics', 138.

19. Michael Wise, 'Three Founder Members of the I.B.G.: R. Ogilvie Buchanan, Sir Dudley Stamp, S. W. Wooldridge, A Personal Tribute', *Transactions of the Institute of British Geographers*, ns 8 (1983), 42–3.
20. LSE OH Fisher, 1978.
21. H. R. G. Greaves, 'William Beveridge by José Harris', *LSE Magazine* (June 1978), 7.
22. Marshall to Hewins, 12 and 17 Oct. 1899, Hewins Papers, 44/226, 44/236.
23. A. W. Coats, 'Marshall and the Early Development of the LSE: Some Unpublished Letters', *Economica*, ns 34 (Nov. 1967), 416.
24. Kadish, *Oxford Economists*, 233.
25. Koot, 'Alternative to Marshall', 3–4.
26. Coats, 'Sociological Aspects', 711.
27. Ronald H. Coase, 'Economics at LSE in the 1930s: A Personal View', *Atlantic Economic Journal*, 10 (Mar. 1982), 34.
28. Robbins, *Autobiography*, 119.
29. Harry G. Johnson, 'Individual and Collective Choice', in Robson, *Man and the Social Sciences*, 13.
30. Hugh Dalton, *Call Back Yesterday: Memoirs 1887–1931* (London: Frederick Muller, 1953), 115.
31. Robbins, *Autobiography*, 126.
32. Coase, 'Economics at LSE', 32.
33. Brian McCormick, *Hayek and the Keynesian Avalanche* (Hemel Hempstead: Harvester Wheatsheaf, 1992), 35.
34. Robbins, *Autobiography*, 19, 20, 110.
35. Shehadi Interview, Hayek.
36. McCormick, *Hayek*, 24.
37. Johnson, 'Individual and Collective Choice', 15.
38. Eveline M. Burns, 'My LSE: 1916–26', *LSE Magazine* (Nov. 1983), 9.
39. Bartley/Lachman interview, 1984, in Shehadi collection, LSE Archive.
40. Beveridge, *LSE*, 94.
41. *PI*, 175.
42. Beveridge, *LSE*, 95.
43. Robbins, *Autobiography*, 129.
44. Shehadi interview, Hayek.
45. Shehadi interview, Kaldor, LSE Archive, early 1980s.
46. Robbins, *Autobiography*, 135.
47. A. W. Coats (ed.), *Economists in Government: An International Comparative Study* (Durham, NC: Duke University Press, 1981), 27.
48. Robert Skidelsky, *John Maynard Keynes*, ii. *The Economist as Saviour 1920–1937* (London: Macmillan, 1992), 377.
49. Ibid. 368.
50. Robbins, *Autobiography*, 153, 154.
51. Skidelsky, *Keynes*, ii. 456–9.
52. McCormick, *Hayek*, 172.
53. Shehadi interview, Kaldor.
54. Johnson, 'Individual and Collective Choice', 15–19.
55. Robbins, *Autobiography*, 133, 154.

56. Johnson, 'Individual and Collective Choice', 22.
57. Michael Newman, *Harold Laski: A Political Biography* (Basingstoke: Macmillan, 1993), 150.
58. Kingsley Martin, *Harold Laski 1893–1950: A Biographical Memoir* (London: Jonathan Cape, 1969), 12.
59. Peter J. O. Self, 'The State versus Man', in Robson, *Man and the Social Sciences*, 66.
60. Robert Dahl, *Modern Political Analysis* (Englewood Cliffs, NJ: Prentice-Hall, 1963), 6.
61. Self, 'State versus Man', 70.
62. André Kaiser, 'Politik als Wissenschaft: Zur Entstehung akademischer Politikwissenschaft in Großbritannien', manuscript of a paper given at the 'Idealismus und Positivismus' colloquium in Werner-Reimers-Stiftung, Bad Homburg, Jan. 1993, 26–7; copy in LSE History.
63. Harold J. Laski, 'On the Study of Politics', in Preston King (ed.), *The Study of Politics* (London: Frank Cass, 1977), 8.
64. Isaac Kramnick and Barry Sheerman, *Harold Laski: A Life on the Left* (London: Hamish Hamilton, 1993), 2.
65. Newman, *Laski*, 77, 87, 86.
66. *CMR* (Michaelmas Term 1950), 34.
67. Prof. Alan Stuart in conversation with the author, 1993.
68. Shehadi interview, Hayek.
69. Shehadi interview, Beales, LSE Archive, early 1980s.
70. Michael M. Postan, 'Time and Change', in Robson, *Man and the Social Sciences*, 31, 32, 37, 35.
71. Negley B. Harte (ed.), *The Study of Economic History: Collected Inaugural Lectures 1893–1970* (London: Frank Cass, 1971), p. xxv.
72. LSE OH Fisher.
73. Shehadi interview, Beales.
74. Robbins, *Autobiography*, 104.
75. LSE OH Fisher.
76. LSE OH Beales, 1973.
77. J. H. Clapham, 'Eileen Power, 1889–1940', *Economica*, NS 7 (Nov. 1940), 353–4.
78. Berg, 'First Women Economic Historians', 322.
79. Postan, 'Time and Change', 35.
80. Berg, 'First Women Economic Historians', 326, 324, 317.
81. *WD* iv. 360.
82. J. R. Williams, R. M. Titmuss, and F. J. Fisher, *R. H. Tawney: A Portrait by Several Hands* (London: Shenval Press, 1960), 3.
83. Harte, *Study of Economic History*, p. xxviii.
84. Williams *et al.*, *Tawney*, 4.
85. R. H. Tawney, *The Acquisitive Society* (London: G. Bell & Sons, 1921), 54; R. H. Tawney, *Equality* (London: George Allen & Unwin, 1964), 80, 56.
86. Williams *et al.*, *Tawney*, 28.
87. Dennis and Halsey, *English Ethical Socialism*, 211.
88. Ross Terrill, *R. H. Tawney and his Times: Socialism as Fellowship* (Cambridge, Mass.: Harvard University Press, 1973), 280.

89. A. H. Halsey, *Traditions of Social Policy: Essays in Honour of Violet Butler* (Oxford: Basil Blackwell, 1976), 257.
90. Terrill, *Tawney and his Times*, 110.
91. Ibid. 13.
92. R. H. Tawney, 'The Study of Economic History', in Harte, *Study of Economic History*, 104, 98, 96, 107.
93. K. Martin, *Father Figures*, 158.
94. Nehru, in Abse, *My LSE*, 25.
95. Kramnick and Sheerman, *Laski*, 251.
96. Dennis and Halsey, *English Ethical Socialism*, 159.
97. 'The Natural Bases of the Social Sciences', July 1925, LSE CF 222/B&C.
98. LSE OH Firth, 1988/9.
99. Helena Wayne, 'Bronislaw Malinowski: The Influence of Various Women on his Life and Works', *Journal of the Anthropological Society of Oxford*, 15 (Michaelmas 1984), 189.
100. Bronislaw Malinowski, *A Diary in the Strict Sense of the Term* (London: Athlone Press, 1967), 4.
101. Wayne, 'Malinowski', 200.
102. LSE OH Firth, 1988/9.
103. Edmund Leach, 'The Epistemological Background to Malinowski's Empiricism', in Raymond Firth (ed.), *Man and Culture: An Evaluation of the Work of Bronislaw Malinowski* (London: Routledge & Kegan Paul, 1957), 124.
104. Michael Young (ed.), *The Ethnography of Malinowski: The Trobriand Islands 1915–18* (London: Routledge & Kegan Paul, 1979), 9.
105. Bronislaw Malinowski, *A Scientific Theory of Culture and Other Essays* (Chapel Hill, NC: University of North Carolina Press, 1944), 150.
106. Edward Evans-Pritchard, 'Notes and Comments', in André Singer (ed.), *A History of Anthropological Thought* (New York: Basic Books, 1981), 198, 199.
107. Raymond Firth (ed.), *Man and Culture: An Evaluation of the Work of Bronislaw Malinowski* (London: Routledge & Kegan Paul, 1957), 1, 2.
108. Beatrice Blackwood Lecture given at Rhodes House, Oxford, 24 Oct. 1990; author's notes.
109. Evans-Pritchard, 'Notes and Comments', 197.
110. Edmund Leach, 'On the Founding Fathers', *Current Anthropology*, 7 (Dec. 1966), 565.
111. Firth, *Man and Culture*, 7–8.
112. Audrey Richards, 'Bronislaw Kaspar Malinowski', *Man* (Jan.–Feb. 1943), 3–4.
113. LSE OH Firth.
114. Ibid.
115. Malinowski to Beveridge, 22 Feb. 1936, Beveridge Papers Supplement, 2/1.
116. Edmund Leach, 'Models of Man', in Robson, *Man and the Social Sciences*, 1972, 167.
117. Robin Fox, *Encounter with Anthropology* (New York: Dell, 1975), 271.
118. PC 6 Feb. 1924, LSE 7/2.
119. Janet Beveridge, 'The Chair of Social Biology 1930–37', 1958, Caine Papers, BLPES.

120. Beveridge, *LSE*, 85.
121. Beveridge, 'Economics as a Liberal Education', 5.
122. 'Origin of Social Biology in the School of Economics', LSE CF 312/E.
123. Beveridge, *LSE*, 87.
124. LSE CF 312/E.
125. 'The Natural Bases of the Social Sciences', LSE CF 222/B&C.
126. Hobhouse to Beveridge, 31 Oct. 1926, LSE CF 101.
127. Mair to Beveridge, Beveridge Papers Supplement, 1/13.
128. Pantin to Mrs Mair, 19 Oct. 1929, LSE SF Hogben.
129. Watson to Beveridge, 4 Feb. 1930, LSE CF 237.
130. Beveridge to Plant, 24 Oct. 1929, LSE SF Hogben.
131. Plant to Beveridge, 15 Nov. 1929, ibid.
132. Hogben to Beveridge, 27 Jan. 1930, ibid.
133. Hogben to Beveridge, 14 May 1930, ibid.
134. *Economica*, Feb. 1931, 1–4.
135. Janet Beveridge, 'The Chair of Social Biology'.
136. Lancelot Hogben, 'The Foundations of Social Biology', *Economica*, 31 (Feb. 1931), 24.
137. Gary Werksey, *The Visible College: A Collective Biography of British Scientists and Socialists of the 1930s* (London: Free Association Books, 1988), 67.
138. LSE OH Brown, 1989.
139. Review quoted by publisher in Lancelot Hogben, *Science for the Citizen* (London: George Allen & Unwin, 1938).
140. Ibid. 10.
141. Blacker, *Eugenics*, 21.
142. Pauline M. H. Mazumdar, *Eugenics, Human Genetics and Human Failings: The Eugenics Society, its Sources and its Critics in Britain* (London: Routledge, 1992), 149.
143. Lancelot Hogben, 'Look Back with Laughter', unpublished autobiography, *c.*1974, collated and ed. G. P. Wells, Hogben Papers, University of Birmingham, A3–A20, 183.
144. 30 Dec. 1937, LSE SF Malinowski.
145. Eugene Grebenik, 'Demographic Research in Britain 1936–1986', in Michael Murphy and John Hobcraft (eds.), *Population Research in Britain*, a supplement to *Population Studies*, 45 (1991) (London: Population Investigation Committee at the LSE, 1991), 6.
146. Blacker, *Eugenics*, 22.
147. Grebenik, 'Demographic Research', 8.
148. Lancelot Hogben, *Political Arithmetic: A Symposium of Population Studies* (London: George Allen & Unwin, 1938), 41.
149. Hogben to Beveridge, 27 Oct. 1937, Beveridge Papers, V19.
150. Gunn office note, 4 Mar. 1931, RF, Folder 934.
151. Beveridge to Stamp, 30 Nov. 1936, Beveridge Papers, V17.
152. Van Sickle to Kittredge, 8 Feb. 1937, RF, Folder 944.
153. SC 25 Feb. 1937, LSE 6/18.
154. Beveridge, *LSE*, 93.
155. Hogben, *Political Arithmetic*, 29.

156. Donald MacRae, 'The Basis of Social Cohesion', in Robson, *Man and the Social Sciences*, 58.

Notes to Chapter 5

1. *CMR* (Michaelmas Term 1933), 5.
2. *PI* 239.
3. N. and J. MacKenzie, *First Fabians*, 406.
4. Newman, *Laski*, 166.
5. Sidney and Beatrice Webb, *Soviet Communism: A New Civilisation?* (London: Longmans, Green, 1944), 927.
6. *PI* 234.
7. Ibid. 184, 247.
8. Beveridge, *LSE*, 52.
9. 'Report of the Committee of Governors on Partisan Propaganda', Governors, 17 July 1923, LSE 4/6.
10. Beveridge, *LSE*, 53–4.
11. Gunn to E. E. Day, 13 Nov. 1931, RF, Folder 935.
12. *PI* 182.
13. Beveridge, *LSE*, 58.
14. Ibid. 54–5, 126.
15. Ibid. 44.
16. *CMR* (Summer Term 1934), 4, 9.
17. Harris, *Beveridge*, 300.
18. Beveridge, 'A Note on the Present Discontents', undated, LSE CF 117/6A.
19. *CMR* (Summer Term 1934), 11.
20. *Student Vanguard* (Feb. 1934), 22; Em.C. 5 Mar. 1934, LSE 6/16.
21. *CMR* (Summer Term 1934), 5, 7, 13–14; Em.C. 5 Mar., 15 Mar., and 22 Mar. 1934, LSE 6/16.
22. Em.C. 22 Mar. 1934, LSE CF 67/11/A.
23. F. W. Pethick-Lawrence to Sir Josiah Stamp, 19 Mar. 1934, ibid.
24. Beveridge, *LSE*, 43.
25. George H. Nash, *The Conservative Intellectual Movement in America since 1945* (New York: Basic Books, 1976), 172.
26. Beveridge, 'Present Discontents'.
27. 'Memorandum: EED, JVS conversation with Sir William Beveridge and Mrs Mair in London, May 22 1934', RF, Folder 939.
28. Kramnick and Sheerman, *Laski*, 323–4; Harris, *Beveridge*, 300.
29. Laski to Beveridge, 19 Apr. 1934, Beveridge Papers Supplement, 2/1.
30. Kramnick and Sheerman, *Laski*, 326–8.
31. *Daily Telegraph*, 12 July 1934; K. Martin, *Laski*, 87; *Daily Telegraph*, 13 July 1934; Harry Henry, 'Sir Ernest Gordon Graham Little, M.P.', *University of London Magazine* (Feb. 1937), 68; Kramnick and Sheerman, *Laski*, 329.
32. Em.C. 19 July, 27 Aug., 5 and 11 Oct. 1934, LSE 6/16.
33. Ibid. 19 and 27 July 1934.
34. *WD* iv. 329.
35. Harris, *Beveridge*, 302.
36. Kramnick and Sheerman, *Laski*, 329: 'Reflections on the "School of

Economics"', LSE CF 116/A,5; Newman, *Laski*, 173.

37. *PI* 234.
38. Robbins, *Autobiography*, 143–4.
39. 'Report of Committee as to German University Teachers', LSE CF 349/1.
40. PC, 10 June 1936, LSE CF 349/A.
41. William H. Beveridge, *A Defence of Free Learning* (London: Oxford University Press, 1959), 2–6; 'Report on the Academic Assistance Fund', PC 7 June 1939, LSE 349/B.
42. Beveridge, *Defence*, 6.
43. R. M. Cooper (ed.), *Refugee Scholars: Conversations with Tess Simpson* (Leeds: Moorland Books, 1992), 92.
44. Gottfried Niedhart, 'Gustav Mayers englische Jahre: Zum Exil eines deutschen Juden und Historikers', in *Exilforschung: Ein Internationales Jahrbuch*, vi. *Vertreibung der Wissenschaften und andere Themen* (Society for Exile Studies, 1988), 100.
45. 'Memorandum . . . May 22, 1934', RF, Folder 939.
46. Beveridge, *Defence*, 12.
47. Em.C. 8 June 1933, LSE 6/16.
48. Robbins, *Autobiography*, 139–40.
49. Shehadi interview, Hayek.
50. Harris, *Beveridge*, 297.
51. Martin Jay, *The Dialectical Imagination: A History of the Frankfurt School and the Institute of Social Research 1923–1950* (Boston: Little, Brown, 1973), 26, 36.
52. Em.C. 14 Feb. 1935, LSE 6/17; Beveridge Papers Supplement, 1/17; Governors, 9 May 1935; letters dated 24 June and 5 Nov. 1936, LSE 4/9.
53. Harris, *Beveridge*, 299.
54. 'German University Teachers', LSE CF 349/1; Beveridge, *Defence*, 3–4.
55. 20 Oct. 1933, Beveridge Papers Supplement, 1/16.
56. R. Edler, 'Learning Law Again', *LSE Magazine* (July 1953), 12.
57. Kittredge to John Van Sickle, 'Memorandum: Prof. Mannheim's Sociological Project', 21 Dec. 1934; Kittredge to Sydnor H. Walker, 16 Nov. 1939, Rockefeller Archive Center, 1.1, 401, Box 73, Folder 967.
58. LSE OH Scharf, 1989.
59. Jones, in Abse, *My LSE*, 34, 41.
60. LSE OH Seear, 1991; Jones, in Abse, *My LSE*, 36; Elizabeth Townsend, 'A Social Science Student 50 Years Ago, Part One', *LSE Magazine* (June 1979), 7; LSE OH Dasgupta, 1988; Elizabeth Townsend, 'A Social Science Student 50 Years Ago, Part Two', *LSE Magazine* (Nov. 1979), 7; LSE OH Seear.
61. Townsend, 'Social Science Student, pt. 1', 7; LSE OH Seear; *The Times*, 19 Aug. 1952.
62. LSE OH Dasgupta.
63. Edler, 'Learning Law Again', 12; LSE OH Seear.
64. Nell McGregor, 'A Student in the Thirties', *LSE Magazine* (May 1971), 13–14; Townsend, 'A Social Science Student, pt. 2', 8.
65. Dr W. M. Philip, 17 Apr. 1993, *The Times Magazine*.
66. Roland Bird, 'The Evening Student', *LSE Magazine* (July 1951), 7.
67. Mike Reddin, 'The Kennedys at LSE', correspondence with author, 10 Dec. 1991.

68. LSE OH Scharf.
69. Townsend, 'Social Science Student, pt. 2', 8.
70. David Hughes Parry, 'E.V.E.: An Appreciation', *LSE Magazine* (July 1954), 5; Kay Lewis, 'Eve Evans', *LSE Magazine* (Nov. 1971), 8.
71. Hughes Parry, 'E.V.E.', 5; DR 1953/4, 13; Lewis, 'Eve Evans', 8; Hughes Parry, 'E.V.E.', 5.
72. Hewins to E. J. Dodson, 6 Oct. 1896, LSE BF Dodson; E.T.R., 'Dodson', *CMR*, NS 3 (Lent Term 1923), 7.
73. Governors, 27 Feb. 1923, LSE 4/6.
74. H. C. Scriven, 'Walter Ernest Wilson', *LSE Magazine* (July 1957), 19; Eve E. Evans, 'Mr. W. Wilson, Clerk of Works: A Tribute on a Notable Occasion', *LSE Magazine* (Jan. 1956), 24.
75. LSE OH Panormo, 1974.
76. *CMR* (Michaelmas Term 1958).
77. LSE OH Brown.
78. *DNB 1941–1950.*
79. *PI* 248.
80. Kittredge to E. E. Day, 12 Nov. 1934, RF, Folder 941.
81. Walker to T. B. Kittredge, 7 Dec. 1934, ibid.
82. N. F. Hall to T. B. Kittredge, 13 Dec. 1934, ibid.
83. Sydnor H. Walker, 31 Dec. 1934, ibid.
84. Kit Jones, 'Fifty Years of Economic Research: A Brief History of the National Institute of Economic and Social Research 1938–88', *National Institute Economic Review*, 124 (May 1988), 37.
85. John Van Sickle, 14 Nov. 1935, RF, Folder 940.
86. K. Jones, 'Fifty Years', 37; Austin Robinson, 'The National Institute: The Early Years', *National Institute Economic Review*, 124 (May 1988), 63.
87. Day to Beveridge, 8 May 1935, and Beveridge to Max Mason, 16 May 1935, RF, Folder 942.
88. LSE, *Review*, 34.
89. Harold J. Laski, *The Dangers of Obedience and Other Essays* (New York: Harper, 1930), 173, 154.
90. Donald Fisher, 'Philanthropic Foundations and the Social Sciences: A Response to Martin Bulmer', *Sociology*, 18 (Nov. 1984), 580–1.
91. Salma P. Ahmad, 'American Foundations and the Development of the Social Sciences between the Wars: Comment on the Debate between Martin Bulmer and Donald Fisher', *Sociology*, 25 (Aug. 1991), 517.
92. John Van Sickle to E. E. Day, 19 Mar. 1934, RF, Folder, 939.
93. Sydnor H. Walker to T. B. Kittredge, 7 Dec. 1934, RF, Folder 941.
94. Robbins, 'A Note on the Director's Reflections', handed to Director at Constitution Sub-Committee, 13 Nov. 1935, LSE 17/5/2.
95. Beveridge, 'Reflections on the "School of Economics"', circulated at Constitution Sub-Committee, 30 Oct. 1935, ibid.
96. Revised Draft Report, Constitution Sub-Committee, Minutes, 6 Nov. 1936, ibid.
97. London School of Economics, *The Working Constitution and Practice of the London School of Economics and Political Science* (London: LSE, 1937), 24, 7.

98. Kittredge, 'Memorandum: Review of Activities of the LSE', RF, Folder 942; *PI* 250.
99. Beveridge to Mrs Rose Dunn-Gardner, 28 Feb. 1937, Beveridge Papers, II.b, 1936, pt. (i).
100. Sydnor H. Walker, 'Memorandum: Conversation, SHW with Sir Wm. H. Beveridge and Mrs J. Mair at the LSE, June 23, 1937', RF, Folder 944.
101. Apr. 1937, Passfield Papers, X.2(i)339.
102. Beveridge, *LSE* 113.
103. DR 1937/8, 17.
104. Walker, 'Memo: June 23, 1937'.
105. *PI* 252–4.
106. Harris, *Beveridge*, 364.
107. Ibid. 440–1, 387, 434.
108. *PI* 349.
109. Mrs Mair to Beveridge, 6 Apr. 1942, Beveridge Papers Supplement, 1/23.
110. Bayley, *The Times*, 25 Nov. 1992; Paul Addison, author's notes on University College, Oxford, 26 Nov. 1992; Harris, *Beveridge*, 471.
111. Bayley, *The Times*, 25 Nov. 1992.
112. Author's notes, 26 Nov. 1992.
113. Bayley, *The Times*, 25 Nov. 1992.

Notes to Chapter 6

1. *WD* iv. 393.
2. Hetherington to Selection Committee, 12 Apr. 1937, Beveridge Papers, II.b.36.
3. Beveridge to Carr-Saunders, 3 May 1937, ibid.
4. Laski, Robbins, Chorley, Plant to Carr-Saunders, 6 May 1937, Carr-Saunders Papers, BLPES, C/2.
5. *WD* iv. 392–3.
6. Kittredge office note, 16 Nov. 1937, RF, Folder 944.
7. Kittredge office notes to Sydnor Walker, 10 Feb. 1938, and S. M. Gunn, 30 Dec. 1938, RF, Folder 945.
8. *WD* iv. 394.
9. LSE application to Rockefeller, Dec. 1938, RF, Folder 945.
10. Kittredge to Sydnor Walker, 28 Feb. 1938, and Sydnor Walker to Carr-Saunders, 6 June 1938, RF, ibid.
11. Governing Body of Peterhouse College Minutes, 25 July 1939, LSE History Box, Reminiscences 1939–45.
12. Hayek, 'LSE', 27.
13. Carr-Saunders to Sir Hubert Sams, in reply to letter 30 Oct. 1945, LSE SF Carr-Saunders.
14. Roy Jenkins, *Portraits and Miniatures* (London: Macmillan, 1993), 143–4.
15. I. G. Patel, 'A Director's Initiation', *LSE Magazine* (Nov. 1984), 3; letter to S. Chapman, 8 Feb. 1993, LSE History Box, Reminiscences 1939–45.
16. Vera Anstey, 'L.S.E. Yesterday, Today and Tomorrow, Part One', *LSE Magazine* (Jan. 1951), 4.
17. B. Bond-Evans, 'The LSE in Cambridge 1941–44', LSE History Box, Reminiscences 1939–45.

18. Anstey, 'L.S.E. Yesterday, pt. 1', 4.
19. Robbins, *Autobiography*, 166–7.
20. Governors, 2 July 1942, LSE 4/10.
21. MacKenzie, in Abse, *My LSE*, 56.
22. Ibid. 50.
23. LSE OH Bohm 1, 1989.
24. LSE OH Griffith, 1989.
25. LSE OH Bohm 1.
26. Dr G. Brand letter to *LSE Magazine*, June 1983, 20.
27. MacKenzie, in Abse, *My LSE*, 49.
28. G.B.S. to Betty Evans, 2 Apr. 1943, LSE History Box, Reminiscences 1939–45.
29. MacKenzie, in Abse, *My LSE*, 49, 51.
30. Barbara Sternberg, 'The View from Colorado', *LSE Magazine* (Nov. 1982), 9.
31. MacKenzie, in Abse, *My LSE*, 53, 56.
32. Ibid. 60.
33. Seaborne Davies, 'To all the Boyz—Greetings', Dec. 1940, LSE R(SR) 1101.
34. J. Hendy letter to *LSE Magazine* (Nov. 1983), 23.
35. D. Hartley letter to *LSE Magazine* (June 1984), 27.
36. Arthur Quiller-Couch, 'Thoughts for Guests', *C.U. Conservative Review*, LSE CF 116/B.
37. Alec Cairncross and Nita Watts, *The Economic Section 1939–1961: A Study in Economic Advising* (London: Routledge, 1989), 29.
38. Ibid. 57.
39. Robbins, *Autobiography*, 176–7, 190.
40. Susan Howson and Donald Moggridge (eds.), *The Wartime Diaries of Lionel Robbins and James Meade 1943–45* (London: Macmillan, 1990), 49, 172.
41. Ibid. 61.
42. Cairncross and Watts, *Economic Section*, 55.
43. Robbins, *Autobiography*, 184–5.
44. Anstey, 'L.S.E. Yesterday, pt. 1', 3.
45. Sternberg, 'View from Colarado', 9.
46. DR 1943/4, 3.
47. GPC, 4 June 1941, LSE 9/2.
48. DR 1944/5, 4.
49. Anstey, 'L.S.E. Yesterday, pt. 2', 2.
50. Watkins, in Abse, *My LSE*, 66.
51. MacKenzie, ibid. 85.
52. Moody, ibid. 106.
53. Anstey, 'L.S.E. Yesterday, pt. 2', 4.
54. Robbins, *Autobiography*, 214, 213.
55. LSE OH Bohm 1.
56. School File, 'History of the LSE 1938–53', LSE CF 116/B.
57. Hayek, 'LSE', 1.
58. Stanley Godfrey correspondence, July 1988, LSE History Box, Reminiscences 1946–59.
59. S. Webb to Director, 2 May 1946, LSE CF 160/14.
60. Kramnick and Sheerman, *Laski*, 481.

61. Ibid. 268.
62. Kramnick and Sheerman, *Laski*, 486–7.
63. Ibid. 517.
64. Ibid. 516–43.
65. Ralph Miliband, 'Harold Laski: By Students and Colleagues', *CMR* (Michaelmas Term 1950), 38.
66. Michael Oakeshott, *Political Education* (Cambridge: Bowes & Bowes, 1951), 7, 22.
67. LSE SF Popper.
68. PC, 22 Oct. 1947, LSE 7/44.
69. Jean Floud in conversation with the author.
70. Robbins, *Autobiography*, 213.
71. LSE OH Halsey, 1988.
72. Colin Crouch and Anthony Heath (eds), *Social Research and Social Reform: Essays in Honour of A. H. Halsey* (Oxford: Clarendon Press, 1992), pp. vi, vii.
73. LSE OH Halsey.
74. A. H. Halsey, 'Provincials and Professionals: The British Post-War Sociologists', *Archives Européennes de Sociologie/European Journal of Sociology*, 23 (1982), 163.
75. Ibid. 152.
76. LSE OH Halsey.
77. Halsey, 'Provincials and Professionals', 166.
78. LSE OH MacRae 1, 1990.
79. LSE OH MacRae 2, 1991.
80. T. H. Marshall, 'Reminiscences', in *Link: LSE Department of Social Science and Administration 1912–1962 Jubilee* (London: LSE, 1962), 8.
81. Richard M. Titmuss, 'Time Remembered', ibid. 4.
82. Robert Pinker, 'Social Work and Social Policy in the Twentieth Century: Retrospect and Prospect', in Martin Bulmer, Jane Lewis, and David Piachaud (eds.), *The Goals of Social Policy* (London: Unwin Hyman, 1989), 95.
83. Kathleen Jones, *Eileen Younghusband: A Biography* (London: Bedford Square Press for the NCVO, 1984), 29.
84. Pinker, 'Social Work', 92.
85. K. Jones, *Eileen Younghusband*, 59.
86. Ibid. 66.
87. *DNB 1971–80.*
88. Margaret Gowing, 'Richard Morris Titmuss 1907–1973', *Proceedings of the British Academy*, 61 (1975), 427, 426.
89. Richard M. Titmuss, *Social Policy: An Introduction*, ed. Brian Abel-Smith and Kay Titmuss (London: George Allen & Unwin, 1974), 150.
90. Ernest Gellner, 'The LSE: A Contested Academy', *THES* (7 Nov. 1980), 13.
91. Nicholas Barr, 'The Phillips Machine', *LSE Quarterly*, 2 (Winter 1988), 308, 315.
92. Ibid. 322.
93. Ibid. 333.
94. *Punch*, 15 Apr. 1953, 456.
95. John Griffith, 'New Directions 8: Law at LSE', *LSE Magazine* (June 1979), 1.
96. McKenzie, in Abse, *My LSE*, 88.

97. Moody, ibid. 113.
98. Bohm, in London School of Economics Alumni Group, *LSE People 1947–1953* (London: LSE, 1987), 35.
99. McKenzie, in Abse, *My LSE*, 102, 93.
100. Moody, ibid. 117.
101. Wheldon, ibid. 138–9.
102. Crick, ibid. 148.
103. Bermant, ibid. 180, 187.
104. Crick, ibid. 158.
105. Minogue, ibid. 167, 170.
106. Kingsley, in LSE Alumni Group, *LSE People*, 32–3.
107. Ibid. 10.
108. Burgh, Ellerbeck, Manley, Wheeler, ibid. 21, 31, 22, 24, 33.
109. DR 1955/6, 10.
110. Carr-Saunders obituary, *The Times*, 8 Oct. 1966.
111. Edmund Carr-Saunders in conversation with the author, 1992.
112. A. M. Carr-Saunders, 'Britain and Universities in Africa', *Universities Quarterly*, 19 (June 1965), 234.
113. A. M. Carr-Saunders, 'English Universities Today', oration delivered at LSE, 11 Dec. 1959 (London: LSE, 1960), 15.
114. Robbins, *Autobiography*, 214.
115. Carr-Saunders to Burgh, 23 Mar. 1957, 'A Letter Re-read', *LSE Magazine* (Nov. 1982), 15.
116. LSE OH Bohm 1.
117. *The Times*, 8 Oct. 1966.

Notes to Chapter 7

1. Sir Sydney Caine, 'The Director Looks Back and Looks Forward', *LSE Magazine* (July 1967), 1.
2. Robbins, *Autobiography*, 270.
3. Caine, 'The Director Looks Back', 3.
4. Ernest Gellner, 'The Panther and the Dove: Reflections on Rebelliousness and its Milieux', in David Martin (ed.), *Anarchy and Culture: The Problem of the Contemporary University* (London: Routledge & Kegan Paul, 1969), 133.
5. Royal Institute of International Affairs, *Documents on International Affairs, 1960*, ed. Richart Gott, John Major, and Geoffrey Warner (London: Oxford University Press/RIIA, 1964), 345.
6. Daniel P. Moynihan, 'The United States in Opposition', *Commentary*, 59 (Mar. 1975), 31–3, 42–3.
7. Peter Strafford, *The Times*, 20 Mar. 1975.
8. Granville Eastwood, *Harold Laski* (London: Mowbray & Co., 1977), 94.
9. Tarlok Singh, 'Vera Anstey's Work for India', *LSE Magazine* (June 1977), 6.
10. Nossiter to S. Chapman, 6 Jan. 1994, LSE History Box, India.
11. Gowing, 'Titmuss', 422.
12. K. Hazareesingh, 'Mauritius LSE Society', LSE History Box, Overseas/International.

13. Emrys Jones, 'New Directions 1: Contemporary Movements in Geography', *LSE Magazine* (June 1976), 1–2.
14. F. J. Fisher, 'The Department of Economic History', *LSE Magazine* (Nov. 1971), 4.
15. Hedley Bull, 'Martin Wight and the Theory of International Relations: The Second Martin Wight Memorial Lecture', *British Journal of International Studies*, 2 (1976), 101.
16. F. J. Fisher, 'Department of Economic History', 3.
17. David Martin, 'The Dissolution of the Monasteries', in Martin, *Anarchy and Culture*, 1.
18. LSE OH Alcock 1, 1993.
19. Basil Yamey, 'Obituary: Professor Sir Ronald Edwards', *LSE Magazine* (June 1976), 8.
20. Arnold Plant, 'Universities and the Making of Businessmen', in Douglas Grant (ed.), *The University and Business* (Toronto: University of Toronto Press, 1958), 520; LSE OH Baxter, 1994.
21. Robbins to Caine, 8 Sept. 1963, LSE CF 782/46/1/A.
22. Coase, 'Economics at LSE', 34.
23. Caine to J. R. Stewart, 1 Aug. 1967, LSE CF 782/46/B.
24. Karl Popper, 'Autobiography of Karl Popper', in Paul A. Schilpp (ed.), *The Philosophy of Karl Popper*, ii (La Salle, Ill.: Open Court, 1974), 91.
25. Popper, conversation with author, 1989.
26. Popper, 'Autobiography', 92.
27. Popper, conversation with author, 1989.
28. Popper, 'Autobiography', 96.
29. B. Levin, *The Times*, 29 Sept. 1990.
30. John Watkins, 'The LSE "School" of Philosophy: A Polish Interrogation, an Interview with John Watkins by Ewa and Andrzej Chmielecka', *LSE Quarterly*, 1 (June 1987), 213.
31. Lakatos obituary, *The Times*, 8 Feb. 1974.
32. John Watkins, 'Imre Lakatos', in *International Encyclopaedia of the Social Sciences Biographical Supplement* (New York: Macmillan/Free Press, 1979), 402.
33. Paul Feyerabend, 'Imre Lakatos', *British Journal for the Philosophy of Science*, 26 (1975), 1.
34. Imre Lakatos, 'Popper on Demarcation and Induction', in Schilpp, *Philosophy of Karl Popper*, i. 260.
35. Karl Popper, 'Replies to my Critics', in Schilpp, *The Philosophy of Karl Popper*, ii. 999.
36. Watkins, 'LSE "School" of Philosophy', 214.
37. LSE OH Alcock 1.
38. Robbins, *Autobiography*, 275.
39. Ibid. 27.
40. Richard Layard and John King, 'Expansion since Robbins', in Martin, *Anarchy and Culture*, 13, 24.
41. Harry Kidd, *The Trouble at LSE 1966–67* (London: Oxford University Press, 1969), 2, 3, 5.

42. Layard and King, 'Expansion since Robbins', 31.
43. Robbins, *Autobiography*, 281.
44. DR 1963/4, 5.
45. John King and Richard Layard, 'The LSE as a Graduate School?', *Universities Quarterly*, 24 (Autumn 1970), 360, 365, 370–3.
46. 'LSE 1964–1981: An Outline of Academic Development', LSE History Box, Major Reports I.
47. Ernest Rudd, 'The Troubles of Graduate Students', in Martin, *Anarchy and Culture*, 68–9, 75–6.
48. Kidd to Borough of Croydon, 16 Oct. 1964, LSE CF 442/22/1/B.
49. 'Recent Changes in Croydon', undated, LSE History Box, Buildings II.
50. Kidd to Claus Moser, 26 Mar. 1965, LSE CF 442/22/1/B.
51. David Kingsley, 'Porters Opinion', *LSE Magazine* (Dec. 1969), 4.
52. Sir Sydney Caine, 'Harry Kidd', *LSE Magazine* (July 1967), 5.
53. LSE OH Alcock 1.
54. Negley B. Harte, *The University of London 1836–1986* (London: Athlone Press, 1986), 260.
55. DR 1963/4, 5.
56. John Ashworth, 'The Case for a "Grand École"', *Lucifer: Magazine for London University*, 1 (Easter 1992), pp. vii.
57. DR 1965/6, 4.
58. DR 1982/3, p. iv.
59. Kidd, *Trouble at LSE*, 6.
60. Bridge to Caine, 19 Jan. 1966, LSE CF 22/33/A.
61. Gellner, 'Panther and Dove', 146.
62. Governors, 11 Dec. 1963, LSE 4/16.
63. *Observer*, 6 Feb. 1966.
64. Carr-Saunders to L. M. N. Hodson, 17 June 1954, LSE SF Adams.
65. Alex Finer, 'Academic Freedom—in Chains', source and date unknown, LSE CF 453/7/2.
66. London School of Economics Students, *LSE's New Director: A Report on Walter Adams* (London: Agitator Pamphlet, 1966), 18.
67. *The Times*, 23 Oct. and 4 Nov. 1966.
68. Kidd, *Trouble at LSE*, 19.
69. Confidential Note on AB, 2 Nov. 1966, LSE CF 453/7/2.
70. 'A Note on the Adams Affair', LSE CF 453/7/2.
71. Kidd, *Trouble at LSE*, 41.
72. Colin Crouch, *The Student Revolt* (London: Bodley Head, 1970), 47.
73. Kidd, *Trouble at LSE*, 45–6, 49.
74. Ibid. 50.
75. Crouch, *Student Revolt*, 49.
76. Kidd, *Trouble at LSE*, 52–3.
77. Crouch, *Student Revolt*, 49.
78. Kidd, *Trouble at LSE*, 53.
79. Crouch, *Student Revolt*, 50.
80. Kidd, *Trouble at LSE*, 57, 59–60.
81. Ibid. 65–6.

82. LSE OH Griffith, 1989.
83. Kidd, *Trouble at LSE*, 73.
84. Tessa Blackstone, Kathleen Gales, Roger Hadley, and Wyn Davies, *Students in Conflict: LSE in 1967*, LSE Research Monograph (London: Weidenfeld & Nicolson, 1970), 1, 163–4.
85. LSE OH Griffith.
86. *Times Educational Supplement*, 24 Mar. 1967.
87. Crouch, *Student Revolt*, 61.
88. *Beaver*, 18 Jan. 1968.
89. Kidd to Caine, 1 June 1967, Adams Papers, BLPES, 7/4.
90. Draft paper on division of secretaryship, LSE SF Adams.
91. Crouch, *Student Revolt*, 65.
92. Caine, 'The Director Looks Back', 3.
93. Lord Bridges, 'Sir Sydney Caine', *LSE Magazine* (July 1967), 4.
94. William Pickles, 'Sir Sydney Caine: An Appreciation', *LSE Magazine* (July 1967), 4–5.
95. Walter Adams, 'The New Director Speaks', *LSE Magazine* (July 1967), 7.
96. Crouch, *Student Revolt*, 67.
97. Editorial, *LSE Magazine* (June 1969).
98. Crouch, *Student Revolt*, 72–5.
99. Ibid. 75.
100. Paul Hoch and Vic Schoenbach, *The Natives are Restless* (London: Sheed & Ward, 1969), 21, 23.
101. Roberts *aide-mémoire* in LSE OH collection, 1994.
102. Crouch, *Student Revolt*, 78.
103. Hoch and Schoenbach, *Natives*, 32.
104. Ibid. 37–8.
105. Ken Minogue, 'Student Militancy at the London School of Economics', unpublished memorandum, Apr. 1969, LSE History, 29.
106. LSE OH Bohm 1, 1989.
107. Hoch and Schoenbach, *Natives*, 55.
108. Crouch, *Student Revolt*, 84–5.
109. Musgrave, 1990, in LSE OH.
110. Hoch and Schoenbach, *Natives*, 9, 6, 65.
111. LSE OH Musgrave.
112. Hoch and Schoenbach, *Natives*, 65.
113. LSE OH Alcock 1.
114. Hansard Official Report, 5th ser., vol. 776, 1968/9.
115. SC 6 May 1969.
116. Judgment of Blackburn Appellate Tribunal, LSE History Box, Troubles.
117. Crouch, *Student Revolt*, 94, 96.
118. *Sunday Times*, 26 Jan. 1969.
119. DR 1968/9, 66.
120. Walter Adams, 'LSE and the New Militancy', *British Universities Annual* (1969), 108–9.
121. Hoch and Schoenbach, *Natives*, 193, 198–9, 194.
122. LSE OH Griffith.

123. Ronald Higgins, 'Troubled Diary: Voices Past and Present', *LSE Magazine* (June 1969), 8.
124. Crouch, *Student Revolt*, 223, 169–70.
125. LSE OH Firth, 1988/9.
126. Gellner, 'Panther and Dove', 140, 137.
127. LSE OH Brown, 1989.
128. Theo Richmond, 'Troubled Diary: Voices Past and Present', *LSE Magazine* (June 1969), 9.
129. DR 1973/4, p. vii.
130. DR 1965/6, 7.
131. London School of Economics, *LSE: The New Library* (London: LSE, 1978), 27.
132. Woledge, 'BLPES', 199.
133. LSE, *New Library*, 4.
134. Crouch, in Abse, *My LSE*, 205.
135. LSE OH Stuart, 1989.
136. Russell *et al., Changing Course*, 289, 295, 279.
137. P. Hennessy, *THES*, 27 Sept. 1974.
138. Walter Adams, 'A Letter from Sir Walter Adams', *LSE Magazine* (Nov. 1974), 2.

Notes to Chapter 8

1. Peter Adams, correspondence with author, 22 May 1975.
2. DR 1974/5, p. vi.
3. Ibid., p. x.
4. Dahrendorf, 'A Centre for Economic and Political Studies in London, Jan. 1976, LSE History Box, British Brookings Debate.
5. Caine, Abel-Smith, and Marshall, in correspondence with author, Jan.–Feb. 1976, ibid.
6. Dahrendorf, 'But Alas! Not Everybody is Equally Noble: Reply to the International Socialists' Pamphlet, The Noblest Director of Them All', ibid.
7. DR 1982/3, p. viii.
8. George Jones, correspondence with author, 1983.
9. DR 1975/6, pp. vi–viii.
10. Dahrendorf, circular to all members of the School, 6 Nov. 1979, as appendix to 'First Steps in a New Situation', AB 5 Dec. 1979.
11. DR 1980/1, p. ix.
12. DR 1985/6, 1.
13. DR 1984/5, p. vii.
14. Ibid., p. x.
15. DR 1987/8, in LSE, *Calendar 1989/90*, 79.
16. DR 1985/6, 4.
17. Patel, 'LSE into the 1990s', attached to DR 1989/90.
18. DR 1985/6, 12.
19. DR 1984/5, p. vi.
20. DR 1987/8, in LSE, *Calendar 1989/90*, 62, DR 1985/6, 1.
21. Patel, 'The Director Writes', *LSE Magazine* (June 1987), 3.

22. DR 1986/7, p. xix.
23. Annual Report 1988/9, 5.
24. DR 1987/8, 5.
25. Oakeshott, *Political Education*, 22.
26. Patel, 'LSE into the 1990s'.
27. 'Goodbye to All That', *Beaver*, 12 Mar. 1990, 8.
28. Ashworth, 'The LSE: A 2020 Vision', delivered 12 Feb. 1991, LSE History Box, Policy Statements.
29. AB 26 May 1965.
30. Noel Annan, *Our Age: Portrait of a Generation* (London: Weidenfeld & Nicolson, 1990), 3.
31. Gellner, 'LSE: Contested Academy', 12–13.
32. *Guardian*, 17 and 30 Mar. 1990.
33. DR 1973/4, p. vii.

BIBLIOGRAPHY

Manuscript collections consulted include:

Adams Papers, BLPES
Beveridge Papers, BLPES
Caine Papers, BLPES
Carr-Saunders Papers, BLPES
Hewins Papers, University of Sheffield
Hogben Papers, University of Birmingham
Mackinder Papers, School of Geography, University of Oxford
Reeves Papers, Alexander Turnbull Library, Wellington, NZ
Rockefeller Archive Center, New York
Shehadi Interviews with 1930s Economists, LSE Archive, BLPES
Steel-Maitland Papers, Scottish Record Office, Edinburgh

The following abbreviations have been used:

AB	Academic Board, LSE
BLPES	British Library of Political and Economic Science
CMR	*Clare Market Review*
COS	Charity Organisation Society
DNB	*Dictionary of National Biography*
DR	Director's Report, LSE
Em.C.	Emergency Committee, LSE
Governors	Court of Governors, LSE
GPC	General Purposes Committee, LSE
LSE	(in source notes) LSE Archive, BLPES
LSE BF	LSE History Biographical File, LSE Archive
LSE CF	LSE Central Filing, LSE Archive
LSE History	LSE History, LSE Archive, BLPES
LSE OH	LSE Oral History, BLPES
LSE SF	LSE Staff File
LSRM	Rockefeller Archive Center, 3.6, Box 3/55, on Laura Spelman Rockefeller Memorial Fund
OP	B. Webb, *Our Partnership*
PC	Professorial Council, LSE
PI	William H. Beveridge, *Power and Influence*
Rem.	LSE, *Reminiscences 1922–1947*
RF	Rockefeller Archive Center, 1.1, 401, Box 71
SC	Standing Committee, LSE
THES	*Times Higher Education Supplement*
UGC	University Grants Committee
WD ii–iv	B. Webb, *Diary*, vols. ii–iv
WL	B. and S. Webb, *Letters*

ABRAMS, PHILIP, *The Origins of British Sociology* (Chicago: University of Chicago Press, 1968).
ABSE, JOAN (ed.), *My LSE* (London: Robson Books, 1977).
ADAMS, WALTER, 'The New Director Speaks', *LSE Magazine* (July 1967).
—— 'LSE and the New Militancy', *British Universities Annual* (1969), 103–11.
—— 'A Letter from Sir Walter Adams', *LSE Magazine* (Nov. 1974), pp. ii–2.
AHMAD, SALMA P., 'Institutions and the Growth of Knowledge: The Rockefeller Foundations' Influence on the Social Sciences between the Wars', Ph.D. thesis (University of Manchester, 1987).
—— 'American Foundations and the Development of the Social Sciences between the Wars: Comment on the Debate between Martin Bulmer and Donald Fisher', *Sociology*, 25 (Aug. 1991), 511–20.
AMERY, JULIAN, *Joseph Chamberlain and the Tariff Reform Campaign: The Life of Chamberlain, vi. 1903–1968* (London: Macmillan, 1969).
AMERY, L. S., *My Political Life, i. England before the Storm 1896–1914* (London: Hutchinson, 1953).
ANNAN, NOEL, 'The Curious Strength of Positivism in English Political Thought', L. T. Hobhouse Memorial Trust Lecture 28, delivered on 7 May 1958 (London: Oxford University Press, 1959).
—— *Our Age: Portrait of a Generation* (London: Weidenfeld & Nicolson, 1990).
ANSTEY, VERA, 'L.S.E. Yesterday, Today and Tomorrow, Part One', *LSE Magazine* (Jan. 1951), 2–4.
—— 'L.S.E. Yesterday, Today and Tomorrow, Part Two', *LSE Magazine* (July 1951), 2–5.
—— *The Economic Development of India* (4th edn.; London: Longmans, Green, 1952).
—— *An Introduction to Economics: For Students in India and Pakistan* (London: George Allen & Unwin, 1964).
ASHBY, ERIC, and ANDERSON, MARY, *Portrait of Haldane at Work on Education* (London: Macmillan, 1974).
ASHLEY, W. J., 'A Survey of the Past History and Present Position of Political Economy: Leicester, 1907', in R. L. Smyth (ed.), *Essays in Economic Method: Selected Papers Read to Section F of the British Association for the Advancement of Science, 1860–1913* (London: Duckworth, 1962).
ASHWORTH, JOHN, 'The Case for a "Grand École"', *Lucifer: Magazine for London University*, 1 (Easter 1992), pp. vii–viii.
BARKER, ERNEST, 'Leonard Trelawny Hobhouse 1864–1929', *Proceedings of the British Academy*, 15 (1929), 536–54.
BARR, NICHOLAS, 'The Phillips Machine', *LSE Quarterly*, 2 (Winter 1988), 305–37.
BASTABLE, C. F., 'Presidential Address to Section F, Economic Science and Statistics', *Report* of the British Association for the Advancement of Science Meeting at Oxford, 1894 (London: John Murray, 1894).
BAXTER, W. T., 'Early Critics of Costing: LSE in the 1930s', in O. Finley Graves (ed.), *The Costing Heritage: Studies in honor of S. Paul Garner* (Harrisonburg, Va.: Academy of Accounting Historians, 1991).
BENTWICH, NORMAN, *The Rescue and Achievement of Refugee Scholars: The Story of Displaced Scholars and Scientists 1933–1952* (The Hague: Martinus Nijhoff, 1953).

BERG, MAXINE, 'The First Women Economic Historians', *Economic History Review*, 45 (May 1992), 308–29.

BERLIN, ISAIAH, *The Hedgehog and the Fox: An Essay on Tolstoy's View of History* (London: Weidenfeld & Nicolson, 1953).

BEVERIDGE, JANET, 'In Retrospect', *LSE Magazine* (July 1951), 8–9.

—— *Beveridge and his Plan* (London: Hodder & Stoughton, 1954).

—— *An Epic of Clare Market: Birth and Early Days of the London School of Economics* (London: G. Bell & Sons, 1960).

BEVERIDGE, WILLIAM H., *John and Irene: An Anthology of Thoughts on Women* (London: Longmans, Green, 1912).

—— 'Economics as a Liberal Education', *Economica*, 1 (Jan. 1921), 2–19.

—— 'The Physical Relation of a University to a City', lecture delivered to the London Society, 16 Nov. 1928, LSE, R(Coll.) Misc. 237 I25.

—— 'The Place of the Social Sciences in Human Knowledge', *Politica*, 2 (Sept. 1937), 459–79.

—— *Power and Influence* (London: Hodder & Stoughton, 1953).

—— *A Defence of Free Learning* (London: Oxford University Press, 1959).

—— *The London School of Economics and its Problems 1919–1937* (London: George Allen & Unwin, 1960).

(Beveridge Oration speeches are published as the annual Director's Reports on the Work of the School.)

BIRD, ROLAND, 'The Evening Student', *LSE Magazine* (July 1951), 7.

BLACKER, C. P., *Eugenics in Prospect and Retrospect* (London: Hamish Hamilton Medical Books, 1945).

BLACKSTONE, TESSA, GALES, KATHLEEN, HADLEY, ROGER, and DAVIES, WYN, *Students in Conflict: LSE in 1967*, LSE Research Monograph (London: Weidenfeld & Nicolson, 1970).

BLOUET, BRIAN W., *Halford Mackinder: A Biography* (College Station, Tex.: Texas A & M University Press, 1987).

BOARDMAN, PHILIP, *The Worlds of Patrick Geddes: Biologist, Town Planner, Re-educator, Peace-warrior* (London: Routledge & Kegan Paul, 1978).

Board of Education, *Reports from those Universities and University Colleges in Great Britain which Participated in the Parliamentary Grant for University Colleges 1908–9*, British Parliamentary Papers, vol. 24, 1910.

BONN, MORITZ H., *Wandering Scholar* (New York: John Day Company, 1948).

BRIDGES, LORD, 'Sir Sydney Caine', *LSE Magazine* (July 1967), 3–4.

British Library of Political and Economic Science (BLPES), *A London Bibliography of Social Science*: being the subject catalogue of BLPES and Goldsmiths, Library of the Royal Statistical Society and the Royal Anthropological Society etc. and University College London, compiled by B. M. Headicar and C. Fuller. 4 vols., 1931; 47 supplements including material from 1929–89.

BROWN, E. H. (ed.), *Geography Yesterday and Tomorrow* (Oxford: Oxford University Press, 1980).

BULL, HEDLEY, 'Martin Wight and the Theory of International Relations: The Second Martin Wight Memorial Lecture', *British Journal of International Studies*, 2 (1976), 101–16.

BULMER, MARTIN, and BULMER, JOAN, 'Philanthropy and Social Science in the 1920s: Beardsley Ruml and the Laura Spelman Rockefeller Memorial, 1922–29', *Minerva*, 19 (Autumn 1981), 347–407.

BURNS, EVELINE M., 'My LSE: 1916–26', *LSE Magazine* (Nov. 1983), 8–9.

CAINE, SYDNEY, *The History of the Foundation of the London School of Economics and Political Science* (London: G. Bell & Sons, 1963).

—— 'The Director Looks Back and Looks Forward', *LSE Magazine* (July 1967), 1–4.

—— 'Harry Kidd', *LSE Magazine* (July 1967), 5.

—— *British Universities: Purpose and Prospects* (London: Bodley Head, 1969).

CAIRNCROSS, ALEC, and WATTS, NITA, *The Economic Section 1939–1961: A Study in Economic Advising* (London: Routledge, 1989).

CANNAN, EDWIN, *History of Local Rates* (2nd edn.; London: P. S. King & Son, 1912).

CANTOR, LEONARD M., 'Halford Mackinder: His Contribution to Geography and Education', MA thesis (University of London, 1960).

CARR-SAUNDERS, A. M., *World Population: Past Growth and Present Trends* (Oxford: Clarendon Press, 1936).

—— 'English Universities Today', oration delivered at LSE, 11 Dec. 1959 (London: LSE, 1960).

—— 'Britain and Universities in Africa', *Universities Quarterly*, 19 (June 1965), 227–39.

—— and WILSON, P. A., *The Professions* (Oxford: Clarendon Press, 1933).

CARSBERG, BRYAN, 'New Directions: Accounting and Finance', *LSE Magazine* (June 1983), 10–11.

CHESTERTON, G. K., *Autobiography* (London: Hutchinson, 1936).

CHORLEY, LORD, 'Beveridge and the LSE, Part One', *LSE Magazine* (Nov. 1972), 5–10.

—— 'Beveridge and the LSE, Part Two', *LSE Magazine* (June 1973), 5–8.

CHURCHILL, WINSTON, *Great Contemporaries* (London: Butterworth, 1937).

CLAPHAM, J. H., 'Eileen Power, 1889–1940', *Economica*, NS, 7 (Nov. 1940), 351–9.

CLARKE, D. A., '"A New Laboratory of Sociological Research": The BLPES', *Journal of Documentation*, 40 (June 1984), 152–7.

CLARKE, PETER, *Liberals and Social Democrats* (Cambridge: Cambridge University Press, 1978).

CLOGG, RICHARD, 'Politics and the Academy', *Middle Eastern Studies*, 21 (Oct. 1985), 1–117.

COASE, RONALD. H., 'Economics at LSE in the 1930s: A Personal View', *Atlantic Economic Journal*, 10 (Mar. 1982), 31–4.

COATS, A. W., 'Sociological Aspects of British Economic Thought (ca. 1880–1930)', *Journal of Political Economy*, 75 (Oct. 1967), 706–29.

—— 'Marshall and the Early Development of the LSE: Some Unpublished Letters', *Economica*, NS 34 (Nov. 1967), 408–17.

COATS, A. W., (ed.), *Economists in Government: An International Comparative Study* (Durham, NC: Duke University Press, 1981).

Coefficients, *Minutes of the Club 1902–06*, LSE, R.(Coll.) Assoc. 17.

COLLINI, STEFAN, *Liberalism and Sociology: L. T. Hobhouse and Political Argument in England 1880–1914* (Cambridge: Cambridge University Press, 1979).

COOPER, R. M. (ed.), *Refugee Scholars: Conversations with Tess Simpson* (Leeds: Moorland Books, 1992).

COREN, MICHAEL, *The Invisible Man: The Life and Liberties of H. G. Wells* (London: Bloomsbury, 1993).

CROUCH, COLIN, *The Student Revolt* (London: Bodley Head, 1970).

—— and HEATH, ANTHONY (eds.), *Social Research and Social Reform: Essays in Honour of A. H. Halsey* (Oxford: Clarendon Press, 1992).

CUNNINGHAM, W. *et al.*, 'Methods of Economic Training in This and Other Countries, Report of the Committee', in *Report* of the British Association for the Advancement of Science Meeting at Oxford, 1894 (London: John Murray, 1894).

DAHL, ROBERT, *Modern Political Analysis* (Englewood Cliffs, NJ: Prentice-Hall, 1963).

DALTON, HUGH, *Call Back Yesterday: Memoirs 1887–1931* (London: Frederick Muller, 1953).

—— *The Political Diary of Hugh Dalton, 1918–40, 1945–60*, ed. Ben Pimlott (London: Jonathan Cape/LSE, 1986).

DAVIS, PATRICK, 'The Troubles: A Chronology with Observations', *LSE Magazine*, (July 1967), 8–19.

—— 'LSE's Coat of Arms', *LSE Magazine* (June 1977), 10–11.

—— and Kingsley, David, 'Troubled Diary', *LSE Magazine* (June 1969), 1–2.

DENNIS, NORMAN and HALSEY, A. H. (eds.), *English Ethical Socialism: Thomas More to R. H. Tawney* (Oxford: Clarendon Press, 1989).

DONNISON, DAVID, 'Taking Decisions in a University' in David Donnison *et al.*, *Social Policy and Administration Revisited* (rev. edn.; London: George Allen & Unwin, 1975).

E.T.R., 'Dodson', *Clare Market Review*, NS 3 (Lent Term 1923), 7.

EASTWOOD, GRANVILLE, *Harold Laski* (London: Mowbray & Co., 1977).

EDLER, R., 'Learning Law Again', *LSE Magazine* (July 1953), 12.

ENSOR, R. C. K., *England 1870–1914* (Oxford: Clarendon Press, 1936).

EVANS, EVE V., 'Mr. W. Wilson, Clerk of Works: A Tribute on a Notable Occasion', *LSE Magazine* (Jan. 1956), 24, 30.

EVANS-PRITCHARD, EDWARD, 'Notes and Comments' in André Singer (ed.), *A History of Anthropological Thought* (New York: Basic Books, 1981).

FEYERABEND, PAUL, 'Imre Lakatos', *British Journal for the Philosophy of Science*, 26 (1975), 1–18.

FIRTH, RAYMOND (ed.), *Man and Culture: An Evaluation of the Work of Bronislaw Malinowski* (London: Routledge & Kegan Paul, 1957).

—— 'Malinowski in the History of Social Anthropology', in Roy Ellen, Ernest Gellner, Gräzyna Kubica and Janusz Mucha (eds.), *Malinowski Between Two Worlds: The Polish Roots of an Anthropological Tradition* (Cambridge: Cambridge University Press, 1988).

FISHER, DONALD, 'The Impact of American Foundations on the Development of British University Education, 1900–1939', Ph.D. thesis (University of California, Berkeley, 1977).

—— 'Philanthropic Foundations and the Social Sciences: A Response to Martin Bulmer', *Sociology*, 18 (Nov. 1984), 580–7.

FISHER, F. J., 'The Department of Economic History', *LSE Magazine* (Nov. 1971), 3–4.

FOX, ROBIN, *Encounter with Anthropology* (New York: Dell, 1975).

FREEMAN, T. W., 'The Royal Geographical Society and the Development of Geography' in E. H. Brown (ed.), *Geography Yesterday and Tomorrow* (Oxford: Oxford University Press, 1980).

FREMANTLE, ANNE, *This Little Band of Prophets: The Story of the Gentle Fabians* (London: George Allen & Unwin, 1960).

GELLNER, ERNEST, 'The Panther and the Dove: Reflections on Rebelliousness and its Milieux', in David Martin (ed.), *Anarchy and Culture: The Problem of the Contemporary University* (London: Routledge & Kegan Paul, 1969).

—— 'The LSE: A Contested Academy', *Times Higher Educational Supplement* (7 Nov. 1980), 12–13.

GILBERT, EDMUND W., 'Seven Lamps of Geography: An Appreciation of the Teaching of Sir Halford J. Mackinder', *Geography*, 36 (1951), 21–43.

—— *Sir Halford Mackinder 1861–1947: An Appreciation of his Life and Work* (London: LSE/G. Bell & Sons, 1961).

GINSBERG, MORRIS, 'The Life and Work of Edward Westermarck', *Sociological Review*, 32 (Jan.–Apr. 1940), 1–28.

GLENNERSTER, HOWARD, with the assistance of Anthea Bennett and Christine Farrell, *Graduate School: A Study of Graduate Work at the London School of Economics* (London: Oliver & Boyd, 1966).

GOFFMAN, ERVING, *Asylums: Essays on the Social Situation of Mental Patients and Other Inmates* (New York: Anchor Books, 1961).

GOWING, MARGARET, 'Richard Morris Titmuss 1907–1973', *Proceedings of the British Academy*, 61 (1975), 401–28.

GREAVES, H. R. G., 'William Beveridge by José Harris', *LSE Magazine* (June 1978), 7.

GREBENIK, EUGENE, 'Demographic Research in Britain 1936–1986', in Michael Murphy and John Hobcraft (eds.), *Population Research in Britain*, a supplement to *Population Studies*, 45 (1991) (London: Population Investigation Committee at the LSE, 1991), 3–30.

Gresham University Commission (Gresham Report), 'The Report of the Commissions appointed to consider the Draft Charter for the Proposed Gresham University in London', British Parliamentary Papers, 1893/4, 31, 24 Jan. 1894.

GRIFFITH, JOHN, 'New Directions 8: Law at LSE', *LSE Magazine* (June 1979), 1.

GROUSE, LIONEL H., 'The Pullets Up Above', *Clare Market Review* (Lent 1949), 32–4.

HADLEY, W., *A Brief History of St Clement Danes Grammar School* (London: St Clement Danes School, 1951).

HALDANE, R. B., 'The Civic University', in *Selected Addresses and Essays* (London: John Murray, 1928), 126–62.

HALDANE, R. B., *An Autobiography* (London: Hodder & Stoughton, 1929).

HALLIDAY, R. J., 'The Sociological Movement, Society and Genesis of Academic Sociology in Britain', *Sociological Review*, 16 (Nov. 1968), 377–98.

HALSEY, A. H., *Traditions of Social Policy: Essays in Honour of Violet Butler* (Oxford: Basil Blackwell, 1976).

—— 'Provincials and Professionals: The British Post-War Sociologists', *Archives Européennes de Sociologie/European Journal of Sociology*, 23 (1982), 150–75.

HARRIS, J., *William Beveridge: A Biography* (Oxford: Clarendon Press, 1977).

—— 'The Webbs, the COS, and Ratan Tata', in Martin Bulmer, Jane Lewis and David Piachaud (eds.), *The Goals of Social Policy* (London: Unwin Hyman, 1989).

HARTE, NEGLEY B. (ed.), *The Study of Economic History: Collected Inaugural Lectures 1893–1970* (London: Frank Cass, 1971).

—— *The University of London 1836–1986* (London: Athlone Press, 1986).

HAYEK, F. A., 'Reflections on the Pure Theory of Money of Mr J. M. Keynes', *Economica* (Aug. 1931), 270–95.

—— 'Reflections on the Pure Theory of Money of Mr J. M. Keynes Continued', *Economica* (Feb. 1932), 22–44.

—— 'The London School of Economics 1895–1945', *Economica* (Feb. 1946), 1–31.

HEILBRONER, ROBERT L., *The Worldly Philosophers: The Lives, Times and Ideas of the Great Economic Thinkers* (5th edn.; Harmondsworth: Penguin Books, 1983).

HENRY, HARRY, 'Sir Ernest Gordon Graham Little, M.P.', *University of London Union Magazine* (Feb. 1937), 66–71.

HEWINS, W. A. S., *The Apologia of an Imperialist: Forty Years of Empire Policy*, i (London: Constable , 1929).

HIGGINS, RONALD, 'Troubled Diary: Voices Past and Present', *LSE Magazine* (June 1969), 8.

HOBHOUSE, HERMIONE, *Lost London: A Century of Demolition and Decay* (London: Macmillan, 1971).

HOBHOUSE, L. T., 'The Roots of Modern Sociology', in *Inauguration of the Martin White Professorships of Sociology, Dec. 17, 1907* (London: John Murray, 1908).

—— *Morals in Evolution: A Study in Comparative Ethics* (7th edn.; London: Chapman & Hall, 1951).

HOCH, PAUL, and SCHOENBACH, VIC, *The Natives are Restless* (London: Sheed & Ward, 1969).

HOGBEN, LANCELOT, 'The Foundations of Social Biology', *Economica*, 31 (Feb. 1931), 1–24.

—— *Mathematics for the Million: A Popular Self-Educator* (London: George Allen & Unwin, 1936).

—— *Political Arithmetic: A Symposium of Population Studies* (London: George Allen & Unwin, 1938).

—— *Science for the Citizen* (London: George Allen & Unwin, 1938).

—— 'Look Back with Laughter', unpublished autobiography *c.*1974, collated and edited by G. P. Wells, held in the Hogben Papers in the Library of the University of Birmingham.

HOLROYD, MICHAEL, *Bernard Shaw, i. 1856–1898: The Search for Love* (London: Chatto & Windus, 1988).

—— *Bernard Shaw, iv. 1950–51 The Last Laugh* (London: Chatto & Windus, 1992).

HOWSON, SUSAN, and MOGGRIDGE, DONALD (eds.), *The Wartime Diaries of Lionel Robbins and James Meade, 1943–45* (London: Macmillan, 1990).

HUGHES PARRY, DAVID, 'E.V.E.: An Appreciation', *LSE Magazine* (July 1954), 5–6.

International Biographical Dictionary of Central European Émigrés 1933–45, vols. i–iii, ed. Herbert A. Strauss and Werner Roder (Munich: K. G. Saur, 1980/1983).

JAY, MARTIN, *The Dialectical Imagination: A History of the Frankfurt School and the Institute of Social Research 1923–1950* (Boston: Little, Brown, 1973).

JENKINS, ROY, *Portraits and Miniatures* (London: Macmillan, 1993).

JOHN, ARTHUR H., *The British Library of Political and Economic Science: A Brief History* (London: LSE, 1971).

JOHNSON, HARRY G., 'Individual and Collective Choice', in William A. Robson (ed.), *Man and the Social Sciences* (London: LSE/George Allen & Unwin, 1972).

JONES, EMRYS, 'New Directions 1: Contemporary Movements in Geography', *LSE Magazine* (June 1976), 1–2.

JONES, KATHLEEN, *Eileen Younghusband: A Biography* (London: Bedford Square Press for the National Council for Voluntary Organizations, 1984).

JONES, KIT, 'Fifty Years of Economic Research: A Brief History of the National Institute of Economic and Social Research 1938–88', *National Institute Economic Review*, 124 (May 1988), 36–59.

KADISH, ALON, *The Oxford Economists in the Late Nineteenth Century* (Oxford: Clarendon Press, 1982).

—— 'The City, the Fabians, and the Foundation of the LSE', undated, unpublished typescript; copy in LSE History.

KAHN-FREUND, OTTO, *Das soziale Ideal des Reichsarbeitsgerichts* (Mannheim: Bensheimer, 1931); reprinted in Th. Ramm (ed.), *Arbeitsrecht und Politik, Quellentexte* (Neuwied: Luchterhand, 1966), 149–210.

—— *Selected Writings* (London: Stevens & Sons, 1978).

KAISER, ANDRÉ, 'Politik als Wissenschaft: Zur Entstehung akademischer Politikwissenschaft in Großbritannien', manuscript of a paper given at the 'Idealismus und Positivismus' colloquium in Werner-Reimers-Stiftung, Bad Homburg, Jan. 1993; copy in LSE History.

KEARNS, GERRY, 'Halford John Mackinder 1861–1947', *Geographers Bibliographical Studies*, 9 (1985), 71–84.

KENDALL, MAURICE, 'The Statistical Approach', *Economica*, NS, 17 (May 1950), 127–45.

KEYNES, JOHN MAYNARD, *Essays in Biography* (London: Macmillan, 1933).

—— *The Collected Works of J. M. Keynes, xvi. Activities 1914–19* ed. Elizabeth Johnson (London: Macmillan/St Martin's Press for the Royal Economic Society, 1971).

KIDD, HARRY, *The Trouble at LSE 1966–67* (London: Oxford University Press, 1969).

KING, JOHN, and LAYARD, RICHARD, 'The LSE as a Graduate School?', *Universities Quarterly*, 24 (Autumn 1970), 360–74.

KINGSLEY, DAVID, 'Moving to Croydon: Retrospect' and 'Porter's Opinion', *LSE Magazine* (Dec. 1969), 1, 4.

KOOT, GERARD M., 'An Alternative to Marshall: Economic History and Applied Economics at the Early LSE', *Atlantic Economic Journal*, 10 (Mar. 1982), 3–17.

KOOT, GERARD M., *English Historical Economics, 1870–1926: The Rise of Economic History and Neomercantilism* (Cambridge: Cambridge University Press, 1987).

KRAMNICK, ISAAC, and SHEERMAN, BARRY, *Harold Laski: A Life on the Left* (London: Hamish Hamilton, 1993).

LAKATOS, IMRE, 'Popper on Demarcation and Induction', in Paul A. Schilpp (ed.), *The Philosophy of Karl Popper, i,* (La Salle, Ill.: Open Court, 1974).

LASKI, HAROLD J., 'The American Political System as Seen by an English Observer', *Harper's Magazine* (June 1928), 20–8.

—— *The Dangers of Obedience and Other Essays* (New York: Harper, 1930).

—— *A Grammar of Politics* (5th edn.; London: George Allen & Unwin, 1967).

—— 'On the Study of Politics', in Preston King (ed.), *The Study of Politics* (London: Frank Cass, 1977).

LAYARD, RICHARD, and KING, JOHN, 'Expansion since Robbins', David Martin (ed.), in *Anarchy and Culture: The Problem of the Contemporary University* (London: Routledge & Kegan Paul, 1969).

—— —— and MOSER, CLAUS, *The Impact of Robbins* (London: Penguin, 1969).

LEACH, EDMUND, 'The Epistemological Background to Malinowski's Empiricism', in Raymond Firth (ed.), *Man and Culture: An Evaluation of the Work of Bronislaw Malinowski* (London: Routledge & Kegan Paul, 1957).

—— 'On the Founding Fathers', *Current Anthropology*, 7 (Dec. 1966), 560–7.

—— 'Models of Man', in William A. Robson (ed.), *Man and the Social Sciences* (London: LSE/George Allen & Unwin, 1972).

LEWIS, KAY, 'Eve Evans', *LSE Magazine* (Nov. 1971), 8.

LISTOWEL, JUDITH, *This I have Seen* (London: Faber, 1943).

London School of Economics, *Brief Account of the Work of the School during Michaelmas and Lent Terms, 1895–6*, LSE 1/1.

—— *Proposed Establishment of a Library of Political Science*, Apr. 1896, LSE 1/1.

—— *Brief Report on the Work of the School since 1895*, 1899, LSE 20/2.

—— *A Brief Description of the Objects and Work of the School*, Mar. 1901, updated to 1910, LSE 1/1.

—— *Memorandum from the Director on the Policy of the School in the Immediate Future*, Finance and General Purposes Committee, 21 Feb. 1908, LSE 5/2.

—— *Memorandum by the Director on the Development of Higher Commercial Education after the War and the Position of the London School of Economics*, *c.*1916, LSE, R(P)fL/10.

—— *The Working Constitution and Practice of the London School of Economics and Political Science* (London: LSE, 1937).

—— *Review of the Activities and Development of the London School of Economics and Political Science (University of London) During the Period 1923–1937*, prepared for the Rockefeller Foundation (London: LSE, Feb. 1938).

—— *Reminiscences of Former Students, Members of Staff and Governors, with other Material Relating to the History of the School 1922–1947*, *c.*1947, LSE, R(SR)1101.

—— *LSE: The New Library* (London: LSE, 1978).

London School of Economics Alumni Group, *LSE People 1947–1953* (London: LSE, 1987), 1–39.

London School of Economics Students, *LSE's New Director: A Report on Walter Adams* (London: Agitator Pamphlet, 1966).

LOUIS, W. ROGER, *In the Name of God, Go! Leo Amery and the British Empire in the Age of Churchill* (New York: W. W. Norton, 1992).

LOWE, JAMES TRAPIER, *Geopolitics and War: Mackinder's Philosophy of Power* (Washington, DC: University Press of America, 1981).

MCCORMICK, BRIAN, *Hayek and the Keynesian Avalanche* (Hemel Hempstead: Harvester Wheatsheaf, 1992).

MCGREGOR, NELL, 'A Student in the Thirties', *LSE Magazine* (May 1971), 13–14.

MCGREGOR, O. R., 'The Social Sciences', in F. M. L. Thompson (ed.), *The University of London and the World of Learning, 1836–1986* (London: Hambledon Press, 1990).

MACKENZIE, NORMAN, 'Socialism and Society: A New View of the Webb Partnership', lecture delivered at LSE, 15 May 1978 (London: LSE, 1978).

—— and MACKENZIE, JEANNE, *The First Fabians* (London: Weidenfeld & Nicolson, 1977).

MACKINDER, HALFORD, 'The Geographical Pivot of History', *Geographical Journal*, 23 (Apr. 1904), 421–44.

—— *Money-Power and Man-Power: The Underlying Principles rather than the Statistics of Tariff Reform* (London: Simkin-Marshall, 1906).

—— 'On Thinking Imperially', in *Lectures on Empire 1906/07* (London: published privately, 1907).

—— *Democratic Ideals and Reality: A Study in the Politics of Reconstruction* (London: Constable, 1919).

MACRAE, DONALD, 'The Basis of Social Cohesion', in William A. Robson (ed.), *Man and the Social Sciences* (London: LSE/George Allen & Unwin, 1972).

MAIR, PHILIP BEVERIDGE, *Shared Enthusiasm: The Story of Lord and Lady Beveridge* (Windlesham: Ascent Books, 1982).

MALINOWSKI, BRONISLAW, *A Scientific Theory of Culture and Other Essays* (Chapel Hill, NC: University of North Carolina Press, 1944).

—— *A Diary in the Strict Sense of the Term* (London: Athlone Press, 1967).

MALONEY, JOHN, 'Marshall, Cunningham, and the Emerging Economics Profession', *Economic History Review*, 2nd ser. 29 (Aug. 1976), 440–51.

MANNHEIM, KARL, *Ideology and Utopia: An Introduction to the Sociology of Knowledge* (London: Routledge & Kegan Paul, 1936).

MARSHALL, T. H., 'Reminiscences', in *Link: LSE—Department of Social Science and Administration 1912–1962 Jubilee* (London: LSE, 1962).

MARTIN, DAVID, 'The Dissolution of the Monasteries', in David Martin (ed.), *Anarchy and Culture: The Problem of the Contemporary University* (London: Routledge & Kegan Paul, 1969).

MARTIN, KINGSLEY, *Harold Laski 1893–1950: A Biographical Memoir* (London: Jonathan Cape, 1969).

—— *Father Figures: A First Volume of Autobiography 1897–1931* (London: Hutchinson, 1966).

MAZUMDAR, PAULINE M. H., *Eugenics, Human Genetics and Human Failings: The Eugenics Society, its Sources and its Critics in Britain* (London: Routledge, 1992).

MEDLICOTT, W. N., *Contemporary England 1914–1964* (London: Longman, 1967).
MIDDLEMAS, KEITH, *Politics in Industrial Society: The Experience of the British System since 1911* (London: André Deutsch, 1979).
MILIBAND, RALPH, 'Harold Laski: By Students and Colleagues', *Clare Market Review* (Michaelmas Term 1950), 24–55.
MINOGUE, KEN, 'Student Militancy at the London School of Economics', unpublished memorandum, Apr. 1969; copy in LSE History.
MOOREHEAD, CAROLINE, *Bertrand Russell: A Life* (London: Sinclair-Stevenson, 1992).
MOYNIHAN, DANIEL P., 'The United States in Opposition', *Commentary*, 59 (Mar. 1975), 31–44.
NASH, GEORGE H., *The Conservative Intellectual Movement in America since 1945* (New York: Basic Books, 1976).
NEWMAN, MICHAEL, *Harold Laski: A Political Biography* (Basingstoke: Macmillan, 1993).
NIEDHART, GOTTFRIED, 'Gustav Mayers englische Jahre: Zum Exil eines deutschen Juden und Historikers', in *Exilforschung: Ein Internationales Jahrbuch*, vi. *Vertreibung der Wissenschaften und andere Themen* (Society for Exile Studies, 1988), 98–107.
OAKESHOTT, MICHAEL, *Political Education* (Cambridge: Bowes & Bowes, 1951).
PARTSCH, JOSEPH, *Central Europe* (London: Heinemann, 1903).
PATEL, I. G., 'A Director's Initiation', *LSE Magazine* (Nov. 1984), 2–3.
—— 'The Director Writes', *LSE Magazine* (June 1987).
PEASE, E. R., *The History of the Fabian Society* (London: A. C. Fifield, 1916).
PICKLES, WILLIAM, 'Sir Sydney Caine: An Appreciation', *LSE Magazine* (July 1967), 4–5.
—— 'When a Director was Secretary', *LSE Magazine* (Nov. 1974), 3–4.
PIMLOTT, BEN, *The Life and Times of Harold Wilson* (London: HarperCollins, 1992).
PINKER, ROBERT, 'Social Work and Social Policy in the Twentieth Century: Retrospect and Prospect', in Martin Bulmer, Jane Lewis, and David Piachaud (eds.), *The Goals of Social Policy* (London: Unwin Hyman, 1989).
PLANT, ARNOLD, 'Universities and the Making of Businessmen', in Douglas Grant (ed.), *The University and Business* (Toronto: University of Toronto Press, 1958).
PLANT, MARJORIE, 'The BLPES', *LSE Magazine* (Jan. 1952), 2–5.
POPPER, KARL, 'Autobiography of Karl Popper', in Paul A. Schilpp (ed.), *The Philosophy of Karl Popper, ii* (La Salle, Ill.: Open Court, 1974).
—— 'Replies to my Critics', in Paul A. Schilpp (ed.), *The Philosophy of Karl Popper, i* (La Salle, Ill.: Open Court, 1974).
POSTAN, MICHAEL M., 'Time and Change', in William A. Robson (ed.), *Man and the Social Sciences* (London: LSE/George Allen & Unwin, 1972).
PRICE, L. L., 'Obituary: William Albert Samuel Hewins', *Economic Journal*, 42 (Mar. 1932), 151–5.
QUALTER, TERENCE H., *Graham Wallas and the Great Society* (London: Macmillan/LSE, 1980).
REEVES, MAUD PEMBER, *Round about a Pound a Week* (London: Virago, 1979).
REEVES, WILLIAM PEMBER, *New Zealand and Other Poems* (London: Grant Richards, 1898).

RENDEL, MARGHERITA, 'How Many Women Academics 1912–76?', in R. Deem (ed.), *Schooling for Women's Work* (London: Routledge & Kegan Paul, 1980).

RICHARDS, AUDREY, 'Bronislaw Kaspar Malinowski', *Man* (Jan.–Feb. 1943), 1–4.

RICHMOND, THEO, 'Troubled Diary: Voices Past and Present', *LSE Magazine* (June 1969), 9.

ROBBINS, LIONEL, 'A Student's Recollections of Edwin Cannan', *Economic Journal*, 45 (June 1935), 393–8.

—— *Autobiography of an Economist* (London: Macmillan, 1971).

ROBINSON, AUSTIN, 'The National Institute: The Early Years', *National Institute Economic Review*, 124 (May 1988), 63–6.

ROBSON, WILLIAM A. (ed.), *Man and the Social Sciences* (London: LSE/George Allen & Unwin, 1972).

ROSENBAUM, E., and SHERMAN, H. J., *M. M. Warburg & Co., 1798–1938: Merchant Bankers of Hamburg* (London: C. Hurst, 1979).

Royal Institute of International Affairs, *Documents on International Affairs, 1960*, ed. Richard Gott, John Major, and Geoffrey Warner (London: Oxford University Press/RIIA, 1964).

RUDD, ERNEST, 'The Troubles of Graduate Students', in David Martin (ed.), *Anarchy and Culture: The Problem of the Contemporary University* (London: Routledge & Kegan Paul, 1969).

RUSSELL, BERTRAND, *The Autobiography of Bertrand Russell, i. 1872–1914* (London: George Allen & Unwin, 1967).

RUSSELL, KIT, BENSON, SHEILA, FARRELL, CHRISTINE, GLENNERSTER, HOWARD, PIACHAUD, DAVID, and PLOWMAN, GARTH, *Changing Course: A Follow-up Study of Students Taking the Certificate and Diploma in Social Administration at the London School of Economics, 1949–73* (London: LSE, 1981).

SCRINVEN, H. C., 'Walter Ernest Wilson', *LSE Magazine* (July 1957), 19–20.

SEARLE, G. R., *The Quest for National Efficiency: A Study in British Politics and Political Thought 1899–1914* (Oxford: Basil Blackwell, 1971).

SELF, PETER J. O., 'The State versus Man', in William A. Robson (ed.), *Man and the Social Sciences* (London: LSE/George Allen & Unwin, 1972).

SEMMEL, BERNARD, *Imperialism and Social Reform: English Social-Imperial Thought 1895–1914* (London: George Allen & Unwin, 1960).

SEYMOUR-JONES, CAROLE, *Beatrice Webb: Woman of Conflict* (London: Allison & Busby, 1992).

SHAW, BERNARD (ed.), *Fabian Essays* (Jubilee edn.; London: George Allen & Unwin, 1948).

—— *Collected Letters 1874–1897*, ed. Dan H. Laurence (London: Max Reinhardt, 1965).

SINCLAIR, KEITH, *William Pember Reeves: New Zealand Fabian* (Oxford: Clarendon Press, 1965).

SINGH, TARLOK, 'Vera Anstey's Work for India', *LSE Magazine* (June 1977), 6.

SKIDELSKY, ROBERT, *John Maynard Keynes, ii. The Economist as Saviour 1920–1937* (London: Macmillan, 1992).

SLOMAN, ALBERT E., *A University in the Making: The 1963 Reith Lectures* (London: British Broadcasting Corporation, 1964).

Sociological Society, *Sociological Papers 1904–06* (London: Sociological Society, 1904–6).

STERNBERG, BARBARA, 'The View from Colorado', *LSE Magazine* (Nov. 1982), 9.

STOCKS, MARY, *My Commonplace Book* (London: Peter Davies, 1970).

SUTHERLAND, GILLIAN, 'The Plainest Principles of Justice: The University of London and the Higher Education of Women', in F. M. L. Thompson (ed.), *The University of London and the World of Learning 1836–1986* (London: Hambledon Press, 1990).

TAWNEY, R. H., *The Acquisitive Society* (London: G. Bell & Sons, 1921).

—— *Religion and the Rise of Capitalism: An Historical Study* (London: John Murray, 1926).

—— *Equality* (London: George Allen & Unwin, 1964).

—— 'The Study of Economic History', in N. Harte (ed.), *The Study of Economic History: Collected Inaugural Lectures 1893–1970* (London: Frank Cass, 1971).

—— *Commonplace Book*, ed. J. M. Winter and D. M. Joslin (Cambridge: Cambridge University Press, 1972).

TERRILL, ROSS, *R. H. Tawney and his Times: Socialism as Fellowship* (Cambridge, Mass.: Harvard University Press, 1973).

THOMPSON, F. M. L. (ed.), *The University of London and the World of Learning 1836–1986* (London: Hambledon Press, 1990).

TITMUSS, RICHARD M., 'Time Remembered', in *Link: LSE Department of Social Science and Administration 1912–1962 Jubilee* (London: LSE, 1962).

—— *Social Policy: An Introduction*, ed. Brian Abel-Smith and Kay Titmuss (London: George Allen & Unwin, 1974).

TOWNSEND, ELIZABETH, 'A Social Science Student 50 Years Ago, Part One', *LSE Magazine* (June 1979), 5–7.

—— 'A Social Science Student 50 Years Ago, Part Two', *LSE Magazine* (Nov. 1979), 7–8.

TREVELYAN, G. M., *British History in the Nineteenth Century and After 1782–1919* (2nd edn.; London: Longmans, Green, 1937).

WALLAS, GRAHAM, 'An Historical Note', *The Students' Union Handbook 1922–23*, LSE CF 116/C.

WATKINS, JOHN, 'Imre Lakatos', in *International Encylopaedia of the Social Sciences Biographical Supplement* (New York: Macmillan/Free Press, 1979), 399–402.

—— 'The LSE "School" of Philosophy: A Polish Interrogation, an Interview with John Watkins by Ewa and Andrzej Chmielecka', *LSE Quarterly*, 1 (June 1987), 213–26.

WAYNE, HELENA, 'Bronislaw Malinowski: The Influence of Various Women on his Life and Works', *Journal of the Anthropological Society of Oxford*, 15 (Michaelmas 1984), 189–203.

WEBB, BEATRICE, *Our Partnership*, ed. Barbara Drake and Margaret I. Cole (London: Cambridge University Press/LSE, 1948).

—— *The Diary of Beatrice Webb, ii. All the Good Things in Life 1892–1905*, ed. Norman and Jeanne MacKenzie (London: Virago/LSE, 1983).

—— *The Diary of Beatrice Webb, iii. The Power to Alter Things, 1905–1924*, ed. Norman and Jeanne MacKenzie (London: Virago/LSE, 1984).

—— *The Diary of Beatrice Webb: iv. The Wheel of Life 1924–1943*, ed. Norman and Jeanne MacKenzie (London: Virago/LSE, 1985).

WEBB, SIDNEY, 'The Provision of Higher Commercial Education in London', in International Congress on Technical Education, *Report* of the Proceedings of the 4th Meeting, held in London, June 1897 (London: Society of Arts, 1897), 205–10.

—— 'Lord Rosebery's Escape from Houndsditch', *The Nineteenth Century and After*, 50 (Sept. 1901), 366–86.

—— WEBB, BEATRICE, 'Reminiscences IV: The London School of Economics and Political Science', *St Martin's Review* (Jan. 1929), 24–8.

—— —— *Soviet Communism: A New Civilisation?* (London: Longmans, Green 1944).

—— —— *The Letters of Sidney and Beatrice Webb, i. Apprenticeships 1873–1892*, ed. Norman MacKenzie (Cambridge: Cambridge University Press/LSE, 1978).

—— —— *The Letters of Sidney and Beatrice Webb, ii. Partnership 1892–1912*, ed. Norman MacKenzie (Cambridge: Cambridge University Press/LSE, 1978).

—— —— *The Letters of Sidney and Beatrice Webb, iii. Pilgrimage 1912–1947*, ed. Norman MacKenzie (Cambridge: Cambridge University Press/LSE, 1978).

—— *Indian Diary*, ed. Niraja Gopal Jayal (Oxford: Oxford University Press, 1990).

WEIGERT, HANS W., *Generals and Geographers: The Twilight of Geopolitics* (New York: Oxford University Press, 1942).

WELLS, H. G., 'The So-Called Science of Sociology', *Sociological Papers*, 3 (1906), 357–77.

—— *The New Machiavelli* (London: John Lane/Bodley Head, 1911).

—— *Experiment in Autobiography: Discoveries and Conclusions of a Very Ordinary Brain (since 1866) ii* (London: Victor Gollancz, 1966).

—— *Ann Veronica* (London: Virago, 1980).

—— *H. G. Wells in Love*, ed. G. P. Wells (London: Faber, 1984).

WERSKEY, GARY, *The Visible College: A Collective Biography of British Scientists and Socialists of the 1930s* (London: Free Association Books, 1988).

WESTERMARCK, EDWARD, 'Sociology as a University Study', *Inauguration of the Martin White Professorships of Sociology, Dec. 17, 1907* (London: John Murray, 1908).

—— *Memories of My Life*, trans. Anna Barwell (London: George Allen & Unwin, 1929).

WHITE, ARNOLD, *Efficiency and Empire* (London: Methuen, 1901).

WIENER, MARTIN J., *Between Two Worlds: The Political Thought of Graham Wallas* (Oxford: Clarendon Press, 1971).

WILLIAMS, J. R., TITMUSS, R. M., and FISHER, F. J., *R. H. Tawney: A Portrait by Several Hands* (London: Shenval Press, 1960).

WILSON, HAROLD, *The Beveridge Memorial Lecture 1966*, given at Senate House on 18 Nov. 1966 (London: Institute of Statisticians, 1966).

—— *Memoirs: The Making of a Prime Minister, 1916–64* (London: Weidenfeld & Nicolson, 1986).

WISE, MICHAEL, 'Man and his Environment', in, William A. Robson (ed.), *Man and the Social Sciences* (London: LSE/George Allen & Unwin, 1972).

WISE, MICHAEL, 'Three Founder Members of the I.B.G.: R. Ogilvie Buchanan, Sir Dudley Stamp, S. W. Wooldridge, A Personal Tribute', *Transactions of the Institute of British Geographers*, NS, 8 (1983), 41–54.

WOLEDGE, GEOFFREY, 'The BLPES', in R. Irwin and R. Staveley (eds.), *The Libraries of London* (2nd rev. edn.; London: Library Association, 1964).

YAMEY, BASIL, 'Obituary: Professor Sir Ronald Edwards', *LSE Magazine* (June 1976), 7–8.

YOUNG, MICHAEL (ed.), *The Ethnography of Malinowski: The Trobriand Islands 1915–18* (London: Routledge & Kegan Paul, 1979).

INDEX

Index compiled by Fiona Plowman

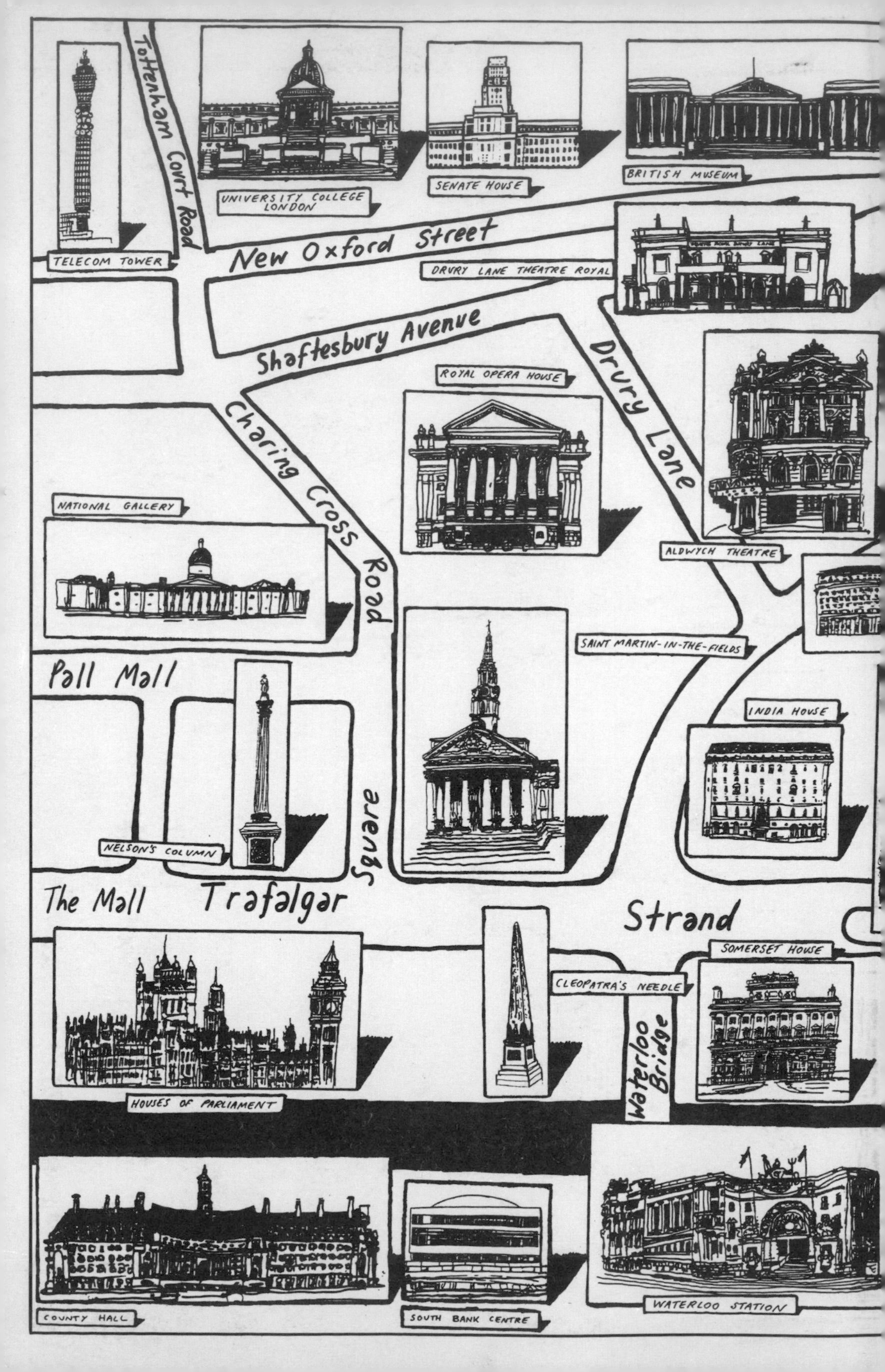

Tottenham Court Road
TELECOM TOWER
UNIVERSITY COLLEGE LONDON
SENATE HOUSE
BRITISH MUSEUM
New Oxford Street
DRURY LANE THEATRE ROYAL
Shaftesbury Avenue
ROYAL OPERA HOUSE
Drury Lane
Charing Cross Road
ALDWYCH THEATRE
NATIONAL GALLERY
SAINT MARTIN-IN-THE-FIELDS
Pall Mall
INDIA HOUSE
NELSON'S COLUMN
Trafalgar Square
The Mall
Strand
SOMERSET HOUSE
CLEOPATRA'S NEEDLE
Waterloo Bridge
HOUSES OF PARLIAMENT
COUNTY HALL
SOUTH BANK CENTRE
WATERLOO STATION